An Historical Who's Who of the
Automotive Industry in Europe

An Historical Who's Who of the Automotive Industry in Europe

JAN P. NORBYE

McFarland & Company, Inc., Publishers
Jefferson, North Carolina, and London

Published with the cooperation of the family and friends
of the late Jan. P. Norbye, who died in 2003 shortly
after completing the manuscript for this book.

LIBRARY OF CONGRESS CATALOGUING-IN-PUBLICATION DATA

Norbye, Jan P.
An historical who's who of the automotive industry in Europe / Jan P. Norbye.
        p.   cm.
Includes index.

ISBN 0-7864-1283-6 (illustrated case binding : 50# alkaline paper)

1. Automobile engineers — Europe — Biography — Dictionaries.
2. Industrialists — Europe — Biography — Dictionaries.   I. Title.
TL139.N67   2006   629.222'092'24 — dc22   2004014710

British Library cataloguing data are available

On the cover, descending: Colin Chapman, Karl Friedrich Benz,
Ferdinand Porsche, Andre Citroën, Armand Peugeot, Enzo Ferrari
(*European Automotive Hall of Fame*). Background map ©2005 PhotoSpin.

Manufactured in the United States of America

McFarland & Company, Inc., Publishers
Box 611, Jefferson, North Carolina 28640
www.mcfarlandpub.com

In memory of
Margaret Norbye
– *J.P.N.*

Table of Contents

Preface

1

An Historical Who's Who
of the Automotive Industry in Europe

3

Index

349

Preface

Jan P. Norbye was born in Oslo on 20 August 1931 and died in Nice on 20 March 2003. A talented linguist, he was fascinated by the process of communication in both spoken and written form. This ability was enhanced by a strong sense of humor; he was a brilliant raconteur. In his work, Norbye was a stickler for accuracy and felt it his duty to set the record straight when errors occasionally appeared in publications covering his specialist area of motor industry history.

Jan's life-long enthusiasm for communication and the motor car industry was triggered at a young age. By five or six, his party piece was the recitation of every make of European car and their engines; he was called in as "Consultant" whenever there was a problem with the family car.

As a child in occupied Norway, he had discovered the importance of communication and his facility for languages. He quickly learned to speak German; taking extra classes with a Norwegian languages teacher for whom he had a profound respect.

Armed with this first success he was eager to progress and when the British arrived in May 1945, Jan was there ready to meet these newcomers. Having made friends with a few of the soldiers stationed near his home, he was taken under the wing of a bluff cockney Army transport Sergeant who not only tutored him in colloquial En-glish, but also taught him to drive. So, by the age of 13, and just reaching the pedals, Norbye was charging around the courtyard of his home in a clapped-out Army jeep. This experience reinforced his passion for cars and the motor manufacturing industry.

On completion of his National Service in the King's Guard (Transport Unit), he left Norway and at the age of 19 went to England to work for HRG Motors. When he returned to Oslo, he took jobs in the car industry with Sultan and B.O. Steen. In 1955 he moved to Paris to work in the Touring Services Department of Esso, France. After his stay in the UK, he had started contributing as a freelance motoring correspondent to the Norwegian publication MOTORLIU, as well as to various English and Continental motor sport magazines and he continued to do this when he worked in France.

During this period of his life, he was in contact with many of the leading motoring personalities of the day: engineers, designers, drivers. These charismatic but often complex characters awakened his interest in the lives and experiences of the early pioneers of the auto industry and he started his archive of photographs together with published (and unpublished) material from many countries. It is this archive which forms the basis of *An Historical Who's Who of the Automotive Industry in Europe*. He intended that it should be illustrated by photographs, but he ran out of time and his illness did not give him the three months he needed to select the pictures.

On leaving Paris in 1958, he went to Volvo in Gothenburg, Sweden, working in the Export After-Sales Department before going to the USA in 1961 to become Technical Editor of *Car & Driver* and later of *Popular Science* magazine. While in the United States, he co-authored several books on American automobile marques with his friend and Detroit editor, Jim Dunne. Norbye also became a founding member of the Society of Automotive Historians in the United States, a privilege of which he was very proud. He was also a member of the Guild of Motoring Writers.

In the early 1970s, he returned to Europe to live in Les Issambres, France, where he went on to write several books on the European and Japanese car industries, together with an industry-acclaimed work on the Wankel engine. As well as continuing his work as a freelance motoring correspondent for worldwide titles, he made a substantive contribution to the definitive three-volume *Beaulieu Encyclopaedia of the Automobile* and was the sole author of the fourth (technical) volume, which has yet to be published. Jan's wife, Margaret, acted as his editorial assistant–cum–typist and as he said on her death in 2001, "Nobody has typed more motoring books than Margaret."

An Historical Who's Who of the Automotive Industry in Europe provides an overview of the major figures in the European motor manufacturing industry from its inception to the present. Exact birth and death dates have been given wherever possible. Clearly there is room for debate on individuals who should have been included or omitted, but that is what Jan would have wanted. His idea was to create a book that would stimulate discussion and ensure that these pioneers and their achievements would not be forgotten by students of the motor industry.

No value judgments are made on these tough, independent-minded people who created the powerful machines of yesteryear and paved the way for today's high-tech industry, but glimpses of their strengths and weaknesses emerge from the page and make for fascinating reading.

Jan is much missed by his family and friends and it is in tribute to his tenacity in trying to complete this book as a reference source for future generations that such extensive efforts have been undertaken by his family, friends and the editors at McFarland to ensure its publication, albeit without the photographs.

Betty Dean
Fall 2005

An Historical
Who's Who of the
Automotive Industry
in Europe

Abarth, Carlo (1908–1979)

Founder of Abarth & Co. in Turin and its managing director from 1949 to 1971.

He was born on November 15, 1908, in Vienna as the son of an official at the Imperial Court. In August 1914, the family moved to Merano in South Tyrol, where grandfather Abarth owned a hotel. Carlo raced bicycles as a boy, then motorcycles, and was apprenticed to a local precision engineering laboratory. The family remained in Merano after the fall of the Hapsburg empire, but Carlo returned to Vienna and found part-time employment in the Degen factory which made frames for bicycles. He also learned to repair motorcycle engines in the Degen shops.

At the age of 19 he designed and built his own motorcycle, using a Villiers engine. He worked for Motorradwerk Max Thun at Traiskirchen and joined the MT racing team. He took full-time employment with Degen in Vienna in 1933 but continued motorcycle racing as a private entrant. He was near Ljubljana in Slovenia in October 1939 when he had a terrible racing accident. He survived and spent the war years there in relative security, repairing cars and trucks, eventually landing a position as plant manager for Ignaz Vok's automotive gas-generator works.

In 1945 he made his way back to Merano and made contact with Piero Dusio, founder of Cisitalia. Abarth was instrumental in getting Dusio the rights to the Porsche Type 360, a 12-cylinder four-wheel-drive racing car, Type 323 (farm tractor) and Type 370 (VW-based sports car). He was named manager of the Cisitalia racing team. On April 15, 1949, he incorporated Abarth & Co. in Turin, financed by the proceeds from the sale of the family hotel in Merano, and risk-capital provided by Armando Scagliarini. Abarth opened shops in Via Trecate, making and selling auto accessories, notably the Abarth low-restriction resonating muffler, and multi-carburetor manifolds for Fiat and other engines. He also took over the Fiat-based Cisitalia models which were renamed Abarth and identified by the scorpion badge. Abarth made a number of speed-record cars which set a total of 131 international class records, and in 1955 he introduced the Abarth 750 production model based on the Fiat 600. Allemano built the roadster bodies for the 750 Spider and 850 Spider. In 1960–61 Abarth made 20 aluminum-bodied Carrera cars for Porsche. The 1000 Bialbero GT, with a twin-cam head on a Fiat block and a Zagato-designed body made in the Abarth factory, which had moved to bigger premises in Corso Marche in 1959, was introduced in 1962. A year later the Abarth Simca 1600 GT Stradale was launched. In 1966 Abarth made its own versions of the Fiat 124 Spider. Racing cars were created for almost every class: Formula 2, Formula 1, Group 6, and more. In all, Carlo Abarth originated 184 different models. He sold Abarth & Co. to Fiat in July 1971, remaining on board as a racing consultant for a period, before retiring to Vienna, where he died on October 24, 1979.

Accles, James George (1867–1939)

Co-founder of Accles-Turrell Autocars, Ltd., and its managing director from 1899 to 1901; joint patent-holder with Charles McRobie Turrell for several automotive devices.

He was of Irish extraction, born in 1867 in Australia, and served a gunsmith's apprenticeship at Samuel Colt's factory in Hartford, Connecticut. He moved to England in 1888 and became a partner in the engineering firm Grenfell & Accles (later, Accles, Limited), with factories on a 35-acre site at Holford Hills, Perry Barr, Birmingham. By 1896 Accles had a steel-tube works with a capacity of 75,000 feet of tubing per week, and a bicycle plant with a weekly output of 600 two-wheelers. Accles also owned factories that made weaving machinery, boot-stitching machines, and aluminum cartridge cases. He became interested in automobiles, formed a partnership with Charles McRobie Turrell, and in 1899 Accles was granted patents for a gasoline-engine ignition system and a carburetor. He sold his interest in the automobile company to

Thomas Pollock, who moved production of the Accles-Turrell car from Perry Barr to Ashton-under-Lyme in 1901. In 1910 he merged his steel-tube business with one of Pollock's enterprises, forming Accles & Pollock, Ltd., at Oldbury. After World War I he resigned to devote himself to experimental and consulting work, but later formed a new industrial enterprise in partnership with G.E. Shelvoke. Due to ill health he retired from Accles & Shelvoke in 1938, and settled in Cornwall where he died within a year.

Ader, Clément (1841–1925)

Managing director and chairman of the Société Industrielle des Téléphones from 1880 to 1925.

He was born on April 2, 1841, at Muret, 20 km from Toulouse, and educated as an electrical engineer. In 1872 he designed an ornithopter and built a tethered test "bird." He began building a steam-powered airplane, the Eole, in 1882, at his own expense, and was awarded a French patent for the design in 1890. The Eole got into the air under its own power on October 9, 1890, at Armainvilliers and stayed aloft for some 50 meters' distance. But the pilot had no means of maneuvering. In 1892 he was commissioned by the Ministry of War to construct a new airplane, the Avion. It failed in preliminary tests and was rebuilt as Avion II, which did not do much better. Avion III, built in 1896, made several long jumps on the Satory plain in October 1897. The Ministry of War lost interest and withdrew its financial support for Ader. His telephone company, on the other hand, prospered and provided funds for his next experiments, this time in automobiles. In 1901 he designed and built a car with a 1½-liter V-twin, and the following year, a new model with a 1.9-liter 90° V-4 engine. The crankshaft was counterweighted. The valves were arranged in F-head formation, with automatic overhead inlet valves. The cylinders were individually cast and bolted to an aluminum crankcase. In 1903 he combined two V-4 engines to make a V-8, and entered the car in the Paris–Madrid road

race. But it was not completed in time and never raced. These cars were built in the shops of the telephone company, rue du 4 Septembre at Levallois. In 1904 he spun off the automobile department as Société Ader and designed new vertical four-cylinder engines in three sizes. He also designed a motorcycle with a single-cylinder engine and shaft drive, and Société Ader produced a line of V-twin and V-4 engines for motorboats. Ader discontinued car production in 1907.

Agnelli, Giovanni (1866–1945)

Secretary of the management board of F.I.A.T. from 1899 to 1901; managing director of F.I.A.T. (later Fiat) from 1901 to 1920; chairman and managing director of Fiat from 1920 to 1925; chairman of Fiat from 1925 to 1943; chairman of Fiat S.p.A. from 1943 to 1945.

He was born on August 13, 1866, as the son of a wealthy lumber merchant at Villar Perosa in Piedmont province. He was educated at a district school in Pinerolo and San Giuseppe College in Turin. From 1884 to 1886 he studied at the Military Academy in Modena, and embarked on a military career. He was serving as a lieutenant in the Savoy Cavalry in Verona in 1892 when he saw a Daimler engine for the first time. It made a deep impression on his mind, and he sought advice from Professor Enrico Bernardi of the University of Padua about the future of the internal combustion engine. What he learned made him decide to leave the military and make a career in the engine business. He returned to Villar Perosa in 1893, taking charge of the family estate, but spent most of his time in Turin, meeting with pioneer motorists, engine makers and car builders. In 1896 he made Turin his home town. His friend Roberto Biscaretti di Ruffia (1845–1940) had two Benz cars which Agnelli examined and drove. With a group of businessmen, bankers and lawyers, he was a co-founder of F.I.A.T.[1] in 1899, formed to take over the small car factory of the Ceirano brothers in Turin. Land was secured on Corso Dante in Turin for putting up a factory designed for the express purpose of building cars.

F.I.A.T. started a racing team to make publicity for its touring cars, and became Turin's biggest car company. In 1906 its finances were in disorder, however. Agnelli paid off the original shareholders and recapitalized the business. He was now Fiat's biggest shareholder. Later that year he made his first visit to the United States, to launch the Fiat car on the American market, and to visit some automobile factories. At home, Fiat took over the Italian branch of Rothschild (coachbuilders) and R. Incerti & C., makers of ball bearings (RIV). He bought shares in local iron works and foundries, and set up Fiat subsidiaries to make marine diesel engines (1907), aircraft engines (1908) and motor boats (1908). He founded Austro-Fiat, which began car production in 1911, and in 1914, Fiat opened a factory for military aircraft. In 1914, Fiat produced 4646 cars, more than half of all cars made in Italy that year. In 1915, Fiat became part of the war economy. The work force increased tenfold in three years, and turnover rose sevenfold. In 1916 a new giant factory site was purchased on Via Nizza in Turin, for construction of the future Lingotto plant, which came on stream in 1923 with capacity for 1200 cars a week (which tripled in the next five years). In 1923 Benito Mussolini, new head of government, arranged for King Vittorio Emanuele to nominate Agnelli as a life-time member of the Senate, which duly passed.

Fiat purchased SPA[2] in 1926 and stopped production of SPA cars, its big factory on the north side of Turin becoming Fiat's main truck plant. Branch factories were set up in other European countries to get around protective barriers. The NSU car factory at Heilbronn was acquired in 1929 to make the Deutsche Fiat, and production of the Polski-Fiat began in Warsaw in 1931. Agnelli backed H.T. Pigozzi in starting Simca in 1934 to produce Fiat-designed cars in France, and a Vickers subsidiary at Crayford, Kent, began assembling Fiat cars in 1936. In 1933 Fiat took over OM[3] of Brescia, and turned it into a truck plant.

Agnelli's political influence, within the industrial sector, was considerable. He was able to obtain cancellation of Ford's building permit for an assembly plant at Livorno and to block Ford's attempted takeover of Isotta-Fraschini in 1929. On the other hand, Mussolini rejected Agnelli's attempt to get control of Alfa Romeo in 1933, as he wanted it under state ownership.

He had placed his shares in Fiat, and a number of other personal investments, into a holding, IFI[4] as early as 1927, under control of the Agnelli family. He stayed. in Turin throughout World War II but had worn himself out, and he died on December 16, 1945.

1. Fabbrica Italiana Automobili Torino.
2. Societa Ligure Piemontese Automobili.
3. Officine Meccaniche (originally Brixia-Züst).
4. Instituto Finanziario Industriale.

Agnelli, Gianni (1921–2003)

Head of Fiat from 1966 to 1996; under his leadership, Fiat became the most important company in Italy and one of the major car companies of Europe.

He was born March 12, 1921, near Turin, Italy. He was the grandson of Giovannni Agnelli (1866–1945), one of the co-founders of Fiat. He studied law at the University of Turin to distinguish himself from his brother, Umberto, and became known as "l'Avvocato" (the lawyer) even though he never practiced law. Agnelli served in the Italian Army in World War II, fighting with a tank regiment on the Russian front, and was twice wounded. On the death of his grandfather in 1945, he became vice-chairman of Fiat. (Agnelli's father, Eduardo, had died in an air crash in 1935.) He became chairman in 1966. During his tenure, he oversaw Fiat's acquisition of Italian companies Alfa Romeo, Ferrari and Lancia as well as foreign truck makers Magirus Deutz and Unic. As honorary chairman, he still had great influence at Fiat until his death. Agnelli had relationships with many key political and financial people in Italy and around the world. Some of those dealings caused controversy and speculation about improprieties relating to Fiat and Agnelli, but he maintained his reputation as one of the most powerful and richest men in Italy. It is said that Khrushchev preferred to talk to him over any politician, as "this lot come and go; you will always be in power."

Agnelli, Umberto (1934–2004)

Managing director of Fiat S.p.A. from 1970 to 1976; vice chairman of Fiat S.p.A. from 1976 to 1993; managing director of Fiat Auto S.p.A. from 1979 to 1989.

He was born on November 1, 1934, in Lausanne, Switzerland, and held a law degree from the Turin University. He was president and chief executive of SAI,[1] Italy's third-biggest insurance company, from 1962 to 1971, also serving as president and chief executive of Fiat-France from 1965 to 1967. In 1968 he took charge of the International Affairs Office of Fiat S.p.A. in Turin, coordinating the group's industrial and commercial operations worldwide. He took command of Fiat Auto S.p.A. when Ghidella left and piloted it safely through a difficult decade. He authorized production of the Fiat Panda, Uno and Croma. He closed Fiat's oldest plant (Lingotto) in 1986 and ordered a vast expansion in Brazil. According to the plan, he was to take over as chairman of Fiat S.p.A. in June 1994, but his older brother decided to extend his term, and Umberto was named chairman of IFI,[2] the family holding, with a 33 percent stake in Fiat S.p.A. and a conglomerate investment portfolio, assuming that office in November 1993.

1. Società Assicurazione Industriale.
2. Istituto Finanziario Industriale.

Ahrens, Hermann (1904–)

Chief engineer of body development, passenger cars, Daimler-Benz AG, from 1937 to 1955; director of the Daimler-Benz AG body engineering department from 1955 to 1971.

He was born on March 14, 1904, at Uslar/Solling and was apprenticed to a local mechanic. Later, he attended technical colleges in Bingen on the Rhine and at Varel in Oldenburg. He began his career with Deutsche Werke in Berlin-Spandau in 1925, a state-owned ex-munitions factory then producing the D-Rad motorcycles and D-Wagen light passenger cars. The company also produced bodies for Brennabor. He began as a body-engineering draftsman and was promoted to designer within two years. In 1928 he joined Horch-Werke AG at Zwickau as head of body development and created a new style for the Horch product, beginning with the 600 and 670 models of 1931, with increased length and flowing fender lines. In 1932 he moved to Stuttgart as body design engineer for Mercedes-Benz, where he continued to modernize the exterior appearance of new models. He applied streamlining techniques selectively, mainly in the rear sections, taking great care to maintain the traditional "face." What he started with the 290 and 350 flourished into creations such as the 500 K and 540 K in 1934. Smaller cars, such as the 230, 170 V, and 320, were styled with sober lines and dimensions, but his V-12 prototypes from 1939–40 were lavish expressions of automotive sculpture as a form of modern art.

He worked at Sindelfingen throughout World War II, building truck cabs and special-purpose bodies for military vehicles. From 1945 to 1949 he was out of the auto industry, designing luggage for Karl Baisch in Stuttgart. He returned to Daimler-Benz AG in 1949 and developed the body shapes and structures for the new 220 and 300 series that went into production in 1951. He designed the company's first unit-construction body for the 180 of 1953, and planned and supervised body design and engineering for the new 220 series that appeared in 1959. He had little to do with the SL models, but started the W-108, W-114/115 and W-116 (S-Class) projects which were taken over by Karl Wilfert.

He retired from Daimler-Benz AG in 1971 and went back to Karl Baisch, where he was active until 1982.

Ainsworth, Henry Mann (1884–1971)

Chief engineer of the Hotchkiss automobile factory at St. Denis from 1909 to 1914; general manager of the Hotchkiss arms and engine factory in Coventry from 1915 to 1923; managing director and chief executive of Automobiles Hotchkiss from 1923 to 1939 and 1946 to 1950.

He was born on August 9, 1884, at Buxton in Derbyshire, into a well-to-do family, and nephew of John James Mannhead of Mann & Overton, auto dealers and coachbuilders in London. He was educated at Parkhurst School in Buxton and held a degree in mechanical engineering from the Manchester School of Technology.

After serving an apprenticeship with Belsize Motors in Manchester, he went to Paris, where his uncle had found him a position in the motor vehicle drawing office of Hotchkiss. He also proved himself an excellent test driver. He got rid of the ball-bearing crankshafts used on earlier models, adopted plain bearings, and put counterweights on the crankshafts. He directed the engineering of the model AB in 1912, the AC-6 and the 5.7-liter four-cylinder model AD. He introduced the one-piece cast-iron four-cylinder block on the Model Y in 1909.

Come 1914 and war in Europe, he was drafted into the Royal Army Intelligence Corps, but soon discharged to set up a factory for making Hotchkiss repeater guns in Coventry. Military orders were canceled at the end of the war, and the factory would have been idled but for Ainsworth's initiative.

He secured a contract from William Morris to build a copy of the Continental engine fitted in the Morris Oxford. In 1920 he designed an air-cooled 90° V-twin, had it mounted in a Morris chassis, and entered it in a hill-climb. The publicity from its racing appearance brought inquiries from BSA who purchased the design and began mounting the V-twin in BSA cars in 1921. He also designed four-cylinder side-valve engines that were installed in Gilchrist and Autocrat cars. He sold the Coventry plant to Morris Motors, Ltd., in 1923 and returned to Paris. He stopped production of all earlier models and began production of the AM, designed in Coventry by himself and Alfred H. Wilde.

The car branch of Hotchkiss prospered as the AM evolved into the AM-2 and spawned the AM-80, but from 1931 onwards, Hotchkiss suffered from high manufacturing costs that had a restrictive effect on demand. Hotchkiss maintained a high profile in international rallies and survived on its reputation. Early in 1940 Ainsworth was recalled to London for military duty, and from 1942 to 1945 he served as technical advisor to the Department of Munitions and Supply, based in Ottawa, Canada. In this capacity he came to know Ward M. Canaday, president of Willys-Overland, who sold manufacturing rights to the Jeep to Ainsworth, who later transferred them to Hotchkiss. He also entertained other car projects with Willys-Overland, but when he returned to Paris, the factory had resumed production of prewar models.

He made an ill-fated attempt to modernize the product by purchasing a license for the J.A. Grégoire prototype, which was never fully developed and had to be abandoned. He retired from Hotchkiss in 1950 and lived many years of comfortable retirement in a high-rent district of Paris until he fell ill, suffered several stays in hospital, and died on January 24, 1971.

Alden, John (1916–)

Chief product engineer of Vauxhall Motors, Ltd., from 1963 to 1971.

He was born in 1916 in Oxford and served an apprenticeship with Morris Garages (MG). He obtained an engineering degree from the Oxford School of Technology and joined Vauxhall Motors, Ltd., in 1938 as a test engineer in the experimental department. In 1944 he was named experimental engineer for Bedford trucks. His next promotion followed in 1959 when he was appointed assistant chief engineer. He became chief engineer on March 1, 1963, and directed the Vauxhall Viva program. He was under increasing pressure from management to cut costs by better coordination with the ongoing programs at Adam Opel AG, and he started the Firenza program, drawing heavily on the engineering of the Opel Manta. On May 1, 1971, he was named technical assistant to the managing director, and in 1974 he was placed in charge of Vauxhall's environmental activities staff. He retired in 1977.

Aldington, Harold J. "Aldy" (1902–1976)

Managing director of AFN Limited from 1929 to 1965.

He was apprenticed to GN[1] at Kingston-on-Thames and in 1922 accompanied Captain A. Frazer Nash in starting production of Frazer Nash cars. He bought the company in 1929 and established a great reputation for the make as a rally car. By 1933 he realized that a more modern product was needed, but could not afford to have one designed-to-order by any well-known firm of consulting engineers. On the other hand, he was very much impressed with the little six-cylinder BMW, and secured manufacturing rights to it. The small AFN factory, Falcon Works at Isleworth, Middlesex, lacked facilities for that, so he imported BMW chassis with right-hand drive, made at Eisenach, and had British bodywork mounted on them. From 1934 to 1939, AFN sold 707 such cars. When World War II broke out, he and his brother Bill[2] volunteered for military service in the RAOC,[3] transferring to the REME[4] on its formation in 1942. He was discharged with the rank of Lieutenant-Colonel in 1946. He had been appointed to the Bristol Aeroplane Company's board of directors in 1940, and was informed of Bristol's plans to start car production. He saw this as an opportunity to revive the Frazer Nash car as well. He was able to visit Munich in 1945, where he acquired a 1940 BMW 328 streamliner. Through the War Reparations Board he was awarded manufacturing rights for the BMW 328 engine and the car models 326, 327 and 328.

In July 1945, Bristol Aeroplane Company Ltd. bought majority control of AFN Limited, and began preparations for building BMW engines and BMW-based cars. Aldington also arranged for the release of Fritz Fiedler to come to England and supervise their engineering. To avoid any risk of marketing conflict, AFN would make two-seaters and Bristol would make four-passenger cars. Yet there was conflict. Aldington was not well liked at Bristol, as he was always proclaiming the superiority of German engineering over British. After a quarrel with Sir George White, chairman of Bristol Aeroplane Co., Ltd., he purchased the AFN shares held by Bristol, which agreed to continue to supply engines for Frazer Nash cars. In 1955 AFN Limited became Porsche distributors, and the last Frazer Nash cars were completed in 1958.

He retired in 1965, due to ill health, and died in 1976.

1. GN were the initials of the two partners, H.R. Godfrey and A. Frazer Nash.
2. William H. Aldington (1901–1980).
3. Royal Army Ordnance Corps.
4. Royal Engineers, Mechanical/Electrical.

Alfieri, Giulio (1924–2002)

Head of the Maserati drawing office from 1954 to 1955; technical director of Maserati from 1955 to 1968; general manager of Lamborghini from 1979 to 1987.

He was born on July 14, 1924, in Parma as the son of a certified public accountant, who hoped he would become a veterinarian. But the boy was interested only in engines. After a high-school education, he enrolled at the Milan Polytechnic in 1944 and graduated in 1948 with a degree in industrial engineering. He began his career with the ship-building yards[1] in Genoa, where he worked on low-speed two-stroke marine diesel engines and steam turbines. In 1949 he joined Innocenti in Milano and designed small two-stroke engines for the Lambretta motor scooters until 1953. He joined Maserati as a test and racing-car engineer in August 1953 and worked on the evolution of the 250/F. He was responsible for the 3500 GT and 5000 GT, the "birdcage" Types 60 and 61 sports/racing cars, and the V-12 inboard/rear engine Type 63. He designed the 1963-model Quattroporte, Type 151 racing coupé, the Indy and Ghibli, and the Bora. In 1968 he stayed in his old office in Via Ciro Menotti, but his title was changed to engine consultant for Citroën. He chopped a planned V-8 to make the V-6 for the Citroën SM and Maserati Merak. When Citroën abandoned Maserati in 1975, it was not long before A. De Tomaso showed up and told him "You have 15 minutes to get out of here."

He went home and handled a variety of engineering assignments from Laverda (farm machinery) and other clients. For a brief period he was president of Honda Italia di Atessa. Patrick Mimran hired him to run the Lamborghini plant at Sant'Agata Bolognese in 1979. He redesigned the 1977 Cheetah prototype, which had a rear-mounted Chrysler V-8, with a front-mounted Lamborghini V-12. He also designed multi-valve cylinder heads for the LP 400 engine and an eight-liter V-12 for off-shore speed boats. He resigned when Lamborghini came under Chrysler's control, and spent his working hours helping his son build up an electronics-and-computer business. He died on March 11, 2002.

   1. Cantieri Navali Riuniti, a subsidiary of Piaggio S.A.

Allard, Sydney Herbert (1910–1966)

Managing director of Allard Motor Company, Ltd., from 1945 to 1960.

He was born on June 9, 1910, as the son of a prosperous building contractor in West London. He was educated at Ardingley public school and St. Saviour's College, Ardingley, spending his free time as a mechanic's helper in a London garage. Later on he worked in a Daimler dealer's service department. He began racing a Morgan, with some success. In 1929 his father bought him a garage in Putney and secured a Ford franchise. He began entering trials with Ford cars, but results were disappointing. In 1936 he built his first Allard Special with Ford V-8 power, and started winning. He got a lot of requests for similar cars, but World War II put paid to any ideas of going into production. He spent the war years operating an Army Auxiliary Workshop at Fulham, London, under contract with the Ministry of Supply. In partnership with his brother Denis and the engineer Reginald J. Canham, he founded a company in 1945 to start building Allard sports cars, and moved into a factory in Clapham High Street in 1946. The cars were noted for their peculiar divided front axle with a transverse leaf spring. Ford power was used exclusively until 1950, when he introduced the J2 with Cadillac or Oldsmobile V-8s and a de Dion–type rear axle. In 1952 he had a work force of 250 and built 500 cars a year. But his basic market was disappearing.

From 1952 to 1956 he offered the smaller Palm Beach with four- or six-cylinder Ford engines, and he built the Clipper three-wheeler from 1955 to 1959. He closed the factory in 1960 and died on April 12, 1966, from cancer.

Allmers, Robert Anton Hinrich (1872–1951)

Managing director of Hansa Automobil-GmbH from 1905 to 1914; managing director of Hansa Automobil-Werke AG from 1914 to 1921; managing director of Hansa Automobil-und Fahrzeug-Werke from 1921 to 1929.

He was born on March 10, 1872, at Absen in Oldenburg as the son of the owner of a daily newspaper. He studied law and social science in Freiburg, Munich and Berlin, and graduated with a law degree from the University of Berlin in 1896. In 1897 he inherited his father's publishing house and printing works, and made his newspaper an influential liberal voice in German national politics. Financed by his father-in-law, Franz Koppen, a local banker and owner of an ironworks, he founded Hansa Automobil-GmbH in partnership with August Sporkhorst at Varel in Oldenburg in 1905. The Hansa cars were a commercial success, and in 1914 he decided to take over Namag[1] of Bremen, but the purchase was financed by a new stock issue which gave Heinrich Wiegand, chairman of Nord Deutsche Lloyd, majority control of the new Hansa company. In 1920 Dr. Allmers took the initiative to set up the GDA[2] as a joint marketing and sales organization for Hansa, Hansa-Lloyd, NAG[3] and Brennabor. The GDA took over 19 dealerships and factory outlets from the manufacturing companies. In 1921 he decided to break up the union with Hansa-Lloyd, and a new Varel-based Hansa company was established with North German bankers as the main shareholders. In 1923 Jakob Schapiro acquired 30 percent ownership of

Hansa Automobil- under Fahrzeug-Werke but sold out to a Dutch holding in 1926. The company ran at a loss in 1927 and 1928, forcing him to negotiate a renewed merger with Hansa-Lloyd on unfavorable terms, trading 120 Hansa shares for 20 Hansa-Lloyd shares. In 1929 both companies were sold to C.F.W. Borgward and Dr. Allmers resigned. He found a staff position with the RDA[4] and moved to Berlin. He was elected chairman of the RDA in 1933, and over the next ten years he did everything possible to resist Nazi state control over the industry. But in 1943, when Albert Speer, Hitler's minister of arms and munitions, placed the RDA under his authority, Dr. Allmers retired to Burg Thurant on the Moselle River. In 1945 the U.S. Army arrested him. He fell ill during the investigation and was sent to a hospital in Reims, where he recovered and obtained his release. He settled in Koblenz where he died on January 27, 1951.

1. Norddeutsche Automobil- & Motoren-AG, Bremen, a subsidiary of Nord Deutsche Lloyd, shipowners with a fleet of ocean liners and cargo vessels.
2. Gemeinschaft Deutscher Automobilabriken (German Automobile Manufacturers' Partnership).
3. Nationale Automobil-Gesellschaft AG, Berlin-Oberschöneweide.
4. Reichsverband Deutscher Automobilindustrie, the national association of automobile manufacturers.

Alzati, Eugenio (1935–)

Deputy director of production for Lancia e C. from 1974 to 1979; managing director of Ferrari Automobili from 1979 to 1985; general technical director of Alfa Romeo S.p.A. from 1985 to 1989; managing director of Officine Alfieri Maserati S.p.A. from 1993 to 1996.

He was born on June 30, 1935, in Turin and graduated in 1965 from the Turin Polytechnic with a degree in aeronautical engineering. He began his career as a production engineer with Fiat at Mirafiori and worked on production programming for the newly erected Rivalta plant. In 1969 he was delegated to Lancia and reorganized production in the Chivasso plant. He laid out the materials flow and assembly processes for the 2000, the Beta, the Gamma, and made the preparations for

the production of the Delta and the Y-10. He was transferred to Maranello in 1979 to clean up the management and administrative problems at Ferrari, and to Arese (Milan) in 1985 to do the same thing on a bigger scale for Alfa Romeo. When Fiat S.p.A. won control of Alfa Romeo S.p.A. in 1987, it was combined with Lancia into a new subsidiary: Alfa Lancia Industriale S.p.A. When the Arese operations were running smoothly, he was dispatched to Brazil in 1989 to sort out the internal troubles of Fiat Automoveis at Belo Horizonte, then producing the 127, the Uno and its Premio, Elba and Fiorino derivatives. He got the Tempra into production in 1992 and began to prepare for the Punto. When he returned to Italy in 1993, Fiat had just taken control of Officine Alfieri Maserati. He was stationed in Modena to evaluate the industrial potential and take steps to put the plant on a profitable basis. He reorganized procurement and rigged up a new (temporary) assembly line. He resigned when Agnelli decided to place Maserati under Ferrari management and left the Fiat organization at the end of 1996. In January 1998, he joined Automobili Lamborghini to supervise engineering, development, purchasing and production, but resigned when it was sold to Volkswagen AG a few months later.

Andreau, Jean (1890–1953)

Designer of Citroën's first six-cylinder engine; chassis engineer with Donnet in 1932–33; director of Chausson's aerodynamic research center from 1934 to 1939.

His family expected him to pursue a military career and he attended St. Cyr military academy. He was attached to the Army's technical services in World War I, devoting all his free time to studies of artillery and aviation. When the war ended, his brain was busy with the problems of the road behavior of automobiles. He applied for a patent for an automatic free-wheel in 1918, and in 1921 started a laboratory for automobile suspension studies in Bordeaux. In 1924 he invented a variable-stroke engine, which was produced in some quantity by Engrenages Citroën. That led to a meeting with André Citroën, who asked him to design a six-cylinder engine. The result was the C.6, first shown in October 1928. He remained with Citroën until 1932, when Jerome Donnet engaged him to design the chassis and suspension for a new front-wheel-drive model. When Donnet declared bankruptcy, he joined Laboratoires Chausson and began scale-model wind-tunnel testing of cars, using a mirror-image system to simulate the roadway. He advised Peugeot on the advantages of streamlining and made suggestions that were first adopted for the 402 sedan and in 1937 he tested a full-scale Peugeot-chassis streamliner which had a drag coefficient of only 0.28 when the industry norm was 0.50.

He also designed streamlined body shapes that were built on Delage and Talbot chassis in 1938–39 and acted as a consultant to the Labourdette coachbuilding firm since 1937. He designed the body for Captain G.E.T. Eyston's Thunderbolt land-speed record car which ran at Bonneville in 1938. During the war he dusted off the drawings of an egg-shaped three-wheeler, made in 1934. That was the basis for the 1947 Mathis VL-333, planned for mass production, but canceled. He designed the DB bodies from 1946 to 1951.

Despite many years of ill health, his spirit remained indomitable to the end. He died in 1953.

Andren, Bertil T. (1916–)

Executive director for car engineering, Ford of Europe, Inc., from 1971 to 1981.

He was born on December 16, 1916, in Milwaukee, Wisconsin, and was a student engineer and project engineer with Chrysler Corporation before continuing his studies at the University of Wisconsin, where he graduated in 1947. He joined Ford Motor Company as a senior project engineer in the Lincoln-Mercury division. In 1955 he was transferred to a corporate engineering office working on international and North American projects, and invented a number of suspension systems. He led a team engaged on the Cardinal project, a small front-wheel-drive car, to be priced below the Falcon, intended for production in the United

States. But Ford Division general manager Lee A. Iacocca was opposed to it, and the Cardinal was given to Ford-Werke AG, where it was developed into the Ford Taunus 12M, making its debut in 1962. Andren was then assistant chief engineer and in 1964 was promoted to engineering and product planning manager for the Ford Motor Company. But in 1967 he was sidelined, and named assistant general manager for product development of Ford Tractor Operations. His next promotion arrived in September 1971, when he moved to Warley, Essex, to take charge of new-model programs for Ford of Europe, Inc. An all-new Escort had been launched in 1967 and the Cortina had been fully redesigned for 1970. The Consul/Granada range was ready for production when he arrived. He started a front-wheel-drive Escort program and supervised its engineering. It went into production at Halewood and Sarrelouis in 1980. He also planned and directed the Sierra program, a modern rear-wheel-drive car with all-independent suspension and a wide choice of engines from 2.0 to 2.8 liter. It replaced the Taunus/Cortina range in August 1982, with final assembly at Dagenham, Genk in Belgium, and Cork in Ireland. He had returned to the United States in 1981 and retired, having reached the age limit.

Appleby, William Victor (1903–)

Chief engineer of engine and gearbox design, British Motor Corporation, from 1953 to 1968.

He was born on August 10, 1903, in Liverpool and educated at Wolverhampton Technical School. He began his career by joining ABC Coupler, Ltd., as a machine operator in 1918, but left to become a draftsman with Sunbeam at Wolverhampton in 1919. After two years in the Sunbeam drawing office, he was transferred to the experimental department, headed by J. S. Irving. He left Sunbeam in 1925, spent two years as a junior draftsman with Vauxhall, and then joined Standard Motor Co. Ltd. as a component designer. In 1928 he went to Bean Industries of Tipton, Birmingham, where he was named chassis design section leader in 1930. In 1931 he joined Thornycroft at Basingstoke as an engine designer for heavy truck applications. He went to Austin in 1934 and worked under T. Murray Jamieson on the supercharged twin-cam 750-cc racing car project. He was transferred to Austin's production-car design office in 1937 and in 1942 was promoted to section leader in the Austin engine design department. By 1952 he was Austin's superintendent of engine design. He had started the A series engine project in 1950. It was mass-produced in several varieties for the A-35, the Minor 1000, the Sprite, and the Mini. In 1961 he was granted a patent for a narrow-angle V-6 with slightly staggered bores, and in 1963 another patent for a narrow-angle V-6 with a common cylinder head and a single overhead camshaft. He designed the E-Series in-line four with chain-driven overhead camshaft and integral gearbox for transverse installation in the Austin Maxi, which was launched in May 1969. But he had retired a year earlier, when the merger talks with Leyland Motors, Ltd., were going on.

Austin, Herbert (1866–1941)

General manager and chief engineer of the Wolseley Tool & Motor-Car Company, Ltd., from 1901 to 1905; managing director of the Austin Motor Company from 1905 to 1914; chairman and managing director of the Austin Motor Company, Ltd., from 1914 to 1941.

He was born on November 8, 1866, at Little Missenden, Buckinghamshire. The family moved to Wentworth in Yorkshire in 1870, and he attended Rotherham Grammar School and Brampton Commercial College. An uncle brought him to Australia in 1882, where he worked in Mephan Ferguson's engineering shop in North Melbourne for two years before becoming works manager of Richard Parks & Co., which imported and installed Crossley gas engines and printing presses in Australia. In 1886 he went to work for Langland's Foundry Company in Melbourne, where he attended classes at Hotham Art School in his spare time. He joined the Wolseley Sheep Shearing Machine Company in Sydney in 1889 and three years later was selected to return to Britain as superintendent of Wolse-

ley's branch factory off Broad Street, Birmingham. In 1895 the company moved into Sydney Works, Alma Street, in the Ashton district of Birmingham, where he designed and built a motor tricycle. He built an improved motor tricycle in 1897, and created the first Wolseley car in 1899 with a front-mounted horizontal single-cylinder water-cooled engine and chain drive to the rear wheels. He set up a motor car department and began production. In 1901 Wolseley sold the motor car department to Vickers Sons & Maxim, which moved production to Adderley Park, Birmingham, and set it up as a separate company, the Wolseley Tool & Motor-Car Company, Ltd. A 10-hp model with a flat-twin engine was added in 1901 and produced up to 1905. Austin resigned in 1905 because the Vickers brothers wanted to make a new car, designed by John D. Siddeley. He won the financial backing of Captain Frank Kayser, director of the Sheffield Steel Company, and Harvey du Cros, Jr., to set up his own company. With two assistants, A.J.W. "Joey" Hancock and A.V. Davidge, he began to design a 25/30 hp car with a vertical front-mounted T-head engine and chain drive in September 1905. Their drawings were displayed at the Crystal Palace Motor Exhibition in London in November 1905. He had found a disused printing works at Northfield, Birmingham, which he bought for £7,750 and retooled for car production, which began in May 1906. Austin produced 147 cars in 1907 and introduced 17 new models from 1906 to 1909, including a six-cylinder model in 1908 and a single-cylinder 7-hp car in 1909. In 1910 Austin made 576 cars and began offering a line of marine engines, including a 380-hp overhead camshaft V-12. Operations were profitable, and in February 1914, the company "went public" with a capital stock of £250,000. Longbridge Works was converted to war production, and the company made a difficult restart in the postwar car business. From 1919 to 1921, only one model was built, the 3.6-liter four-cylinder 20 hp. He tried to sell the company to General Motors in 1919–20, but the offer was far too low. He then turned to Henry Ford but failed to start serious negotiations. In March 1921, the company was placed in receivership.

He withdrew from its daily affairs, obtained a personal loan of £200,000 from Lord Leigh (1885–1938) of Stoneleigh Abbey, who later became chairman of the Standard Motor Co., Ltd., and went to work at his home at Lickey Grange, Bromsgrove, with just one assistant, Stanley Edge, on the design of a new car for the mass market. Seeing how successful Renault had been with two-cylinder models, Austin was in favor of a parallel-twin. But Edge said it was too crude, and the engine should have four cylinders, like the Peugeot Quadrilette. Hancock and Engelbach had rushed the Austin 12 with a 1.6-liter four into production for 1922, and by March 1922, the company was out of receivership. Still, only 2600 cars were made at Longbridge that year. Austin returned with his economy car project, and gave it to the company against a royalty on every Austin Seven it sold. It came on the market in 1923, priced at £165, compared with £299 for the two-seater Morris Cowley. In 1926 over 25,000 cars were turned out. Austin had tried to sell out to General Motors again in 1924–25, but refused the £1.3 million offer. He next proposed a merger with Morris Motors, Ltd., but W.R. Morris turned a deaf ear. In 1926–27 Austin and Morris fought for control of Wolseley Motors, Ltd., but Morris won. Austin responded by broadening his model lineup. A new six-cylinder 20 was introduced in 1926 followed by a six-cylinder 16 in 1927. Austin took part in the planning of every new model, but played little part in the design work. A.J.W. Hancock was the chief designer, John Clarke the head of the engine drawing office, and George Parker designed and engineered the bodywork. By 1930 Longbridge had 17,000 workers, 6,000 machine tools, and produced 60,000 cars a year. The 10 and the new 12/4 appeared as 1932 models. By 1934 the Ten outsold the Seven, which was finally replaced by the 8 in March 1939. The Austin 14 came on the market in October 1934 and replaced the 16 for 1936. Herbert Austin had been knighted in 1917 and in 1936 became Baron Austin of Longbridge.

Birmingham University gave him an honorary doctor's degree in law in 1937. In his lifetime, his name was on 180 patents. He died on May 23, 1941, after several months of pneumonia and a heart attack.

Axe, Royden (1937–)

Styling director of Chrysler UK from 1970 to 1976; director of styling, BL Light Medium Cars Division, from 1981 to 1987; design and concept director of Austin-Rover from 1987 to 1990.

He was born in 1937 in Lincolnshire and began sketching cars at the age of six. In 1959 he joined the Rootes Group for a four-year car body-engineering course, and upon graduation was invited to work in the styling studio. He was named chief stylist for the Rootes Group in 1967. After he had designed the 1967 Sunbeam Rapier, he was transferred to Chrysler's styling center at Highland Park, Michigan. In 1981 Harold Musgrove brought him back to England to design new Rover cars. He designed the Rover 800 and the 200/400 series. He resigned from Rover in 1990 to form Design Research Associates, Ltd., at Warwick. In 1994 he co-designed with Graham Hull, the Rolls-Royce design chief at Crewe, the Bentley Java prototype.

Bache, David (1928–)

Styling director of Rover from 1955 to 1982, Austin from 1975 to 1982, Triumph and Jaguar from 1978 to 1982.

Born in 1928, he joined the Austin Motor Company at Longbridge as a student engineering apprentice in 1948 and attended evening classes at the Birmingham College of Art and Birmingham University. He worked under Dick Burzi in the Austin design studio until 1953 when Maurice Wilks invited him to Rover. His ideas were far too radical for Maurice Wilks, but he was allowed to raise the deck lid on the P4 for 1955 and add wraparound tail lights. Two years later Wilks approved his proposal for new front fenders on the P4.

His first big job at Rover was the P5 body, though it was based on a Pinin Farina prototype on the P4 chassis. Marketed as the Rover

3-Litre, the P5 went into production in 1959. He styled the Rover T4 gas-turbine car, which set the pattern for the P6 Rover 2000 introduced in 1963. He wanted to copy the Ghia-built Chrysler Dart prototype, but was forced to hang on a grille and remove the tailfins.

He designed an angular coupe as a Rover-Alvis P6 BS for British Leyland in 1966–67, which was condemned as a potential rival for Jaguar. He also designed a two-door "Gladys" hatchback on the Rover 3500 platform, built by Harold Radford in 1967, that never went anywhere. Inspired by the Maserati Ghibli, he designed the Rover SD1 (3500) launched as a four-door hatchback in June 1976.

In 1975 he was given styling responsibility for all British Leyland cars, just in time to put his personal touch on the Austin Metro. He modernized the Jaguar XJ and Daimler Sovereign for 1979 and did the preliminary outlines for the XJ-40 project before resigning in 1982, to open his private design-consulting practice.

Baguley, Ernest Edward (b. 1863)

Technical director of the Ryknield Motor Company, Ltd., from 1902 to 1906; manager of the BSA[1] motor vehicle department from 1907 to 1911; managing director of Baguley Cars, Ltd., from 1911 to 1921.

He was born in 1863 at Burton-on-Trent and served an apprenticeship with steam-locomotive builders Hawthorne, Leslie. He began his career as an engineer with Bagnall's of Stafford, but left start his own business. In February 1902, he established the Ryknield Motor Company, Ltd., at Burton-on-Trent and in 1903 began producing 10/12 hp car with a vertical-twin engine. It was soon supplemented by one-ton commercial vehicles based on the same chassis and power train. A 15-hp three-cylinder and 24-hp four-cylinder models followed in 1905. He left Burton-on-Trent in 1906 and the following year found a position with BSA at Small Heath, Birmingham, where he designed a range of cars. The 2½-liter 14/18 and the 4.2-liter 18/23 had four-cylinder L-head engines, and the 5.4-liter 25/38 was closely copied on the Itala that won the Peking-to-Paris trek in 1907, with a T-head four-cylinder engine. After the BSA merger with Daimler in 1910, F. Dudley Docker dismissed him as being "incapable of organization." He returned to Burton-on-Trent and purchased his old factory, now operating as the Ryknield Motor Company, Ltd., and transformed into Baguley Cars, Ltd. He began making a 15/20 hp car which was near-identical with the BSA 14/18, powered by an L-head four, with a four-speed gearbox and worm-drive rear axle. In 1915 he took over the Salmon Motor Co. and moved production of its Ace light car into the Baguley works. The Ace had an 850-cc four-cylinder engine, two-speed gearbox, and chain drive. Car production was suspended during World War I, and in 1920 he tried to revive the old 15/20 model but failed. In 1922 he converted the factory to produce Drewry electric railcars, which became the basis for the start of Baguley-Drewry, Ltd., where he was active for the remaining years of his career.

1. Birmingham Small Arms Company, Ltd.

Bahnsen, Uwe (1930–)

Chief designer, Ford of Britain's advanced and interior design studio from 1967 to 1970; chief designer of the Ford-Werke AG exterior styling studio from 1970 to 1976; vice president of design, Ford of Europe, from 1976 to 1986.

He was born on May 30, 1930, in Hamburg and graduated from the Academy of Fine Arts, Hamburg, in 1950. He spent the next seven years in advertising and graphic design, during which time he got involved with motor racing and designed the body for a racing car. That led to his appointment at Ford-Werke AG in 1958 as a member of the design staff, then headed by Wesley P. Dahlberg. He was transferred to Dearborn as design manager in 1962, returning to Cologne in 1963 as design executive for exterior design and design engineering. He worked on the Granada and the second-generation Capri, and led the design team for the 1975 Escort. He created an entirely new Ford style for the 1982 Sierra, followed up by the 1985 Scorpio. He resigned from Ford of Europe, Inc., in October 1986 to become chief of education at the Art Center

College of Design in Lausanne, a position he held until his retirement in May 1995.

Baily, Claude Walter Lionel (1902–1988)

Chief engine designer of Jaguar Cars, Ltd., from 1948 to 1952; executive engineer from 1952 to 1960; executive director of power unit design from 1960 to 1968.

He was born in 1902 at Twickenham as the son of John Baily, maker of furniture for coaches. He was educated at Richmond Hills School, Surrey, and the Henry Thornton School at Clapham Junction. He went on to study mechanical engineering at the Regent Street Polytechnic in London and joined the Anzani Engine Company, Ltd., in London in 1918 as an apprentice, and became an engine draftsman for Anzani in 1922. When the Anzani factory closed in 1926, he worked successively for Sunbeam, Villiers, and Henry Meadows for short periods. In 1928 he joined Morris Motors (Engines) Ltd. in Coventry, rising to the rank of chief draftsman and later, assistant chief designer. He designed the side-valve Morris 8 engine with a three-bearing crankshaft, which went into production in 1934. He resigned from Morris in 1938 and worked with Electrolux, then with Rotol, before joining SS Cars, Ltd., as a draftsman in 1940. Under W.M. Heynes's direction, he designed overhead-valve versions of the four- and six-cylinder Standard engines for installation in Jaguar cars. In 1945 he worked on a number of experimental engines, such as the four-cylinder twin-cam XF and the four-cylinder splayed-valve XG. The XJ was tested in both four- and six-cylinder versions before being redesigned as the XK-80 (four) and XK-120 (six). He was responsible for the larger versions (3.8-liter in 1958, 4.2-liter in 1964) as well as the short-stroke derivative 2.4-liter of 1955. He went into retirement in 1968, settling near Lake Coniston in Westmorland, where he died on February 15, 1988.

Barbarou, Marius J.B. (b. 1876)

Chief designer of the Clément-Bayard and Clément-Gladiator, 1899–1902; chief engineer of Benz & Cie in 1902–03; chief engineer of Delaunay-Belleville from 1903 to 1914; technical manager of Lorraine-Dietrich from 1914 to 1924; technical director of Lorraine-Dietrich from 1924 to 1930; managing director of Lorraine-Dietrich from 1930 to 1936.

He was born at Moissac (Tarn-et-Garonne) in southwest France on October 28, 1876. He designed and built a car in his own garage in 1898, a light car with an air-cooled V-twin engine.

Clément & Cie engaged him in 1899 to prepare a range of new models, beginning with a chain-drive 7 hp single-cylinder model and a shaft-driven 12-hp two-cylinder car. Both had front-mounted engines. He also designed the Gladiator models, including powerful racing cars, for Clément.

Benz's business manager Julius Ganss brought him to Mannheim at the end of 1902 to put new ideas into Benz cars. But Barbarou was to have little influence there, for the Benz-Parsifal range of four models (8/10, 10/12, 12/14 and 16/20 hp) launched in 1903 were mainly the work of Fritz Erle. Barbarou returned to Paris in 1903 at the invitation of Delaunay-Belleville, an old, established engineer firm at L'Atéliers de l'Hermitage, Saint-Denis (Seine) which had been building experimental cars since 1901.

Barbarou introduced shaft drive on the 1905 16-hp model, but chose chain drive for powerful 24-hp and 40-hp four-cylinder models. His six-cylinder 15 CV of 1907 set all French makers of luxury cars on a new course. By 1909, Delaunay-Belleville was offering three six-cylinder models from 2½ to 11.8 liters size. From 1911 onwards, all models were shaft-driven.

In 1914 he left Saint-Denis for Lorraine-Dietrich at Argenteuil and designed some remarkable aircraft engines, such as a double-overhead-camshaft in-line six with splayed valves, and single-overhead-camshaft V-8 and V-12 engines of up to 400 hp. In 1920 he even created a 1000-hp W-24 engine that was built and tested but not marketed.

The B-26 15 CV car, introduced in 1920, was powered by a six-cylinder 3445-cc overhead-valve engine with two spark plugs per cylinder and a crossflow head. The first version was rated at 72 hp, but the B-26 Sport, introduced in 1924, claimed 90 hp. It won the 24-hour race at Le Mans in 1925 and 1926. Barbarou also designed a four-cylinder derivative which was built from 1922 to 1929. The 15 CV was replaced by a 20 CV L-head model in 1930, but demand was very slack and the company, an affiliate of the Société Générale Aéronautique since 1930, made its last car in 1935. The following year Barbarou joined Hispano-Suiza as an aircraft engine designer, going into retirement in 1940.

Barber, John Norman Romney (1919–)

Finance director of Ford of Britain from 1962 to 1965; finance director of British Leyland from 1971 to 1973; deputy chairman and sole managing director of British Leyland from 1973 to 1975.

He was born in 1919 and began his career as a clerk in the Welfare Department. He was promoted to department head in the Welfare Office in 1953 and left government service in 1955 to take a position with Ford of Britain, working in market analysis, forward planning, and financial control. He was in charge of cost-accounting for the Ford Cortina, which became Ford's most profitable car. He left Ford in 1965 to join Associated Electrical Industries and masterminded its merger with the General Electric Co. He was offered the chairmanship of GEC, but gave in to Sir Donald Stokes's invitation to join British Leyland which took over British Motor Corporation on January 17, 1968. He applied Ford methods to cost-accounting the Morris Marina and the Austin Allegro, acting also as the group's chief business strategist. He deliberately priced Austin and Morris cars above Ford and Vauxhall, to reap higher profits. But the Allegro failed to meet its market-share target, and the profits never materialized. He aimed the Rover SD-1, appearing in 1976, at a higher price bracket than its P5 (3-Litre) predecessor, and invested heavily (£31 million) in new plant and equipment at Solihull, Birmingham. But

British Leyland suffered heavy losses, and he was given a lot of blame for the mounting debts. He was forced to resign in 1975.

Barenyi, Bela (1907–1997)

Pre-development engineer in the Mercedes-Benz advanced engineering section from 1939 to 1972.

He was born on March 1, 1907, at Hirtenberg near Vienna and received a technical education. He was a trainee engineer with Steyr in Vienna and later worked on the engineering staff of Austro-Fiat. As early as 1925 he published proposals for modern economy cars, some with rear engines, others with front-wheel-drive. One of his designs made prior to 1931 anticipated the basic ideas that went into the Volkswagen and were credited to Dr. F. Porsche. Patent litigation started in 1951 challenging the validity of some VW patents was resolved three years later when the German patent office ruled in Barenyi's favor.

In 1934 he spent a few months with Adler, and then became an agent for Clemens August Vogt, a patent lawyer who represented Chrysler in talks over licenses for "Floating Power" engine suspension. He went to Berlin to work for the Society for Technical Progress, and spent some time in Paris as head of its French Brand Office. After a brief stint with the Belgian State Railways in Brussels, he went to work for Norton (motorcycles) in Birmingham.

He joined Daimler-Benz AG in 1939, and in 1940 he proposed a crash-safe vehicle structure with rigid floor, non-formable passenger compartment and progressively collapsing zones front and rear, built-in rollover bar, side-impact protection, full-perimeter bumper system, and a collapsible steering column.

His safety-car ideas were backed up by further patents from 1947 and 1951. Some elements of his inventions were used for the Mercedes-Benz 180 sedan of 1953, and in the 220 series of 1959. By the time of his retirement in 1972, his name was mentioned on no less than 2500 patents. He settled in Vienna, where he died on May 30, 1997.

Bartlett, Charles John (1889–1955)

Managing director of Vauxhall Motors, Ltd., from 1930 to 1953, and its chairman from 1953 to 1954.

He was born in 1899 at Bibury, Gloucestershire, and after a high-school education went on to take a course in business methods and accounting at Bath Technical College. He began his career as a clerk with General Motors, Ltd., at the Hendon depot, then busy receiving and preparing numbers of Canadian-built Buick chassis for the British market. He rose in rank to head of the accounts department and in 1926 was named general manager of General Motors, Ltd., which had acquired Vauxhall Motors, Ltd., in 1925. He was transferred to Vauxhall in 1930. The company was operating at a loss, with sales of only 1278 cars in 1929. He brought out a line of Bedford vans and trucks in 1931, with sales or more than 11,000 units the first year. He revitalized the car business, topping 4000 units in 1931, and set Vauxhall on its new course with the introduction of the Cadet late in 1930. He relied heavily on commercial and technical guidance from Detroit (and even from Opel) and by 1938, production topped 60,000 vehicles.

During World War II, he directed Vauxhall's activities to great effect, building three sizes of military trucks to a total of 250,000, plus 5640 Churchill tanks and vast numbers of engine parts for Bristol and aircraft components for De Havilland, A.V. Roe, and Hawker.

When civilian production resumed, he coped with material shortages, militant labor unions, a national economy in decline, and the need to develop new export markets. As early as 1947 Vauxhall's car production was back up to 30,000 units. He was able to maintain Vauxhall in fourth rank among Britain's car makers, behind the Nuffield Organization, Austin, and Ford, though constantly challenged by the Rootes Group and Standard-Triumph. He retired in 1954 and died in 1955.

Bashford, Gordon (1917–1991)

Design engineer with Rover from 1935 to 1940; head of advanced car engineering for Rover from 1946 to 1972.

He first came to Rover as an office boy in 1931 and then worked with Alfred Herbert in Coventry to get machine-tool experience. Returning to Rover in 1935, he worked in the drawing office under Robert Boyle and became an expert on suspension systems. During World War II he became involved with gas-turbine engine research and the development of the Meteor tank engine. He designed the new chassis for the P3 with its underslung rear end and the P4 chassis for 1950 with its full-length frame and rear-axle kickup. He also designed the mounting system for the gas-turbine installation in the T-1 vehicle in 1949–50, and was in the center of the Land Rover development from 1945 to 1950. His ideas were not expressed to any extent in the 1958 P5 (3-Litre) but blossomed in the 1963 P6 (2000) with its front coil springs mounted horizontally against the bulkhead and its unique version of a de Dion type rear end. He designed the Range Rover frame, front and rear axle suspension, and started the design of the suspension systems for the SD-1 program before his retirement in 1972.

Basshuysen, Richard van (1932–)

Head of NSU engine testing from 1965 to 1975; chief engineer of Audi engine development from 1976 to 1985; executive engineer for the Audi V-8 from 1985 to 1988; responsible for all Audi power unit and vehicle development, upper-class models, from 1988 to the end of 1989.

He was born on February 29, 1932, and studied power-unit engineering in college. He joined Aral (motor fuels) in its Bochum laboratories in 1955, came to NSU in 1961 and worked on Wankel engine development under Walter Froede.

He had a central part in the development of the Ro-80 twin-rotor engine which went into production in 1967 and a long series of improvements up to 1976. He worked on a number of piston-engine programs under Franz Hauk's direction, and broadened his scope into complete vehicle concepts. He led some of the most crucial new-car programs for Audi until he came into conflict with F. Piech and resigned in 1990. He became publisher of the

MTZ[1] and ATZ,[2] technical monthly magazines on power-unit and automobile engineering.

1. Motor-Technische Zeitung.
2. AutomobilTechnische Zeitung.

Bastow, Donald (1909–1989)

Assistant to the technical director of Lagonda from 1944 to 1947; chief engineer of Jowett Cars, Ltd., from 1952 to 1954; assistant chief engineer of Coventry Climax in 1955–57.

He was born on June 8, 1909, at Bolton-le-Sands, and educated at Dulwich College and University College, London. He worked as a post-graduate trainee with Daimler in Coventry from 1929 to 1932 and then joined Rolls-Royce as a junior draftsman on chassis design. Soon he became a section leader in the car design office and drew up the independent front-suspension systems for the Phantom III, the Wraith, and the Bentley Mark V. In 1941–42 he designed the suspension system for the Cromwell tank. He went to Lagonda in 1944 as a technical production engineer and became W.O. Bentley's assistant. He worked on calculations for the postwar car and handled Bentley's liaison with the manufacturing engineers. In 1947 he accompanied W.O. Bentley to an office at Weybridge, working under a consultancy contract with Armstrong-Siddeley. When that arrangement ended in 1950 he signed up with BSA Group Research as manager of the engine and mechanism laboratory, which he ran for two years.

He went to Jowett at Bradford, Yorkshire, in 1952 and worked on the Jupiter R4, and planned a new-generation range of cars, but found himself out of a job when Jowett stopped car production in 1954.

He went to Metalistik Ltd. as technical assistant to the managing director, and then joined Coventry Climax as assistant to Wally Hassan. In 1956 he became chief engineer of Coventry Climax electric vehicles (fork-lift trucks), and a year later joined Birfield Engineering. He invented and patented a modified Rzeppa constant-velocity universal joint, eliminating the locating pin by changing from a toroidal to an elliptical ball track. This was tremendously significant, for it brought down the Rzeppa joint's cost to the point of making its application on low-priced front-wheel-drive cars economically feasible. Without it, the BMC Mini might never have come into being. The Birfield-Rzeppa joint, in fact, started a whole wave of front-wheel-drive car projects throughout Europe. When Birfield was taken over by Hardy-Spicer in 1959, he was named chief engineer of Hardy-Spicer. He resigned in 1963 and spent a year with Dobson & Barlow, Ltd., and then two years with TMM (Research) Ltd. He opened a consulting engineer's office in Cheshire in 1966, moving to Woodland End at Mells in Somerset two years later, with a drawing office in his private home.

Beach, Harold (1911–)

Chief engineer of Aston Martin from 1953 to 1964; technical director of Aston Martin Lagonda, Ltd., from 1964 to 1974; technical consultant to Aston Martin Lagonda, Ltd., from 1975 to 1978.

He was born in 1911 in London and began his career with Barker & Co., coachbuilders in North Kensington, starting in the machine shop, later working in the drawing office. After five years with Barker, he joined Beardmore as a draftsman at their Clapham works. In 1937 he joined James Ridlington's JR Engineering to work on the design, engineering and construction of racing cars and speed-record vehicles. In 1939 he became a design engineer with Garner Straussler Mechanization at Park Royal, specialists in all-wheel-drive vehicles of all sizes, primarily for military purposes. One of his first tasks was the installation of two Ford V-8 engines with a common drive line in a special vehicle. He remained with Garner Straussler throughout the war, and in 1946–47 was associated with its tractor projects for civilian markets. He left in 1950 when he was signed up as a draftsman by Aston Martin at Feltham, Middlesex. He was assigned to a DB2 replacement project. He worked in close association with Professor Eberan and assisted him in designing the DB3S chassis and Project 117, intended for a future Lagonda V-12. In 1953 he started work on Project 114, intended as a replacement for the DB2/4 Mark III. The initial version had a perimeter frame, 3-liter Lagonda engine, and de Dion rear suspension. It was redesigned with a platform chassis when John Wyer chose Carrozzeria Touring for body design and construction, and a coil-sprung rigid rear axle. The first DB4 prototype ran in July 1957 and production began in September 1958 in the former Tickford factory at Newport Pagnell. The DB5 also had Touring's "Superleggera" body construction, but for the DB6 of 1965, Beach adopted conventional body construction, so as to permit a cabriolet version. The DBS, arriving in 1968, was an entirely new project with a fabricated steel structure and a de Dion rear end. The DBS V-8 was added a year later.

Beaumont, Mario Revelli De *see* Revelli De Beaumont, Mario

Becchia, Walter (b. 1896)

Technical director of Automobiles Talbot from 1925 to 1940; engine design engineer with Citroën from 1941 to 1964.

He was born on March 31, 1896, at Casale Monferrato. His technical studies were interrupted in October 1916 when he was called up for military service. He was assigned to the Italian air force and flew airships until the end of World War I. Immediately after his discharge in October 1919, he joined Fiat as a member of the engineering staff. He became a member of G. C. Cappa's team of racing car designers and helped create the Fiat 804 which won the French Grand Prix at Strasbourg in 1922. That provoked Louis Coatalen to lure two young engineers away from Fiat: Becchia was assigned to Talbot-Darracq at Suresnes, and Bertarione to Sunbeam in Wolverhampton. Becchia designed the four-cylinder twin-cam 11-liter Talbot for 1923 and its supercharged derivative for 1924. The following year Bertarione arrived at Suresnes to design future racing cars, and Becchia was placed in charge of touring-car engineering. His first non-racing machine went into production in 1927 as the M.67 11 CV 2-liter six. He fol-

lowed up with the P.75 17 CV 3-liter six in 1928 and the H.78 22 CV 3.8-liter straight-eight in 1929. In 1932 he adopted independent front suspension for all models, with a transverse leaf spring. When Tony Lago took over the company in 1934, he maintained Becchia in his functions. The T-120 3-liter six arrived in 1935, with a T 120 Sport version as the first hemi-head Talbot having splayed valves operated from a single camshaft by a clever arrangement of pushrods and rocker arms. The T 150 followed in 1936, with a 4-liter six having seven main bearings, marketed in five lines: Baby, Major, Master, Lago Special and Lago Super Sport. In 1940 the factory was retooled for aircraft-engine production, and Becchia felt insecure about the company's future. He left and signed up with Citroën in January 1941. They showed him a water-cooled flat-twin from 1936, intended for a rural-transport economy car. He redesigned it for air-cooling and developed it into a 9-hp 375-cc power unit for the 2 CV introduced late in 1948. He designed a water-cooled flat-six, intended for the DS-19, but it fell so far short of its performance targets that it was abandoned. He worked on a twin-cam 1100-cc four-cylinder unit for a potential sports coupe using the Panhard CT 24 body shell, but it came to nothing.

Beckett, Terence Norman "Terry" (1923–)

Manager of Ford of Britain's new product planning department from 1955 to 1966; chairman of Ford of Britain from 1976 to 1980.

He was born in 1923 in Wolverhampton and studied engineering at the Wolverhampton and South Staffordshire Technical College. He was on military duty throughout World War II, and in 1946 began studies at the London School of Economics, graduating in 1950. He joined Ford as a trainee and within a year was named personal assistant to Sir Patrick Hennessy. In 1954 he was transferred to the product planning department and became its manager a year later. He was responsible for planning the Cortina, a low-cost compact sedan of simple, conventional construction, at a time when the parent company in Dearborn and Ford-Werke AG in Cologne were developing a project of similar size, but of novel concept, with front-wheel-drive and a V-4 engine. The Cortina went into production at Dagenham in 1962 and for the next 20 years, it accounted for about half of Ford of Britain's profits. In 1966 he moved over into marketing and was named Ford of Britain's sales director in 1969. He became chairman of Ford of Britain in 1976 and watched the development of the front-wheel-drive Escort which was introduced in 1980. He left Ford that year to become director-general of the Confederation of British Industry, and retired from that office in 1987.

Behles, Franz (1928–)

Chief engineer of vehicle development with Audi NSU Auto Union AG from 1971 to 1974; director of BMW car concepts and engineering coordination from 1981 to 1987.

He was born in 1928 in Dresden as the son of an executive of Wacker-Chemie (electro-mechanical works), educated in local schools and called up for military service in the Luftwaffe in 1944. He went right back to school in 1945 and graduated from the Lessing College at Hoyerswerda in Saxony in 1946. After escaping to the West, he held several trainee jobs in engineering shops until 1948, when he was admitted to the Munich Technical University. He graduated in 1952 and began his career as a draftsman in the motorcycle engineering office of BMW in 1953. A year later he was able to join Daimler-Benz AG where he worked successively on racing-car engines, passenger-car chassis, vehicle pre-development, and body engineering. In 1952 he won a doctor's degree in engineering from the Munich Technical University, and a year later he was detailed to Auto Union (part-owned by Daimler-Benz AG) as a section chief for the pre-development of new vehicles. He was placed in charge of a drawing-office department in 1965 and reached director's rank in 1970.

In 1973 he was given complete charge of Audi vehicle development, with the additional responsibility of body shape and architecture. He directed the Audi 50 project (which became the first VW Polo) and the Audi 100, made from 1977 to 1982.

When the Audi engineering department underwent an organizational revolution in 1974, he became head of vehicle engineering, body shape and architecture, with supervisory duties over vehicle development. In 1977 he became a member of the strategy work group and in 1978 a member of the product planning committee. In 1980 he was named head of project management and pre-development. He was aiming for the title of technical director, which Ludwig Kraus had promised he would inherit on his retirement. But the powers-that-be in Wolfsburg brought in Ferdinand Piëch for that post, and Behles resigned in protest. He went to BMW, arguing relentless for the adoption of front-wheel-drive, only to be disappointed, time after time. He retired in 1987.

Bellanger, Robert (1884–1966)

Managing director of SA des Automobiles Bellanger Frères from 1913 to 1925.

He was born in 1884 in Paris, into a wealthy family owning industrial property. He had a general and commercial education, and became interested in automobiles as a teenager. He did not care about economy cars; he was only attracted to high-powered luxury cars. In 1907 he set up an agency for Westinghouse cars in rue Marbeuf, just off the Champs-Elysées, quickly adding the Delaunay-Belleville franchise. About 1910 he decided to make his own cars and contracted with Société des Automobiles Grégoire of Neuilly to produce them to his specifications. The first Bellanger, Type 70, appeared in 1912 with a four-cylinder Daimler sleeve-valve engine, four-speed gearbox, Lanchester worm-drive, and Riley wire wheels. With the financial backing of his brother Marcel, and to a lesser extent, Pierre and René, he took over the Grégoire works and founded his own company on July 21, 1913. Car production was not completely interrupted by World War I, but the Bellanger family had a big plant on Boulevard Dixmude in Paris which received contracts for aircraft components and industrial equipment, and took over boat-building yard[1] at St. Malo on the channel coast. The Neuilly plant delivered some armored vehicles. Postwar production began with Model A1, a relatively modest proposition powered by a 3-liter four-cylinder Briscoe engine, which was to find a market as a taxicab in London as well as Paris. Bellanger made its own engines for the luxury models, a 30-hp four and a 50hp V-8. The latter was aimed at the very top of the prestige market, but it may have lacked quality, for only 10 were built. The A1 remained in production through 1923 and the 30 hp model until 1925, when the Neuilly factory was closed. It was leased to Peugeot in 1926 and sold to Rosengart in 1927. Robert Bellanger decided to go into politics and was elected to the Chamber of Deputies (lower house of parliament) in 1927, later becoming a Senator. In 1928 he signed a contract with de Dion–Bouton for production of a new Bellanger car, the B1 with a 1328-cc four-cylinder side-valve engine. It was identical with the 8 CV de Dion–Bouton, and a small series was produced at Puteaux. In 1929 he obtained a long-term lease on the Chateau de Brégançon on the French Riviera, a fortress from about 1000, and spent a fortune on restoring it. In 1937 he established a real-estate business[2] for use of the Boulevard Dix-mude factory, still family property. He spent his retirement years in Paris and died there in June 1966.

   1. Chantiers Navales de l'Ouest.
   2. Société Immobilière et Financière pour l'Industrie Automobile.

Bellentani, Vittorio (1906–1968)

Technical director of Officine Alfieri Maserati from 1946 to 1955.

He was born on March 6, 1906, in the Emilia-Romagna province of Italy, began working as a mechanic and became a skillful draftsman. He designed the 175-cc Mignon motorcycle for Vittorio Guerzoni of Modena in 1931, and as a reward Guerzoni sent him to study engineering in Switzerland. He returned with a diploma in 1936, but his sponsor was

no longer in the motorcycle business. He worked in a variety of jobs in and around Modena before he met Enzo Ferrari in 1939, and was hired to help build the Auto-Avio roadster. He stayed with Ferrari throughout World War II, working as a machine tool designer. In 1946, he joined Maserati as technical director, with responsibility for the touring cars (A6G 1500 and derivatives). His authority also included the racing cars, which were Massimino designs. In 1952 Bellentani lured Colombo away from Alfa Romeo to design a new Grand Prix car (which became the 250/F in 1954), upon which Maserati lost Massimino's services. In 1953 Bellentani outlined the 150 S four-cylinder engine and began testing engines in speed boats before they were seen on the race tracks. That's how the V-8's and the big 3500 GT six engines were developed.

He left Maserati at the end of the 1955 racing season and promptly signed up with Ferrari, where he designed four-cylinder 3- and 3½-liter engine as well as new 3½ liter V-12. He assisted Alfredo Ferrari in the design of the 1½ liter "Dino" V-6 and assisted Andrea Fraschetti on racing car projects.

In 1960 he left Ferrari to set up his own machine shop in Modena. Later, he also operated a small aluminum foundry and an accessory business. Seriously hampered by ill health, he more or less retired before death came on March 26, 1968.

Bensinger, Jörg (1936–)

Development engineer with Audi NSU Auto Union AG from 1972 to 1987.

He was born in 1936 in Stuttgart as the son of a Daimler-Benz engineer, W.D. Bensinger, and studied engineering, graduating in 1960. After a period as a trainee at Untertürkheim, he joined Porsche as an experimental engineer. He came under Helmut Bott's orders and did valuable research work on road-holding and vehicle dynamics. Leaving Porsche in 1972, he went to Audi at Ingolstadt. In the final months of 1976 he was a member of an Audi test team on a winter-driving expedition to Finland. A number of experimental front-wheel-drive cars were being tested, but it was noticed that for running errands, going to lunch, etc. everybody preferred the ungainly low-powered off-highway four-wheel-drive VW Iltis they had taken along. Bensingen discussed this with another member of the expedition, Walter Treser, and they filed a report making a strong recommendation for a four-wheel-drive Audi passenger car. The report came to Piëch's desk very quickly. His reaction was favorable, and the four-wheel-drive project, went on the drawing board in February 1977. The first Audi quattro prototype was ready for testing in November 1977. The production model was first shown in March 1980 as a low sports-coupé with a 200hp turbo-intercooled 2144-cc five-cylinder engine.

Audi planned a series of 400 quattro, period. Eleven years later, 11,500 had been built and sold. In 1984, Audi began making four-wheel-drive versions of other models: 90 quattro, 80 quattro, and 200 Turbo quattro. Bensinger left Audi at the end of 1987 to work on advanced engineering projects for Uni-Cardan AG, the German subsidiary of GKN[1], makers of visco-couplings, constant-velocity universal joints, shafts and differentials.

1. Guest, Keen, Nettlefolds.

Bensinger, Wolf-Dieter (1907–1974)

Powertrain design and development engineer with Daimler-Benz AG from 1943 to 1972.

He was born in Baden-Württemberg in 1907 and educated as a mechanical engineer, graduating in 1931. He began his career with DVL[1] in Berlin as a research engineer for aircraft-engine control systems. Over the next 12 years he conducted experiments in several areas of aircraft engine design, aid in 1940 was delegated to take charge of the Wankel technical center at Lindau, then engaged in the development of rotary valves. He left DVL in 1943, returned to Stuttgart, and joined Daimler-Benz AG as an aircraft-engine test engineer. In 1945 he led preliminary studies for a new generation of passenger-car engines which came to fruition in 1951 when the sin-

gle-overhead-camshaft 2.2 and 3-liter sixes went into production at Untertürkheim.

His responsibilities increased to include advanced power systems, clutch and gearbox. When Daimler-Benz acquired a license for the Wankel rotary-combustion engine in 1960, the tasks of evaluation, estimating its potential, proposing changes, and planning its development and applications, fell on Bensinger. He designed single- and twin-rotor Wankel engines in 1961–62, put them on test, and began solving their problems. At the same time, he developed the 230 SL, 300 SE aid 250 S and 250 SE engines, and designed a four-cylinder Heron-head[2] engine for the 1965 Audi 60.

In 1969 he began design work on three- and four-rotor Wankel engines that were mounted in the C-111 experimental streamliner which won world-wide attention and fueled the "Wankel fever" then raging in the auto industry. He retired in 1972 but remained a consultant to Daimler-Benz, and gave lectures at Stuttgart University. His health was failing, however, and he died on June 17, 1974. Daimler-Benz AG then phased out its Wankel program after direct costs of DM 100 million.

1. Deutsche Versuchsanstalt für Luftfahrt = German Test Establishment for Aviation.
2. Bowl-in-piston-crown combustion chamber, with flat-faced cylinder head, patented by Samuel D. Heron.

Bentley, Walter Owen (1888–1971)

Chief engineer of Bentley Motors, Ltd., from 1919 to 1931; chief engineer of Lagonda Motors, Ltd., from 1935 to 1947.

He was born on September 16, 1888, as the son of a businessman at St. John's Wood in London, and educated at Clifton College, leaving at the age of 16. He was apprenticed to the Great Northern Railway works at Doncaster, and stayed on at Doncaster for another two years as an engineer on steam engines. Returning to London in 1910, he joined the National Motor Cab Company as assistant to the works manager. In 1912 he set up a car dealership in partnership with his brother Horace (1886 —1967) with franchises for Buchet,

La Licorne and DFP, and show-rooms in a mews off Baker Street. He joined the Royal Navy Voluntary Reserve in 1914 and became attached to the technical department of the Navy Air Service. He maintained liaison with many suppliers, such as Sunbeam and Rolls-Royce, and was delegated to Gwynne's of Hammersmith, London, who were building Clerget aircraft engines under license. He designed the Bentley Rotary I and the enlarged Bentley Rotary II aircraft engines, but the war ended before they could go into production.

In 1919 he opened a small drawing office on Conduit Street and rented a workshop in New Street Mews, Upper Baker Street, Marylebone, where he began construction of the first Bentley car.

The 3-Litre Bentley was displayed at Olympia in October 1919, and he moved production into a small factory in Oxgate Lane, Cricklewood. The first deliveries were made in 1921, with a five-year guarantee. The engine was a long-stroke four-cylinder unit with a single overhead camshaft. Four-wheel-brakes were adopted in 1923. The 3-Litre, in several varieties, was the only model up to 1926 when the 6½-Litre "Big Six" was added. A 160-hp version sold as the "Speed Six" arrived in 1928, along with the 4½-liter four (conceived as the six with one cylinder chopped off at each end). The six-cylinder 8-Litre was first seen at the London Motor Show in 1930 when the company was heading for financial ruin. W.O. tried to sell it to Napier, and negotiations were still going on when Rolls-Royce stepped in and bought it from the receivers in 1931. He was asked to stay on as a consulting engineer, but his acceptance led to a most unhappy period in his life. He claimed to be "held hostage" by Rolls-Royce and treated like a "dangerous ex-enemy" until his exit in 1935. A.P. Good engaged him to revitalize the Lagonda model range, and he brought out the V-12 Rapide in 1937. He led the design team that created the postwar Lagonda 2½-Litre, with a twin-cam six, cruciform frame and all-independent suspension. He left Lagonda when David Brown took it over and became a consulting engineer. From 1947 to 1952 he

did a lot of work for Armstrong-Siddeley, notably for the 1953 Sapphire engine.

He retired to Shamley Green near Guildford, Surrey, where he died on August 13, 1971.

Benz, Carl Friedrich Michael (1844–1929)

Technical director, Benz & Cie, Rheinische Gas-Motorenfabrik from 1883 to 1903.

He was born on November 25, 1844, in Mühlburg as the son of a railway engine driver. He attended local schools the Karlsruher Gymnasium, and was admitted to the Karlsruhe Polytechnic in 1860. He began his career in 1864 with the KMG[1] locomotive works and two years later went to work in the machine shops of Johann Schweitzer, Sr. in Mannheim. He moved into Schweitzer's drawing office in 1867 but left two years later to join the Benckiser brothers in Pforzheim, makers of scales and bridge builders. In 1871 he went into partnership with August Ritter in setting up a mechanical workshop in Mannheim, becoming its sole owner in August 1872.[2] Five years later he began experimental work on gas engines, and his first two-stroke engine patent was dated August 4, 1877. His two-stroke engine first ran on New Year's Eve, 1879, and he was granted a new patent on July 11, 1881. Financed by nine partners, local businessmen and bankers, he started Gasmotorenfabrik Mannheim in 1882, but the enterprise collapsed in internal disputes within months. One of the points of conflict was Benz's intention of making an automobile, while his partners were just looking for a return on their investment. By this time, Benz had also realized that his stationary two-stroke engine was not suitable for purposes of automotive traction, and he began experimenting with four-stroke engines.

In 1883 he was joined by Friedrich Wilhelm Esslinger and Max Caspar Rose in starting Benz & Cie, Rheinische Gasmotorenfabrik, in Mannheim. The statutes of incorporation included a clause opening the door for the development of an automobile. The first four-stroke Benz engine was tested in 1884. It was a horizontal single-cylinder unit with spark ignition, putting out ¼ hp at 250–300 rpm, with a weight of 96 kg. He made a better one a year later, delivering 0.9 hp at 400 rpm, and mounted it in a three-wheeled carriage of his own conception. He never contemplated installation of his engine in an existing horse-drawn carriage. A patent for this "gas engine vehicle" was dated January 29, 1886. The Benz automobile made its first test run on July 3, 1886. The company then began producing a line of four-stroke industrial engines. Over the following years, Benz made several three-wheeled automobiles, but no real production was undertaken. Esslinger and Rose were opposed to wasting money on these experiments, and withdrew from Benz & Cie in 1890. Benz managed to refinance the company with Julius Ganss (1851–1905) and Friedrich von Fischer (1845–1900) as partners. It was Emile Roger of Paris, licensed to produce Benz engines in France, who steered Benz to the concept of a four-wheeled automobile. Benz began prototype construction in 1891 and started series production of the Benz Vélo in 1893, followed by the bigger Victoria.

After that, Benz confined himself to the automobile drawing office, while his partners handled all other business. In 1896 Benz & Cie produced its 500th car, and the Kontramotor (horizontally opposed twin) went into production. The 1000th Benz car was built on May 1, 1897. Yet it was becoming evident to his partners that Benz had limited imagination about progress in car design. At the same time, the market for industrial gas engines went into a steep decline. To raise capital, Benz's partners reorganized the enterprise as a joint stock corporation, Benz & Cie, Rheinische Gasmotorenfabrik AG, in 1899, and in 1901 Julius Ganss put Georg Diehl and Fritz Erle into a separate office to create a line of modern automobiles. Hedging his bet, he also hired Marius Barbarou from Paris, with his own assistants, gave them their own office at Mannheim, and the same task. Carl Benz resigned in protest on April 23, 1903, but retained his seat on the board for another year.

In 1905 he assisted his sons Richard and Eugen in setting up an automobile factory at Ladenburg on the Neckar. Officially, the co-founders were Eugen and his father, as Richard was still working in the Benz & Cie drawing office, and did not move to Ladenburg until 1908. Carl and Eugen designed, tested and developed a conventional 10/22 PS four-cylinder car, resembling the contemporary Adler, NAG, Dürkopp and others, and production got under way in 1909. Carl Benz retired in 1912, but the make C. Benz Söhne endured until 1926. Carl Benz died on April 4, 1929 in Ladenburg.

1. Karlsruher Maschinenbau Gesellschaft.
2. Carl Benz, Mannheim Iron Foundry and Mechanical Workshop.

Bercot, Pierre (1903–1991)

Joint managing director of SA André Citroën from 1950 to 1958; president and managing director of Citroën from 1958 to 1970; honorary president of Citroën from 1970 to 1977.

He was born in 1903 in Paris and educated as a lawyer.

After getting his law degree, he studied economics and linguistics, and was awarded a degree from the National School of Oriental Languages. In 1937 he applied to join Michelin and was hired. Pierre Boulanger had him transferred to Citroën to work in cost-accounting and calculation of salaries and wages. He graduated to positions of increasing responsibility in Citroën's manufacturing and administrative offices, and in November 1950 was named joint managing director, with Antoine Brueder holding the same title. After the launching of the DS-19 in October 1955, he understood that Citroën lacked the long-term viability needed to stay afloat on its own and began searching for a French partner. He made feelers to H.T. Pigozzi of Simca, but saw his efforts nullified by Chrysler's entry as a big Simca shareholder in 1963. He immediately began making advances to Peugeot and arrived at a written agreement on cooperation in the areas of purchasing, technological research, and manufacturing, but it was never signed. It was annihilated by political intrigues, and Peugeot linked up with Renault in a similar set of schemes.

He arranged Citroën's takeover of SAAE Panhard & Levassor in 1965, set up joint ventures with NSU (Comobil and Comotor) and bought control of Officine Alfieri Maserati S.p.A. He organized Citroën's purchase of Berliet (trucks and buses), perhaps only to prevent it from falling prey to Volvo. He was not in at the beginning of the negotiations to merge Citroën with Fiat, however, which took place between Giovanni Agnelli and François Michelin, but became involved later on, at the level of industrial cooperation. He was instrumental in frustrating the merger plans, which were cancelled in 1973. He authorized turning the experimental Maserati-powered front-wheel-drive sports car into an SM production model, and cleared the way for the CX to replace the aging D-Series. He was well aware of Citroën's failure to fill the vast gap between the two-cylinder 2 CV and the D-series, and took steps to fill it, bringing out the GS in 1970. But it was not enough. He retired on November 24, 1970 and died in Paris in April 1991.

Bergmann, Sigmund (1851–1927)

Chairman of Bergmann Elektrizitäts-Werke AG from 1900 to 1927.

He was born on June 9, 1851, in Mühlhausen and attended local schools up to the age of 16. He went to work for a maker of electrical instruments, went to America about 1875 and started a joint venture with Thomas A. Edison of Menlo Park, New Jersey. He returned to Germany in 1891 and established Bergmann & Co. in Berlin, makers of electrical equipment. Three years later he set up another company, Bergmann Elektromotor und Dynamo Werke AG. The two were merged in 1900. He set up an automobile department in 1905 and began production of Fulgura battery-electric cars and vans in 1907. In 1908 he secured a license for the Métallurgique and took over the Deutsche Metallurgique Gesellschaft in Cologne, and set up for car production in a branch factory at Berlin-Halensee. Production began in 1909, with most of the components brought in from Belgium, gradually achieving an all–German content. The Bergmann-Métallurgique cars were powerful, expensive machines, but the 1913–14 range also included a small 6/18 as well as the 14/40 and the majestic 7.3-liter 29/70. The factory was converted to build trucks during World War I, and after the war, L'Auto-Métallurgique refused to renew the license. Bergmann ceased production of motor cars in 1922, switching to battery-electric propulsion, and specializing in post-office vans. He sold his company to Siemens-Schuckert in 1926 and died in 1927.

Berliet, Marius Maximin François Joseph (1866–1949)

Owner and director of Marius Berliet, Ingénieur Constructeur, from 1901 to 1917; chairman of SA des Automobiles Marius Berliet from 1917 to 1921; managing director of SA des Automobiles Marius Berliet from 1921 to 1929; chairman of SA des Automobiles Marius Berliet from 1929 to 1949.

He was born on January 21, 1866, at Croix-Rousse, Lyon, as the son of a textile-mill owner. He broke off his formal schooling at 15 to become a weaver's apprentice, and at 19 joined the parental shops. He spent his free time designing a car, which he built with the aid of two mechanics in 1895. It was a narrow four-wheeler with two seats in tandem (passenger in front), and a rear-mounted horizontal single-cylinder engine. His second car, built and patented in 1898, had a bench seat for two, and a steering wheel. He offered the design to Audibert, Lavirotte & Cie, who rejected it. In 1899 he opened a workshop in rue de Sully, Lyon, and hired a few workers. They built 11 cars with two-cylinder engines that year. In 1902 he bought Audibert, Lavirotte & Cie, and moved into their spacious plant at Lyon-Monplaisir. New models were patterned on the Mercedes, and built in 10/16 and 24-hp sizes. The model range evolved, and in 1905 Berliet sold a license to the American Locomotive Company, which provided funds for plant expansion. Berliet made 450 cars in 1906. The first Berliet truck was built at Monplaisir in 1907, with series production since 1910.

The first six-cylinder Berliet car came on the market as a 1910 model and in 1914 Berliet offered a model range from 10 to 36 hp. Orders for war trucks, ammunition, and eventually tanks, enabled Berliet to buy a vast tract of land at Venissieux for an additional plant. The company became a joint stock corporation on December 12, 1917. The first postwar Berliet car was a copy of the wartime Dodge, designated VB. But Berliet used inferior steel in certain critical applications, and the VB gave Berliet a reputation for poor quality. A consortium of bankers took control in 1921, while the engineers prepared a new car, the VF four-cylinder 12/16 hp model which was successful. Six-cylinder 10 CV and 11 CV cars were introduced in 1927. In 1929 Marius Berliet sold properties he owned in Cannes, repurchased the bonds issued in 1921, and regained control. He shifted the emphasis from cars to trucks, discontinued the six-cylinder cars at the end of 1932, and simplified the range of four-cylinder models. The last model, made in 1938–39, had a Berliet chassis with a Peugeot 402 body. In November 1939 the French government requisitioned the Berliet works, but in June 1940, the Germans moved in. For four years Berliet fought to maintain truck production for the civilian market in the face of German orders for war trucks. He was arrested on September 4, 1944, and accused of collaboration with the Germans, and given a two-year prison sentence by a court in Lyon in June 1946.

The court also ordered the confiscation of all his property, only to find he didn't own anything, having put all his assets into a family trust in July 1942. He was given sick leave to a clinic in Cannes, where he underwent several operations for intestinal cancer. He died in Cannes on May 17, 1949.

Bernardi, Enrico Zeno (1841–1919)

Head of engine design for Miari, Giusti & C. from 1896 to 1899; technical director of Societa Italiana Bernardi from 1899 to 1901.

He was born on May 20, 1841, in Verona as the son of a medical professor and educated in local schools up to the age of 15. He then began private studies, to prepare for the university. A brilliant student, he delivered a thesis on the "inherent memory of hydraulic apparatus" at the Veneto Institute of Science, Letters and Art in 1857. He studied mathematics at the University of Padua from 1859 to 1863, earning a doctor's degree in mathematics. He stayed on at the University as assistant professor in geodesics, hydrometry, rational mechanics and experimental physics until 1867, when he moved to Vicenza as professor in physics and mechanics at the Industrial Technical Institute[1]. In 1870 he was elected member of the Academy of Science at the University of Padua, and from 1879 to 1915 held the title of extraordinary professor of hydraulic, thermal and agricultural machines in the University's school of Applied Engineering. In 1882 he designed and built a single-cylinder horizontal gas-engine, "Pia" which he patented and saw going into series production in Padua[2]. In 1884, he built a toy tricycle driven by a "Pia" engine. It had a single front wheel, which steered, and belt drive to the left rear wheel only. It was in 1885 that he began in earnest to design an automotive engine.

In 1891–92 he advised Giovanni Agnelli on the future of the internal combustion engine, and in 1893 installed one of his "Lauro" engines on a trailer, with belt drive to the rear wheel of his son's bicycle. He displayed a three-wheeled car in Padua in 1894, with the engine under the bench seat and drive to the single rear wheel. The two front wheels were tiller-steered, but his second car had a steering wheel. His first four-wheeled prototype was built in 1898. He obtained a new internal-combustion engine patent in 1896 and a spark-ignition patent for the use of platinum point in the contact breaker in 1898. Several sizes of Bernardi engines were produced from 1896 to 1899 by Miari, Giusti & C. in Padua. He then founded his own company to manufacture three- and four-wheeled light cars, but the business was liquidated in 1901. He continued design work on engines and cars but stopped in 1906 to pursue other interests. In 1911 he plunged into photographic studies and conducted research on color photography and three-dimensional effects. He retired from teaching in 1915 and moved to Turin, where his son was working. He died on February 21, 1919, in a private clinic in Turin.

    1. Istituto Tecnico Professionale Industriale di Vicenza.

    2. Societa Veneta per Imprese a Costruzioni Pubbliche di Padova.

Berriman, Algernon Edward (b. 1882)

Chief engineer, Daimler of Coventry, from 1912 to 1927.

He was born in 1882 and received a general and technical education. As a premium apprentice with the C. A. Parsons Company, he worked on steam-turbine generators. He had also ten years of experience in automotive engineering when he joined the Daimler Company as chief engineer in 1912. For two years he occupied himself more with product quality than with innovation. During World War I he organized the production of Gnome air-cooled radial aircraft engines in the Daimler works.

In 1919 he became a member of Daimler's board of directors and also held an administrative position with the BSA[1] parent organization. He designed new chassis for the first postwar models, which had 5- and 7½-liter six-cylinder sleeve-valves engines dating back to 1912.

He brought out a four-cylinder 20-hp 3.3-liter in 1921, only to replace it a year later with a 21-hp 3-liter six on the same chassis. He designed two small sixes for 1923, a 1542-cc 12 hp, and a 2167-cc 16 hp model. The lineup proliferated until Daimler was producing 13 basically different chassis in 1924. That was uneconomical and he reduced it to five chassis by 1926. In 1925 he modernized the sleeve-valve engine by changing from cast-iron to lightweight steel sleeves, fitting bigger ports, and adopting aluminum pistons. The model range had a remarkable spread for a small-volume manufacturer. At the low end was an 1872-cc 16/55 six. The next step up was a 2648-cc 20/70 six, topped by a 3568-cc 25/85 six, a 5764-cc 30/90 six, and an 8458-cc 45-hp six. At the Olympia Motor Show in October 1926,

Daimler unveiled the Double-Six with a V-12 engine made up of two 25/85 units at a 60° angle, making a 7136-cc 50-hp power plant — silent, well-proven and dead reliable. A 30 hp 3744-cc Double-Six created by combining two 16/55 units was added during 1927. Berriman resigned from Daimler when L. B. Pomeroy arrived in 1928, worked as a consulting engineer in the Coventry and Birmingham areas, and joined the Pressed Steel Company in 1939.

    1. Birmingham Small Arms Company, Ltd.

Bertarione, Vincent (1892–1962)

Chief engineer of Automobiles Talbot from 1925 to 1928; chief engineer of Hotchkiss from 1928 to 1961.

He was born on January 27, 1892, in Italy, and held a degree in mechanical engineering. He joined Fiat as a draftsman in 1920 and a year later was assigned to the racing-car drawing office. Louis Coatelen brought him to Sunbeam in 1922, and he was placed in charge of creating a new Grand Prix car for 1924. Its engine, understandably, bore a strong resemblance to the Type 404 Fiat twin-cam two-liter six. He was transferred to Paris in 1925 and designed the Talbot Type DUS touring car with an overhead-valve 2Y2-liter engine. He abandoned the use of torque-tube drive and cantilever springs in favor of an open, double-jointed propeller shaft and semi-elliptic leaf springs for the 1927 M 67L and M 75L. He led the design of the 1485-cc straight-eight supercharged racing Talbot which earned the nickname "invincible" during the 1927 season.

H.M. Ainsworth invited him to Hotchkiss in 1928, where he brought out the 3-liter AM 80 in less than six months. For 1932, he replaced the torque-tube with an open propeller shaft on the AM 80, and developed a sportier AM 80S. For 1933 to AM 80 was renamed 617 and the AM 80S became 620. He replaced the AM 2 with the more modern 411 and 413. He had an independent front suspension system ready in 1934, but Ainsworth preferred the rigid axle. He developed the 620 into a 686, a champion rally car and designed a little side-valve engine for the 1937 Amilcar Compound.

During the war, he designed the 704, with rear-axle drive, as a replacement for the front-wheel-drive Compound, but it was never put into production. After the war, he developed the Artois four-cylinder and Anjou six-cylinder models, and was finally allowed to adopt independent front suspension in 1949. They were discontinued in 1954, after which he concentrated on the design of light and medium-duty trucks, with gasoline and diesel engines. He retired at the end of 1961 and died in 1962 during an operation for intestinal cancer.

Bertelli, Augustus Cesare (1890–1979)

General manager of Alldays & Onions, Ltd., from 1919 to 1923; chief engineer of Aston Martin from 1928 to 1936.

He was born on March 23, 1890, in Genoa to Florentine parents who settled in Britain when he was a baby. He was educated at Cardiff Grammar School and Cardiff Technical College, and served an apprenticeship with the Dowlais Iron and Steel Works, taking evening classes at University College, Cardiff. He wanted to see his native land, arrived in Turin at the age of 20 and worked as an experimental engineer with Fiat. On occasion, he went as riding mechanic to Felice Nazzaro in road races. He went back to Britain in 1911 and became manager of Howell's Garage in Cardiff. In the early part of World War I he was an engineer with the Hendon Experimental Aircraft Engine Department, and later joined Grahame-White Aviation. In 1918 he became works manager of Alldays & Onions, Ltd., in Birmingham, with promotion to general manager a year later. He was in charge of producing the Enfield-Allday LC 11 four-cylinder 10/20 hp car since 1921, but left in 1923 to build a racing car. He joined the Burt-McCollum (sleeve valve engine) company in 1924, and spent all his spare time at Brooklands. In association with William Somerville Renwick and the financial backing of Lord Charnwood, he bought the assets of Lionel Martin's enterprise in 1926 and designed an overhead-camshaft 1496-cc four-cylinder engine for the 1927 Aston Martin. He moved

production from South Kensington to Feltham, Middlesex. He stayed on when the company was sold to L. Prideaux Brune in 1931, and again to Sir Arthur Sunderland in 1932 but resigned in 1936.

He worked as a consulting engineer with Coventry Simplex, Rover, and Armstrong-Siddeley until 1939, when he joined High-Duty Alloys, one of Lord Kenilworth's companies. After World War II he organized Templewood Engineering Company, which he directed for the rest of his career.

He retired about 1965 and died in 1979.

Bertetto, Adrien (1923–)

General manager of industrial operations, Régie Nationale des Usines Renault, from 1975 to 1980; director of Renault car-product planning and automotive research and development from 1980 to 1985.

He was born in 1923 and held an engineering diploma from the Ecole des Arts et Métiers. He continued his studies and graduated from the Ecole des Moteurs in 1948. He began his career with the Régie Nationale des Usines Renault in 1948 and spent the next four years on experimental work with automatic transmissions. With valuable input from Gaëtan de Coye de Castelet, Renault's director of Applied Research, he developed the Transfluide automatic transmission which became optional on the Frégate in 1957. He held the title of chief engineer for automatic transmissions from 1956 to 1961. He became director of Applied Research in 1961 and held that office until 1968, when he was transferred to the office of industrial planning. He then served as manager of the central office of methods from 1969 to 1975. From 1980 onwards he worked on advanced studies, made evaluations of new ideas and proposals, conducted research into material substitutions, and made recommendations based on feasibility studies. In 1958 he was placed in charge of robot and automation systems, working in liaisons with Renault's ACMA and RETI machine tool and industrial-engineering subsidiaries. He retired in 1989.

Bertodo, Roland (1933–)

Director of power train engineering, Austin-Rover Group from 1985 to 1987; product engineering director of Austin-Rover from 1987 to 1989.

He was born on April 28, 1933, in London, and studied at the Hendon College of Advanced Technology and the Willesden Technical College. From 1949 to 1954 he was an engineering apprentice with De Havilland Aircraft Company. He spent four years in a research establishment and was head of the test laboratory of Bristol Siddeley Engines from 1958 to 1966. For the next four years, he worked on diesel engine research for General Electric.

He joined the Perkins Engine Group at Peterborough in 1970, beginning as reliability manager, later holding the titles of test engineering manager and systems engineering manager before being named director of product engineering in 1976. After a year as engineering director, he left Perkins in 1979 to become managing director of Dexion, Ltd. He came to Austin-Rover in 1985 and took on a program to win engine-technology independence from Honda. That led to the K-series 16-valve four-cylinder engine family and a line of derivatives. In January 1989 he was appointed strategic planning director of Austin-Rover, but went to Italy in 1990 as managing director of Rover Italia. He resigned in 1993 to open an engineering-consultant office at Palazzo in Piedmont in collaboration with Telos Management Consultants of Milan.

Bertone, Nuccio (1912–1997)

Chief executive of Carrozzeria Bertone; considered one of the greatest nurturers of design talent in history.

He was born July 4, 1912, as Giuseppe but was always called Nuccio. The son of Giovanni Bertone (1884–1972), founder of the famous Italian coachbuilder in 1921, Nuccio began working with Bertone in 1933 and began the transformation of a small family workshop into a famed manufacturer of prototypes. The company survived the war by building ambulances and military vehicles and

Nuccio took charge of Bertone in 1945. Asked to design a successor to the Disco Volante by Alfa Romeo in the 1950s, Bertone created the first of the designs that came to be known as the BAT (Berlina Aerodinamica Technica) cars on the Alfa 1900 Sprint chassis. Under Bertone's leadership, Carrozzeria Bertone produced numerous landmark sports cars, including the Lamborghini Miura, Espada and Countach; the Alfa-Romeo Giulietta Sprint and Montreal; the Iso Grifo; Lancia Stratos; and the Fiat 850 Spider and Dino Coupe. The Bertone design team was also responsible for several high-volume cars, including the Daewoo Espero, Simca 1200S Coupe, Volvo 780 Coupe and the Citroën BX, ZX, Xantia, XM and Berlingo. Nuccio was less of a stylist than an entrepreneur, but he hired many famous stylists, including Franco Scaglione, Giorgio Giugiaro and Marcello Gandini. Nuccio offered his own design observations to his staff through the years and was still active in the business up until his death on February 26, 1997, at the age of 82. Bertone was succeeded as chairman by Paolo Caccamo, and the company continues to flourish today.

Besse, Georges (1927–1986)

President and managing director of the Régie Nationale des Usines Renault in 1985–1986.

He was born on December 25, 1927, at Clermont-Ferrand, and attended the Ecole Polytechnique from 1948 to 1950. He entered the Ecole des Mines in Paris in 1950 and graduated in 1954. He began his career in the pits of the Bazailles ore mines and later served as a staff engineer of the Béthune Mines. He was attracted to nuclear energy and served from 1955 to 1958 as joint industrial director of France's Atomic Energy Commission. In 1958 he became its managing director and in 1965 he was named head of USSI, contractor for building the nuclear power plant at Pierrelatte.

He joined the Alsatian-Atlantic nuclear power group within Alcatel and served as president of Alcatel until its merger with CIT. After a period as joint managing director of CIT-Alcatel, he became chairman of Eurodif in 1974, with responsibility for uranium enrichment at the Tricastin plant. From 1976 to 1982 he led Cogema, France's main supplier of nuclear matter.

When the French government nationalized Pechiney, the giant aluminum producer, in 1982, he was chosen as president of Pechiney and endowed with a big budget for its recovery. Three years later the government watched with consternation as Renault's losses mounted, and put Besse in charge of turning the Régie around in January 1985. He took drastic steps and started its recovery, but was assassinated on a Paris sidewalk by terrorists on November 17, 1986.

Bettmann, Siegfried (1863–1951)

Chairman and managing director of the Triumph Cycle Company, Ltd., from 1900 to 1933; vice chairman of Triumph Limited from 1933 to 1936; chairman of Triumph Engineering, Ltd., from 1936 to 1939.

He was born on April 18, 1863, in Nuremberg as the son of a law clerk and estate manager. He was educated privately and at preparatory school in Nuremberg but ended his formal education at 16. He worked as an office boy in Antwerp and later as a clerk with a mail-order business in Paris. He arrived in London in November 1884, settled in Islington, and held a succession of low-paid office jobs until he joined Singer Sewing Machine Company in the export department. He lost his position when the expected order flow failed to materialize, and obtained a parental loan to set up a bicycle shop in London. He contracted with the Coventry Cycle Company for a supply of Bettmann bicycles, which he sold in the home market, and with William Andrews in Birmingham and Hillman, Herbert & Cooper in Coventry for export models which carried a Triumph badge. In 1887 he went into partnership with Mauritz J. Schulte, who made plans for making their own Triumph bicycles. They took over a former ribbon factory in Much Park Street, Coventry, where bicycle production began in 1891. With fresh capital staked by local businessmen, they formed New Triumph Cycle Company in 1897. The "New" was deleted in 1900. They

began experimenting with motorcycles in 1902, using Minerva engines, and started production in 1905. From an output of 533 Triumph motorcycles in 1906, they reached 3,000 in 1909. Bicycle production continued, and motorcycle assembly was transferred to a newly purchased plant on Priory Street, Coventry, in 1907. The partnership with M.J. Schulte broke up in 1919, and Bettmann began thinking about a sideline in cars. In 1921 the Triumph Cycle Company, Ltd., purchased the idle plant of the Dawson Car Company, Ltd., in Clay Lane, Stoke, Coventry, where 65 Triumph cars were turned out over a two-year period. Car production was moved to Priory Street in 1923, and the new 10/20 TLC model put on the market at £430. The 13/35 followed in 1925 and the 15/50 in 1926. A new 15 replaced all earlier models for 1927, to be supplemented by the K-series Triumph 8 in 1928. But the company lost money from 1930 to 1933. The motorcycle branch was sold to J.Y. Sangster, and Bettmann lost his title. In 1936 he took over the rights to the Triumph bicycle, but success eluded him, and he sold out to Raleigh Cycle Company of Nottingham in 1939. He retired and died on September 23, 1951.

Bez, Ulrich (1943–)

Head of advanced vehicle engineering for BMW from 1982 to 1984; director of BMW Technik GmbH from 1984 to 1988; director of research and development for Porsche from 1988 to 1991; vice president in charge of engineering and development for Daewoo Motor Company from 1993 to 1998; became chief executive officer of Aston Martin in 2000.

He was born on November 7, 1943, in Stuttgart and attended local schools. He worked as a trainee with Porsche from March to September 1965, then studied aircraft technology at the University of Stuttgart from October 1965 to December 1971. He rejoined Porsche in 1972 to work in research and development at the new Weissach technical center and ended up as head of vehicle research. He went to BMW in 1982 and prepared the Z-1 prototype in 1985–86.

In August 1988 he returned to Porsche, replacing Helmut Bott. In the 1991 season, Porsche was under contract to supply Formula 1 racing engines to a Footwork-sponsored team. In the first six races entered, not one of their cars finished. Bez was made a scapegoat and resigned in October 1991. His last project for Porsche was Type 993, a planned replacement for the 911.

He went to Korea in October 1993 to direct the creation of a new model range for Daewoo, returning to Germany in September 1998 to become chairman of A. Friedrich Flender of Bocholt, a transmission subsidiary of Babcock-Borsig AG. In January 2000 he went to Ford as a business consultant, as Ford was then preparing to make a bid for Daewoo Motor Co. On July 5, 2000, Ford named him chief executive of Aston Martin.

Biggs, Theodore James (18??–19??)

Chief designer of Arrol-Johnston, Ltd., from 1909 to 1912; chief engineer of Humber, Ltd., from 1912 to 1914.

He was a prominent racing cyclist in his youth, which led to permanent employment by the Raleigh Cycle Co., Ltd., of Nottingham. He became a motorcycle engineer and in 1904–05 designed a Raleigh car with a 16-hp Fafnir engine, but only one prototype was made. In 1906 T.C. Pullinger invited him to join Humber at Beeston as a design engineer. The company was then in the throes of moving car production into a factory taken over from the Centaur Cycle Company in Folly Lane, Stoke, Coventry, where big expansions were planned. He designed the racing cars for the 1908 Tourist Trophy with their four-inch-bore, four-cylinder overhead-camshaft engines. As the Beeston works were closed in 1908, he was transferred to Coventry, but in April 1909 accompanied T.C. Pullinger to the Underwood Works at Paisley, where he designed the 11.9-, 15.9- and 23.9-hp Arrol-Johnston models. Returning to Humber, Ltd., in April 1912, he immediately revised the Eleven and Fourteen, and designed the Humberette, with tubular frame, air-cooled V-twin, 3-speed gearbox and shaft drive, and

rack-and-pinion steering. He also designed a new Humber Ten for 1914, and resigned in November 1914 in order to take a better-paid position with F.E. Baker, precision engineers at Kings Norton.

Bignan, Jacques (18??–19??)

Partner in and managing director of Bignan & Picker from 1910 to 1919; managing director of Automobiles Bignan from 1919 to 1926.

He began his career at the turn of the century by preparing engines for speed boats and met Charles Picker, pioneer of the hemi-head twin-cam type of engine. They formed a partnership to start engine production in 1910 and opened a factory in rue du Chemin de Fer in Courbevoie in 1911. Bignan power units were primarily marine engines, but also found their way into some cars, notably the Crespelle from 1912 onwards. Picker designed a very successful single-overhead-camshaft 2½-liter unit in 1913, but the best-selling Bignan engine was a side-valve 1½-liter. In 1914 Bignan & Picker moved into roomier premises in rue de Normandie in Courbevoie and announced a range of four-cylinder engines from 7 to 40 (tax) hp and a six-cylinder 60 (tax) hp unit. Bignan and Picker went separate ways during the war. Picker formed a link with Société Janvier, engine manufacturers, and Bignan graduated from the Ecole Supérieure d'Aéronautique. Bignan also set a consulting engineers' office with A. Granjean and Emile Reno. He still had the factory at Courbevoie and decided to become a car manufacturer. The first Bignan Sport of 1919 had the pre-war T-head 1½-liter engine mounted in a new chassis supplied by the Ateliers de La Fournaise of Saint-Denis. He wanted to offer a 2-liter model and began an association with Némorin Causan, who agreed to design a 2-liter engine, but happened to have complete drawings for a new 3-liter on hand. Bignan rushed it into production and it went into the 18 CV Type 132C of 1920, quickly followed by a 3½-liter version. Ateliers de La Fournaise made the chassis. Concurrently, Automobiles Bignan was offering an 1100-cc car which was a disguised AL-Type Salmson. In 1924 Bignan bought the ex–Grégoire factory at Poissy and moved assembly operations there. Causan designed two versions of the 2-liter Bignan engine, the base model with a single overhead camshaft and valves splayed at 10°, and a super-sports version with desmodromic valves. Only two test cars were made with the second engine. In 1924 Causan added two cylinders to the base unit to form a 3-liter six, and also made a Cozette supercharger installation on the 2-liter. All along, Bignan let his enthusiasm guide his actions, neglecting financial control, and in 1926 his company was bankrupt and taken over by Société La Cigogne. With incredible resilience, he began offering a nice, sporty car he named Celtic, produced in 7 CV and 10 CV versions by S.A. Ariès which had a lot of excess capacity. The Celtic was made from 1926 to 1929. In 1927 Bignan was also selling SCAP–powered four-cylinder EHP[1] cars with Bignan-badged radiators. His final attempt was the 1930 Bignan-MOP, powered by a 2½-liter straight-eight SCAP engine.

1. Etablissements H. Précloux, La Garenne-Colombes.

Binks, Charles (1864–1922)

Technical director of Charles Binks, Ltd., from 1900 to 1905; technical director of New Leader Cars, Ltd., in 1905–06; chief engineer of Roydale Engineering Company, Ltd., from 1907 to 1909.

He was born in 1864 at Bootle near Liverpool and was educated at York. He served an apprenticeship with a bicycle manufacturer at Birkenhead, and then set up his own cycle shop. In 1900 he moved to Apsley, Nottingham, and became an agent for Darracq automobiles. He also designed motorcycles, and in 1903 obtained a patent for a four-cylinder motorcycle, moved into the Whitehall Works, and began production. He also designed a 10/12 four-cylinder car, and introduced the Leader in 1905. He was building six cars a week when the plant was destroyed by fire in 1905. He found new backers to restart the business as New Leader Cars, Ltd., but disagreements arose and he preferred to leave. He formed a partnership with A.E. Learoyd to set

up Roydale Engineering Company, Ltd., in former textile mills owned by the Learoyd family, Trafalgar Works, Leeds Road, Huddersfield, Yorkshire. He designed the Roydale cars, with four-cylinder T-head engines, 3.2-liter 18/22 and 3.9-liter 25/30, with dual ignition, Binks carburetors, and shaft drive. Production began in the spring of 1907 but he resigned in 1909 when he moved to Salford to design engines for Sir W.H. Bailey & Co. Independently, he opened a factory at Eccles to make Binks carburetors (renamed Amal in 1930). He caught pneumonia and died in 1922.

Bionier, Louis (1898–1972)

Chief body engineer of Panhard from 1929 to 1967.

He was born in 1898 at Alfortville, suburb of Paris, and left school at 11, when his father died, to help his mother feed the family. Later he served an apprenticeship in a local machine shop, and joined Panhard & Levassor as a machine-setter in 1915. During the years that followed, he took evening classes at the Ecole Pratique, held a variety of positions in the factory, and became a department manager. In 1927 he was named project engineer for touring cars, convertibles, and racing cars, and two years later was promoted to chief engineer of body construction, which also allowed him to show his talent as a designer. He designed the 6 DS and 8 DS bodies for 1930, created the Panoramique in 1933 and the unit-construction Dynamic in 1936.

He engineered the aluminum frame-and-body for the 1946 Dyna Panhard, whose production was sub-contracted to Facel-Metallon. In 1948 he designed the experimental Dynavia to see how far aerodynamics-without-compromise could be taken on a car, and some of its lessons were applied in the Dyna-54, a six-passenger sedan with an all-aluminum body shell (supplied by Chausson). He also designed the 24 CT and 24 BT of 1963–64. When the Panhard company came under Citroën's control, he designed the body for the Citroën Dyane, produced from 1967 to 1984.

He retired in 1968 and died in 1972.

Birkigt, Marc (1878–1953)

Technical director of La Hispano-Suiza, Fabrica de Automoviles in Barcelona from 1904 to 1919; majority owner and technical director of S.A. des Automobiles HispanoSuiza of Bois-Colombes from 1923 to 1939.

He was born on May 6, 1878, in Geneva as the son of a tailor. Orphaned at the age of 11, he was raised by his grandmother. At 15, he began his studies at the Ecole de Mécanique in Geneva, revealing a rare sense for physics and kinematic mechanisms. His professors were so impressed with the young man that they arranged for a scholarship for him at the Federal Technical University[1] in Zurich. But his grandmother had grown too old, and her means too limited, to support him, so he was forced to refuse it. He had to work for a living and found employment in the machine shops of Piccard & Pictet in Geneva. His formal education had ended shortly after his 18th birthday. One of the customers he came to know at work was a Swiss engineer named Bouvier who represented a Spanish company belonging to Emilio La Cuadra. Bouvier offered Birkigt the opportunity of going to Barcelona, where La Cuadra had just sold his electric power station in order to set up a factory for making electric railroad locomotives, and work as a design engineer. He accepted, and led the engineering team that created their first electric locomotive in 1898–99. It proved despairingly underpowered and the project was given up. When Birkigt expected a reprimand, or possibly the sack, he swiftly steered the conversation with La Cuadra away from locomotives and suggested that they should make automobiles instead.

La Cuadra asked for detailed plans, and then agreed. Both single-cylinder and parallel-twin powered cars appeared in 1901, made to Birkigt's designs. Some cars were sold, but at a loss. La Cuadra sold the automobile factory to J. Castro, who kept Birkigt on and even gave him the title of technical director. New Castro cars with two- or four-cylinder engines were produced in 1903, but the business was unprofitable, and the company went bankrupt in 1904. The assets were sold to Don Damian Matéu, who refinanced and renamed it La Hispano-Suiza, with Birkigt as a partner and technical director. For the first time, he was given a proper budget for design and development. He became more daring in his drawing-board work and began applying for patents, the first one in 1905, for his system of combining the engine, clutch and gearbox into one sub-assembly, carried in a sub-frame consisting of two flexibly mounted cross-members. He built up an engineering staff which proved quite prolific, completing designs for no less than 35 different cars over a ten-year period.

Due to strong international demand for the T.15 "Alfonso XIII" Don Damian decided to set up a branch factory in France, and from 1909 to 1914 Birkigt alternated between Paris and Barcelona. In 1915 he designed a V-8 aircraft engine that became famous, and was produced under license by 13 other French companies including Peugeot and Chenard-Walcker. It was also made by three Italian companies, by Wolseley in Britain, by Wright-Martin in the United States, by Mitsubishi in Japan and The National Arsenal in Russia, for an aggregate total of nearly 50,000 units. After World War I, Birkigt spent most of his time in Paris, developing the H6 series into one of the world's top-quality prestige cars. He also designed several V-12 aircraft engines and launched his J.12 car with V-12 power in September 1931. Car production at Bois-Colombes ceased in 1938 and Birkigt returned to Barcelona in 1940, and spent the war years there. When the company was nationalized, he decided to retire, and in 1947 moved to Versoix on Lac Léman, where he designed new high-precision pistols and rifles, and relaxed by sailing. He died of cancer on March 15, 1953.

1. Eidgenössische Technische Hochschule.

Bitter, Erich (1933–)

Managing director of Bitter Automobile GmbH & Co. K-G from 1971 to 1986; president of Bitter & Co. Automobile GmbH and the Bitter Automobile Company from 1986 to 1991.

He was born in 1933 at Schwelm near

Düsseldorf as the son of a bicycle-shop owner. Always a keen cyclist, he was 16 when he built his own racing bicycle and started racing. He made a living as a door-to-door Coca-Cola salesman, but found he could do better by buying, fixing and selling cars. He stopped peddling to race motorcycles in 1954, and three years later bought an old Citroën which he entered in local rallies and races. When he got an NSU Prinz in 1959, he began winning. He became an NSU dealer in Schwelm, and then began racing a Volvo PV 544, which earned him the Volvo franchise. In 1961 he also raced an Abarth and in 1962 a Jaguar 3.8 saloon. In 1962 he set up Rallye-Bitter as a sales outlet for accessories, racing uniforms, and tuning kits. He also became a Saab dealer and in 1964 signed up as sole importer of Abarth and Intermeccanica cars. In 1968 he began racing a Group 5 Opel Rekord, but in May 1969 he had a bad accident driving an Abarth, which put a stop to his racing career. He made plans for a Bitter grand-touring car, and asked Bob Lutz about buying Opel platforms for it. Lutz agreed to supply shortened Opel Diplomat platforms with a Chevrolet V-8 and Turbo Hydra-matic transmission. Bitter made some styling sketches, and GM's Charles M. Jordan and David R. Holls gave definitive shape to the Bitter CD coupé. He contracted with Karl Baur Karosseriefabrik in Stuttgart to make the body shells. Assembly space was found in the Leple works at Heilbronn. Production began in 1973 and sales went well until fuel prices doubled in 1974. His operation had a break-even point of 250 cars a year, but with sales at half that level, he had no hope of making a profit. Still, 390 Bitter CD coupés were completed by 1979. Then Opel, having discontinued the Diplomat, could no longer supply the platform. He began planning a new car, based on the new Opel Senator, with a 3-liter six and de Dion rear suspension. He went to Michelotti for the styling, which was retouched by Henry Haga and George Gallion of Opel. Body shells were to be made by Carrozzeria Ocra in Turin, with final assembly by Steyr-Daimler-Puch AG in Graz. Production of the Bitter SC coupé

began in 1981, followed by a four-door sedan in 1984 and a cabriolet in 1985. After the first run of 50 coupé bodies, the contract for body-shell production was switched to Maggiora in Turin. By 1986, 600 Bitter SC cars had been built. He invented a soft-top coupé and sold the idea to Opel, and Karl Baur produced 2000 Aero-Kadetts from 1968 to 1978. In 1983 he had an idea for a roadster based on the Opel Manta, but all plans were canceled when Opel stopped building the Manta. Creditors, mostly suppliers and former personnel, sued his company in 1986 and it was liquidated. He made his comeback through Bitter Automobile of America, Inc., a sales organization set up in 1983 in partnership with Lee Miglin, a Chicago real-estate developer. They formed a new holding, Bitter Automobile Company, based in Santa Monica, California, with Samuel Tucker, a former vice president of Volkswagen of America, Inc., as chairman. It had a production subsidiary, Bitter & Co. Automobile GmbH of Schwelm. Erich Bitter made plans for new cars based on the Opel Omega platform, Type 3 appearing in 1990 and Type 4 in 1990. Bodies were styled by CeComp of Turin, and Steyr-Daimler-Puch AG was contracted for final assembly. Bitter spoke of 3000 cars a year, but only prototypes were built. His last venture was called the Bitter GTI and appeared in 1998 as a supersports model based on the Lotus Elise, powered by a Chrysler V-10. The prototype was built in the shop of Toine Hezemans in Eindhoven.

Bizzarrini, Giotto (1926–)

Owner and president of Autostar, Livorno, from 1962 to 1963; owner and president of Prototipi Bizzarrini from 1963 to 1966; president of Bizzarrini S.p.A. from 1966 to 1969.

He was born on June 6, 1926, in Livorno and studied at the university of Pisa, graduating in 1953 with degrees in mechanical and industrial engineering. He began his career with Alfa Romeo, working in the experimental department from 1954 to 1957. That year Enzo Ferrari engaged him to work in quality-control and testing, and a year later promoted him to head of the experimental department for

Gran Turismo cars. He was project engineer for the TR1/60 and developed the chassis for the 250 GTO. He also designed a small-car chassis for a project Ferrari had undertaken for Beretta,[1] famous makers of pistols and shotguns, who were eager to branch out into sports cars. After Beretta changed its mind, Ferrari arranged the sale of the little "Mitra" to Niccolo de Nora, who produced it in Milan as the ASA[2] 1000 from 1962 to 1966. He left Ferrari in November 1961, along with Carlo Chiti and Romolo Tavoni, to join Count Volpi's ATS[3] in Bologna, but left within months to set up his own consulting practice. He became technical consultant to Renzo Rivolta, who was about to start production of an Italian version of the Gordon car for the American market. In 1962–63 he designed a 3½-liter V-12 engine for Lamborghini, renamed his business, and moved into bigger shops in Via Lulli 1 in Livorno. He designed the Grifo sports coupé, powered by a Chevrolet V-8, and began small-scale production in 1963. In 1966 he sold the Grifo name to Renzo Rivolta, and renamed his own car, GT Strada. In 1966 he "went public" and sold shares in his company, which introduced the GT Europa, powered by a four-cylinder Opel Rekord S engine. He created the P 538 roadster and the Bizzarrini Manta. His company was declared bankrupt in 1968 and the last cars were made in 1969. He became involved with American Motors who wanted a car to compete against Ford's De Tomaso Pantera, designed and built two chassis, and was offered a production contract, which was canceled in 1972.

1. Fabbrica d'Armi Pietro Beretta, Gardone Val Trompia.
2. Autocostruzioni S.p.A., Via San Faustino, Milano.
3. Automobili di Turismo a Sport S.p.A., Via Altabella, Bologna.

Black, John Paul (1895–1965)

Joint managing director of the Hillman Motor Company, Ltd., from 1921 to 1929; joint managing director of the Standard Motor Company, Ltd., from 1929 to 1934; sole managing director of the Standard Motor Company, Ltd., from 1934 to 1953; chairman of the Standard Motor Company, Ltd., from 1953 to 1954.

He was born on February 10, 1895, at Kingston-on-Thames, as the son of a municipal clerk. He studied law at London University and began his career in a London lawyer's office. At the start of World War I he volunteered for military service, was assigned to the Royal Naval Voluntary Reserve and was on a ship taking part in the evacuation from Gallipoli on Suvla Bay in 1916. He was transferred to the Royal Tank Regiment and fought in France in 1917, being discharged in 1919 with captain's rank. He worked for a brief period with a London firm of solicitors, joined Hillman Motor Co., Ltd., in 1920 and moved to Coventry. During his years with Hillman, the company had a one-model policy, the 11-hp cars from 1919 being upgraded to a 14-hp car for 1926. When it came under control of Rootes, Ltd., in 1928, he was also given a seat on the boards of Humber, Ltd., and Commer Cars, Ltd.

He was engaged by Reginald W. Maudslay to direct a general restructuring of the Standard Motor Company, Ltd., and became its sole managing director in 1934. He got impressive results, for in 1930, Standard had made only 6000 cars, and in 1936, the company built (and sold) 34,000 cars.

In 1936 he was an eager supporter of the government's "Shadow Scheme" to built aircraft and aircraft-engine factories with public funds, and assign operations to private industry. He also tried to buy a controlling-minority stake in SS Cars, Ltd., but failed. In September 1939, he was named chairman of the government's Joint Aero Engine Committee. His company operated Shadow factories in Banner Lane, Coventry, and on Fletchamstead Highway, and the Canley works were converted to war production. Standard Motor Company, Ltd., assembled 1000 De Havilland Mosquito fighter/bombers, built Oxford trainer planes and fuselage sections for the Bristol Beaufighter, and manufactured huge numbers of Claudel-Hobson aircraft carburetors and cylinder barrels for Bristol engines, as well as a quantity of complete Bristol Hercules engines. In November 1944, Standard purchased the Triumph Motor Company, Ltd., for £77,000 from Thomas Ward & Co. Ltd., sold the Clay Lane plant and prepared to move Triumph car production into the Canley works. He negotiated a contract to produce Ferguson tractors equipped with Standard engines and leased the Banner Lane factory for tractor production, which began in 1946. He also leased the factory on Fletchamstead Highway which he turned into a gearbox plant and spare part depot. He authorized new Triumph sedan and sports models but gave top priority to the Standard Vanguard, which went into production in the summer of 1947 with body shells by Fisher & Ludlow. During 1950 alone, the Canley plant turned out 50,000 Vanguards. The little Standard 8 was added to the program in 1953, a few months behind the launching of the Triumph TR2 roadster. But he was erratic in many business decisions, and lost the confidence of his executives and engineers. They forced him to sign a letter of resignation in January 1954. He retired to his farm at Llanbedr, Merioneth, in Wales, and died at Cheadle Hospital, Cheshire, on December 24, 1965.

Blackwood Murray, Thomas (1871–1929)

Technical director of the Albion Motor Car Company, Ltd., from 1899 to 1929.

He was born in 1871 at Biggar in Lanarkshire and educated at George Watson's College, Edinburgh. He enrolled at Edinburgh University and graduated with a degree in mechanical engineering. He obtained his first patents in 1891, covering an automatic engine-speed governor, and soon became known for research and development work on low-tension ignition systems, electrical generators, motor fuels and airflow. His first positions in industry were with King, Brown & Company, electrical engineers, and subsequently, with Kennedy of Glasgow. He became associated with Sir William Arrol's Mo-Car Syndicate, where he met and befriended Norman Osborne Fulton. In partnership with N.O. Fulton he founded the Albion Motor Car Company,

Ltd., in December 1899, with shops in Finnieston Street, Glasgow. He designed the first Albion car, Type A-1, with a 2-liter flat-twin mounted under the seat and chain drive to the rear axle. It had tiller steering and solid tires. He adopted the steering wheel for Type A-2 in 1901, and the engine was bored out to 2.7 liter size. Type A-3 was a light truck with the same engine and a wide choice of wheelbases and load capacities. For the 16-hp model, he adopted the vertical in-line four, mounted in front, behind a square radiator. It evolved into the 24/30, and was replaced by a new 2½-liter 15-hp model in 1912. In 1913 the board of directors voted to discontinue car production and develop a line of heavy trucks and buses. Production had been transferred to roomier premises at Scotstoun outside Glasgow in 1903, and this plant turned out some 2500 military transport vehicles during World War I. In tribute to his services to Scottish industry, the Edinburgh University conferred a doctor's degree in science upon him. Dr. T. Blackwood Murray directed the design of all Albion vehicles up to the time of his death in 1929.

Blatchley, John (1915–)

Chief body designer of Rolls-Royce, Ltd., from 1944 to 1970.

He was born in 1915 and began his education in a Jesuit boarding school at Chesterfield. When he was 12, he was found to suffer from rheumatic fever, and spent the next three years without schooling. This period steered him towards his future career, as he spent all his time making sketches of cars. In 1930 his father thought he was well enough to go back to school and enrolled him at Oxford. But he took no interest in studying and flunked the exams. He made his father understand that all he cared about was to draw cars, and the two agreed that the best way to become a professional designer was to get a solid background in engineering. Thus motivated, he did well at the Chelsea School of Engineering and later graduated from Regent Street Polytechnic. In 1935 he was engaged by J. Gurney Nutting as a junior designer under A.F. McNeil, who admired his imagination and appreciated his sureness of line. He handled difficult assignments with sober artistry and in 1937, when McNeil left, he was named chief designer. During the next three years, he designed some spectacular Bentleys, but also some bodies for Alvis and Lagonda chassis. He was rejected for military service because of a heart flutter, and was sent to Rolls-Royce at Hucknall, Nottinghamshire, where he designed engine pods and other sheet-metal parts for aircraft. Rolls-Royce engineers began preparing new cars for the postwar market in 1943 and he became styling designer for Rolls-Royce and Bentley cars. He was able to concentrate strictly on the creative side, for H.I.F. Evernden handled liaison with the body engineers and the coachbuilders. Yet it was known in the organization that Blatchley never designed anything without knowing how it could be made one of the benefits of his technical training. He styled the Bentley Mk VI and Rolls-Royce Silver Dawn, whose bodies were produced by Pressed Steel, Ltd., at 50 percent the unit cost of a Park Ward body. He made the styling design for the Silver Cloud and S-Type, compromising the traditional looks as well as the demands of fashion. He was handed an even more difficult task when the company decided to adopt unit body construction for the 1965 Silver Shadow. He had to fit the body within compact dimensions without losing its Rolls-Royce identity altogether. He designed a convertible version, the Corniche, which went into production in 1971, a year after his retirement to Hastings in 1970.

Boillot, Jean (1926–)

Managing director of Automobiles Peugeot from 1979 to 1980; chairman and chief executive of Automobiles Peugeot from 1980 to 1983; chairman of PSA Peugeot-Citroën from 1984 to 1990.

He was born on February 6, 1926, at Jougne (Doubs) as the son of a hotel-keeper and educated in high schools at Besançon and Pontarlier. He went to Paris to study Law at the Faculté de Droit and also won a CPA[1] diploma. He began his career in 1951 by joining Crédit Lyonnais, one of France's big three state-owned banks, and in 1954 was hired by Peugeot, in Paris, to work in sales and service. He was president of Peugeot Canada from 1961 to 1963. Upon his return to France, he served as regional sales manager at Nantes from 1964 to 1966, and national sales manager of Automobiles Peugeot from 1967 to 1974. He was general sales manager (including exports) from 1974 to 1976, when he was given a seat on the board of directors. In 1980 he directed the blending of the Talbot (ex–Chrysler) sales organization with, Peugeot's and ordered a Talbot version (Samba) of the Peugeot 104 to be built at Poissy. In 1983 he was sidelined by Jean-Paul Parayre as vice chairman of the Automobiles Peugeot board, but in 1984 he was rehabilitated by Jacques Calvet, with responsibility for the complete PSA[2] automobile division. He examined the ex–Chrysler C-28 project and ordered it re-engineered so as to share a maximum of Peugeot 205 components. It went into production as the Peugeot 309 at Poissy in September 1985.

He allocated funds to double production capacity of the SMAE[3] engine works at Tremery in Lorraine, and ordered the phasing-out of Citroën and ex–Chrysler engines. He started planning a replacement for the Citroën CX as a sister model of the Peugeot 605, due to take over from 505. But the Citroën had to be a front-wheel-drive car, and the Peugeot engineers preferred rear-wheel-drive for a car of that size and power. For cost-considerations and other reasons, the two could not be reconciled. He prevailed on the Peugeot engineers to go to front-wheel-drive for the 605, which added to the delay of getting it to the production stage. The Citroën XM was launched in March 1989 and the Peugeot 605 in September 1989. Both had serious quality problems, and the XM was a failure in the market place.

He took early retirement in March 1990.

1. Centre de Préparation aux Affaires.
2. Peugeot, Société Anonyme, established in 1975 as holding for all the former operations and subsidiaries of Peugeot and Citroën.
3. Société de Mécanique Automobile de l'Est.

Bollée, Amédée, Jr. (1867–1926)

Director of the Usines Bollée at Le Mans from 1887 to 1926.

He was born on January 30, 1867, at Le Mans as the son of Amédée Bollée, Sr. (1844–1917), noted steam-vehicle pioneer. He grew up in the parental works, becoming familiar with foundries and forges, machine tools and machine construction, from childhood onwards. He helped his father design the Mail Coach, built in 1885 for the Marquis de Broc, and then designed a steam-powered three-wheeler which weighed 650 kg and reached a speed of 40 km/h. Armand Peugeot inspected it closely, but no business was actually concluded with Peugeot. Amédée Bollée understood as well as Armand Peugeot did, that making light-weight high-speed cars meant farewell to steam power. Consequently, he began to design a four-stroke internal-combustion engine in 1886. The first test engine was operational within a year. In 1888–89 he designed and built a three-cylinder rotary engine for airship propulsion, but it had cooling problems and was discarded. A few years later he turned his attention to the two-stroke engine and in 1895 built a two-stroke engine equipped with a primitive type of supercharger, for which he received a patent. He had begun to design an automobile in 1894 and in 1896 received a patent for a new four-stroke engine. It was water-cooled, with a condenser (not a radiator), the combustion chambers were hemispherical, with bowl-in-crown pistons, and a jet-nozzle carburetor. The horizontal parallel-twin weighed 150 kg and delivered 6 hp at 600 rpm. It was mounted transversely in the nose of the frame, with belt drive to a central differential, and an individual drive shaft to each rear wheel.

When Adrien de Turckheim saw the car in 1897, he purchased the manufacturing rights for De Dietrich and tooled up to make cars at Niederbronn. By 1899 the Usines Bollée were producing an engine a day for De Dietrich. Amédée Bollée, Sr., retired that year, leaving his elder son in command. He remodeled the factory to produce cars in series. In 1899, he designed and built the "Torpilleur" with a four-cylinder horizontal rear-mounted engine having dual carburetors, and chain drive. The frame was "underslung" (below the axles) and the speed-boat-like open body had a slanted windshield. Bollée sold the "Torpilleur" design to De Dietrich also. His first production car was the 1901 Model D, with a 10-hp rear-mounted horizontal twin. In 1902 he obtained a patent for a fully balanced parallel-twin with a third "slave" cylinder, and in 1903 he was experimenting with fuel-injection. In 1904 he adopted vertical front-mounted four-cylinder engines and began building some very expensive, powerful touring cars, such as the 1907 Model E whose 6.3-liter engine put out 45 hp at 1000 rpm. In 1910 he was awarded a patent for a hydraulic zero-lash valve lifter, and in 1912 the elegant 4-liter Model F featured a hydraulic vibration damper on the crankshaft. The last E was made in 1919 and the last F in 1921. He saw no market for his kind of car in the postwar world, and converted the plant to make piston rings and other high-precision auto parts. He died unexpectedly on December 13, 1926.

Bollée, Léon (1870–1913)

Director of Etablissements Léon Bollée from 1896 to 1913.

He was born on April 1, 1870, in Le Mans as the second son of Amédée Bollée, Sr. (1844–1917), and as a boy spent more time in the parental engineering works than he did at school. He was only 18 when he invented a clever calculating machine, capable of direct multiplication, consisting of more than 3000 parts. For this invention, he won a gold medal from the French Science Academy and a visit from Thomas Alva Edison, "wizard of Menlo Park." It steered him into a profession as designer of cash registers, machines for ticketing, dating and numbering, and even a cigarette-rolling machine. He was also a keen cyclist and used a tricycle for his personal transport. But his doctor told him the exercise might be too much for his weak heart, which made him think of motorizing his tricycle. By 1895, it had evolved into something quite different, a voiturette with two front wheels, seating for two in tandem, and a single-cylinder horizontal air-cooled engine with belt drive to the rear wheel. So many people wanted to buy one that he arranged for its production by Diligeon & Cie of Albert (Somme) in 1896. H. J. Lawson purchased the British manufacturing rights and Herbert Austin copied its main elements for his Wolseley tri-cars. In 1897 he designed and built a four-wheeled light car with a steel-tube frame, no springs, and a front axle steered as a unit from a central pivot. The air-cooled single-cylinder engine was mounted in front, with belt drive to the rear axle. In spite of its obvious drawbacks, Alexandre Darracq bought the manufacturing rights to it. The licenses brought him a lot of money, and he bought land at Les Sablons outside Le Mans for a spacious factory. He updated the design he had sold to Darracq, with a backbone frame and independent front suspension, put the engine at the rear end, and began production. In 1903 he adopted the Mercedes layout and began building a 20-hp four-cylinder model with chain drive. It was redesigned with shaft drive in 1906, a year when he introduced his first six-cylinder model. He sold a license for some of these cars to the Worthington Automobile Company of New York City. He had so much factory space he was able to offer facilities for airplane construction to Wilbur and Orville Wright, who spent a lot of time at Le Mans in 1907–09. In 1910 he began producing a 12-hp four and an 18-hp six, both with shaft drive. His production figures far out-stripped those of the Usines Bollée, makers of his brother's cars.

But Léon Bollée's heart gave out when business was at its best and he died on December 16, 1913.

Böning, Alfred (1907–1984)

Chief engineer of BMW cars and motorcycles from 1945 to 1968; department head for the development of BMW chassis and transmissions, cars and motorcycles, from 1968 to 1972.

He was born on October 13, 1907, in Salerno to German parents who brought him to Esslingen at the age of ten. He graduated from high school in Esslingen and spent a year as a draftsman with an engineering firm in Stuttgart. From 1927 to 1930 he studied at the Esslingen Technical College,[1] and after grad-

uated joined NSU as a design engineer. He left NSU for BMW in November 1931 and became a motorcycle designer. He also served as assistant to the head of motorcycle testing. In October 1937 he was placed in charge of frame design, transmission and sidecars. He put better brakes on all models, designed the frames for the R 12, the R 5 and the military sidecar combination R 75 with drive not only on the rear wheel but also the sidecar wheel. In 1943 he was transferred to the aircraft-engine department and worked on tests of rockets and turbojet engines. In 1945 he was placed in charge of the drawing office for cars and motorcycles, and head of vehicle development. He designed the new R 50 motorcycle and a small car, Type 331, with a 750-cc motorcycle engine, which was turned down by the board. He designed the 501 chassis with torsion-bar suspension, and directed the design of the V-8 engine for the 502, 503 and 507. He redesigned the Isetta with BMW power, and created the 600 and 700 minicars with rear-mounted motorcycle engines. Fritz Fiedler assumed the duties of new-car planning and pre-development after 1957, and in 1960 a separate engine design office was set up under Alex von Falkenhausen. Böning designed the R-27 single-cylinder motorcycle for 1960 and developed the R 50 and R 60. In 1968 he was given a new title and full responsibility for car and motorcycle chassis and transmission. He made important contributions to the 3.0S and 3.3L, as well as the original 5-series models of 1974 and the 3-series cars of 1975. He retired on October 31, 1972, but remained a consultant to BMW for another two years. He died on February 15, 1984.

1. Höheren Maschinenbauschule Esslingen.

Bonnet, René (1904–1983)

General manager of Société Deutsch-Bonnet from 1947 to 1961; chairman of Société des Automobiles René Bonnet from 1961 to 1965.

He was born on December 27, 1904, at Vaumas (Allier) near Vichy as the son of a carpenter cabinet-maker. His formal schooling ended at the age of 11 when the teachers of his local school were drafted into military service.

He worked in his father's shop and other carpentry firms. While on military duty in the Toulon naval base; he had an accident when diving, and was sent to Hôpital Bouville at Berck on the Channel coast, where he was to remain for three years. He spent his time reading books on engineering. In 1927 he returned to Vaumas, living an aimless existence. That changed in 1929, when his recently widowed sister asked him to come to Champigny-sur-Marne and run her late husband's Garage Pouhez. He then discovered his true calling. The business grew so fast that he needed more floor space, and in 1932 he bought a cartwright's shop from Madame Deutsch, another widow. He became friends with her son Charles, a student and a car enthusiast. He became a Citroën dealer, operating as the Garage du Marché, in Champigny. In 1936 Bonnet and Deutsch joined forces to prepare and race cars based on the front-wheel-drive Citroën. The first one, with a stylish roadster body, was ready in 1938. The called it the DB. The garage survived World War II as a repair business, and in 1944 the partners revived their racing projects.

From 1945 to 1949 they built seven DB cars, mostly roadsters and one single-seater. The last one was a sports coupé, for unlike Deutsch, Bonnet had dreams of making touring cars in series production. In 1948 he arranged with Antem to supply a number of bodies for a GT coupé designed by Mirhelotti for Stabilimenti Farina. When the Michelin family heard about it, they told the management of Citroën not to sell power trains to DB. They turned to Paul Panhard, who was quite willing to sell them their air-cooled flat-twin New models were designed and tested, beginning with a 500-cc single-seat racer, a cabriolet, and a short-wheelbase coupé. An aluminum-bodied coupe was shown in October 1952 and Antem supplied bodies for 30 roadsters. The HBR 5 coach was launched in 1953 with a plastic body made by Chausson and found a ready market. Unlike Deutsch, Bonnet had misgivings about the Panhard engine, which he felt was "overstressed." As early as 1954 he built a car with a four-cylinder Renault engine installed in an inboard rear location., The contract with Chausson expired in 1957, but an alternative supplier was found by setting up a joint venture, Société Plastique des Vosges at Contramoulin–St. Léonard, and production of the HBR 5 coach continued. When it was discontinued in 1961, 660 units had been built. The Le Mans cabriolet was added in 1953, with the Panhard engine and front wheel drive, and a plastic body made by a joint venture with a Matra subsidiary, GAP (Générale Automobile Plastique) at Salbris.

Matra also arranged for assembly facilities in a disused Norman textile mill at Romorantin. In 1961 Bonnet broke up the partnership with Charles Deutsch and cancelled the engine-supply contract with Panhard.

The Le Mans was redesigned with a front-mounted Renault engine and a radiator end in 1962. René Bonnet introduced the Missile cabriolet and Djet coupé, both with inboard/rear-mounted Renault engines. In 1965 he sold his company to Matra, but continued to manage the Garage du Marché at Champigny. He was killed in a traffic accident near Epernay while driving to his country house at Connantré (Marne) on January 13, 1983.

Bono, Gaudenzio (1901–1978)

General manager of Fiat S.p.A. from 1946 to 1955; managing director of Fiat S.p.A. from 1955 to 1969; vice president of Fiat S.p.A. from 1969 to 1974.

He was born in 1901 in Turin and held a degree in mechanical engineering. He joined Fiat in 1923, working mainly in the area of technical management up to 1931, when he was appointed vice director of SPA (the truck and military-vehicle division). He served as managing director of SPA from 1939 to 1946, when he was transferred to the corporate headquarters, with primary responsibility for the car division. He directed the reconstruction of the plants which had suffered considerable war damage, and had his say in the planning of new models, where to make them and in what numbers. He steered the 1400 and the 1100/103 into production, and pushed for a modern successor to the 500C. In this period he also lectured on special technonologies for automobiles at the Turin Polytechnic, and in 1957 was named president of the Galileo Ferraris National Electrotechnical Institute in Turin.

He promoted Fiat's return to the six-cylinder car market in 1959 and authorized a number of new projects including the 124 sedan, coupe and spider. He also helped prepare Fiat's industrial expansion into southern Italy, leading to new plants at Cassino, Termoli, and the vast proving grounds at Nardo.

He retired in 1974 and died in Turin in 1978.

Bönsch, Helmut Werner (1907–)

Director of BMW car planning, cost analysis and quality control from 1958 to 1963; director of BMW marketing and product planning from 1968 to 1973.

He was born on October 27, 1907, and studied engineering in Berlin. From 1926 to 1940 he occupied technical positions with Siemens, Deutsche Shell, and Mannesmann. He survived World War II and became a director of Kronprinz AG, wheel manufacturers in Solingen. In 1958 he moved to Munich and took charge of cost analysis, product planning and quality control for BMW. In 1963 he was named technical liaison manager, which meant coordinating the requirements of BMW's export markets, looking at the trends, and advising the car-planning staff and development engineers. He had considerable influence on the type of cars BMW was to build, and played a central role in the creation of the BMW 2000 series, the 1600-2, the Formula 2 racing engine, and the six-cylinder Bavaria series (2500/2800) sedans.

He retired from BMW in 1973.

Borgward, Carl Friedrich Wilhelm (1890–1963)

Managing director of Goliath Werke Borgward & Co. GmbH from 1928 to 1931; managing director of Hansa Lloyd & Goliath Werke Borgward & Tecklenborg GmbH from 1931 to 1937; chairman of Hansa-Lloyd-Goliath Werke Carl F.W. Borgward from 1937 to 1945; chairman of Goliath Werke GmbH from 1948 to 1963; chairman of Lloyd Maschinenfabrik GmbH (later Lloyd Motoren Werke GmbH) and Carl F.W. Borgward GmbH from 1949 to 1963.

He was born on November 10, 1890, in Altona, suburb of Hamburg, as the son of a coal merchant. He was apprenticed to a machine shop in Altona, and graduated after two years at a technical college[1] in Hamburg. He began his career with an engineering company[2] in Hanover and took evening classes at Hanover Technical University. After two years in Hanover he joined Schnellhass & Druckenmüller in Bremen as an engineer. In 1914 he was sent for training with the Pioneer Corps in Hamburg, and then to the western front in 1915. Within months, he was wounded in action and discharged. He then joined Carl Francke in Bremen as an engineer for hydrogen gas apparatus on airships. By 1919 he had saved enough to become a partner in Bremen Reifenindustrie, a small works that made some tires, but mainly kitchen tools and farm implements. They started manufacturing radiators in 1920 and won a big contract from Hansa-Lloyd. A year later the company became Bremer Kühlerfabrik Borgward & Co.

In 1921 they made two-passenger test car, designed by Fritz Kynast, powered by a two-cylinder two-stroke engine, and shelved it. In 1924 Borgward invented the Blitzkarren ("lightning cart"), a delivery vehicle with 250 kg payload capacity. It had a single front wheel steered from a bicycle saddle behind the cargo box. The engine was mounted alongside, with belt drive to the rear axle. Very quickly, demand far outstripped production. To finance a new factory, he went into partnership with Wilhelm Tecklenborg (1882–1948), a businessmen from a Bremen shipbuilding family. A new Blitzkarren appeared in 1925, with a front axle below the cargo box, and a motorcycle rear-half, marketed with the Goliath trademark. Bigger Goliath vehicles soon followed. Bremer Kühlerfabrik was renamed Goliath Werke in 1928, and Borgward purchased Bremer Carrosseriewerk vormals Louis Gärtner AG. The Goliath Rapid three-wheeled pickup truck with a cab was launched in 1929. That year Borgward also took control of Hansa-Lloyd Werke AG of Bremen-Hastedt and Hansa Automobil-Werke AG in Varel. They were consolidated with Goliath in 1931. In 1932 he began producing a three-wheeled car designed by Herbert Scarisbrick, with a 5.5-hp single-cylinder two-stroke engine, as the Goliath Pionier. Four-wheeled cars with rear-mounted two-stroke Ilo engines appeared in 1934 as the Hansa 400 and Hansa 500. His first "real" car was the Hansa 1100, with a four-cylinder engine and all-independent suspension. It was joined by a six-cylinder Hansa 1700 in 1935.

In 1937 he bought Tecklenborg's parts, becoming sole owner, and in 1938 began construction of a new auto plant at Sebaldsbrück. It produced 3-ton trucks and a half-track artillery tractor in World War II. In 1945 the plants were 75 percent destroyed, and Borgward was interned by the U.S. occupation forces in Lager Ludwigsburg under investigation for his part in "war crimes." He was released in 1948, blameless. He founded separate Lloyd, Goliath and Borgward companies. Production of the Lloyd LP 300 and Goliath GP 700 got under way in 1950, preceded by the Borgward Hansa 1500. He devoted himself to constant product renewal and a relentless broadening of the model range, neglecting accounting and legal matters, which bored him. His industrial empire collapsed in 1961 and he went into retirement. He died in Bremen of heart failure on July 28, 1963.

1. Höhere Maschinenbauschule, Hamburg.
2. Louis Eislers Eisenkonstruktionen, Hanover.

Bott, Helmuth (1925–1994)

Head of Porsche vehicle testing on road and track from 1962 to 1971; director of product development for Porsche from 1971 to 1984; executive vice president of Porsche from 1984 to 1988.

He was born on August 23, 1925, at Kirchheim/Teck as the son of an electrical engineer responsible for the local power stations. He was yanked out of school and sent to the Russian front to fight with the 23rd Panzer Division before he was 17, becoming one of the Wehrmacht's youngest lieutenants before he was wounded and sent back to a hospital in Germany in November 1944. In 1945 he was making a living as a schoolteacher, while looking for educational opportunities for himself. He joined Daimler-Benz AG as a trainee in October 1947 and won a scholarship to Stuttgart Technical University, stretching his income by working part-time as a draftsman for Bosch. When he got his degree, Daimler-Benz named him experimental engineer. He went to Porsche in 1952 as an assistant production engineer but was soon transferred to chassis development. He cured the aerodynamic lift and instability problems of the 356 and revamped the swing-axle rear suspension to make its behavior less capricious.

Though mainly concerned with production cars, he also designed the bodies for the 908 and 917 racing cars. He took personal responsibility for developing the 911 Carrera 4 x 4 and the 959. He laid out the Weissach test tracks and led the 928 through its development phase.

From 1984 onwards he had a say in the company's policy-making and continued to work long hours. He retired on September 30, 1988, and died in his home at Münsingen-Buttenhausen on May 14, 1994.

Boulanger, Pierre Jules (1885–1950)

General manager of Citroën from 1935 to 1938; president of Citroën from 1938 to 1950.

He was born on March 10, 1885, at Sin-le-Noble (Nord). He went through grade school and high school, volunteering for military service in 1904 and was assigned to the First Battalion of Balloonists[1] at Satory where he met Marcel Michelin and the two became friends. Four years later, he sailed for America, traveled west, and worked for some time on a ranch. Next he held a position with a utility company, and then went to Vancouver as a draftsman with a firm of architects. In 1911 he founded a construction company in Canada, which he led for three years before returning to Paris in 1914. He was drafted into the Air Force[2] as a corporal, and in 1915 was placed in charge of an observation squadron. By 1918 he was a technical advisor to the Ministry of

Armament, and was discharged with captain's rank in 1919.

He called on his friend, Marcel Michelin, who introduced him to his uncle Edouard. He was immediately engaged in an administrative position at the Clermont-Ferrand tire factory, and by 1922 had become a close collaborator of Edouard Michelin, the technical genius of the family. In the years that followed, he became the right-hand man of Pierre Michelin, son of Edouard. When Michelin's biggest customer, Citroën, fell behind on its payments in 1934, Pierre Michelin sent Boulanger to the Quai de Javel and all the other Citroën plants to make an assessment of their worth.

Citroën's fate became a highly politicized issue. The labor unions fought for the jobs of 19,000 workers and lawyers took action on behalf of creditors, mainly suppliers. It was solved by Michelin's acquisition of the Citroën assets (and debts), and Pierre Michelin's appointment as president of Citroën, with Boulanger at his side. It was Boulanger who ordered the engineers to develop a very small economy car, which resulted in the creation of the 2 CV many years later.

He was named president of Citroën when Pierre Michelin was killed in a road accident on December 29, 1937. Boulanger reorganized the company along military lines, introducing the Michelin family's addiction to secrecy at all levels, but particularly in technical matters. He also provided a generous budget for technical research, but had little or no ability to bring experimental projects into actual production. The arrival of the 2 CV in 1948 was an exception. But he never set a deadline for the successor model to the old 11 CV or the 15-Six, both overdue by the time of his death, in a road accident near Le Vernet on November 11, 1950, driving a 15-Six H with an experimental four-speed gearbox.

1 Premier Bataillon d'Aérostiers.
2 Armée de l'Air.

Bouton, Georges-Thadée (1847–1938)

Co-founder of Société de Dion–Bouton & Trépardoux in Paris in 1887; technical director of Société de Dion–Bouton & Cie from 1894 to 1927.

He was born on November 22, 1847, at Montmartre, Paris, as the son of an artist painter and his wife who taught music. After elementary schooling in Paris, he began an apprenticeship with a mechanic named Dubourg in 1862, and in 1867 was engaged as an engineer by the Mediterranean Forges and Worksites[1] at Le Havre, and worked in many phases of shipbuilding. Returning to Paris in 1869, he went into partnership with his brother-in-law, Charles Armand Trépardoux (husband of his sister Eugénie), still a student, who graduated from the Ecole des Arts et Métiers of Angers with an engineering diploma in 1871. They opened a workshop and began making scientific instruments for clients such as Ducretet, Bourbouze, and others. They also built various kinds of mechanical toys, including model steam engines.

They sold a model steam engine to Count de Dion in 1882, and he joined their partnership in 1883. That year, Trépardoux designed their first steam tricycle, patented a fast vaporization boiler, and began to design a steam car with front wheel drive and rear-wheel steering, which was built in 1886. They produced a small number of steam tractors and passenger vehicles, and invented the tractor-semitrailer combination in 1893.

Bouton and Count de Dion turned their attention to internal-combustion engines in 1890, and in 1893 Trépardoux (1855–1920) left the partnership and had no further contact with the auto industry. Bouton designed a small spark-ignition engine which was capable of rpm speeds twice to three times higher than the Daimler and Benz engines of the time and began production. It was built in many sizes and powered the de Dion–Bouton tricycles from 1895 to 1900, the Vis-à-Vis four-passenger "voiturette" from 1899 to 1904 and the Populaire from 1902 through 1911. A patent for a parallel-twin was issued to de Dion–Bouton in 1902, and Bouton's first four-cylinder model went into production in 1905. The company also made many models with V-8 engines from 1909 to 1923. After that, the engineering of their products failed to keep up with the progress made by its rivals, the sales curve dropped, and management began selling off real-estate to pay its debts. Bouton's functions were terminated in 1927 but he was connected with the disposition of the company assets up to 1933.

He died on October 31, 1938, in Paris.

1. Forges et Chantiers de la Méditerranée.

Bowden, Benjamin G. (ca. 1897–)

Chief designer of Hoyal Body Building Corporation from 1928 to 1936; body engineer and chief designer of Humber, Ltd., from 1936 to 1944; founder of Allen-Bowden Ltd. in 1945.

He was born about 1897 in London and began his career as a junior designer with Clement Talbot Ltd. where he spent three years. He became a designer with the Chelsea Motor Carriage Builders Company. From 1923 to 1928 he held positions with the Albany Carriage Company, Park Ward, H.J. Mulliner, and James Young, and then became chief designer with Hoyal. Hoyal's main business was to produce small series of special bodies on Morris, Austin, and Daimler chassis, but also built some bodies on American-make chassis. He joined the Rootes Group in 1936 and established firm styling themes for Humber and Hillman. He restyled the Hillman Minx twice, the Humber Snipe and Pullman, and a body for the Hillman 14. He met Donald Healey towards the end of 1944 and designed a very modern coupé for the Healey chassis that was then on the drawing board. His Healey body was put in production by Elliott & Sons of Reading in 1946. Allen-Bowden Limited was an industrial-design studio which failed to attract enough business to make it viable and was closed some two years after its foundation.

In 1947 he went to the United States as an independent design consultant, and did a lot of work for the Ford Motor Company in the 1948–52 period. In 1952 he set up his own industrial design studio in Toledo, Ohio, and worked for the Schwitzer Corp., Simmons Steel, and other clients. In 1955 he joined Murray Corporation in New York, and eventually became vice president of engineering and research for the Schwitzer Division of Wallace-Murray Corporation. Moving to Lake Worth, Florida, in May 1972, he won a permanent contract for defense-work designs from General Dynamics, which kept him busy till the end of his long career.

Boyle, Robert W. (1900–)

Chief engineer of Morris Motors (Engine Branch) Ltd. from 1936 to 1938; assistant chief engineer of Rover from 1938 to 1957; executive director of engineering of Rover from 1957 to 1961.

He was born in 1900 and trained as an electrical engineer. He began his career as a draftsman under Georges Roesch with Clement Talbot Ltd. in London. He joined Rover in 1929 as an assistant to Maurice Wilks, but left in 1933 to go to work for Morris Motors as chief experimental engineer. He was put in charge of all Morris engine design in 1936, but returned to Rover in 1938 as assistant chief engineer. He remained with Rover throughout World War II and was the principal engineer for the P4 (1949–63), the 3-Litre P5 and the 1963 P6 (Rover 2000).

He retired from Rover in 1964 to set up his own consulting-engineer office.

Bracq, Paul (1933–)

Chief designer of interior styling, Automobiles Peugeot, from 1974 to 1996.

He was born in 1933 and grew up in the Gironde area. He was just a small boy when he began repainting the toy cars his father gave him. Later he used plasticine to alter their shape. At the age of 12–13 he was making balsa-wood scale models of cars. He studied art and graduated from the Ecole Boulle in 1952. He went to see Philippe Charbonneaux who hired him in 1953, but was called up for military service. Due to the war in Algeria, he was kept in uniform until 1957. Karl Wilfert brought him to his studios at Sindelfingen in 1957. The pagoda roof on the Mercedes-Benz 230 SL was mainly his idea. He was the principal stylist on the 220 SE coupe and convertible, and worked on the W-114 and W-115. The 600 limousine was basically his design. He returned to France in 1966 and became chief designer of Brissonneau & Lotz, where he designed the Turbotrain for Alsthom and a proposed Simca coupe that was nixed.

In 1969 he joined BMW in their Munich studios and drew the prototype for the 1600 TI. He designed the Turbo concept car and the first 3-series and 5-series models. He joined Peugeot in 1974 and styled the interiors for the 505, the 205, 405 and 605. He retired to Cauderan near Bordeaux in 1996.

Bradshaw, Granville Eastwood (1889–1969)

Chief designer of ABC[1] from 1910 to 1920; consulting engineer to Belsize Motors, Ltd., from 1921 to 1924.

He was born in 1889 and went to work as draftsman with Star Engineering Company in Wolverhampton in 1908. In 1910 he signed up with W.L. Adams, owner of the Aeroplane Engine Company at Redbridge, Southampton, which they reorganized as the All-British Engine Company. The Southampton location was very good for selling marine engines, but too far from the aircraft industry, so in 1911–12 ABC leased premises at Brooklands where T.O.M. Sopwith and Harry Hawker had test-flight depots and flying schools. His first ABC aircraft-engine was a 40-hp water-cooled in-line four, and in 1912 he also designed an air-cooled flat-twin 500-cc motorcycle engine, which was produced at Brooklands and fitted in the ABC motorcycle, made by a subsidiary, ABC Road Motors, Ltd., starting in 1913. Late in 1914 ABC had to vacate its base at Brooklands on order from the Royal Aircraft Establishment, and moved to Hersham, near Walton-on-Thames. Bradshaw was given an engine-design contract from the Royal Aircraft Establishment and drew up two sizes of water-cooled 90° V-8s. The larger (100-hp) unit was put in production by Armstrong-Whitworth. His next design was an air-cooled radial eight-cylinder unit which was given the name Dragonfly, and was made in considerable numbers by Beardmore Aero Engines,

Ltd., Crossley Motors, Ltd., Belsize Motors, Ltd., F.W. Berwick & Co., Ltd., and Wright Aeronautical Corp. of Dayton, Ohio. In 1918–19 he designed a small-wheeled motor scooter with its 125-cc single-cylinder engine mounted horizontally above the rear wheel, put on the market as the Skootamoto. He also designed the ABC light car with an overhead-valve flat-twin 1100-cc air-cooled engine, but deliveries had just begun in November 1920 when the All-British Engine Company declared bankruptcy. The assets were taken over by A. Harper, Sons & Bean, Ltd., who formed ABC Motors (1920) Ltd. to continue production at Hersham. Bradshaw withdrew at the time of the sale and approached Belsize Motors, Ltd., with a new car design, which went into production in 1921 as the Belsize-Bradshaw. The V-twin overhead-valve 1.3-liter engine had a combination of air- and oil-cooling, produced by Dorman Engine Company to Bradshaw's design. About 1000 of them were built up to 1925. He joined BSA[1] and designed a succession of air-cooled flat-twin motorcycle engines. He opened a technical consulting practice, worked on military projects during World War II and about 1951–53 was busy designing engines for motor scooters. He died in April 1969.

1. Birmingham Small Arms Company, Ltd.

Braess, Hans Hermann (1936–)

Director of technical research for Dr. Ing. h.c. F. Porsche K-G from 1977 to 1980; director of BMW's science and technology center from 1980 to 1996.

He was born in 1936 and graduated with a degree in mechanical engineering from Hanover Technical University in 1960. He stayed on for another two years in Hanover for postgraduate studies on motor vehicle design and construction. He began his career as a chassis designer with Ford-Werke AG in Cologne and won a doctor's degree in engineering for his thesis on Steering Behavior and Vehicle Dynamics. He became a professor at Munich Technical University and in 1966 took his class to visit the Porsche factory at Zuffenhausen, which resulted in an invitation from Helmut Bott to take a position in the Porsche organization. Braess was interested, but stalled for time. In 1969 he called Bott to suggest they hire his best-ever student, Helmut Flegl, which was done. And in 1970 Braess joined Porsche as a member of the scientific research section. He became head of basic concepts and pre-development in 1975 and director of technical research two years later. He invented a new rear suspension geometry, self-adjusting roll-steer effects in accordance with torque magnitude and torque reversals in the driving (rear) wheels, which became the "Weissach axle" of the Porsche 928.

He returned to Munich in 1980 to take charge of BMW's science and technology center. His influence on subsequent BMW production cars lies not in actual designs, but in the basic principles of vehicle layout, disposition of the masses, and suspension geometry for wheels that steer as opposed to (or combined with) wheels that drive. From 1992 onwards BMW allowed him to go on lecture tours to the technical universities of Stuttgart, Dresden and Munich. He retired from BMW in 1996.

Branitzky, Heinz (1929–)

Business manager of Dr. Ing. h.c. F. Porsche K-G from 1972 to 1976; chief financial officer and deputy chairman of Porsche from 1976 to 1988; chairman of the Porsche management board from 1988 to 1992.

He was born on April 23, 1929, in Zölz, Upper Silesia, and graduated with a diploma in business administration from the Berlin Free University in 1952. He began his career in the office of business qualifications and tax advice of a government cooperative[1] in 1952, becoming manager of his department in 1955. In 1960 he joined Carl Zeiss (optical systems) at Oberkochen as financial planning manager. He escaped to the West in 1965 and joined Porsche in October 1965 as finance manager. He was promoted to business manager in March 1972 and chief financial officer in November 1976. He succeeded Peter Schutz as chairman in February 1988 when the company was headed for financial straits. He organized

a new stock issue to raise $150 million and laid off 1000 workers. In the face of shrinking sales, he replaced the 924 with the 944 and authorized erection of a new body plant for the 911 and 928 at Zuffenhausen. He presided over the introduction of the Carrera 4 all-wheel-drive model and the 911 Speedster. But the sales curve went down and down, and losses mounted. He was replaced in September 1992 by Wendelin Wiedeking.

1. Berliner Treuhandvereinigung.

Brasier, Charles-Henry (1864–1941)

Technical director of Société d'Electricité et d'Automobiles Mors from 1896 to 1901; technical director of Société du Trèfle à Quatre Feuilles from 1901 to 1904; managing director of Société des Automobiles Brasier from 1904 to 1906; chairman of Société de Construction d'Automobiles le Trèfle à Quatre Feuilles from 1906 to 1909; chairman of Société des Automobiles Brasier from 1909 to 1926.

He was born on March 9, 1864, at Ivry-la-Bataille as the son of a gardener in the service of the Laporte family on their estate. He studied engineering at the Ecole des Arts et Métiers in Chalons-sur-Marne and graduated in 1883. He began his career as a draftsman with a railroad company[1] working on equipment and traction machinery, and joined Mors in 1886, working under the orders of Henri Perrot. He designed a coke-fueled Mors steam car in 1887, and a steam-powered tricycle built by Mors for Roger de Montais. It was in 1895 that Emile Mors decided to produce automobiles with internal combustion engines, and Brasier designed a V-4 engine with a gear-driven oil-pressure pump that was installed in the 1896 Mors dogcart. He also applied for a patent on flat-twin engines of 3 CV and 5 CV in 1896. He designed the racing cars that made the Mors name famous, taking part in the 1897 Paris–Trouville, the 1898 Marseille–Nice and 1898 Paris–Amsterdam road races. In 1899 he adopted the in-line four-cylinder engine, mounted in front, with chain drive to the rear axle, first a 16 CV racing car and then a 10 CV touring car. He left

Mors in October 1901, went into partnership with Georges Richard and designed four types of light car with 8 CV and 12 CV front-mounted vertical parallel-twin engines, adding a 10 CV twin in 1902 and four-cylinder 16 and 24 CV models in 1904. The cars were named Georges Richard up to 1903, and Richard-Brasier in 1903–04. The partnership broke up in 1904 because Georges Richard wanted to mass-produce low-priced cars, while Brasier "was only interested in racing" and designed only expensive cars. Brasier took over the factory in rue Galilée at Ivry-Port to make Brasier cars, still wearing the four-leafed clover badge of the Richard Brasier. They were big, powerful, expensive cars, and after profitable operations in 1906, the company was on the brink of ruin in 1907. It was refinanced by a group of Swiss investors, who put a banker, Jacques Osmond, on the board of directors, holding the purse strings. The factory expanded into Quai d'Ivry at Ivry-Port, and from 1909 to 1914 the Brasier name won new fame for its speed-boat engines. Under Swiss influence, Brasier began building a two-cylinder 10/12 in 1911 and an 11 CV four-cylinder car in 1912. In 1915 the factory was requisitioned for military work and produced 6000 Hispano-Suiza V-8 aircraft engines in four years. He made a careful re-entry on the car market with the four-cylinder 18/30, adding a 12 CV family car in 1923. No longer present on the race track, and lacking any claim to outstanding quality, the Brasier name lost its gloss, and its market share went into a terminal decline. Brasier sold the company to Jean Chaigneau in 1926 and went into retirement.

1. Chemins de Fer d'Orléans.

Breitschwerdt, Werner (1927–)

Director of passenger-car body development, Daimler Benz AG from 1973 to 1978; management board member for technical affairs from 1978 to 1983; chairman of the Daimler-Benz AG management board from 1983 to 1987.

He was born on September 23, 1927, in Stuttgart and graduated from the Stuttgart Technical University in 1951 with a degree in electrical engineering. He worked as an assistant professor in electrical equipment at Stuttgart Technical University in 1952–53 and joined Daimler-Benz AG on April 16, 1953, as a body test engineer at Sindelfingen. In 1960 he was named section leader of production car body engineering, and in October 1963, department head in the car-testing department at Sindelfingen. On May 1, 1965, he became chief department head of car-body testing and was made director of the car-body testing department on July 15, 1967. He was promoted to deputy director of car-body engineering in July 1971 and director of car-body development in July 1973. He was given the additional responsibility of styling on January 1, 1974. He directed body design and engineering for the W-123 and W-124 mid-range models and supervised the planning for the W-126 S Class. As chairman, he was in complete agreement with the purchase of MAN's half-stake in MTU[1] for $225 million in 1985 but argued against purchases such as AEG[2] and Dornier. He wanted all expansion confined to the corporation's core activities of cars, trucks, buses, special purpose vehicles, and engines, while some other directors saw better growth opportunities in diversification into unrelated industries. They were stronger politically and prevailed, forcing him to hand in his resignation in July 1987. The board maintained an office for him at Untertürkheim as technical advisor until 1997.

1. Motoren and Turbinen Union.
2. Allgemeine Elektrizitäts-Gesellschaft.

Brillié, Eugène (1863–1941)

Co-founder of Gobron & Brillié in 1897; designer of Gobron-Brillié motor vehicles from 1898 to 1903; designer of Schneider buses and trucks beginning in 1905, and chief engineer of the Schneider battle tank of World War I.

He was born in Paris on May 8, 1863, and graduated from the Ecole Centrale des Arts et Manufactures in 1887. He began his career as a railway engineer but developed an interest in the internal-combustion engine and its fuels (gasoline, alcohol, benzole and naphtha) and filed a number of patent applications. In 1897 he went into partnership with Gustave Gobron to produce automobiles powered by his patented opposed-piston engine (two pistons per cylinder, with con-rods extending to both ends, lever-and-rod linkage to a single output shaft). Gobron-Brillié racing cars were prominent from 1898 and won the Light Vehicle Class in the 1902 Circuit des Ardennes. At Ostend on April 10, 1903, Louis-Emile Rigolly set a new world land-speed record with a Gobron-Brillié at 166.66 km/h.

Later that year, Brillié severed his connection with Senator Gobron and went to work for Ateliers Schneider at Le Havre. He devoted his career to heavy-duty vehicles from then on, but kept right on inventing. In 1907 he obtained a patent for an automatic spark-advance device.

He withdrew from business activity in the early twenties and died in 1941.

Brough, George (1881–1970)

Managing director of Brough Superior Cars, Ltd., from 1935 to 1939.

He was born in 1881 at Nottingham as the son of W.E. Brough, an engineer who made his own car in 1899 and produced motorcycles of original design from 1908 to 1926. George grew up in his father's shops on Vernon Road, Basford, Nottingham, and it became his ambition to build a better machine than his father did. He set up his own shop in Haydn Road, Nottingham, in 1921 and began producing the Brough Superior, mainly powered by air-cooled V-twin JAP, MAG, or Matchless engines of 680 or 750 cc. In 1934 he built a sports car for his own use. Encouraged by the interest it sparked, he decided to exploit the market for an exclusive touring car of very pretentious styling. He contracted with the Hudson importers for a supply of complete chassis, and with Atcherley for coachwork. He retrofitted the chassis with hydraulic brakes and reinforced the rear suspension with transverse torsion bars and radius rods. Engines for the Alpine Grand Sports were equipped with a Centric low-pressure supercharger. In May 1938 he displayed a new

model with a Charlesworth body, and a Lincoln-Zephyr V-12 engine. He also introduced a new and more powerful motorcycle, the 996-cc Golden Dream. The shops undertook precision-engineering work during World War II, and he never resumed production of motorcycles or cars. He died in February 1970.

Brown, David (1904–1993)

Owner of Aston Martin from 1946 to 1972 and Lagonda from 1947 to 1972.

He was born on May 10, 1904, in Huddersfield, Yorkshire, as the son of Frank Brown and grandson of David Brown (1843–1903) who founded a company to make gears and patterns in 1860. He was educated privately and at Rossal School in Yorkshire. In 1921 he was apprenticed to the family firm of David Brown & Sons (Huddersfield) Ltd., who had produced the Dodson car from 1910 to 1913 and the Valveless from 1912 to 1915. His apprenticeship took him through machine shops, foundries, pattern shops and drawing offices in British factories, and also to South Africa, where the company was installing gears in gold mines, and the United States, to get a look at American production methods. In 1926 he was appointed foreman of a worm-gear department in the Huddersfield works, and became assistant works manager in 1927. He was named general manager of the Keighley Gear Company in 1928 and joined the board of David Brown & Sons (Huddersfield) Ltd. in 1929. He became its joint managing director in 1932 and sole managing director in 1933. In 1935 he set up David Brown Tractors, Ltd., and produced Ferguson-Brown tractors from 1936 to 1939. He paid £20,500 for the assets of Aston Martin, Ltd. In 1946 and £52,500 for those of Lagonda Motors, Ltd., in 1947. He determined the general makeup and character of the products and provided a realistic racing budget. He took over Tickford, Ltd., coachbuilders at Newport Pagnell, in 1955 and moved all car production there in 1957. In a vast consolidation, he became chairman of David Brown Holdings, which controlled David Brown Gear Industries, David Brown Tractors, Ltd., Vosper-Thornycroft shipbuilding, plus some smaller industrial subsidiaries. Aston Martin and Lagonda were merged in 1958 and never made a big profit. But losses began mounting in 1969 and reached crisis proportions in 1972, leading to the sale of Aston Martin Lagonda, Ltd., to Company Developments, Ltd., and David Brown Tractors, Ltd., to the J.I. Case Division of Tenneco, Inc. When Vosper-Thornycroft was nationalized in 1978, he retired to Monaco. He died in Monaco on September 6, 1993.

Brown, George W.A. (b. 1881)

Chief engineer of Clement Talbot, Ltd., from 1911 to 1913; chief engineer of Sir William Beardmore & Co., Ltd., from 1914 to 1920.

He was born in 1881 and educated at Charterhouse School and Glasgow University, where he became known as a brilliant mathematician. He began his career as a design engineer with Humber, Ltd., in 1903. Herbert Austin engaged him in 1907 to design a Grand Prix racing car. He responded with a six-cylinder 9.7-liter machine with chain drive (one was also made with shaft drive). In 1911 G.P. Mills, works manager of Clement Talbot, Ltd., in London hired him to prepare a speed-record car for Percy Lambert. It was basically a 25-hp 4½-liter six with special pistons, bigger carburetor, and a longer-geared rear axle. It was timed at 109.43 miles per hour for a flying-start lap of Brooklands race track. In 1913–14 he also modernized the chassis of all Talbot models with semi-elliptic leaf springs and an open propeller shaft in place of the former transverse leaf spring and torque-tube drive. In 1913 he designed a three-wheeled cyclecar for the Coventry Motor Company (later Coventry Premier, Ltd.) and in 1914 Sir William Beardmore brought him to his shipbuilding yards at Dalmuir on the river Clyde, to organize the production of Austro-Daimler aircraft engines in the Underwood Works at Paisley. He also designed several cars for the postwar market. One was the Beardmore 15/20, built at Paisley from 1919 to 1928 as a taxicab and family car. Another was the Beardmore 10/18 which had an advanced four-cylinder 1½-liter engine with splayed overhead valves operated by a single overhead camshaft (using the layout of the Austro-Daimler aircraft engine). The same cylinder-head configuration was also used for a six-cylinder 2.6-liter engine in a new chassis which went into production in Dumfries as the Arrol-Johnston Victory in 1919. Sir William Beardmore had been chairman of the New Arrol-Johnston Co. Ltd. and its biggest shareholder since 1904. The 10/18 hp Beardmore was assembled at the Victoria Works, Anniesland, which Beardmore had acquired for war production in 1914. But both the 10/18 and the Victory had quality problems, due to a lack of development, and were replaced for 1921, the 10/18 by a simpler Eleven designed by A. Francis and the Victory by the prewar 15.9 hp model designed by Theo Biggs.

Brown, R.J. "Tom" (1907–)

Chief engineer of Morris Motors (Engines Branch) Ltd. in 1939–40 and from 1945 to 1948.

He was born in 1907 and educated at Birmingham Central Technical College. He began his career with BSA[1] in 1930 and joined Wolseley Motors, Ltd., in 1931 as a development engineer. He became an engine development engineer with Austin Motor Company, Ltd., in 1933 and in 1936 joined Morris Motors (Engines Branch) Ltd. in Coventry as a design engineer. He was named assistant chief engineer, under A.V. Oak for Morris engines in 1937 and succeeded "Vic" Oak as chief engineer two years later. In 1940 he was "on loan" to the Ministry of Supply and spent the war years as director of machine tool production. Upon his return to Oxford, he led the design and development of a single-overhead-camshaft in-line 2.2-liter six. The camshaft drive and valve train were full of intricate refinement, but his engine was criticized as being "structurally weak." It was installed in the Morris Six and Wolseley 6/80, built from 1948 to 1954. He made a four-cylinder version, scaled down to 750 cc, as a candidate for powering the Mosquito — the project that led to the Morris Minor in 1948.

But Alec Issigonis, the brain behind the Mosquito, had sketched out a side-valve flat-four. Brown put it on his drawing board, and also provided overhead-valve alternatives of 800 cc and 1100 cc, with the radiator mounted behind the engine. But both were too wide to fit between the front wheels of the Mosquito. Finally, after much testing, the side-valve flat-four was also rejected, partly due to vibration problems and partly due to excessive manufacturing-cost estimates. It was decided to use the old (1935) Morris eight L-head engine. Brown quickly drew up an overhead-valve version, but Lord Nuffield vetoed it. Feeling that he was wasting his time with Morris, Brown resigned on August 31, 1948, and went to Sheepbridge, major suppliers of engine parts and other hardware, as managing director of the parent company and chairman of its 16 subsidiaries.

1. Birmingham Small Arms Company, Ltd., was then producing three- and four-wheeled cars with front wheel drive, designed by F.W. Hulse.

Bruhn, Richard (1886–1964)

Managing director of Auto Union AG from 1932 to 1945; managing director of Auto Union GmbH from 1949 to 1958.

He was born on June 25, 1886, at Flekeby in Holstein as the son of a farm worker. His schooling was part commercial, part technical. He began learning to be an electrician, but was apprenticed to a mechanical workshop. Later on he studied business management and joined AEC[1] in Berlin, and was only 24 years old when AEG sent him to England as manager of the London branch. He served in the German Navy in World War I, and afterwards studied political science at the University of Kiel, graduating in 1921 with a doctor's degree. He spent five years as an executive with a Kiel-based company, and then joined Hugo Junkers in Dessau in 1927, serving as a director of both the aircraft and the engine branches. In 1929 he took over the Pöge Electrical Works in Chemnitz, and became acquainted with J.S. Rasmussen, who persuaded him to take over the management of the company[2] that made DKW motorcycles and cars.

In 1930 the State Bank of Saxony held huge debt claims against several car makers in the region, including DKW, Audi and Horch. The bankers approached Dr. Bruhn, asking him to submit a report on the situation, He invented a scheme for an industrial cooperative, whose share capital would be controlled by the bank. It was incorporated in 1932 as Auto Union AG, the members being Audi, DKW, Horch and Wanderer.

He directed Auto Union AG so as to maximize the synergy between the members, sharing plant capacity and undertaking engineering assignments for each other. It became one of the pillars of Germany's auto industry in the Nazi period, behind Adam Opel AG and Daimler-Benz AG. He escaped to the West in 1945, joined forces with Carl Hahn (1884–1961) the former sales director of Auto Union AG, and others, in making plans for a fresh start. They founded Auto Union GmbH in Ingolstadt in 1949 and also acquired a plant in Düsseldorf. They started production of DKW motorcycles, light vans, and passenger cars.

He guided the company's affairs until 1958, when it fell into Fritz Flick's control, and left when Flick put a stop to motorcycle and light van production.

1. Allgemeine Elektrizitäts-Gesellschaft (General Electrical Company).
2. Zschopauer Motorenwerke AG.

Bugatti, "Jean" Gianoberto Carlo Rembrandt Ettore (1909–1939)

Design and engineering executive of Automobiles E. Bugatti from 1930 to 1939; director of Bugatti's operations at Molsheim from 1936 to 1939.

He was born on January 15, 1909, in Mülheim on the Rhine. as the first son of Ettore Bugatti, and spent his early childhood at Molsheim in Alsace. He was six years old when the family moved to Milan, where he attended local schools. After a brief stay in Paris in 1918–1919, he continued his education in Strasbourg schools and in his father's factory. Everywhere, his nature and interests were at odds with the curriculum, and he put paid to any ambitions the family had nurtured about university studies. Brimming with artistic talent, he had also inherited his father's intuitive understanding of mechanisms and physics, and was perfectly at home in the factory. He also enjoyed testing the cars and proved an excellent driver.

In 1930 he was given a number of responsibilities at the factory and took over the body-design tasks. He was also instrumental in changing Bugatti's engine architecture, from one to two overhead camshafts, from vertical valves to widely splayed valves. Some of his body designs were of great beauty, while others were stronger in their originality than actual harmony of lines. He created the Coupé Napoléon on Type 41 chassis, a masterpiece of extravagance and several commonsense open and closed bodies for Type 49 and Type 50.

In 1933 he designed the roadster body for Type 55, the quintessential sports car of its period, and made preliminary sketches for the Ventoux and Stelvio body styles which Gangloff produced on Type 57 chassis. "Jean" also designed the racing car bodies, notably the 1937 Grand Prix single-seater and the Type 57 streamliners that won the 24-hour race at Le Mans in 1937 and 1939. The styling theme from Type 55 was taken to extreme limits in the Atalante and Atlantic coupés.

The Bugatti team drivers had great respect for "Jean"'s skill and mastery at the wheel, but his father forbade him to compete in races. Still his life came to an untimely end on the evening of August 11, 1939, when he crashed into an apple-tree when swerving at very high speed to avoid a cyclist who had somehow strayed on to the rest road that was closed to traffic.

Bugatti, Ettore Arco Isidoro (1881–1947)

Founder of Automobiles E. Bugatti in 1909 and its owner and leader until 1940.

He was born on September 15, 1881, in Milan, as the son of an artistic cabinet maker, Carlo Bugatti, whose talents included sculpture, painting, silversmithing and architecture. The family had been artists for generations and Ettore enrolled at the Brera school of Fine Arts in Milan only to find that his instincts, talents and interests steered him towards mechanical inventions and design. Finally his father agreed to letting him start an apprenticeship with Prinetti & Stucchi, bicycle manufacturers who also built motor tricycles under a de Dion–Bouton license. In 1898 he designed a dual-engine version of the tricycle which the company agreed to build. But they balked when he proposed a four-wheeled car with four separate engines grouped at the rear end.

In 1899 he showed drawings of new cars to the Counts Gulinelli who agreed to back him and founded the Bugatti-Gulinelli Motor Co. in Milan. Its first product had a four-cylinder overhead-valve engine and a four-speed gearbox. It won an award at the International Exhibition in Milan in 1901 and a medal from the Automobile Club de France. It also led to an invitation from De Dietrich of Niederbronn in Alsace to join them as head of the motor vehicle design office. Since he was not yet of age, his father co-signed the contract.

He lacked basic qualifications for leading an engineering office and was dismissed in 1904. That same year he obtained backing from E.E.C. Mathis to design the Hermès car and serve as technical director of the Société Alsacienne de Constructions Mécaniques in Graffenstaden. The Hermès had quality problems and Mathis sacked him. Bugatti tried his luck with Gasmotorenfabrik Deutz of Cologne, and Arnold Langen hired him in 1907. Bugatti designed the Deutz car, basically a "new, improved" Hermès, plus an all-new model for 1909.

Also in 1909 he bought property at Molsheim in Alsace, including a mansion as living quarters for his family, a former dye works which he converted into a mechanical workshop, plus stables and kennels. The first Bugatti cars were manufactured at Molsheim in 1910. He became financially secure when he sold a license for Type 12, a small car with an 856-cc four-cylinder engine, to Peugeot. The Bugatti name won fame with the 1327-cc Type 13 and Type 14, the Gaillon (named for winning a hill-climb).

He fled from Molsheim when World War I broke out, and Count Zeppelin provided him with a safe-conduct to travel to Milan with his family, baggage, and two racing cars.

In October 1914, he went to Paris and volunteered his services to France. For a year and a half he rented a workshop at Puteaux but in 1916 obtained the use of offices, laboratories, and factory space belonging to the Duc de Guiche, in rue Chaptal at Levallois, where he developed the 500-hp U-16 (two vertical blocks of eight) aircraft engine. He sailed for America at the end of 1917 to organize its production in the Duesenberg works at Elizabeth, New Jersey, and returned to Paris near the end of the war in 1918.

He returned to Molsheim in January 1919 and began work on Types 22 and 23, for which he sold licenses to Crossley in England, Diatto in Italy, Rabag in Germany. He also sold a special suspension-system license to Panhard in Paris. Bugatti broke into Grand Prix racing with Type 35 in 1924 and met with great success. Type 43 of 1927 was a road-going sports version of the 35, with outstanding performance. Type 41, the "Bugatti Royale" was built to outpower and outluxurize any prestige car in the world-but only seven units were built.

In his very richest hours, Bugatti spent more time breeding horses and riding them, maintaining the kennels, and designing household utensils than he did with the cars. He became intrigued with the French trend to high-speed railcars and designed one powered by four straight-eight Type 41 engines. For its production, he added a vast new hall at Molsheim which dwarfed the car factory.

In 1936 the railcar-factory workers joined France's general strike and occupied the car factory, putting a stop to car production. Bugatti was so emotionally wounded that he left Molsheim and took up residence in Paris. He set up two drawing offices in Paris and opened an experimental workshop in rue du Débarcadère, Levallois, coordinating their activities with the planning being done at Molsheim. All Bugatti activity at Molsheim ended abruptly in September 1939, at the declaration of war against Nazi Germany. All the best machine tools were shipped to rented factory space in Bordeaux (which was badly damaged by the RAF in a bombardment raid in November 1940). The Molsheim factory was occupied by the Germans who quickly arranged for its purchase at their own price, less than half of its real value.

Privately, Bugatti bought the De Coninck yards on the Seine at Maisons-Lafitte and started to design a high-speed torpedo-boat. In 1942 he purchased Automobiles La Licorne from the Lestienne family, with a big factory at 5–17 rue Mathilde in Courbevoie, and designed Bugatti Types 72 and 73 as prototypes for postwar Licorne models.

The Molsheim establishment was confiscated by the French authorities in 1945 as "enemy property" and placed under the Administration of Estates office. Bugatti sued to recover the property, and the case was settled in his favor in July 1947.

His plans for La Licorne were never implemented in part due to the delays over the rights to the Molsheim factory, and in part because he was aging and not in good health. He died in the American Hospital at Neuilly on August 21, 1947, and was buried at Ermenonville.

Bullock, William Edward (1877–1968)

Managing director of Singer & Co., Ltd., from 1919 to 1936.

He was born on March 14, 1877, at Handsworth, Birmingham, as the son of a smith, and was educated at Smethwick Technical School. He joined Dennison & Wigley of Handsworth as a toolmaker's assistant at the age of 14 but left in order to get experience in electrical engineering. When his old firm was reorganized as the Wigley-Mulliner Engineering Company, he returned to Handsworth, and moved to Coventry when the company bought a new factory there. In 1905 Wigley-Mulliner Engineering Company became the core of the newly founded Coventry Ordnance Works, and he was named works manager. He joined Singer & Co., Ltd., in 1908 as works manager of the Canterbury Street plant in Coventry. When the new Singer Ten was introduced in 1912, all earlier Singer cars were phased out, and production increased to 1350 cars in 1913. The factory was converted to make munitions at the start of World War I, and in 1919 Bullock quickly got the Singer Ten back into production. Under his management, Singer & Co., Ltd., grew at a fast pace. He arranged the takeover of Coventry Premier in 1920 and Coventry Repetition Company in 1922. A six-cylinder Singer 15 was given an overhead-valve engine. In 1925 he negotiated the purchase of the Sparkbrook Manufacturing Company, Ltd., of Payne's Lane, Coventry, and converted the plant from motorcycles to car production. Singer made 9000 cars in 1926. In 1927 Singer purchased the bankrupt Calcott Brothers, Ltd., and remodeled its Coventry works into a service center and spare parts depot. Also in 1927 Bullock made a deal with BSA[1] to take over a seven-story factory building in Coventry Street, Small Heath, Birmingham, which was reequipped and became Singer's main assembly plant. He then disposed of leased factory space on Brewery Road and Waverley Road in Coventry. The Singer Junior with an overhead-camshaft 847-cc four-cylinder engine came on the market and sustained a rising sales curve in a market that shrank considerably in 1929–31. Bullock multiplied his initiatives, speeded up the model-renewal cycle, and engaged talented engineers such as H.C.M. Stevens, A.G. Booth and Leo Shorter. But the company's profit margin got narrower, and the balance sheet turned to losses in 1934 and 1935. In 1936 Singer & Co., Ltd., was on the verge of bankruptcy, and Bullock approached Spencer Wilks of Rover with a merger plan. Wilks agreed, but the Rover board of directors turned it down. Bullock failed to win re-election to the Singer board and resigned. He joined a company that made Aron fire pumps, but retired in 1939 and spent World War II working on a family-owned farm.

1. Birmingham Small Arms Company, Ltd.

Burney, Sir Charles Dennistoun (1889–1968)

Managing director of Streamline Cars, Ltd., from 1928 to 1933.

He was born in 1889 as the son of Sir Cecil Burney, admiral of the Royal Navy, and succeeded to his baronet's title in 1929. He was a tireless inventor with a broad field of interests. During World War I he invented the paravane for mine-sweeping and became a commander in the Royal Navy. In 1924 he was placed in charge of the team which produced the British Airship R-100 at Howden, Yorkshire. He wanted to apply aircraft construction principles to automobile bodies, and set up a company in rented premises at Maidenhead in 1928. He chose a layout with an outboard/rear engine installation and an extremely spacious interior. The body was built up around a duralumin cage with fabric cover and a profile derived from airship shapes. The first prototype had a four-cylinder Alvis engine. Later ones had Beverley-Barnes straight-eight, Armstrong-Siddeley six, or Lycoming straight-eight engines. Only 12 cars were built. His idea was to sell the manufacturing rights, not complete cars, and aroused some interest from E.W. Hives of Rolls-Royce, though the outcome was negative. In 1933 he sold a license to Crossley Motors, Ltd., and closed the Maidenhead shops. He turned his mind to aeronautical inventions for military purposes (rockets and flying bombs) and scientific advances for civilian applications, such as the use of sonar for detecting schools of fish. He retired to Bermuda about 1952 and died there in November 1968.

Bussien, Richard (1888–1979)

Technical director of Voran Automobilbau AG from 1926 to 1932; deputy chief engineer of Ford-Werke AG from 1934 to 1940; chief engineer of Ford-Werke AG from 1940 to 1951.

He was born on November 18, 1888, in Leipzig and educated as an engineer. He began his career as a motoring journalist and author of technical books. In 1917 he became editor of the annual *Automobil-Technisches Hand-*

buch, which Edmund Rumpler had first published in 1901 under the title *Automobil-Technischer Kalender*. He became an advocate of front-wheel-drive and in 1926 set up a consulting-engineer's practice with facilities for prototype construction in Kleiststrasse, Berlin. His Voran 4/20 prototype had independent front suspension with upper and lower transverse leaf springs and double-joined drive shafts. Its main features were patented (DRP 456,925). He also designed a Voran 5/25 and a Voran 6/30, and lobbied the German auto industry to sell licenses for their production. He created the GMJ front-wheel-drive prototype in 1926 and in 1928 was associated with Ing. Garsstka in the design of a big front-wheel-drive bus for ABOAG. In 1932 NAG took over the entire Voran establishment, and he designed the NAG–Voran 220 which appeared at the Berlin Auto Show in February 1933. It had a central-tube frame and all-independent suspension, an air-cooled 1½-liter flat-four engine and unit-construction body. He left the company when NAG discontinued car production and signed up with Ford in 1934.

Busso, Giuseppe (1913–1999)

Assistant chief engineer of Alfa Romeo S.p.A. from 1948 to 1954; manager of the Alfa Romeo drawing office from 1954 to 1966; vice director of product development for Alfa Romeo from 1966 to 1969; director of Alfa Romeo product engineering from 1969 to 1972; central vice director of the Alfa Romeo engineering staff from 1972 to 1973; joint central director of the Alfa Romeo engineering staff from 1973 to 1977.

He was born in 1913 in Turin and held an expert's diploma in industrial engineering. He joined Fiat in the aircraft section and worked on calculations and checking. He moved to Milan in 1939 and became an assistant to Grazio Satta in the design of racing cars. At the start of World War II he was assigned to aircraft-engine design under Wilfredo Ricart, and from 1943 to 1945 he was a member of a small team working on postwar car projects in Albergo Belvedere above Lake Orta. In 1946–48 he was on loan to Ferrari at Maranello, but returned to the Portello works and was given responsibility for all mechanical organs for the 1900, the Giulietta, and the 2000. He designed the five-speed gearboxes and the steering gear and linkages for all production models. From 1966 onwards he became more and more involved with product planning and liaison with marketing and service. He had a supervisory role in the making of the Montreal and the Alfetta, and participated in the preliminary studies for the V-6. He retired in 1977 and died in Milan in 1999.

Calvet, Jacques Yves Jean (1931–)

Vice president of the Peugeot SA board from 1982 to 1984; chairman of Peugeot SA from 1984 to 1997.

He was born on September 19, 1931, at Boulogne-sur-Seine, as the son of Louis Calvet, philosophy teacher at the Lycée Henri IV. He graduated with a law degree from the Lycée Henri IV. He graduated with a law degree from the Lycée Jeanson de Sailly in Paris, where he was known for his phenomenal memory. He continued his education by earning a degree in finance and economic science from the Institute of Political Studies in Paris and a diploma from the E.N.A.[1] His career began with an appointment to the Court of Accounts (a watchdog over government spending) in 1957. Two years later he was made a "head of mission" in the Ministry of Finance and in 1970 cabinet chief for Valéry Giscard d'Estaing, French minister of Economy and Finance. In January 1974 he was named joint managing director of the state-owned Banque Nationale de Paris, and assumed its presidency in 1979. The left-wing Mitterrand government pushed him out in 1981.

He had little knowledge of cars and was not much of a driver, but Peugeot engaged him in June 1982 in an observatory capacity. Within two months he was named head of Automobiles Peugeot and vice president of the parent organization, Peugeot SA.

In January 1983 he stopped all cooperation with Matra and sold Peugeot's block of Matra shares. He was vocal in his criticism of the Maastricht treaty and wanted France to put up tariff barriers against Japanese imports.

He pursued an accelerated model-renewal program that brought to market the Citroën AX in May 1986, Peugeot 405 in June 1987, Citroën XM in June 1989 and Peugeot 605 in November 1989, Peugeot 306 in January 1993 and Citroën Xantia in November 1992. In February 1989 he railed against the government that Renault should give up its 50 percent share in the Française de Mécanique, the giant engine plant at Douvrin, to Peugeot to make up for years and years of state subsidies to Renault, while Peugeot had always been a net taxpayer. His dynamism revitalized the entire French auto industry and its suppliers. He retired when reaching the statutory age limit in September 1997.

1. Ecole Nationale d'Administration, spawning ground for government officials and civil servants.

Camusat, Maurice (1880–1953)

Technical director of Delaugere & Clayette from 1920 to 1923; technical director of Usines Léon Bollée from 1923 to 1928.

He was born in 1880 in the Mayenne and graduated in 1902 from the Ecole des Arts et Métiers at Angers. He began his career with the Compagnie des Chemins de Fer de l'Ouest, went to Paris and after a brief period with Renault, joined Lacoste & Battmann, major suppliers to the auto industry, who also made some cars (Lacoba). In 1905 he was engaged by Peugeot as an engineer in the Levallois plant, where he was soon named assistant chief engineer. He left Peugeot in 1910, went to England and spent three months with Daimler in Coventry before going to Argyll in Glasgow, where he was a member of Henri Perrot's staff for 18 months. Late in 1912 he joined Pipe in Brussels and designed a number of new models. Returning to Paris in 1914, he worked for the next six years under Michelat's orders at Delage.

In 1920 he joined Delaugère & Clayette in Orléans, makers of conventional but overpriced cars. When they closed their engine plant, he moved to Léon Bollée in Le Mans and modernized their product. When the company was sold to W.R. Morris, he stayed on, working mainly on the production side. In 1928 he joined Citroën as a design engineer and became a specialist on gears, transmissions and axles for all kinds of vehicles. He served as a plant engineer and was eventually placed in charge of prototype construction. He retired in 1945 and died on January 2, 1953.

Canavese Giovanni (1927–)

Manager of Fiat Auto's car-design office from 1980 to 1982.

He was born on March 26, 1927, at Garessio and studied engineering at the Turin Polytechnic. He joined Fiat in 1952, working as a service engineer in the truck-and-bus department. In 1957 he was transferred to the passenger-car side, assigned to the export department with the title of product coordinator for South America. Starting in 1968, he spent three years with the Fiat rally team, preparing competition versions of the Fiat 125 sedan and the 124 Spider. From 1972 to 1974 he was project manager for the Fiat 131, forced to switch goals halfway because of the petroleum crisis and quadrupling of motor-fuel prices. The 131 therefore began life as an economy car, but high-performance models appeared in 1978. In 1980 Paolo Scolari promoted him to manager of the car design office, just as Project 154 (Lancia Thema and Fiat Croma) was put on the drawing board. He also directed the design and engineering of Projects 159–160 (Fiat Tempra and Tipo) and 176 (Fiat Punto) before retiring in 1992.

Cantarella, Paolo (1944–)

Managing director of Fiat Auto S.p.A. from 1990 to 1996; chief executive officer of Fiat S.p.A. beginning in 1996; chairman of Fiat Auto and Iveco NV.

He was born in 1944 at Varallo Sesia in Piedmont and graduated from the Turin Polytechnic with a degree in mechanical engineering. He began his career as an executive with an automotive components supplier firm and arrived in the Fiat orbit in 1977. He became manager of the Fiat group's intersectorial coordination department and in 1980 was named assistant to the managing director of Fiat S.p.A. Three years later he became managing director of Comau, the Fiat group's machine tool and automation affiliate, where he led a forceful expansion phase over six years. He came to Fiat Auto S.p.A. as head of the supplies and distribution division in 1989. He instituted wide-sweeping reforms in the production setup, closing the Desio (ex–Autobianchi) plant in 1991 and the Chivasso (ex–Lancia) plant in 1992. In 1992 he announced a program to develop and introduce 18 new models by the year 2000 at a budgeted cost of $32 billion. The cars arrived on schedule, but the Bravo/Brava and Marea, crucial mid-market models, failed to reach their sales targets. He spearheaded Fiat's return to niche-markets with the Fiat Coupé, Fiat Punto cabriolet, and the Fiat Barchetta.

Cappa, Giulio Cesare (1880–1955)

Chief engineer of Aquila Italiana from 1905 to 1914; design engineer with Fiat from 1914 to 1924.

He was born on April 28, 1880, in Voghera in Lombardy and studied engineering at the University of Turin. In 1904 he designed and built a motorcycle with a water-cooled single-cylinder engine, and the following year he was engaged by Marquis Pallavicino di Priola to design cars named Aquila Italiana. He made technical history in 1906 by creating a six-cylinder overhead-valve engine with aluminum pistons, mounted with the clutch and gearbox on an aluminum platform lodged in the chassis frame. He stayed on after the Marquis was killed in a collision with a train on a level crossing, and the company was taken over by Marsaglia Bank, and designed powerful, high-quality cars. Aquila Italiana cars won the Mont Ventoux hill-climb in 1912, took second place in the 1913 Targa Florio, and won the 1913 Parma–Poggio race.

In 1914 he joined Fiat as a design engineer and was responsible for military vehicles and aircraft engines. In 1921 he designed the four-cylinder 1½-liter engine for Type 803 and in 1922 the twin-overhead-camshaft 1991-cc six for Type 804, which won the French Grand Prix at Strasbourg. He designed the overhead-camshaft four-cylinder engine for the 509

economy car and the big six for the 519 luxury car.

He left Fiat in 1924 to set up office as an independent consultant. One of his first masterpieces was a 1000-hp 120-cylinder aircraft engine for Lorraine-Dietrich. He designed several production models for Itala, the 61, 65 and 75, which were midsize six-cylinder touring cars. He also designed the Itala 11 racing car with a 1049-cc V-12 engine and front wheel drive. One was built, but never raced. In 1928 he designed a light military vehicle for Ansaldo, which was later produced in some quantity by OM.[1] He received engine-design contracts from Ernesto Breda and Enrico Piaggio, and acted as a consultant to Alfa Romeo, CEMSA,[2] and Bugatti. In 1935 he designed and patented a three-cylinder horizontal diesel engine with opposed pistons, a modular concept that could also be built as a six- or 12-cylinder unit.

1. Officine Meccaniche, Brescia.
2. Costruzione Elettro-Meccaniche Saronno (a Caproni affiliate).

Carden, John Valentine (1889–1935)

Chairman of the Carden Engineering Company, Ltd., from 1913 to 1922; technical consultant to Vickers-Armstrong, Ltd., from 1921 to 1935.

He was born in 1889 at Farnham, Surrey, into a noble family and privately educated. He set up an engineering shop on the family estate, where he made his first prototype cars and airplanes. He searched for the ultimate in lightweight construction, and his 1913 Carden single-seater combined wooden unit-body construction with an inboard/rear-mounted single-cylinder JAP engine. In 1914 he moved into more spacious premises in Somerset Road, Teddington, and started production. An improved derivative with a V-twin engine was sold to Ward & Avey, Ltd., who took over the Teddington plant in 1916 and produced it as the AV Monocar. During World War I, he worked almost exclusively on aviation projects, but in 1919 a cyclecar with side-by-side seating for two and a V-twin engine was ready.

He sold the design to Edward A. Tamplin, designed a new Carden with a similar chassis configuration but a 700-cc horizontal parallel-twin driving the rear axle, and started production in a factory at Ascot, making up to 40 cars a week in 1921–22. He sold his company to Arnott & Harrison, Ltd., who moved production to Hythe Road, Willesden Junction, north London, in 1923. He designed a lightweight military vehicle with crawler-tracks for go-anywhere mobility, which Vickers put into production as the Carden-Lloyd. A Carden-Lloyd Bren-gun carrier followed, using an 85-hp Ford V-8 engine and Horstmann suspension. He was killed in an airplane crash at Tatsfield, Surrey, in December 1935.

Caspers, Albert (1933–)

Executive director, Ford of Europe's engineering and vehicle manufacturing group from 1990 to 1993; vice president, Ford of Europe, product development and production, from 1993 to 1998.

He was born in 1933 at Lissendorf in the Eifel mountain range and obtained an engineering diploma from the Aachen Technical University. He began his career in 1958 as a plant-engineering trainee with Ford-Werke AG in Cologne, and in time won a succession of promotions in manufacturing and product engineering within Ford of Europe, Inc. In 1983, he was named director of body and assembly operations, Ford of Europe, with responsibility for the Cologne and Sarrelouis plants in Germany, Genk in Belgium, Almusafes in Spain, Dagenham, Halewood, Langley and Southampton in the United Kingdom. He became vice-president of Ford of Europe's Manufacturing Operations in 1989, taking on the additional duties of product development in 1993.

He retired in 1998.

Cattaneo, Giustino (1881–1973)

Technical director of Isotta-Fraschini from 1907 to 1938.

He was born in August 1881 at Caldogno, and studied at the Vicenza Professional Institute. In 1900–01, he assisted Enrico Bernardi in his engine experiments and patent applica-

tions. Next, he designed the 1903–04 Florentia two-cylinder cars. In 1904, Roberto Züst engaged him to design a powerful touring car, which went into production in a Milan factory. The Züst cars impressed a lot of people, including Cesare Isotta and Oreste Fraschini, who invited him to design their automobiles and aircraft engines. He created the big Isotta-Fraschini AN and BN types with four-cylinder engines from five to eight liters displacement, followed by the 11-liter Type CN. He also designed some small-engine models, such as the four-cylinder 1.4-liter FEN and the 2.2-liter HNC. In 1910 he introduced four-wheel brakes and the IF-6 model with a six-cylinder 70-hp 9.5-liter engine.

He spent the years of World War I on aircraft-engine design. After the war, he created the Tipo 8A car, a giant deluxe model with a straight-eight 6-liter twin-carburetor engine which grew in size over the years to 7.3 liters, its output rising from 80 to 110 hp. When Mussolini came to power in Italy, Isotta Fraschini was given big contracts for aircraft engines, and Cattaneo designed the Asso 500 twin-overhead-camshaft 27.7-liter V-12, followed by a W-18 in 1930.

Car production was curtailed in 1932 and abandoned in 1935. He remained in his post, concentrating on aircraft engines, while the car factory was converted to build diesel engines, trucks, buses, and military vehicles.

He retired in 1938 and maintained residence in Milan until his death in 1973.

Causan, Némorin (18??–19??)

Racing engine designer with Delage in 1906–08; racing car designer with Grégoire in 1909–12; aircraft engine designer with Panhard & Levassor in 1914–15; founder of Automobiles Causan in 1922.

Armed with an engineering diploma from the Ecole des Arts et Métiers, he began his career in the Clément-Bayard drawing office in 1903. In 1907 he fitted aluminum pistons in a Delage racing engine and invented a system of desmodromic valve operation. He designed a five-speed gearbox with overdrive fifth which Louis Delage used in his 1911 four-cylinder voiturette racer. He put a two-speed rear axle on the 1911 Grégoire racing voiturette and for 1912 an 85-hp 3-liter four with four horizontal valves per cylinder. The V-8 and V-12 aircraft engines he designed for Panhard & Levassor also featured four valves per cylinder.

In 1915 he went to the United States and a year later joined the Chalmers engineering staff in Detroit. He returned to Paris in 1919 and designed a racing engine for Jacques Bignan, revising and complicating his desmodromic valve gear. In 1920 he designed an 1100-cc overhead-valve four-cylinder for the M.A.S.E.[1] cyclecar and a similar unit for Louis Lefèvre, maker of La Perle cars, who asked him to design a 1500-cc in-line six as well.

He established Automobiles Causan with shops in Route de la Révolte, Levallois-Perret, and designed a 350-cc two-cylinder supercharged engine, mounted in a cyclecar with two-speed gearbox. A few of them were produced in 1924 by d'Aux in Reims. He designed overhead-camshaft six-cylinder engines for the Lestienne family, makers of La Licorne cars, and the Bucciali brothers. In 1927 he designed a 1500-cc V-8 with dual superchargers for Vernandi, which did not go into production. His V-16 3-liter design was not even built as a prototype. In 1928 he was busy with an opposed-piston two-stroke engine for Dr. Etchegoin's racing motor boat. His career ended unexpectedly when he died while traveling in Egypt.

1. Manufactures d'Armes, Saint-Etienne.

Cavalli, Carlo (1878–1947)

Technical director of Fiat from 1919 to 1928.

He was born in 1878 in Val Vigezzo as the son of a local magistrate, and he took up the study of law to please his parents. But he never worked in a law office, for after graduation, he taught himself the rudiments of mechanical engineering. He joined Fiat in 1905 as a draftsman, and in 1908 was named head of the drawing office. He was responsible for the 1908 Fiat 1 Taxi, the company's first four-cylinder unit with a one-piece block. It also had a giant one-piece casting that constituted an engine sump as well as a gearbox casing. He also designed the 20/30 hp Type 3 and 30/45

hp Type 4 introduced in 1910, and in 1911 was promoted to assistant to the Technical Director. He created the Fiat Zero, Fiat's first model to be marketed as a complete car with open-tourer coachwork. It was also Fiat's first attempt at mass production, and over 2000 Zero's were built from 1912 to 1915. He also led the design of Fiat's S 57/14B racing cars, which in 1916 were equipped with four-wheel mechanical servo-assisted brakes.

During World War I he directed the design of the 15 Ter and 18B military trucks and a 40-ton armored vehicle. He also received a patent for his method of mounting a one-piece crankshaft on ball-bearings and began experiments with supercharging. The postwar Fiat 501 was his design. A roomy four-passenger sedan with a 23-hp 1.4-liter engine, it was an immediate success in export markets as well as in Italy. Fiat produced more than 45,000 of them in less than seven years. The 505 and 510 were also his creations, conventional in concept. The 510 had a 46-hp six-cylinder engine and led to the 510 S of 1920, a rakish torpedo with four-wheel brakes. He modernized the prewar Taxi and prepared a luxury car with a V-12 engine (but its production was canceled). It was downscaled into a 4.8-liter Type 519 six, which was good for 110–115 km/h. He designed the straight-eight supercharged engine for the Type 805 racing car of 1923 and created Fiat first "small" car, Type 509 of 1925, with a 990-cc overhead-camshaft engine designed by Bartolomeo Nebbia. It was the first Fiat to be produced at Lingotto, and over 90,000 were built in a five-year period.

He retired in 1928, primarily for health reasons, to his native valley, and died at dawn on October 1, 1947, at Santa Maria Maggiore near Domodossola.

Ceirano, Giovanni (1865–1948)

Managing director of Giovanni Geirano Junior & Co. in 1904; managing director of Fabbrica Junior Torinese di Automobili in 1905–06; managing director of SCAT[1] from 1906 to 1917; managing director of SA Giovanni Ceirano Fabbrica di Automobili from 1919 to 1924; chairman of SCAT from 1924 to 1931.

He was born in 1865 in Turin and was associated with his older brothers G.B. and Matteo in manufacturing Welleyes bicycles in Turin. They built the Welleyes automobile in 1898–99 but sold its design and all materials on hand to F.I.A.T. The brothers then founded Fratelli Ceirano and began building Ceirano cars. They split up in 1903 when Matteo went off to start making Itala cars and Giovanni left to build Junior cars. G. B. Ceirano kept the old factory going, but changed the car's name to Rapid, and the corporate title to Société Torinese Automobili Rapid (STAR). The range of Junior cars included a single-cylinder model, a twin, and a four-cylinder 16/20. He left in 1906 to set up SCAT in Turin as makers of powerful high-performance cars, later adding lower-priced models to add volume. SCAT was building 400 cars a year by 1912. During World War I SCAT was building military trucks and aircraft engines under Hispano-Suiza license. But in 1917 he sold his holding in SCAT to Italian investors associated with Hispano-Suiza. It took him less than two years to start a new company in his own name, making cars and commercial vehicles with the Ceirano name plate. This enterprise was mainly financed by Count Belli di Carpenea, and the first model was a 20-hp touring car with a 2.3-liter four-cylinder side valve engine. It was followed by the 2.6-liter CA2 sports car and the big 3-liter CS2H. He used some of the profits from Ceirano sales to buy SCAT stock, and in 1923 he retrieved control of SCAT. He merged the two companies and the SCAT name plate disappeared from the products. But car sales were declining by 1927, and he compensated by building up the production of trucks and buses. Heading for bankruptcy in 1928, he formed an alliance with the Fiat Consortium, which opened the way for sharing production facilities with Ansaldo and Ceirano in Turin and OM in Brescia. The last Ceirano car was a Type 150 S made in 1930. He resigned in 1931 and in 1932 the company was swallowed up by SPA.[2] He died in 1948.

1. Società Ceirano di Automobili Torino.
2. Società Ligure Piemontese di Automobili.

Ceirano, Matteo (1870–1941)

Joint managing director, Fratelli Ceirano in 1901–03; managing director of Matteo Ceirano e C. in 1903–04; technical director of Società Piemontese Automobili Ansaldi-Ceirano from 1906 to 1908; technical director of the Società Ligure Piemontese Automobili SA from 1908 to 1918.

He was born in 1870 in Turin and had his early training in the bicycle industry. In 1901 he joined his brother Giovanni Battista Ceirano in making cars powered by single-cylinder Aster and de Dion–Bouton engines in a shop on Corso Vittorio Emmanuele in Turin. In 1903 they added a four-cylinder model with an engine designed by Aristide Faccioli, but later that year Matteo left to start his own business in Via Guastalla, Turin, to make the Itala car. His fledgling company was taken over by bankers from Genoa who reorganized it as Itala Fabbrica Automobili with works in Via Petrarca, Turin, and he walked out. In 1906 he went into partnership with Cavaliere Michele Ansaldi, a well-connected Turin industrialist to build the SPA car in a large (12,000 m²) plant at Barriera Crocetta on the outskirts of Turin. A year later they had 300 workers and built 300 cars. In 1908 their company merged with FLAG[1] which was short on technical resources but strong on capital. The corporate headquarters were moved to Genoa, while the FLAG plant at La Spezia was closed and production concentrated at Barriera Crocetta. The first SPA light military truck appeared in 1909, and an important number of trucks was turned out during World War I, while another section of the plant made 200-hp aircraft engines. Matteo Ceirano was appointed Cavaliere di Lavoro in 1915. He retired from business in 1918.

1. Fabbrica Ligure di Automobili, Genova, began producing the FLAG car under Thornycroft license in 1905.

Cena, Bruno (1943–)

Chief engineer and product manager for Fiat Auto's Platform D cars from 1996 to 2000; president and chief executive officer of Carrozzeria Bertone beginning in June 2000.

He was born on November 4, 1943, in Turin and graduated from the Turin Polytechnic in 1970 with a degree in aeronautical engineering. Later that year he joined Fiat as an assistant to the director of car development and testing. An excellent driver, he also introduced the first instrumented handling tests at Fiat in 1971, making it possible to objectively quantify minute differences in vehicle response and road behavior. From 1974 to 1977 he was in charge of Fiat's performance development department, and then took part in a Fiat Research Center project to study hybrid power systems. In 1978 he was assigned to the Lancia test department and given responsibility for road-testing and chassis development of the Thema, until its launch in the summer of 1984. In 1986 he was named manager of the Lancia test department, and headed the development of the Dedra 4WD and Delta HF Integrale. From 1991 to 1996 he was in overall charge of testing for Fiat, Lancia and Alfa Romeo. Then he was given responsibility for the development and quality of a new car family, platform D, beginning with the Alfa 156 and Fiat Marea in 1997, continuing with the Lancia Lybra in 1999. He left in April 2000, with Fiat's consent, to join Bertone.

Chaffiotte, Pierre (ca. 1918–)

Chief engineer of engine design, Régie Nationale des Usines Renault, from 1970 to 1983.

He was born about 1918 and graduated from the Ecole des Arts et Métiers of Cluny. He also held a diploma from the Ecole Nationale des Moteurs. He joined Hispano-Suiza in 1945 as an aircraft-engine designer. When the company purchased a license for Hercules diesel engines, he worked on diesel-engine testing and development, and was later part of the team that designed the V-12 military-vehicle diesel engine. He joined Renault in 1970 to take charge of passenger-car engine programs and directed Renault's adoption of electronic ignition, fuel injection, turbocharging, five-speed gearboxes, and diesel-engine development. His last contribution was the F8M engine, Renault's first unit with a belt-driven overhead camshaft, which went into production in 1984. He retired in 1983.

Chapman, Anthony Colin Bruce (1928–1982)

General manager of Lotus Engineering Company from 1952 to 1959; chairman of Lotus Cars, Ltd., from 1959 to 1980; chairman of Group Lotus Companies Plc from 1968 to 1982; chairman of Lotus Engineering from 1980 to 1982.

He was born on May 19, 1928, at Richmond, Surrey, and studied engineering at University College, London. He learned to fly with the University Air Squadron and served in the RAF, but not as a pilot. He began his career as a development engineer with British Aluminium Company, Ltd., and in 1947 built a trials ("mudplugging") special using a 1930 Austin Seven chassis. In 1950 he went into partnership with the brothers Michael and Nigel Allen who offered him space in their garage at Muswell Hill, North London. In 1952 he founded his own company and moved into a factory in Tottenham Lane, Hornsey, London. He designed the Lotus Mk 6, his first production model, the stark roadster Mk 7, and planned the Lotus Elite with its unit-construction plastic body. It was built from 1957 to 1963 with a total of 998 units. He created the Elan roadster with its steel backbone frame and plastic body, making a total of 14,457 units from 1963 to 1973. Lotus car production had been moved to Delamare Road, Cheshunt, Berkshire, in 1959, but it became too small and in 1965, car production moved into a new factory on the former Hethel airfield near Wymondham, Norfolk. The first new car from Hethel was the Lotus Europa with an inboard/rear-mounted Renault 16 engine, making a total of 9230 units from 1966 to 1975. Lotus produced over 3000 cars in 1972, and Chapman decided to start making his own engines, beginning with a twin-cam 2-liter unit for the Lotus Elan, Europa and the Elan 2, plus the Jensen-Healey. When the Jensen contract was canceled, it meant a £15 million loss for Group Lotus. A new Elite coupé was introduced in 1974 but its production cost was 45 percent over budget! It was supplemented by a lower-priced Eclat in 1975, but total production fell to 655 cars that year. After a long delay due to lack of funds to buy tooling, the Esprit went into production in 1976. In November 1978, Chapman won a $17.5 million contract to develop the De-Lorean sport coupé for production in a government-sponsored factory at Dunmurry, Northern Ireland. Chapman then turned his back on production cars and in 1980 set up Lotus Engineering as a technical consultancy and subsidiary of Group Lotus. He received an order for engines and chassis development for the Sunbeam-Lotus from Talbot Motors Co., Ltd. (ex–Chrysler-UK), and in June 1981, signed an agreement for technical cooperation with Toyota Motor Company. The books of Group Lotus were deep in red ink and in the first week of December 1982, American Express refused to renew its credit line, originally granted in 1977. Colin Chapman died at his home, Ketteringham Hall, Norfolk, from a heart attack during the night of December 15-16, 1982.

Charles, Hubert Noel (1893–1982)

Design engineer of MG Car Company, Ltd., from 1930 to 1935; chief engineer of MG cars from 1935 to 1938; development engineer of Austin Motor Company, Ltd., from 1943 to 1946.

He was born on November 22, 1893, at Barnet, Hertfordshire, and educated at Highgate School and University College, London. Early in 1914 he was a member of a racing team preparing a Triumph motorcycle equipped with fuel-injection for a speed-record attempt, an effort interrupted by World War I in which he served as a mechanic in the Royal Navy. In 1915 he was transferred to the Royal Flying Corps as an engineer, working on the development of engines for fighter planes, and was discharged with the rank of captain. He joined Zenith carburetors as a sales engineer in 1919 and signed up with Automotive Products in 1921, working in the technical branch of the sales department. He went to Morris Motors, Ltd., in 1924 as a technical assistant in the production department, where he met Cecil

Kimber and started moonlighting for MG. At Kimber's request, he was delegated to the Abingdon factory in 1930. Using a maximum of Morris parts, he developed a family of overhead-camshaft MG engines and helped develop the J-Type Midget, the K-Type Magnette, the F-Type Magna and the P-Type Midget. He was solely responsible for the chassis of the 1935 R-Type racing car, with a fork-shaped backbone frame and all-independent torsion-bar suspension. When Lord Nuffield sold the MG Car Company, Ltd., to Morris Motors, Ltd., he was given a new office at Cowley. He worked on the suspension systems for the Morris Ten and Twelve, and prepared the engine for the TA-series MG. He left the Nuffield organization in 1938 and became chief engineer of Rotol Airscrews, where he developed a new reversing gear and a blade-feathering device. In 1943 he signed up with Austin Motor Company, Ltd., and moved to Longbridge as a development engineer. He designed the independent coil-spring suspension for the postwar models beginning with the A-110 Sheerline and the A-40 Devon and Dorset. He was the principal engineer for the Sheerline and Princess limousines, and his name is listed as inventor or co-inventor of two engine patents assigned to Austin in 1946. He resigned in 1946 and became a consultant to Cam Gears, Ltd., best known for its steering gears, where he assisted Joe Craig in the development of an exhaust back-pressure controlled carburetor for Norton motorcycles. He retired about 1961, settled at Teddington, Middlesex, and died in an Oxford hospital on January 18, 1982.

Charron, Ferdinand "Fernand" (1866–1928)

Chairman of CGV[1] from 1901 to 1908; managing director of Société Anonyme des Anciens Etablissements Clément-Bayard from 1909 to 1911; chairman of Automobiles Alda from 1912 to 1919.

He was born in 1866 at Angers and never had much formal education. He learned to ride high-wheel bicycles at the age of 15 and raced them with great success. Out of 45 races he entered in 1886, he won 32. Having married a daughter of Adolphe Clément, he went to work in the Clément bicycle factory and by 1893 was director of their Paris sales branch. In 1895 he became interested in automobiles and started a dealership (CGV) in partnership with Emile Voigt and Léonce Girardot, selling many makes including Panhard-Levassor, Peugeot and de Dion–Bouton. He was at the head of CGV when they began building their own cars, designed by Girardot, and throughout its existence. He sold CGV to British investors in 1908, and returned to Clément-Bayard as chief executive. But he wanted full independence and in 1911 he bought the ENV factory at Courbevoie, and began building Alda automobiles, probably designed by Girardot. The factory was requisitioned in 1914 and put under Farman management to produce airplanes. After World War I, Henry Farman bought the factory to build Farman cars. A few hundred cars bearing the Alda badge were produced from 1919 to 1921.

1. Charron, Girardot et Voigt.

Cheinisse, Jacques (1934–)

Product planning manager of Société des Automobiles Alpine Renault from 1970 to 1976; head of high-class car product planning with Régie Nationale des Usines Renault from 1976 to 1979; head of medium- and high-class car product planning with Renault from 1979 to 1984; deputy manager of Renault product planning from 1984 to 1987; director of Renault product management from 1987 to 1995.

He was born in 1934 at Le Havre. He went to work for Jean Rédélé as a salesman for Alpine cars in 1963 and became an instructor for the Alpine rally team drivers. He was named sales manager for Alpine in 1966 and manager of the Alpine racing team in 1968. He worked on the evolution of the Alpine A-310 until 1976. He was chief planning engineer for the R-25 which went into production in October 1983 and was mainly responsible for the R-21 concept, in which some engines were mounted transversely and others longitudinally. He laid down the basic specifications of the R-19 which replaced the R-9 and R-11 in

1988, and the first Clio, launched in 1990. The Laguna of 1994 began as an upscaled R-19. In September 1995 he was given a new assignment to study international markets for Renault cars. He retired in 1999.

Chenard, Ernest Charles Marie (1861–1922)

Co-founder of Chenard-Walcker and its chief executive officer from 1899 to 1922.

He was born on July 7, 1861 at Nanterre into a branch of an old family from Pithiviers with traditions in pewter pots. He had some technical education and in 1883 became a draftsman with the Western Railways.[1] In 1888 he set up his own shop in rue de Normandie at Asnières to sell bicycles. He soon was building both bicycles and tricycles On June 26, 1897, he applied for a patent covering a fore-carriage with a passenger's seat, steered by handlebars, which could instantly convert a tricycle into a quadricycle. He mounted single cylinder Aster and de Dion–Bouton engines on the rear frames of both tricycles and quadricycles, and sold a license to the British Motor Syndicate, earning regular royalties.

On January 19, 1899, he formed a partnership with Henri Walcker to produce motor vehicles, beginning with motor quadricycles. They made their first front-engine automobile (with belt drive) in 1900 and began producing their own 4 CV single and 8 CV parallel-twin engines. Their first four-cylinder model was a 20/24 CV model appearing in 1902, with shaft drive. Business grew fast, and in 1906 Chenard-Walcker went public, majority control being held by the Banque Charles Victor in Paris. The following year the company moved into a very large factory at Gennevilliers.

In 1912 Chenard-Walcker took over the Société Parisienne de Carrosserie Automobile of Neuilly, and moved it to a site adjacent to the factory at Gennevilliers in 1921. During World War I, Chenard-Walcker made shells, mostly, but also a number of Hispano-Suiza V-8 engines, and filled subcontract orders from Renault and Salmson. Passenger-car production was resumed in 1919 with considerable success, but Ernest Chenard died from a heart attack on May 28, 1922.

1. Compagnie des Chemins de Fer de l'Ouest.

Chenard, Lucien (1894–1971)

Technical director of Chenard-Walcker from 1912 to 1923; member of the Chenard-Walcker management board from 1923 to 1936.

He was born on January 17, 1894, at Asnières as the son of Ernest Chenard and held an engineering degree from the Institut Industriel du Nord. He was a much-decorated officer in World War I, winning the Legion of Honor, Military Medal and Croix de Guerre. Upon joining his father's business in 1919, he created the methods department and began cost-accounting. He took a keen interest in the product and directed the engineering staff, notably in the design of the overhead-camshaft four-cylinder 3-liter model made from 1921 to 1927, and the 1100 and 1500 tank-bodied sports cars made from 1924 to 1937.

He held top responsibility for production as well as the products, beginning in 1923. By 1927 Chenard-Walcker had 2700 workers and produced 100 cars a day. He was involved in the plans for combining the functions of purchasing and production with Delahaye, which resulted in some crossbreeding. The 1928 Delahaye 107 M was also sold as a 10 CV Chenard-Walcker, slightly disguised with its own radiator shell and other exterior details. In return, the 9 CV and 16 CV Chenard-Walcker models were also produced in Delahaye versions. Delahaye built its 14 CV Type 108 for both sales organizations, while Chenard-Walcker produced the 9 CV (Delahaye 109) and 12 CV (Delahaye 110) with both trademarks. He shared no blame at all in the personality conflicts which led to the severance of the alliance in 1931.

He supervised the wholesale product-renewal of 1933–34, when all older engines were taken out of production and replaced by new four-, six- and V-8 power units. The Aigle series had modern styling and independent front suspension. Lucien Chenard also called in J. A. Grégoire as a consultant when he decided to build a series of front-wheel-drive models in 1934–35. By that time it became all too evident that manufacturing costs were way out of line, and prices desperately uncompetitive. Sales volume dwindled and he resigned from the board when Chenard-Walcker declared bankruptcy on June 13, 1936.

He remained active in company affairs until 1943 when he left to work on gas-turbine research for Bethenod. He became a consultant to Simca and was again associated with J. A. Grégoire when Simca planned to mass-produce the two-cylinder economy car developed for and sponsored by Aluminium Français. In 1945 he took over a big auto repair shop at Courbevoie where he spent the rest of his career.

He died in 1971.

Citroën, André (1878–1935)

Managing director of Société Nouvelle des Automobiles Mors from 1908 to 1913; director of André Citroën Ingénieur-Constructeur from 1919 to 1924; chairman and chief executive of SA André Citroën from 1924 to 1935.

He was born on February 5, 1878, in Paris as the son of a diamond merchant, and educated at the Lycée Condorcet and the Ecole Polytechnique, graduating in July 1900. He bought a patent for a herringbone gear system on a visit to his mother's family in Poland in 1900, and on his return to Paris joined Frères Hinstin, makers of locomotive parts at Corbeil. In 1902 he headed a new department of Hinstin for making herringbone gears, which was spun off in 1905 as Société des Engrenages Citroën, Hinstin & Cie.

He joined the automobile industry in 1908 by accepting an offer to lead the newly founded company making Mors cars in Paris and raised production from 125 cars a year to 1200 in a span of five years. In 1913 he acquired majority control of the gear company and reorganized it as Société d'Engrenages Citroën. In August 1914, he was called up for military service and served as a Captain in the French Artillery. In a few weeks of front-line duty he saw a boundless market for munitions, which were always in short supply. Through connections from the Ecole Polytechnique, he approached the commander of the Artillery and the minister of arms production with a plan to produce 20,000 shrapnel-filled shells a day. The City of Paris provided a suitable factory site under a 99-year lease, and on the strength of the government contract, he was able get financing for plant construction and equipment. Production began at mid-year 1915 with an output of 10,000 shells a day, which doubled within mere months. During 1917 the daily output climbed to 50,000 shells, and in 1918 the government asked Citroën to sort out the production problems at the Roanne Arsenal. He accomplished that before the war ended. He had no further government orders for his factory in Paris, but he already had a plan. He would make automobiles. He engaged several engineers who prepared several prototypes, from cyclecars to luxury limousines. He chose the medium-size touring car, and the first 10 CV Type A was completed on June 4, 1919. At a time when most makers built chassis only, Citroën offered a complete car at an amazingly low price. He also broke new ground in marketing techniques in 1921 by starting a rent-a-car company, with optional lease contracts. In 1924 he also set up a taxicab company, changed his company into a joint stock corporation, and made his second trip to the United States, 12 years after his first visit to America.

For every car Renault built in 1925, Citroën made two. He was the uncontested leader of the French motor industry, and the most daring in his business methods. But he took a lot of profits out of the company and wasted a fortune on gambling and a lavish life-style. By 1927 the company was in financial straits, and Banque Lazard agreed to save it only on condition of having a seat on the board of directors. He was no longer master in his own office.

He visited the United States again in 1931 and decided on his biggest gamble yet. He would build a revolutionary front-wheel-drive car, so different from earlier models that it demanded a complete remodeling and retooling of the factories the body shops at St. Ouen, the Grenelle works which made front and rear axles, the gearbox plant at St. Charles, and the final assembly halls at the Quai de Javel.

The industrial investments coming on top of the design and development cost of the new car, which went into production on March 7, 1934, broke the company's financial back. The biggest creditor, Michelin, supplier of wheels and tires, organized a rescue operation, but at its own price, which was to put Michelin executives in all key positions at Citroën.

Andre Citroën was told not to come to his former office any more. His failure in business ruined his health, and he was taken to the Clinique Georges Bizet in Paris, where he died from gastric ulcers and a tumor on July 3, 1935.

Claveau, Emile (1892–1974)

Designer of Claveau cars from 1923 to 1956; head of the Bureau d'Etudes et Recherches Techniques from 1936 to 1956.

He was born in 1892 in the Touraine, studied architecture at the Academie des Beaux-Arts, and opened an architect's cabinet in Joué-les-Tours. He was eminently successful and made a small fortune. About 1921 he became interested in the architecture of automobiles and began design work on a car in 1923. The Autobloc Claveau was first shown in 1926 in two body styles, open roadster and four-passenger torpedo. The engine was an air-cooled side-valve flat-four, constructed mainly of Alpax, mounted in an inboard/rear position, permitting a short-front chassis. It had all-independent suspension with sliding pillars and enclosed coil springs. The roadster weighed 670 kg and the torpedo 720 kg. In 1927 he displayed the LR streamliner with a four-passenger two-door fuselage-style body and cycle fenders for all four wheels. It was given extensive press coverage but he was never able to sell manufacturing rights for it. He read books on engine design and in 1930 made a water-cooled two-cylinder two-stroke unit with a third pumping cylinder. He proposed to mount it in a small front-wheel-drive car with slab sides, and showed drawings of three models in 1930: a two-passenger roadster and coupé and a four passenger coupe. By

1932 his prototypes were shown with mock-ups of a side-valve 90° V-4 1125-cc engine, but the only running prototype was powered by a production-model vertical in-line four, mounted behind the final drive, and the gearbox in front, with the clutch at the nose of the gearbox. He opened a technical consultant's office in Paris in 1936 but ended up doing more housing and building architecture than car design until the end of World War II. In 1946 he displayed the Claveau Descartes, a 5-meter long streamlined four-door sedan with a Duralinox unit-construction body and a smooth undertray. The engine was an Alpax-block hemi-head 2.3-liter V-8 with a five-speed gearbox, driving the front wheels. It was called "futuristic" but did not prove to be prophetic in its layout, shape, or selection of materials. The last Claveau prototype appeared in October 1955, and was an economy sedan with DKW power train and rubber springs. But once again, it was a design with no future.

Clegg, Owen (b. 1877)

Chief engineer of the Rover Company, Ltd., from 1910 to 1912; chief engineer of SA Darracq from 1913 to 1920; managing director of Automobiles Talbot from 1920 to 1933.

He was born in 1877 in Leeds, Yorkshire, and served an apprenticeship with Kitson, locomotive engineers, where he later became assistant works manager. He joined Wolseley in 1904, but did not get to work on the cars, for the management of Vickers, Ltd., transferred him to Glasgow as manager of a Clydeside naval yard. Eager to work in automobile engineering, he left Glasgow in September 1910 to become works manager of the Rover Company, Ltd., in Coventry, where he designed the 2.2-liter four-cylinder Rover 12. He linked up with Darracq in March 1912, and designed new 15- and 20-hp models with single-sleeve-valve engines. In 1913 he moved to Paris and supervised Darracq's production of aircraft engines during World War I. He designed the Talbot Type A, produced from 1920 to 1923 with a 4.6-liter side-valve V-8 engine. He also created the 1921 Type B with its 1505-cc overhead-valve engine, designed by Louis Coatalen. Then followed the 16 CV Type V 20 which evolved into the 18/36 in 1923, and the 8 CV and 12 CV four-cylinder models of 1922. As Owen Clegg, now a prominent shareholder in STD Motors, Ltd.,[1] assumed the duties of management and policy-making, Louis Coatalen as group technical director gave the Suresnes factory an impressive engineering staff with men like Edmond Moglia, Walter Becchia, Vincent Bertarione and (briefly) Georges Roesch. From 1925 onwards, the group's racing activities were centered at Suresnes under Coatalen's direction. Clegg's primary interest was powerful touring cars and prestige cars. In 1926 he discontinued the production of all four-cylinder Talbot models to bring out a new range of sixes, with eight-cylinder models in the pipeline. He reckoned that with a race-winning reputation, he could sell such Talbot products with a high profit margin. The racing cars often won, but the luxury cars did not sell well, and led to a decline in the company's financial situation. The Talbot chassis were modernized in 1932, with elegant new styling, but the company was on the way to ruin. When STD Motors, Ltd., was liquidated in 1934, the Suresnes branch was sold to A.F. Lago, and Owen Clegg retired.

1. Sunbeam-Talbot-Darracq Motors, Ltd.

Clément, Gustave-Adolphe (1855–1928)

Chairman and managing director of SA des Etablissements Clément-Bayard from 1903 to 1919.

He was born on September 22, 1855, at Pierrefonds (Oise) and educated at Compiègne and Crepy-en-Valois. He was accepted by the Ecole des Arts et Métiers in Paris, but had been orphaned, and his guardian sent him out to work as a grocer's apprentice. He ran away, on his bicycle, and found work in a mechanic's shop. He bought a new bicycle and began racing, with great success. In 1878 he was able to open his own bicycle shop and later started a school for cyclists. He began building bicycles, for which he found a ready market, and in 1894 established Société Anonyme des Vélocipèdes Clément in Paris. To secure his supplies of frames, cranks, and other components, he organized a plant[1] at Mézières in 1895. A year later he merged his bicycle company's with Darracq's to form Clément, Gladiator & Humber (France) Ltd., becoming its managing director. Production was concentrated in the former Darracq works at Pré-Saint-Gervais. In 1897 he founded A. Clément & Cie at Levallois-Perret and bought a ready-made car design from A.C. Krebs, who soon afterwards joined Panhard & Levassor. The car was called Clément-Panhard and produced in the Avenue d'Ivry factory, while Clément's company handled the sales. By this deal, Clément became a member of the board of Panhard & Levassor and its chairman in 1899. Concurrently, the Pré-Saint-Gervais factory had begun making both motorcycles and automobiles, some with a Gladiator name plate, others badged Clément. Chassis exported to Britain were renamed Talbot. In 1902 the London agents bought manufacturing rights and a joint venture, Clément-Talbot Ltd., was founded. In January 1903, he resigned from his office with Panhard & Levassor, and nine months later he severed his ties with Clément, Gladiator & Humber (France) Ltd. He added the name Bayard to his own (with authorization from the Ministry of Justice) and founded a company to make the Clément-Bayard automobile at Levallois-Perret, with many parts from Mézières. Production began in 1904 with an annual output of 1200 cars. In 1906 he sold a license for the Clément-Bayard to the Diatto brothers in Turin. Clément-Bayard started aircraft-engine production and he sponsored the construction of both airships and airplanes, including the famous Santos Dumont "Demoiselle." He retired in 1919, settled on the French Riviera, and died on May 10, 1928, from a stroke occurring while driving his car in Paris traffic.

1. Fonderies et Ateliers de Mécanique de la Macérienne.

Coatalen, Louis Hervé (1879–1962)

Chief engineer of Sunbeam Motor Car Co., Ltd., from 1909 to 1913; joint managing director of Sunbeam from 1913 to 1920; technical director of STD Motors, Ltd.,[1] from 1920 to 1931.

He was born on September 11, 1879, in Concarneau, Finistère, and attended college in Brest. He graduated from the Ecole des Arts et Métiers of Cluny with a degree in mechanical engineering. After military service in the French Army, he began his career with Panhard & Levassor, then joined Clément, whom he left for a position with de Dion–Bouton. In 1900 he went to England and became the assistant of Charles Crowden, who made cars at Leamington Spa. A year later he joined Humber in Coventry as chief engineer for motor cars and masterminded the move to a new factory in Folly Lane, Stoke, Coventry, which came on stream in 1908. He had left Humber in 1907 to prepare the high-performance Hillman-Coatalen touring car. He went to Wolverhampton in 1909 as chief engineer of Sunbeam, revitalized the model range with the help of ex–Humber engineers, and laid out a new factory in Upper Villiers Street in Wolverhampton on a site which the company had acquired earlier. He started experimental work on aircraft engines in 1913 and Sunbeam offered some remarkable multicylinder units to the British aircraft industry during World War I. In 1920 Sunbeam was taken over by A. Darracq & Co. (1905) Ltd., fresh from its acquisition of Clément Talbot, Ltd., and the three were merged into STD Motors, Ltd. He appointed chief engineers for Talbot and Darracq (renamed Talbot) but ran the Sunbeam engineering department himself. He launched Sunbeam into Grand Prix racing and land-speed record attempts. He transferred the racing-car activity to Talbot at Suresnes in 1926, and from then on spent most of his time in Paris. The 1927 Talbot Grand Prix car was most successful, and a 1000-hp Sunbeam special powered by two V-12 aircraft engines set a new world land speed record at over 200 miles per hour. But STD Motors,

Ltd., was losing money on its production cars, and had to declare bankruptcy in 1931. Coatalen was dismissed, as were his fellow directors. He bought control of the French Lockheed company (hydraulic brakes) and became a director of KLG.[2] His technical imagination turned to the needs of the moment, as he watched Hitler's Germany prepare for war. He wanted to give the French aircraft industry a better engine and designed a remarkable V-12 diesel engine with a centrifugal supercharger running at ten times crankshaft speed. It delivered 575 hp at 2000 rpm, but was not put into production. He spent the war years in Britain, returning to Paris in 1945, where he picked up the reins of the Lockheed business. He began toying with disc-brake designs, and was still inventing improvements in 1958.

He died on May 23, 1962, in Paris.

1. Sunbeam-Talbot-Darracq Motors, Ltd.
2. Kenelm Lee Guinness (spark plugs).

Colombo, Gioacchino (1903–1987)

Design engineer with Alfa Romeo, S.p.A. from 1935 to 1939; chief engineer of Auto Avio Costruzioni Ferrari S.p.A. from 1940 to 1949; research engineer with Alfa Romeo S.p.A. from 1950 to 1952; project engineer with Officine Alfieri Maserati S.p.A. in 1952–53; technical director of Automobiles E. Bugatti from 1954 to 1957.

He was born on January 9, 1903, at Legnano and attended schools in his home town. In 1920 he joined Stabilimenti Franco Tosi in Legnano as a student engineer on low-speed diesel engines and steam turbines. In 1923 he submitted a design for a two-stage piston-type supercharger to Alfa Romeo, which led to an invitation to join the company's engineering staff. He worked almost entirely on racing car projects, under the guidance of Vittorio Jano, and in 1932 was delegated to Scuderia Ferrari which had taken over the Alfa Romeo racing team. He was an assistant to Luigi Bazzi in the creation of the Bimotore racing cars, one with two eight-cylinder engines side by side (and two propeller shafts), the other with an eight-cylinder engine in front and a second one in

the tail. He returned to Alfa Romeo in 1935 and worked on the 8C 2900 sports models and the Tipo 158 "Alfetta" racing car. He returned to Modena in 1939 to assist Enzo Ferrari, and spent World War II on the design of machine tools. In 1945 Enzo Ferrari asked him to design a V-12 engine and a chassis for it. That became the Ferrari Tipo 166, first seen in 1947. He designed all Ferrari cars up to 1949 when he returned to Alfa Romeo. In 1952 Vittorio Bellentani lured him back to Modena as project manager for the Maserati A6 GCS, a sports model powered by the 2-liter six from the Grand Prix car. He designed the engine for the 250/F Grand Prix car, but left Modena in October 1953 to go to Molsheim and design new racing and sports cars for Bugatti. He created the Type 251 Grand Prix car with a 2½-liter straight-eight mounted transversely behind the driver, and a suspension system with laterally interconnected springs. It was raced once (without success) in 1956. He returned to Italy in 1957 and designed helicopter engines for MV Agusta at Verghera until 1970. During those years he also made some engine designs for OSCA and Abarth. He opened a consulting-engineer's office in 1970 and continued to advise his clients on engine design.

Constantinesco, George (1881–1965)

One of the most prolific inventors of his time; developed improvements in internal combustion engines and aspects of engines such as carburetors, fuels and transmission elements.

George, or Gogu Constantinescu (as the name would be spelled in the country of his birth) was born in Craiova, Romania, on October 4, 1881. He arrived in London in 1920 and within three years had applied for 18 British patents related to improvements in internal combustion engines. He became a naturalized British subject in 1916 and earned considerable fame and fortune with his invention of the synchronized machine gun which could fire bullets through the revolving propeller blades of a fighter aircraft. He formulated the theory of sonics which related to the transmission of power by vibrations. At

45, he was named one of seventeen world leaders in science and technology by the *Graphic*. After World War I, Constantinesco had an idea for a low cost "people's car" that would offer high performance and low fuel consumption by using a cheap 500 cc single cylinder two-stroke air-cooled engine together with his unique torque converter transmission which would eliminate the conventional gearbox and clutch. The car created quite a stir at the London and Paris Motor shows in 1925 and excited many articles in the world press. General Motors acquired a license to build the car in 1926, but its advantages were not marketable at that time and it was never massproduced. Constantinesco continued to develop practical applications of his theories until he had a fall from which he never recovered shortly after his 84th birthday. At his death, Constantinesco had 133 British patents in the fields of automobile engineering, fluid power, mechanical transmissions and others.

Cordiano, Ettore (1923–1995)

Head of Fiat's advanced engineering department from 1958 to 1966; director of all Fiat production-model programs from 1966 to 1983.

He was born on June 29, 1923, in Turin and graduated from the Turin Polytechnic in 1947. After postgraduate studies, preparing a thesis on motor transport, he joined Fiat in 1947 and was assigned to the calculations office. In 1950 he arrived in the car design office under Montabone and worked on several projects including the 100 (Fiat 600) and 110 (Fiat Nuova 500). In 1956 he was transferred to Deutsche Fiat AG at Heilbronn to prepare new concepts and experimental cars. His department often worked on 20 different projects simultaneously. In 1958 he returned to Turin and began outlining the future of Fiat's model range. He foresaw the general adoption of front-wheel-drive on Fiat cars. Project 109 was loosely based on one of his experimental front-wheel-drive cars from Heilbronn, and evolved into the Autobianchi Primula, going into production in 1965.

He was also responsible for the Fiat 130 chassis, with all-independent suspension and rear wheel drive and the Fiat 135 (Dino) sports models with Ferrari-based V-6 engines. He personally directed the engineering of three front-wheel-drive cars scheduled for mass production, the 1969 Fiat 128, 1971 Fiat 127, and 1978 Fiat 138 (Ritmo). He started the Type 4 program, planned as a common platform for the Lancia Thema, Fiat 154 (Croma), Alfa Romeo 164, and the Saab 9000, but retired in 1983, before production startup, and set up a consulting practice in Turin.

Cornacchia, Felice (1930–)

Head of Fiat's passenger-car body engineering department from 1972 to 1976; chief engineer of the car product planning office from 1976 to 1983.

He was born in 1930 at Moncalvo near Asti and studied at the Turin Polytechnic where he graduated with a degree in mechanical engineering in 1953, picking up an aeronautical engineering degree in 1954. He began his career with Fiat as a tool development engineer and later moved into body engineering. In 1957 he was transferred to the experimental department of Deutsche Fiat AG in Heilbronn where he created the Fiat Austria prototype with front-wheel-drive and two alternative power units, one an air-cooled Steyr flat-twin, the other a small water-cooled four based on the Fiat 600. Returning to Turin in 1959, he was chief body engineer on a number of Fiat models, including the 1100 D, the 124, the 128 and the 130. He was promoted to head of the passenger-car body engineering office and vice-director of Fiat, working in close liaison with Enzo Franchini on safety and crashworthiness research. In 1976 he was named head of product planning, and started the Regata and Ritmo programs. In 1983 he became executive assistant to the technical director, Paolo Scolari, with responsibilities covering the full range of Fiat products. He retired from Fiat in 1995.

Cosmo, Joseph De *see* De Cosmo, Joseph

Costin, Francis Albert "Frank" (1920–1995)

Technical director and general manager of Costin Research and Development from 1960 to 1975; chief designer of Timothy Research, Ltd., from 1987 to 1982; chief engineer of Costin Drake Technology from 1982 to 1987; chairman of Costin Engineering, Ltd., from 1987 to 1995.

He was born on June 6, 1920, at Harrow, Middlesex, and educated at Harrow Weald College. In 1937 he joined General Aircraft at Feltham as a mechanic, and was soon promoted to the drawing office. From 1940 to 1943 he worked with Airspeed, Ltd., in the design office at Portsmouth, and then joined Supermarine Aviation, where the final versions of the Spitfire were then under construction. He was a project design engineer with Percival Aircraft, Ltd., from 1946 to 1948, and was technical director of Sanderson and Costin of Newbury from 1948 to 1951. He joined the De Havilland Aircraft Company, Ltd., in 1951 as aerodynamic flight test engineer in the experimental department at Christchurch, transferring to the Chester plant in 1953. He met Colin Chapman and designed a sports car body for him in his spare time, maintaining his position with De Havilland until 1958. He designed the Lotus Mark 9 body and the body for the 1957 Vanwall racing car. He also designed a bulbous but aerodynamically efficient coupé body for the Maserati 450 S of 1957. In 1959 he created a shape and a plywood structure for Jem Marsh of Speedex Castings and Accessories, Ltd., at Luton. It went into production in 1960 as the Marcos GT. For the next 15 years he had a lucrative sideline in the design of speedboat hulls, but he kept on developing the all-plywood body concept. Auto Project 5 was a small coupe with a DKW engine, and Auto Project 9 was a sports-racing prototype with a Lotus engine. He designed a number of Jaguar-powered roadsters for Brian Lister, and an off-road vehicle for J.C. Bamford. Auto Project 14 was a racing car financed by Roger Nathan and led to a proposed production car, the Amigo, in 1970, using Vauxhall power. The project was kept alive until 1980. He died on February 5, 1995, at his home in Ireland.

Costin, Michael Charles "Mike" (1929–)

Director of Lotus Cars, Ltd., from 1956 to 1962; director and chief deputy engineer of Cosworth Engineering, Ltd., from 1962 to 1990.

He was born on July 10, 1929, at Harrow, Middlesex, and educated at Salvatorian College and Harrow School. In 1946 he was apprenticed to De Havilland's aeronautical technical school at Hatfield, and following military service in the RAF, went to work for De Havilland Aircraft Company, Ltd., as a design draftsman of test rigs for airplane systems. He left Hatfield in 1953 to join Colin Chapman at Lotus Engineering, Edmonton, as head of forward development. He led the staff which produced the designs for the Lotus 18, 19, 20, 21, 22, 23 and 24. He was a cofounder with Keith Duckworth of Cosworth Engineering, Ltd., in 1958, but remained with Lotus until his contract expired in 1962. From the time he joined Cosworth Engineering, Ltd., on a full-time basis, his career blended with Keith Duckworth's. He retired in 1990.

Cotal, Jean

Inventor of transmission improvements and founder of the French firm Cotal.

Jean Cotal developed transmission improvements and founded a company for the production of his inventions. In the early 1930s, Cotal built a pre-selector gearbox with an electromagnetic clutch. Cotal elements such as the Cotal half-automatic electromagnetic 4-speed planetary transmission were used in high-class cars in the 1930s, including Delage, Delahaye and Talbot-Darracq.

Cousins, Cecil (1902–1976)

Works manager of the MG Car Company, Ltd., from 1930 to 1935; assistant general manager of MG Car Company, Ltd., from 1935 to 1939; production manager of the Nuffield Organization's Abingdon plant from 1945 to 1952, and the BMC Abingdon plant from 1952 to 1955.

He was born in 1902 at Oxford and served a four-year apprenticeship in a local mechanic's shop. In 1920 he went to work for Cecil Kimber at Morris Garages, and in 1922 Cecil Kimber transferred him to his "special" department as a technical advisor on both product and production matters. He was good at organizing the work and was promoted to works manager in 1930, and from 1935 to 1939 he was assistant to John Thornley, the general manager. He got the MG TC back into production in 1946 and accommodated the assembly of the RM-series Riley models at Abingdon in 1949. He organized the production of the TD, the TF and eventually, the MGA and MGB. In 1955 he was reprimanded for the poor quality of the Riley Pathfinder assembled at Abingdon, and demoted. He stayed on at Abingdon in a subordinate capacity until his retirement in 1967.

Craig, Alexander (1870–1935)

Works manager of Humber, Ltd., from 1896 to 1900; consulting engineer to Humber, Maudslay, Lea-Francis and Singer in 1902–03; works manager of the Standard Motor Co., Ltd., from 1903 to 1907; managing director of Maudslay Motor Company, Ltd., from 1907 to 1935; and chairman of Maudslay Motor Company, Ltd., from 1910 to 1935.

He was born in 1870 in the Midlands and educated at Masons College, Birmingham. After serving an apprenticeship with the Crewe Locomotive Works, he joined the shipbuilding firm, Dennys of Dumbarton, as an electrical engineer in 1892. Moving to Coventry in 1896, he laid out the first factories that produced Humber motorcycles and automobiles. In 1900 he set up an independent consulting office in Hertford Street, Coventry, and signed contracts to lay out a motor-vehicle factory for Maudslay at Parkside, Coventry, and to design cars for Maudslay and Lea-Francis. The first Maudslay had a three-cylinder horizontal 3383-cc overhead-camshaft engine mounted in an underfloor position, with a three-speed gearbox and shaft drive to the rear axle. The Lea-Francis was the same car, with some dimensional differences, and a slightly

bigger (3680-cc) engine. He designed the first Standard of 1903, with a 6-hp short-stroke single-cylinder horizontal underfloor engine, and in 1904 designed the first Singer, with a horizontal parallel-twin mounted under the front seat. He designed a 4.8-liter three-cylinder model for Maudslay in 1904 along with two six-cylinder cars of 6.8-liter and 9.7-liter size. He also designed the Standard 12/15 hp with its underfloor parallel-twin engine, and a succession of other Standard models, culminating with the six-cylinder 15 hp car of 1907. He was named managing director of Maudslay Motor Company, Ltd., in 1907, which did not stop the directors of Rover from calling on him in May 1909 to update their single-cylinder 8-hp model. He led the Maudslay company through World War I and back into the civilian market in 1919, but discontinued its car production in 1923 to expand into a strong line of commercial vehicles. The change in business emphasis gave him a free hand to renew the contact with the Rover Company. He designed the Rover 14/45 with its single-overhead-camshaft engine in 1925 and the 16/50 in 1927. He accepted a seat on Rover's board of directors in 1926 and served as its chairman in 1931–32.

Critchley, James Sidney (1865–1944)

Works manager of the Daimler Motor Syndicate, Ltd., from 1893 to 1895; general manager of the Daimler Motor Company, Ltd., in Coventry from 1896 to 1900; chief engineer of Crossley Brothers, Manchester, from 1904 to 1906.

He was born in Bradford, Yorkshire, in 1865, and educated at Bradford Grammar School. He began his career as a patent agent with John Waugh, M.I.C.E. of Bradford, and later joined T. Green & Sons, Smithfield Iron Works at Leeds, who were specialists in electric streetcars. He grew into a competent industrial engineer, with wide knowledge of tooling and machinery, production methods, and factory layout. F.R. Simms hired him in 1893 to run a shop under a railway arch near Putney Bridge, London, installing Daimler engines in motorboats.

His first Daimler car designs were based on Panhard-Levassor practice, with components from German, French and British suppliers. In 1897 he organized engine production in the Motor Mills at Coventry. He designed a four-cylinder Daimler engine in 1898, and a year later he dealt with the problem of a stock of 55 4-hp Cannstatt-built V-twin engines that no one wanted, by designing a car to suit them. That was the Light Daimler, not a great car, but a commercial necessity. During the time that followed, he felt a lack of support from the management and resigned on June 8, 1900.

He joined the British Electric Traction Company in London as a motor specialist, retaining a stake in the Daimler Motor Company, Ltd., and a seat on the Daimler board until 1902, when British Electric Traction merged with Brush Engineering company and planned to build motor vehicles. Dissatisfied with the indecision and inertia of management, he welcomed an invitation from Charles Jarrott and William Letts, prominent car dealers in London, to design a high-performance touring car. He drew up a 22-hp T-head four-cylinder machine in 1903. Jarrott and Letts then talked the Crossley Brothers, gas-engine makers in Manchester, into building it. The engine was supplied by Mutel in France and the frame was ordered from a Belgian supplier. Assembly took place in the Crossley works in Pottery Lane, Openshaw, Manchester. He also designed the 28-hp and 40-hp models that followed, and switched from chain-drive to shaft drive on the 1906 models. In 1905 he also designed an engine for the Leyland-Crossley omnibus, built to order from the London General Omnibus Company.

He designed the Critchley-Norris steam vehicles of 1908 and spent the rest of his career in non-automotive industrial positions. He retired in 1934 and died in 1944.

Crook, Thomas Anthony Donald "Tony" (1923–)

*Managing director of Anthony Crook Motors, Ltd., from 1947 to 1960; president and chief ex-*ecutive officer of Bristol Cars, Ltd., beginning in 1961.

He was born on February 16, 1923, and grew up in Lancashire. He was educated at Clifton College and studied engineering at Cambridge University, joining the RAF in the middle of World War II. After a turn of duty in the Near East, he was stationed at an airfield in Lincolnshire where he made contacts in the local motor trade. Upon his discharge from the RAF, he went to work for a garage company that belonged to Raymond Mays and Peter Berthon. He began a parallel career as a racing driver which ended with an accident at Goodwood in 1955, but he had also become a Bristol dealer at Caterham Hill, later adding the Fiat and AC franchises.

When Bristol Aeroplane Company merged with Hawker-Siddeley in 1959, the car division was put up for sale. Crook purchased it, in partnership with Sir George White. He brought out the Bristol 407 with a Chrysler V-8 engine in 1961, leading to a succession of Chrysler-powered models. He became the sole owner of Bristol Cars, Ltd., when White retired in 1973. He introduced the Beaufighter in 1980 and the Beaufort, Britannia, and Brigand in 1982. When the Blenheim appeared in September 1993, it was the first new Bristol in 11 years. Bristol made about 100 cars a year, and in 2001 Crook was planning a new Fighter with Chrysler's V-10 engine.

Crossley, William John (d. 1911)

Chairman of Crossley Brothers, Ltd., from 1866 to 1910; chairman of Crossley Motors, Ltd., in 1910–11.

The two Crossley brothers, Frank W. and William J., traveled to Cologne in 1866 and acquired manufacturing rights for Otto & Langen's Deutz gas engines. They tooled up their Openshaw, Manchester, works for gas engine production in 1869. At first they adhered strictly to the German design, but later began to innovate. In 1886 the Crossley brothers used rotary valves in both vertical and horizontal gas engines. They won a good reputation for their workmanship and quality. Frank died in 1897, and William led the company as sole owner. He went into politics and was elected Member of Parliament (Liberal Party). He also held a seat on the board of directors of the Manchester Ship Canal and was declared a "Freeman" of the City of Manchester. He got his start in the automobile industry because the Jarrott & Letts agency in London wanted a British-built car of the Mercedes type, and engaged J.S. Critchley to design it. William Letts and Charles Jarrott took the designs to Crossley, asking if he could undertake its production. It was a 22-hp car with a four-cylinder 4.6-liter T-head engine running up to 900 rpm. The first ones were built in a separate shop on City Road, Hulme, Manchester, in 1903–04. Production was started at Openshaw to build the 1904 28-hp model, followed by a 20-hp car in 1905. A 40-hp model designed by G.W. Iden was added in 1906. In 1908 he began production of a 4½-liter 20/25 model designed by A.W. Reeves and G. Hubert Woods, adding a 2.3-liter 12/14 in 1910. That year he spun off the automobile branch from the gas engine company, founded Crossley Motors, Ltd., and transferred car production to a factory in Napier Street, Gorton, Manchester. He died in 1911.

Crouch, John William Fisher "Billy" (1883–1952)

Managing director of Crouch Cars, Ltd., from 1912 to 1915; chairman of Crouch Motors, Ltd., from 1915 to 1928.

He was born in 1883 in Coventry and attended local schools. He joined Daimler Motor Company, Ltd., as a trainee in 1899, became a draftsman, joined Captain H.H.P. Deasy as an engineer in 1906. He began designing his own car in 1910, a light three-wheeler with a rear-mounted V-twin. His father sold his hat-making business at Atherstone and financed the construction of the first Crouch Carettes. In 1912 he moved into a small factory at Tower Gates, Bishop Street, Coventry, and before the end of the year a prototype 8-hp four-wheeler was ready, also powered by a rear-mounted V-twin. In 1913 four-wheeled Crouch cars won Gold Medals in the London-to-Land's End and London-to-Man-

chester trials. During World War I the factory was fully engaged in aircraft-engine work, but production of the V-twin model resumed in 1919. It was redesigned for 1922 with the engine mounted in front and shaft drive but taken out of production at the end of 1923. The Crouch 12/24 was introduced as a 1923 model with a four-cylinder 12-liter Anzani engine and by 1925 he was turning out 10 or 12 cars a week. He was chronically in debt and the cars suffered from lack of progress. He was never able to put a differential in the rear axle, and never adopted four-wheel-brakes. He went bankrupt when payment was never received for 200 cars, made to order and shipped to Australia in 1928.

Cudell, Max (ca. 1860–19?)

Managing director of Cudell & Co. K-G from 1898 to 1906.

He was born about 1860 and spent his early career in a succession of business appointments. In 1897 he secured manufacturing rights for de Dion–Bouton single-cylinder engines and motor tricycles, and a year later set up his own company with a factory in Aachen (Aix-la-Chapelle) and began production. He added the four-wheeled Vis-à-Vis model in 1899, with a 402-cc single-cylinder engine having two flywheels. It became available with 520- and 700-cc engines in 1901–02, using two-speed gearboxes with a reverse. Seeking to develop his technology faster than de Dion–Bouton's, he hired Paul Henze and Carl Slevogt to create new two- and four-cylinder models in 1903. The 2½-liter four-cylinder 16/20 appeared late in 1904 with a four-speed gearbox and a choice of propeller-shaft or chain drive. The technically magnificent 6.1-liter 35/40 Phönix was the star of the Berlin auto show in 1905, but it was made at very high cost. Cudell had expanded too fast for his resources and taken initiatives without a budget, and the company was declared bankrupt. With great acumen, he formed a partnership with Leo Kolbing in Berlin to start Cudell Motoren GmbH. Marine engine production and the stock of spare parts for Cudell cars were transferred to Berlin. The Berlin plant also began making compressed-air starters in 1907. The Aachen plant became the property of a new corporation, Aktiengesellschaft für Motor- und Motor-fahrzeugbau vorm. Cudell & Co. which disposed of the cars under construction and the ones completed by 1907. Then it was liquidated, paying 5 cents on the dollar. Nothing if not resilient, Max Cudell appeared as managing director of Deutsche Automobilbank AG of Berlin-Charlottenburg in 1920, a credit organization for the auto industry and its suppliers, founded by Joseph Vollmer, Max Cudell, plus two bankers, a country-estate proprietor, and a lawyer. But it fell victim to Germany's hyperinflation in 1923.

Cunningham, Alexander A. (1926–1996)

Managing director of Adam Opel AG from 1974 to 1976; chief executive of General Motors European Operations from 1976 to 1978; group executive for GM Overseas Group from 1978 to 1986.

He was born in 1926 in Sofia, Bulgaria, to a British father and Bulgarian mother. The family fled to Britain when World War II broke out, and he served in the RAF. After the war, he went to America and began his GM career with Frigidaire in 1948, graduated from the GM Institute in 1952, and was assigned to GM Overseas Corporation in 1953. He was transferred to Adam Opel AG in 1964. He authorized the Ascona program and had the Rekord made over for 1967. He also gave the green light for the Senator and approved the adoption of front-wheel-drive for the Kadett and Ascona. He supervised the coordination of Vauxhall with Opel, but was called back to the United States in 1980 as group executive for Body and Assembly. In 1984 he was named executive vice president for North American Automotive Operations. He was headed for the presidency of General Motors, but was thrown off course by a heart attack in 1985. He retired the following year and died on September 5, 1996, in Tustin, California.

Cureton, Thomas (ca. 1863–1921)

Managing director of the Sunbeam Motor Car Company, Ltd., from 1905 to 1914; joint managing director of the Sunbeam Motor Car Company, Ltd., from 1914 to 1918; chairman of the Sunbeam Motor Car Company in 1918–19.

He was born about 1863 in Northamptonshire and educated at Rugby School. He began an apprenticeship with John Marston, Ltd., of Wolverhampton in 1877. Ten years later, Marston began making Sunbeam bicycles, and Cureton moved into the bicycle department. By 1896 he was works manager of the Sunbeamland Cycle Company. His mind was increasingly occupied with automobiles, and in 1899 he submitted a well-reasoned memo to John Marston, arguing the case for building cars. He designed and built a prototype with a vertical single-cylinder engine in 1899, and a second test car with a two-cylinder engine the following year. But the board of directors withheld their approval for his plan. It was not until 1901 that the company could start making cars, but the product was not of Cureton's design. It was a design by Mabberley Smith with four wheels in lozenge formation, a side-saddle bench seat, and the driver's seat behind, with tiller steering to the single front and rear wheels. The front-mounted single-cylinder de Dion–Bouton engine had chain drive to the two side wheels. It was not produced in the bicycle works, but in a separate plant on Upper Villiers Street in Wolverhampton, at Cureton's recommendation. Some 300 Sunbeam-Mabley units were made up to 1904. At T.C. Pullinger's suggestion, Marston made arrangements with Marius Berliet to copy his 12-hp car, and supply engines from Lyon. Cureton was in charge of the production. In 1905 the motor car department was set up as a separate corporation, with Cureton as managing director, and the factory began producing a powerful 16/20 model designed by Angus Shaw. In 1909 Cureton engaged Louis Coatalen as chief engineer, leading to a total product renewal. He directed the company's aircraft-engine production during World War I and in 1918 succeeded John Marston as chairman. But little more than a year later he retired due to ill health and died in July 1921.

Cutler, Gerard Mervyn (b. 1902)

Technical manager and chief engineer of the Car Division, Armstrong-Siddeley Motors, Ltd., from 1951 to 1953; technical engineer for medium cars for the Rootes Group from 1953 to 1956; chief engineer of long-term projects for the Rootes Group from 1956 to 1960.

He was born in October 1902 in Wolverhampton and educated at Alex Hunters Preparatory School and Wolverhampton School. He served an apprenticeship with the Sunbeam Motor Car Company, Ltd., from 1931 to 1935. He was a draftsman on aircraft engines with the Bristol Aeroplane Company, Ltd., from 1935 to 1938, and an aircraft engine design engineer with Armstrong-Siddeley Motors, Ltd., from 1938 to 1941. He worked as an experimental engineer on engines and transmissions for Humber, Ltd., during World War II and in 1946 joined Armstrong-Siddeley Motors, Ltd., as chief development engineer in the Car Division. In 1948 he went to Morris Motors, Ltd., as deputy chief engineer for engines (under Tom Brown) but returned to Armstrong-Siddeley in 1951. He directed the Sapphire program and designed the 3.4-liter six with splayed overhead valves. Returning to the Rootes Group in 1953, he prepared a new-generation Humber Hawk for 1957, and then started a small-car project, assisted by Mike Parkes and Tim Fry. It evolved into Project Apex, which finally materialized in May 1963 as the Hillman Imp. He had left the Rootes Group in 1960 to become chief technical executive of Automotive Products, Ltd., and retired in 1966.

Daimler, Gottlieb Wilhelm (1834–1900)

Chief engineer of Gasmotorenfabrik Deutz from 1872 to 1882; technical director of Daimler Motoren Gesellschaft from 1890 to 1895.

He was born on March 17, 1834, as the son of a baker in Schorndorf, Württemberg, served an apprenticeship with a gunsmith, and went to Stuttgart to study at the district trade school.[1] He began his career in 1853 as a mechanic with Werk Grafenstaden in Alsace, where steam-locomotive production was taken up in 1856. When he was offered the post as manager of the locomotive branch, he declined, feeling the need for further schooling. After three years at the Stuttgart Polytechnic Institute, he returned to Grafenstaden and served as manager of the locomotive branch from 1859 to 1861. He went to England to broaden his steam-engine experience and held engineering positions with Smith, Peacock & Tannet of Leeds, Roberts & Company in Manchester, and the machine tool works of Sir Joseph Whitworth in Coventry. He returned to Germany in 1863 as chief engineer of Metallwarenfabrik Straub & Sohn, and two years later took over the management of a machine factory owned by a religious institution[2] at Reutlingen. In 1869 he left Reutlingen to become factory manager of Maschinenbau-Gesellschaft Karlsruhe, and during 1871 he began negotiations for a position with Gasmotorenfabrik Deutz in Cologne. Eugen Langen engaged him as chief engineer, effective August 1, 1872. He got a thorough grounding in two-stroke internal combustion engines, and took part in Deutz's experimental work on four-stroke gas engines, for which Deutz obtained a patent in 1877. He resigned from Deutz in 1882 to set up his own mechanical workshop in Cannstatt, on Stuttgart's east-end doorstep. Here, with Wilhelm Maybach as his assistant, he developed a petroleum-fuel four-stroke engine, for which a patent (No. 28,022) was issued in 1833. They built a vertical single-cylinder engine which was installed in a kind of bicycle in 1885. It was covered by a patent (No. 36,423). A bigger engine was mounted in a horse-drawn carriage in 1885, with belt drive to the rear wheels and a steering linkage to the front axle. Daimler's idea was to make engines, not to build automobiles. He was awarded a patent for a motorboat (No. 39,357) in 1886. His cars and boats were made to demonstrate the feasibility of such applications, and attract buyers. In 1888 he sold a license for his patents to William Steinway of Long Island City, New York.

He sold an engine license to Panhard & Levassor in 1889. He went into partnership with F.R. Simms to promote his patents and sell Daimler engines in the British market in 1893. In 1887, Daimler engine production was moved into a new factory at Seelberg, with 20 workers. Daimler's narrow-angle V-twin went into production in 1889 and the first four-cylinder unit, a marine engine, was built in 1890. He also built up a steady business in engines for railcars, With wealthy partners, three industrialists and a banker, he founded Daimler Motoren Gesellschaft on November 28, 1890. His partners were not interested in automobiles, they were out to make a profit from the engine business. Yet a number of Daimler cars were built, generally on a one-off basis, from 1891 onwards. The first Daimler truck appeared in 1894. Car production did not begin until 1897, and then on a small scale, with the Phoenix model which had a front-mounted engine. A year later, Daimler began making engines for airships.

In 1895 Daimler ceded his title to Wilhelm Maybach, but continued to work as a director of the company until his death on March 6, 1900.

1. Königlishe Landesgewerbeschule.
2. Maschinenfabrik des Bruderhauses.

Daimler, Paul Friedrich (1869–1945)

Technical director of Austro-Daimler from 1902 to 1905; technical director of Daimler Motoren Gesellschaft from 1907 to 1922; chief engineer of Horch-Werke AG from 1923 to 1928.

He was born on September 13, 1869, in Karslruhe, as the elder son of Gottlieb Daimler. He attended neighborhood schools in Stuttgart, and worked part-time in his father's shop in Cannstatt. He held an engineering degree from the Stuttgart Technical University and joined Daimler Motoren Gesellschaft as a design engineer in 1897. He made complete drawings for a light car in 1899, but it never went into production because the plant was being revamped to build 40 cars of a new Maybach design ordered by Count Emil Jellinek. Has ambitions frustrated at Cannstatt,

he was pleased to accept a transfer to Austro-Daimler in Wiener-Neustadt. The cars produced there, however, were not his designs, but adaptations of various Mercedes models designed by Wilhelm Maybach. It was Emil Jellinek who intrigued his way to have him replaced by Ferdinand Porsche, and in 1905 Paul Daimler returned to Stuttgart. He was eager to have Maybach's title, and built up a list of pretexts why Maybach had to go. He reached his goal in 1907, and won the board's approval by the chance that a car of his design won the French Grand Prix in 1908.

Following the example of the British Daimler Company in Coventry, he embraced the Knight-patented sleeve-valve engine and introduced the 10/30, 16/40, and a 16/45 for 1910, in parallel with cars having poppet-valve engines. He designed the 16-valve four-cylinder 4½-liter Grand Prix racing car for 1914, and created aircraft engines that were produced in considerable numbers during World War I. He initiated experiments with superchargers and developed a system that went into production on two postwar Mercedes sports cars. But in 1921 he fell out with the supervisory board, because they wanted to produce profitable cars, and he was blamed for having a fixation on high performance. He went to Horch at Zwickau in Saxony in 1922 and designed two cars with twin-cam straight-eight engines, the 3.4-liter Type 305 going into production in 1927 and the 4-liter Type 375 arriving a year later. They were not successful, and he was forced to resign from Horch.

Dallara, Giampaolo (1939–)

Chief engineer of Automobili Lamborghini from 1963 to 1968; chief designer of De Tomaso Automobili from 1968 to 1972.

He was born in 1939 in Varano De'Melegari near Parma and graduated with a degree in aeronautical engineering from the Milan Polytechnic. He was awarded 96 points out of a maximum of 100 for his thesis on "The Construction of a Two-Seater Jet Plane for Training." His professor had done some consulting work for Ferrari and recommended Dallara to the Commendatore. He worked at Maranello from January 1960 to April 1961 as an assistant to Carlo Chiti in the racing department, making stress calculations, vibration analyses, and suspension-geometry studies. At the instigation of his cousin, Giulio Alfieri, he went to Maserati in May 1961 and worked on Tipo 64 sports/racing car with an inboard/rear-mounted V-12 engine, and the Tipo 151 coupé with its front-mounted V-8 engine. He joined Lamborghini in June 1963 and designed the chassis for the 350 GT. In 1965 he invented the Lamborghini Miura, with its transversely mounted inboard/rear V-12 engine. He got restless, wanting to get back into racing, and signed up with A. De Tomaso in September 1968. He designed the De Tomaso F 2 car for 1969 and the Ford V-8 powered F 1 car for 1970. He updated the Mangusta into a Pantera, and left De Tomaso in 1972. He became a consulting engineer for Iso Rivolta, returned to his native village and opened a small factory, and got a contract from Lancia to develop the Stratos for race-and-rallying. He built and developed the Lancia 037 chassis in 1980–82. His radically streamlined sports car from 1973, with the Fiat 128 power train mounted between the rear wheels, remained in the prototype stage. He built a series of Dallara F 3 racing cars with Alfa Romeo engines, beginning in 1982. In 1988 he built the first BMS[1] racing car for Giuseppe Lucchini, owner of the Scuderia Italia, and for 1989 designed a new BMS F 1 car with a Ford-Cosworth V-8 engine. He made the Dallara F-190 for Scuderia Italia in 1990, installing the Cosworth DFR V-8 engine. He mounted Judd V-10 engines in his 1991 cars and obtained Ferrari V-12 engines for the F-192 a year later. In 1990–92 he designed a new chassis with a tubular steel frame for a new Iso Rivolta Grifo sports coupé which never got into production. He continued as one of the main builders of F 2 and F 3 racing cars.

1. Brescia Motor Sport.

Dangauthier, Marcel (1911–)

Technical director of Peugeot from 1956 to 1976.
He was born in 1911 and held an engineer-ing degree from the Ecole Centrale des Arts et Manufactures. He also graduated from the Ecole Nationale des Moteurs.

He began his career with Peugeot in 1936 as a trainee engineer at Sochaux and stayed on as a draftsman. In 1945 he became a design engineer in Peugeot's technical center at La Garenne–Colombes, and was promoted to director of the technical center in 1954. He succeeded Henri Dufresne as technical director on May 1, 1956, and took charge of the 404 development. He steered Peugeot into front-wheel-drive and directed the 204 program from beginning to end, following up with the 304. He was also behind the concept of the 504 and 604 rear-wheel-drive cars with all-independent suspension. He started the 505 project after completing the 104, and semi-retired in 1976, taking full retirement in 1977.

Daniels, Jack (19??–)

Chief design and development engineer of the Nuffield Organization from 1945 to 1949; chief development engineer of Austin-Morris from 1974 to 1977.

He was born in Oxford and attended the Central School, Oxford, where he was noted for being good at wood-working and technical drawing. He sat for an exam to join the Great Western Railway, but his headmaster told him about somebody building sports cars at Cowley, so he went to see them, discovering the beginnings of MG. He went to work for Cecil Kimber in 1927, on the staff of H.N. Charles alongside Adrian Squire and W.S. Renwick. They created every new MG for a ten-year span, from the 8/33 Midget to the J-Type Midget, from the 18/80 Sports Six to the 2-Litre SA, from the J-Type Midget to the PA, from the Magna 12/70 to the Magnette K3 and Magnette N. The team was transferred to Morris Motors in 1936, where he worked in experimental engineering until 1939, when he was put into a military-vehicle design office. One of their projects was a reconnaissance vehicle with a unit-construction body, another was an 85-ton tank. After the war, he led the development of the new Morris Oxford and Oxford Six. He also became involved with the radically new Morris Minor. As early as 1950 he proposed three new designs as potential replacements for the Minor, but all were rejected. One survived, however, and went into production in 1958 as the Wolseley 1500, followed by a Riley version. He had a hand in the front-wheel-drive experiments made at Morris in 1954, but left in 1956 to join Austin at Longbridge. He worked as a test engineer on the Hydrolastic suspension from 1957 to 1961, led the chassis studies for the Austin 1800 and the 3-Litre, the Allegro and the Princess.

He retired in 1977.

Daninos, Jean Clément (1906–2001)

President of Facel from 1939 to 1962.

He was born on December 2, 1906, in Paris, into a family of Greek origin. He began drawing cars when he was a small boy. It was a passion that never left him. He was educated at the Jeanson de Sailly and Buffon schools and graduated with an engineering degree from the Ecole des Arts et Métiers in Paris. He also held a degree from the Scarborough School of Mechanical Engineering. A former schoolmate, Jacques Bingen, whose sister had married Andre Citroën, enabled him to join Citroën in April 1928. He began as a body-engineering trainee, was transferred to the methods department, and became director of special vehicles. He worked on the C6 roadster and designed the coupé and cabriolet versions of the 11 "Traction," but left Quai de Javel when Michelin took over in 1935. He went to Morane-Saulnier, working on wing design and stainless steel applications, and was co-designer of the Morane-Saulnier 405 fighter plane. In 1937 he founded his own company, Métallon, making a range of kitchen cabinets and stainless steel sinks. In 1939 he established the Forges et Atéliers de Construction d'Eure-et-Loire (Facel) at Dreux, with a factory at Amboise. During World War II, Facel absorbed Métallon and made aircraft engines and components. Daninos, however, fled to the United States in 1940, where he worked for General Aircraft Equipment, Inc. He also invented an ice-cube machine that proved highly profitable. He returned to Paris in 1944 and recovered control of Facel-Métallon. He won a contract for the all-aluminum Dyna-Panhard unit body, and made 45,000 of them. Facel-Métallon also produced over 15,000 coupé and roadster bodies for Simca from 1947 to 1961, and won a Ford contract to build the bodies for the Comète. He designed the Cresta body for the Bentley chassis, and built 17 of them. Facel-Métallon produced the bodies for the Delahaye VLR, and supplied steel pressings for Somua truck cabs, Vespa motor scooters, and Massey-Ferguson tractors. The Facel Vega existed as a concept in 1950. He had a prototype built in England, with a six-cylinder Hotchkiss engine, and offered the design to Hotchkiss, Delahaye, Salmson and Talbot, without meeting much interest. In 1954 he decided to produce it himself, with a Chrysler V-8 engine and Chrysler automatic transmission. For the next seven years, it was France's only prestige car. He added the lower-priced Facellia in 1959, but the company was failing financially, and he had to give up control of Facel in 1962. In 1965 he tried to make a deal with Rover to assemble a new Facel at Solihull, but the French government vetoed it. He lived an active retirement life, remaining alert and lucid till the end.

He died from cancer on October 13, 2001.

Darl'Mat, Emile (1892–1974)

Parisian concessionaire who produced a famed series of sports cars on the Peugeot 203 chassis.

Breton-born Emile Darl' Chechmate (commonly spelled as Darl'Mat) became associated with Peugeot as an agent in 1923. His first concept was the Eclipse — a hardtop convertible with the top being swallowed by the trunk. His next idea was to propose a light bodied sports car with the chassis and engine made by Peugeot. This car was serviced by the Peugeot network and was intended to have a reasonable sales price. Georges Paulin designed the aluminum body and Marcel Pourtout, a very well-known French coach builder, was responsible for production. The engine was tuned by Peugeot and mounted on a short chassis. Four models were available: a roadster, a convertible, a coupe, and a race car almost identical to the roadster (the last model). A trio of Peugeot Darl'Mat 302 Roadsters competing in the 24 Hours of Le Mans in 1937 finished 7th, 8th and 10th. In the 1938 Le Mans 24 Hours, the Darl'Mat 402 achieved 5th place overall and won the 2 Liter class. Between 1950 and 1954 Darl'Mat commissioned 250 Peugeot 203 saloons with restyled bodies and tuned engines. He remained active in the industry until his death in 1974.

Darracq, Alexandre (1855–1931)

Founder of A. Darracq & Cie in Paris; partner in the British-owned firm of A. Darracq & Co. (1905) Ltd. from 1905 to 1912.

He was born on November 4, 1855, in the Bigorre area of the central Pyrenees and began his career as a draftsman in the Arsenal at Tarbes. Moving to Paris in search of broader opportunities, he joined Hurtu & Hautin, sewing-machine manufacturers, as an engineer. When Hurtu & Hautin began making bicycles, he obtained first-hand experience in that line of business. He left Hurtu & Hautin to start making bicycles of his own design in a shop on Boulevard Saint Martin in Paris, distinguished himself as a bicycle racer, and in 1891 associated himself with Jean Aucoc to produce Gladiator bicycles in a factory at Pré-Saint-Gervais.

In 1894 he started production of Perfecta motorcycles, secured manufacturing rights to a car designed by Léon Bollée in 1896, and began production of Darracq cars at Suresnes in 1897.

His company fell under British control in 1902, but he stayed on as head of the Darracq car operations in France until 1912. In 1919 he sold all his interests in the automobile industry, bought an excellent property in Monte Carlo, and spent the rest of his life in retirement.

He died in 1931.

Dawson, Alfred J.

Works manager of the Hillman Motor Car Company, Ltd., from 1910 to 1914 and 1916–18; managing director of the Dawson Motor Car Company, Ltd., from 1919 to 1921.

He served an apprenticeship with the Great Western Railway in its Wolverhampton shops and worked as an engineer with British Thomson-Houston. He was a production engineer with the Daimler Company from 1906 to 1910, when William Hillman engaged him as works manager. He designed the 1913 Hillman 9-hp car with its side-valve 1327-cc four-cylinder engine, which was well received on the market and probably saved the company from bankruptcy. He was works manager of the Hotchkiss ordnance factory in Coventry in 1914–15 but returned to Hillman, and updated the 9-hp car into the "Peace" model. Late in 1918 he left to produce his own car. With the financial backing of Malcolm Campbell, whose agency also handled distribution of the Dawson, he moved into a factory on Clay Lane, Stoke, Coventry, which had engine-manufacturing capability. The Dawson engine was a 1.8-liter overhead-camshaft four-cylinder unit, and he set out to make a quality car at a reasonable price. The 1919 two-passenger coupe was listed at £450. But he — and Campbell — could never sell enough cars to reach break-even point, and he raised prices, which further inhibited the demand. Nonpayment of suppliers led to lawsuits and receivership in 1920, and the factory closed after no more than 65 Dawson cars had been completed. Dawson emigrated to Australia.

Dawtrey, Lewis Henry (1900–)

Chief engineer of the Standard Motor Company, Ltd., from 1952 to 1955; chief engineer of research and development for Standard cars from 1956 to 1963 and Triumph cars from 1956 to 1969.

He was born on April 15, 1900, in Halifax, Yorkshire, and educated at schools in Halifax and Dewsbury before going on to Coventry Technical College in Coventry. He joined Humber, Ltd., in 1919 as a trainee engineer, served as a draftsman with Humber from 1921 to 1923 and a technical designer with Humber from 1923 to 1931. He left Humber to join Standard and held the position of technical assistant to the technical director from 1931 to 1940. He was technical executive of Standard from 1940 to 1945, organizing war production. From 1945 to 1949 he held the title of technical engineer and was deeply involved with the Vanguard project as well as the Triumph Renown and Mayflower. He led the Standard 8 project, with its derivatives, and prepared the second-generation Vanguard. From 1956 onwards he had only minor influence on the shape of new Standard Triumph products, but he made valuable contributions to ensure that the Triumph 1300 and 2000 were passably "bug" free when launched in 1963. He retired in 1969.

Day, Graham (1933–)

Chairman of the Rover Group from 1986 to 1991.

He was born on May 3, 1933, in Halifax, Nova Scotia, as the son of a British-born former stockbroker. He was educated at Queen Elizabeth High School and Dalhousie University, graduating in 1956 with a degree in law. He spent eight years with a local law firm before joining Canadian Pacific in 1964, where he gained experience in the hotel business, in shipping, vessel procurement, and running a truck fleet. He went to Britain in 1971 as chief executive of Cammell Laird Shipbuilders, and from 1975 to 1977 he was vice president of H.M. Government's organizational committee for shipbuilding. From 1977 to 1980 he lectured on maritime transport at Dalhousie University in Halifax. He worked for three years as president of Chantiers Davie at Lauzon, Quebec, and vice president of Dome Petroleum, returning to Britain in 1983 as managing director of the state-owned British Shipbuilders. In just over two years, he closed 30 shipyards and cut the payroll from 60,000 to 9,000. He was appointed chairman of British Leyland on May 1, 1986, and lost no time in renaming it the Rover Group, on July 10, 1986. As he had done with shipbuilding, he started dismantling the unwieldy enterprise. In 1987 he disposed of Unipart in a management buyout, sold Leyland Buses to Volvo, Leyland Trucks and Freight-Rover to DAF. What remained of the Rover Group, he sold to British Aerospace in March 1988 for £150 million. He left in 1991.

Deasy, Henry Hugh Peter (1866–1947)

Chairman and chief executive of H.H.P. Deasy & Co. from 1903 to 1908; chairman and joint managing director of the Deasy Motorcar Manufacturing Company, Ltd., from 1906 to 1908.

He was born in 1866 in Dublin as the son of the Right Honorable Rickard Deasy, Lord Justice of the Irish Appeals Court. He joined the Army in 1888 and was discharged with the rank of captain in 1897. He spent the next three years on explorations in Tibet and China. Upon his return to Ireland, he discovered automobiles, and moved to London in search of opportunity. In 1903 he founded a company with a repair shop and a small coachbuilding works in London, and became agents for the Martini. He developed a scheme to use Martini's factory at St. Blaise, Switzerland, to produce a 24-hp touring car to sell for £500. To this end, he set up a finance company, Hills-Martini, Ltd., which bought control of the Martini works in 1906. Acting out another scheme, he took over the factory at Parkside, Coventry, formerly occupied by the Iden Motor Company, and engaged E.W. Lewis to design a range of high-powered touring cars to wear the Deasy name. His loose management came in for strong criticism from the board of directors, which demanded his resignation before the end of 1908. He returned to Dubin, entered local politics, and ended his career as Lord Justice Baron Deasy, Irish Court of Appeals.

De Beaumont, Mario Revelli *see* Revelli De Beaumont, Mario

Deconinck, Rémi (1950–)

Head of technical services, Régie Nationale des Usines Renault, from 1980 to 1985; chief engineer of Renault low-line product planning from 1985 to 1989; chief engineer of Renault advanced product planning from 1989 to 1995;

director of product, Renault SA from 1995 to 2001; senior vice president, Renault SA product planning beginning in 2001.

He was born in 1950 and graduated from the Institut Supérieur des Matériaux et de la Construction Mécanique with a diploma in third-cycle mathematics and physics. He joined Renault in 1977 as a gearbox designer and was assigned to new-car product planning two years later. He was responsible for the concept and development of the R-19 which went into mass production at the Douai plant in May 1988, and led the planning of the 1990 Clio and 1992 Twingo. he was personally responsible for the Avantime and the "Next" hybrid-electric experimental car. He also directed the planning and development of the 2002 Mégane.

De Cosmo, Joseph (b. 1863)

Head of the car design office, Fabrique Nationale d'Armes de Guerre (FN) from 1900 to 1903; managing director of De Cosmo & Cie from 1903 to 1908.

He was born in 1863 in Italy and schooled as an engineer. He went to Egypt in 1881 to take charge of a steam-locomotive repair shop, and on his return to Italy went to work in a torpedo factory. He was named superintendent of the Navy Arsenal in Naples, but moved to France in 1890. In 1893 he was working for Gautier-Wehrlé on steam-vehicle design in Paris, and in 1894 he was assisting Emile Delahaye in the construction of his first car in Tours. He designed and patented a motorcycle, and in 1896 went to Coventry to work for George Singer. In 1899 he found out that the firearms manufacturer FN of Herstal-lez-Liège was eager to start a sideline in automobiles. The roots of FN go back to the reign of Erard de la Marck (1505–1538) when arms manufacturing began in Liège. He joined FN and took charge of car design and engineering. The first FN had a front-mounted two-cylinder engine, two-speed gearbox and chain drive to the rear wheels. In 1902 he created the FN 14 CV four-cylinder 4-liter shaft-driven touring car, but resigned from FN in 1903 to start his own company, with a factory in rue

de la Vieille Montagne, Liège. The first De Cosmo car was a 24/30 CV four-cylinder model, followed by a 30/35, both with three-speed gearbox and shaft drive. The 24/30 was also exported to England, where the Wilkinson Sword Company, Ltd., of Acton, London, sold it with a Wilkinson badge. He also made truck and bus chassis, and engines for motorboats. In 1906 he introduced a 45/55 CV car with a six-cylinder 6-liter engine. The factory closed in 1908.

De Dietrich, Eugène (b. 1844)

Chairman of the Société De Dietrich & Cie of Niederbronn in Alsace from 1881 to 1904.

He was born on October 9, 1844, at Niederbronn as a descendant of Domenicus Josephe De Dietrich, mayor of Strasbourg and founder of the De Dietrich Iron Works at Jägerthal in 1684. He attended local schools and graduated from the Karlsruhe Forestry Academy. He joined the family business in 1867, assuming administrative duties in running the forges at Grafenweier, the steel mills at Reichshoffen, and the machine works at Niederbronn. He started a branch factory for making railway rolling stock at Lunéville in Lorraine in 1880, producing important numbers of freight cars but also some passenger carriages. In the 1890s the Alsatian aristocracy became pioneer motorists in a big way, and Eugène De Dietrich proved to be an enthusiastic driver. In 1896 he decided to manufacture cars in the De Dietrich factories, and purchased a license from Amédée Bollée, Jr. The first De Dietrich cars were built at Reichshoffen in 1897, with engines supplied by Amédée Bollée of Le Mans. In 1899 he also signed a license agreement for a simple car with a single-cylinder engine with Alexis Vivinus, and in 1902 he engaged the young Ettore Bugatti to come to Niederbronn and design a range of four-cylinder cars. Two years later he became disenchanted with the automobile business, dismissed Ettore Bugatti, and led the enterprise into broader fields of industrial and durable consumer goods.

de Dion, Wandonne de Malfiance (Jules-Félix Philippe Albert) (1856–1946)

Co-founder of the Société de Dion–Bouton & Trépardoux in Paris in 1887; chairman of de Dion–Bouton & Cie from 1894 to 1927.

He was born on March 9, 1856, at the Chateau de Maubreuil, Carquefou near Nantes in the Loire estuary, as the scion of a noble family with roots in Brabant, son of the local mayor and member of the regional council. No record of any formal education has come to light. He went to Paris as a spoiled brat with too much spending money and became notorious as a high-society playboy. His friend the Duke of Mornay asked him one day if he could find some new gadgets for a party gag, and he went to a shop on the Boulevard des Italiens where he had bought useless trinkets before. When his eyes fell on an exquisite miniature steam engine, he forgot all about his errand and bought the steam engine. Later he tracked down its makers, two mechanics in a small shop in rue Pergolèse, named Bouton and Trépardoux. He offered them higher pay than they were making if they would work for him.

That was the start of a historical partnership which was to have an impact on the evolution of the automobile. They began making stationary steam engines in 1882 and produced their first steam-powered quadricycle in 1883. For the first time, Count de Dion had found a worthy cause and a useful purpose in life. His family, which had tolerated his earlier life as a wastrel and favorite of the gossips, chose this occasion to disown him and block his access to funds which legally belonged to him, for industrial investment. Suddenly he had to work for a living, becoming totally dependent on the steam engine and vehicle business for an income. He worked hard, stayed out of debt, and did not go and beg the family for help. Eventually they relented, and he could once again draw freely on his capital. He studied engineering and became convinced that the internal combustion engine was superior to steam power for vehicular applications. Experimental spark ignition engines were designed, built and tested from 1890 onwards, and the first marketable high-speed engine was built in August 1894, putting out ½ hp at 1500 rpm from a single 143-cc cylinder. A year later, they put a 1.5 hp 275cc engine in the frame of a Humber tricycle, and tooled up to produce motor tricycles.

In 1897 de Dion–Bouton had 200 workers and the old factory had no room for expansion. Count de Dion secured the financial backing of Baron de Zuylen de Nyevelt de Haar and moved into spacious buildings on both sides of rue Ernest in the Paris suburb of Puteaux in 1898. The company hired another 200 workers, and by 1900 the payroll totaled 950.

He inherited the title of Marquis de Dion from his father who died in 1901. He had already gone into politics and was elected deputy from the Loire Atlantique. From 1901 onwards the factory also made four-wheeled automobiles and in 1908 de Dion–Bouton established a taxicab company in Paris.

The Marquis won reelection to the Chamber of Deputies in 1910 and was active also in motor-industry and motoring associations. Still he was an effective leader of the de Dion–Bouton company, which had 3000 workers in 1911. His interest in cars and the car business waned after 1918, and in 1923 he was elected to the Senate. He lost control of the company when it was declared bankrupt in 1927. He was never again to hold office in the auto industry, but continued his political career up to 1940 when, in a flourish of patriotism, he refused to go to Vichy and denied his vote to Marshal Pétain, plenipotentiary head of state during the Nazi occupation of France.

He died on August 19, 1946.

De Freminville, Charles *see* Freminville, Charles de

de Jong, Sylvain (1868–1928)

Joint managing director of Minerva Motors, SA from 1903 to 1906, managing director of Minerva Motors SA from 1906 to 1921, chairman of Minerva Motors SA from 1921 to 1928.

He was born on January 5, 1868, in Amsterdam and had a general education. He first tried his hand at journalism but decided to go into business, and moved to Antwerp in 1883.

In 1889 he opened a bicycle agency in Antwerp, and in 1892 traveled to England to see how his suppliers produced them. He then began to make Mercury bicycles in Antwerp in partnership with his brother Jacques. In 1896 he made a tour of the American bicycle industry. On his return, Sylvain and Jacques resigned from the Mercury Cycle Company and established themselves as S. de Jong & Cie in Antwerp, introducing the Minerva bicycle. The work force in their shops on rue Jacob grew from an initial 40 men to 200 within two years. In 1899 they became dealers for de Dion–Bouton engines, and also began manufacturing a small clip-on engine as auxiliary power for bicycles. Bigger Minerva engines appeared in 1900, and they introduced the Minerva motorcycle. That year they also built a test car with a 6-hp two-cylinder engine and chain drive. They prospered as motorcycle makers and also supplied air-cooled single-cylinder engines to Cottereau, Gobron, Opel, Adler, Enfield, Humber, Quadrant and Seidel-Naumann. A new prototype car was ready in 1902, and Minerva Motors SA was founded in 1903 in partnership with David Citroën, who had set up a British sales organization for Minerva motorcycles and took office as joint managing director. That partnership ended in 1906 and Sylvain de Jong became sole managing director with Jacques de Jong as technical director.

They bought a factory site at Berchem outside Antwerp, and began car production there in 1904, offering a 10-hp twin, a three-cylinder 15-hp, and a 20-hp four-cylinder model. The 15 and 20 were eliminated in 1905 to be replaced by a 14-hp four-cylinder model, which soon grew to 16-hp size. A new 22-hp four-cylinder car followed in 1906, the first shaft-driven Minerva. The first six cylinder was a majestic 40/60 introduced in 1907. In this period he aimed for a self-contained factory with a 1200-strong work force making its own carburetors, radiators, batteries and wheels. In 1908 the de Jong brothers bought a license for the Knight sleeve-valve engine and phased out all poppet-valve car engines. The 1910 model lineup included 15-, 28- and 38-hp engines. Motorcycle production, which had peaked at 1200 a year in 1907, was discontinued in 1915. By 1913 Minerva was building 3000 chassis a year and supplying sleeve-valve engines to Mors and Mathis.

In 1918 the brothers went back to Amsterdam together with a technical staff, and rented space in the Spyker plant, where they prepared new models. Two luxury cars, 20-hp four-cylinder and 30-hp six-cylinder, were introduced in 1919, followed by a 2-liter four-cylinder 15 CV model in 1922. In 1923 Minerva Motors SA took over SAVA[1] and its factory was remodeled as a repair and service shop. In 1924 they took over the Auto Traction company which had been building tractor and semi-trailer combinations under license from Chénard & Walcker, and introduced a line of Minerva commercial vehicles. In 1925 they set up a body plant at Mortsel, enabling Minerva to offer complete cars and not just chassis. Under the pressure of intensified competition, Sylvain de Jong approached FN[2] in 1927 with a plan for a joint industrial program, which was never implemented. He had taken ill in 1926 and died on April 29, 1928, in Antwerp.

1. Société Anversoise pour la Fabrication de Voitures Automobiles.

2. Fabrique Nationale d'Armes de Guerre, Herstal-lez-Liège, makers of a range of medium-priced family cars.

Delagarde, Louis (1898–1990)

Project engineer with Panhard & Levassor from 1931 to 1965.

He was born in 1898 at Vitry-le-Frangois and served in the French Artillery from 1917 to 1918. He studied mechanical engineering at the Ecole Centrale des Arts et Manufactures at Chalons-sur-Marne from 1919 to 1921. He began his career with Panhard & Levassor as a draftsman in 1921. He worked mainly on engine development and made valuable contributions to the use of lightweight materials for all reciprocating parts in the sleeve-valve engine. He was part of the development team for Panhard's first diesel engines, and in 1931 he began to design the AM 175 light armored vehicle. He designed the chassis elements for the Dynamic sedan of 1936, with torsion-bar independent suspension. In 1939 he undertook the design of a multi-wheeled armored vehicle carrying a 37-mm machine gun. In 1942 he began development work on an air-cooled flat-twin of 350 cc, which evolved into the 610-cc power unit of the front-wheel-drive 1946 Dyna Panhard. In 1946 he started the design of the EBR eight-wheeler, an armored vehicle powered by an air-cooled flat-12 engine, conceived as six Dyna engines grouped around the same crankshaft. The company built about 1200 EBR vehicles. He designed the power trains and chassis for the Dyna 54, PL 17, and the 24 CT. In 1965 he was transferred from car engineering to the military vehicle division SCMPL,[1] serving as its head until 1975. He retired to Bourg-la-Reine, dedicating his time to technical research. He died there in 1990.

1. Société de Constructions Militaires Panhard-Levassor.

Delage, Louis (1874–1947)

Managing director of Automobiles Delage from 1905 to 1935.

He was born on March 22, 1874, in Cognac as the son of an assistant station master with the local railroad system, and attended local schools. After three years of technical studies at the Ecole des Arts et Métiers at Angers, he graduated in 1893 (only 19!) with a degree in mechanical engineering. He began his career in the public-works sector in Bordeaux, where he reached foreman's level. Taking a deep cut in salary, he signed up with Turgan-Foy, an engineering company at Levallois-Perret just outside Paris. He worked himself up to the title of chief draftsman, and then went on to Peugeot, where he became head of the experimental shop and chief chassis tester. He set up his own company at Levallois in 1905, in partnership with Augustin Legros. While they were designing the first Delage car, the factory made money by making parts for Helbé and subcontracting for others. The Delage car

went into production in 1906, and by 1907 they had an 87-man work force. In 1911 he purchased a large property at Courbevoie and moved Delage production there. The payroll swelled to 350 in 1912, and production capacity topped 1000 cars a year. During World War I the Delage factory made shells and other war materiel, providing ample cash flow to prepare new cars for the postwar market. The 4½-liter six-cylinder CO was introduced before the end of 1919, followed by the 3-liter four-cylinder DO a year later.

Up to 1928 Delage had a Grand Prix racing team, with some remarkable cars, the last being a 11-liter twin-cam supercharged straight-eight. The range of production models expanded, mainly up-market, culminating with the magnificent D8 straight-eight 4-liter arriving as a 1930 model. At the same time, the D6–17 replaced all the previous sixes, but was supplemented with the D6–11 for 1932. But Delage was losing money, and had to declare bankruptcy in 1935. Louis Delage retired, but remained in Paris, though he was no longer a rich man, and his health began failing.

He died on December 14, 1947.

Delahaye, Emile (1843–1905)

Owner of Emile Delahaye, Ingénieur-Constructeur, from 1879 to 1897; director of Emile Delahaye & Cie from 1897 to 1901.

He was born on October 16, 1843, in Tours as the son of a tapestry worker. After attending local schools, he studied mechanical engineering at the Ecole des Arts et Métiers in Angers, graduating in 1869. He began his career far from home, in the drawing office of Ateliers J. F. Cail & Cie in Denain, near France's Belgian border. The Cail enterprise was then a giant of the railway locomotive, railway equipment, and steam traction-engine industry. He served as a military engineer on the Ardennes front in the Franco-Prussian war of 1870–71, and happened to be in Brussels when the war ended. He knocked on the door of Cail's Belgian agents, Halot & Co., and went to work in their factory. He returned to his native Tours in 1875 and went into partnership with Louis Brethon, who had a foundry business there. In 1879 Delahaye bought out his partner and branched out from foundry work into engine production. He designed the four-stroke gas- and petroleum-fuel engines himself. The next step was to make a complete automobile. He designed a horizontal parallel-twin and a belt-drive transmission, which he installed in the rear end of the chassis. Production began in 1896, and Delahaye drove one himself in the Paris-Marseille-Paris road race, finishing tenth in the overall classification. He felt that he needed new partners to expand the business, and was well aware that his age was beginning to tell. He began negotiations with Léon Desmarais and Georges Morane, who had a large factory right inside Paris, on rue du Banquier. He sold his company to them, but the cars kept the Delahaye name, and he served as a director of the enterprise until retiring to Saint Raphael on the French Riviera in 1901. He died at Vouvray on June 1, 1905.

Delamare-Deboutteville, Edouard (1856–1901)

Internal combustion engine pioneer and builder of a prototype automobile in 1884.

He was born on February 8, 1856, as the younger son of a cotton-spinnery owner, in Rouen. He was educated at Lycée Corneille and the Rouen Commercial College.[1] In 1878 he went to work in the parental mills, now run by his older brother. He began to make improvements in the machinery and invented a universal machine capable of serving as a lathe, drill, miller and planer. He began drawing up an engine in 1881. It was a four-stroke unit with pre-compression of the mixture, mechanically operated inlet and exhaust valves, and spark ignition. A horizontal single-cylinder 22-hp test engine was ready in 1883. Assisted by Léon Malandin, he motorized a tricycle, and then they built a two-cylinder 8-hp engine which they installed in a horse-drawn wagon. This engine featured sealing rings on the pistons, a throttle-controlled carburetor, and an exhaust muffler. The wagon was equipped with a steering linkage for the front axle, a progressive-action clutch and a

differential between the rear wheels. It was extensively tested on the roads around Fontaine-le-Bourg in Normandie. Then the engine was taken out and the vehicle scrapped. He continued to design and develop industrial engines, hoping to replace steam power in the textile mills with a plurality of small, industrial engines. At the same time, he designed bigger and bigger engines, receiving a Gold Medal in Paris in 1889 for 100-hp unit and installing a 250-hp engine in the Grands Moulins de Pantin 1894. In cooperation with Powel & Matter in Rouen, he created the 90-hp Simplex engine in 1896, and also began collaborating with John Cockerill at Liege, who developed 200-hp and 600-hp Simplex engines. He gave up his engine studies to pursue other interests, such as studies of local bird life and shellfish farming. He died at the Chateau de Mont-Grimont on February 17, 1901.

1. Ecole Supérieure de Commerce de Rouen.

Demel, Herbert (1955–)

Product planning and development engineer of Audi from 1990 to 1993; Audi board member for technical affairs from 1993 to 1994; chairman and chief executive officer of Audi AG from 1994 to 1997.

He was born in Austria in 1955 and studied mechanical engineering at Vienna Technical University, graduating in 1981. His career began with Robert Bosch GmbH where he worked on anti-blocking brakes and anti-wheelspin systems from 1984 to 1986, following by three years on engine-management systems and direct-injection diesel-engine equipment.

He moved to Ingolstadt in 1990 and took charge of car development in March 1993. He took a particular interest in the A3, first shown in July 1995 and put in production late in 1996. He also guided the development of the A8 which went into production at Neckarsulm in 1994.

After three years as chairman of Audi AG, he was named chairman of VW do Brasil on July 1, 1997, a company with five plants and an annual output of 665,000 vehicles.

De Seze, Guy *see* Seze, Guy De

Desgouttes, Pierre (1874–1955)

Technical director of Automobiles Cottin & Desgouttes from 1906 to 1922.

He was born on February 26, 1874, in Lyon, and educated as an engineer. He began his career with Audibert & Lavirotte in 1896, where he met Marius Berliet. In 1901 he assisted Marius Berliet in the design of his 22-hp four-cylinder car. He set up his own factory on Place du Bachut, Lyon-Monplaisir, in 1904, operating as Société d'Automobiles Pierre Desgouttes et Cie, and began producing a big four-cylinder 24/40 touring car in 1905. The following year Cyrille Cottin joined him as a partner, refinanced the company, and took the title of chairman. Desgouttes designed an excellent 18/22 touring car, and for two years also offered a huge 50/70, both with chain drive. He adopted shaft drive for the new 9 CV in 1912, and the whole range followed. His first six was a 45-hp model introduced in 1909. The factory began building tourist buses in 1907 and made important numbers of military trucks during World War I. Car production resumed in 1919 with the pre-war 14 CV, 23 CV and 32 CV models, plus a more modern 18 CV car.

He prepared the overhead-valve 2.6-liter 12 CV family car for 1922, and resigned.

De Silva, Walter Maria (1951–)

Chief designer of the Alfa Romeo styling center from 1986 to 1998; chief designer of Seat SA since 1999.

He was born on February 27, 1951, at Lecco on the eastern extension of Lake Como and joined Fiat as a designer in 1972, working under the guidance of Sergio Sartorelli and Giampaolo Boano. He contributed to the styling of the Ritmo and the Regata, and left to work with an industrial design firm. He returned to Fiat a few years later and in 1986 was placed in charge of the Alfa Romeo styling center. He led the design teams that created the Proteo and Nivola prototypes, and was responsible for the design of the Alfa 145, 146, 156 and 166 production-car bodies.

He left the Fiat group in 1999 to become styling director of Seat SA in Barcelona, a subsidiary of Volkswagen.

De Tomaso, Alejandro (1928–2003)

Chairman of De Tomaso (Modena S.p.A.) Automobili from 1965 to 1995; managing director of Officine Alfieri Maserati from 1975 to 1989; managing director of S.p.A. Nuova Innocenti from 1979 to 1990.

He was born on July 10, 1928, in Buenos Aires as the son of a wealthy building contractor and grandson of a mason from Naples who emigrated to Argentina. He dropped out of school at 15 to go to work as a farmhand at one of the Ceballos[1] estates. With some like-minded friends, he started an anti–Peron newspaper but left the editorial profession in 1950 to go road-racing with an old Bugatti. In 1955 he was part of another anti–Peron movement whose leaders' named soon appeared on the police "wanted" list. He was flown out to Uruguay on a small plane and made his way to Italy. He found work as a racing mechanic with OSCA[2] in Bologna and became the owner of a second-hand OSCA M/T-4 roadster which he raced with some success. In 1958 he went to Modena and set up his own business. He built his first car, a Formula 2 racer with a modified Alfa Romeo engine, in 1980. In 1963 he designed the Vallelunga sports car with a backbone frame fabricated from sheet steel and an inboard/rear-mounted Ford Corsair engine. By 1965 he had a factory on Via Peri in Modena, and founded his company. He scaled up the Vallelunga chassis to take a Ford V-8 engine, creating the Mangusta. Financed by Rowan Controllers, Inc.[3] of Red Bank, New Jersey, he bought Carrozzeria Ghia S.p.A. in 1967 and secured control of Carpozzeria Vignale S.p.A. in 1969. The Mangusta was re-designed as the Pantera in 1969. Ford Motor Company president Lee A. Iacocca imagined a big market for it in the United States and in 1970 paid Rowan Controllers, Inc., $4.3 million for 80 percent ownership of De Tomaso Automobili, Ghia and Vignale. The spacious Vignale factory was refurbished for production of the Pantera body shell and final assembly of the Pantera. He erected a new plant on Via Emilia Ovest just outside Modena where he began production of the Deauville four-door sedan in 1971 and the Longchamp coupe a year later, both with Ford V-8 power. In 1970–72 he purchased two Italian motorcycle manufacturers, Benelli at Pesaro and Moto Guzzi at Mandello Lario. The Benelli purchase was made with the financial backing of GEPI[4] and the Moto Guzzi acquisition was bankrolled by IMI.[5] He made plans for a "Panterina" with an air-cooled Benelli in-line six mounted transversely behind the seats, but it was never built. In 1973 Ford Motor Company bought full ownership of Ghia (now including Vignale) and stopped Pantera production in 1974. De Tomaso added the Pantera to the products of the Modena plant. In 1975 the Officine Alfieri Maserati was in liquidation. He prevailed upon Romano Prodi, president of GEPI, to finance a takeover. Prodi became chairman of Maserati, with De Tomaso as managing director. He stopped production of the Bora and the Merak, but continued the Khamsin. He created the Kyalami by putting the Maserati 4136-cc V-8 in the Longchamp. Leyland Innocenti S.p.A. was put in liquidation in November 1975. With a capacity of 40,000 cars a year in its factories at Lambrate, Milan, it was a bigger piece of industrial property than De Tomaso had ever touched. He talked it over with Romano Prodi and GEPI stepped in with a $140 million rescue plan. He continued production of the Innocenti Mini, converted parts of the Lambrate works to motorcycle production, and moved the final assembly of Maserati cars from Modena to Lambrate. In 1981 the Maserati Biturbo replaced all other models. The Mini became available with 2- and 3-cylinder Daihatsu engines (made under license by Moto Guzzi). In 1986 he won a contract with Chrysler to build 10,000 Maserati-labeled convertibles over five years, and introduced a new four-door Maserati. Deliveries to Chrysler fell far short of schedule and Chrysler canceled the deal on September 1, 1988. That started the decline of De Tomaso's affairs. Fiat S.p.A. and Mediobanca took a 49 percent stake in Maserati in 1989 and a 51 percent stake in Nuova Innocenti in 1990. Maserati assembly was taken out of Lambrate and put back in Modena. De Tomaso sold his motorcycle companies in 1989. On January 22, 1993, he suffered a cerebral hemorrhage from which he never made a complete recovery, and on July 12, 1995, he gave up all his business functions. He died on May 21, 2003.

1. Ceballos was his mother's maiden name.
2. Officine Specializzate Costruzioni Automobili, owned by the Maserati brothers.
3. Founded by A.L. Haskell, a former GM executive, and father of Isabel De Tomaso. A Rowan subsidiary, De Tomaso Industries, Inc., was set up in 1970 to coordinate the activities of De Tomaso Automobili, Ghia S.p.A., Vignale S.p.A., De Tomaso of America, Inc., of Livonia, Michigan, and Autosports Products, Inc., also in Livonia.
4. Gestione a Participazioni Industriali, established in 1970 with IMI subscribing 50 percent of the capital, and smaller stakes by IRI (Istituto per la Ricostruzione Industriale) and ENI (Ente Nazionale Idrocarburi) the state-owned petroleum giant.
5. Istituto Mobiliari Italiano, a state-owned credit organization.

de Turckheim, Adrien G. *see* Turckheim, Adrien G. de

Deutsch, Charles (1911–1980)

Engineering director of DB sports cars from 1936 to 1962; creator of the 1962–64 CD-Panhard and the 1965–67 CD-Peugeot.

He was born on September 6, 1911, at Champigny-sur-Marne as the son of a modest cartwright and wagon builder. A good pupil in school, he also spent a lot of time in the parental shop, learning the use of tools. He made pocket money by painting license plates for cars. He was in his early teens when his father died, forcing him to share part of every day between homework and helping his mother run the business. He was accepted by the Ecole Polytechnique and graduated in 1930 with a degree in road-and-bridge construction. He went on to France's national school of bridges and highways[1] and the school of applied engineering.[2] In 1935 he went to work for the government administration of bridges and highways, but the only available post was that of inspector of navigation on the Seine River. He accepted, and sold the parental shops to a local mechanic and Citroën dealer, René Bonnet. Together they made the first Citroën-based DB sports/racing car in 1936.

Continuing in his government post, he found time to study at the Ecole Supérieur d'Electricité. In 1947 Deutsch and Bonnet chose the Dyna Panhard as the basis for a new generation of sports cars. He left the inland waterways administration in 1951 and became attached to the liquid fuel depot service of the department of public works, designing new DB cars in his spare time. From 1957 to 1963 he served as joint managing director of the Trapil[3] pipeline corporation. The partnership with René Bonnet broke up in 1962 and he designed a radically new CD-Panhard car. In 1964 he became president of a liaison group for equipment used in the petroleum, petrochemical and national gas industries, concurrently heading a company engaged in automotive research and petroleum-transport technical office.[4]

His next car, the CD-Peugeot of 1965, used the 204 power train in an inboard–rear transverse installation and the most advanced aerodynamics (by Robert Choulet). He founded SERA[5] which became consultant to Matra, Porsche, Ligier, Alfa Romeo and Lola Engineering. He designed the CD-GRAC with a four-cylinder Alfa Romeo engine, but the project was abandoned in 1966. Ten years later he prepared a small single-seater based on Renault components, aimed at starting a new racing class to train the youngest drivers. He never completely retired, but fell ill and died on December 7, 1980.

1. Ecole Nationale des Ponts et Chaussées.
2. Ecole d'Application du Génie.
3. Société du Transport Pétrolier par Pipeline.
4. Société d'Etude Pétrolière Marine.
5. Société d'Etudes et Réalisations Automobiles.

Dick, Alick Sydney (1916–1986)

Managing director of the Standard Motor Company, Ltd., from 1954 to 1959, managing director of Standard-Triumph International, Ltd., from 1959 to 1961.

He was born on June 20, 1916, as the son of a medical doctor at Massingham, Norfolk. He was educated at Chichester High School and the Dean Close School at Cheltenham, and served an apprenticeship with the Standard Motor Company, Ltd., from 1934 to 1938. He was invited to stay on and was named chief buyer for Standard's Shadow Factory in Banner Lane, which was making Bristol Hercules aircraft engines. In 1942 he was placed in charge of production control for both the motor vehicle plant at Canley and the Banner Lane plant. In 1945 he was promoted as John Black's personal assistant, and was given a seat on the board of directors in 1947. New titles followed — first, assistant managing director and then deputy managing director. In 1953 he negotiated the sale of Banner Lane to Massey-Harris for £12½ million. On Sir John Black's retirement in 1954, he smoothly took over his office. He started merger talks with executives of the Rover Company, Ltd., but George Farmer cut them off. In 1954 he arranged the purchase of Mulliner, Ltd., of Birmingham, and Hall Engineering (Holdings) Ltd. In 1956 he secured ownership of Bean Industries, Tipton, Birmingham, to get their foundries for Standard, and took over Alford & Alder, Ltd., of Hemel Hempstead, makers of chassis components and sub-assemblies. He brought modern management to Standard-Triumph but failed to master cost-accounting and plunged the company deeply into debt. He successfully negotiated its takeover by Leyland Motors, Ltd., in 1961, only to be purged in the reorganization that ensued, losing his office to Stanley Markland.

Diétrich, Eugène De *see* De Diétrich, Eugène

Di Giusto, Nevio (1953–)

Head of the Lancia design studio from 1985 to 1988; chief engineer of the Lancia body design office from 1988 to 1990; head of Fiat's styling center in 1991; assistant to Mario Maioli in 1992; director of Fiat Auto S.p.A. body engineering from 1993 to 1997; director of Fiat, Alfa Romeo and Lancia platform development from

1997 to 2001; director of product engineering, Fiat Auto S.p.A. beginning in 2001.

He was born on July 1, 1953, at Magnano in Riviera, Udine province, and obtained a doctor's degree in aeronautical engineering from the Turin Polytechnic in 1977. He delivered an experimental thesis on "Winged structures obtained by aluminum extrusion," and was engaged by Fiat as a research engineer. His first task was to propose alternative materials for body structure and mechanical components. From 1979 to 1982 he was head of the wind tunnel at Fiat's Orbassano research center, and in 1982 he was placed in charge of Vehicle Architecture and Innovation as a long-term planning function. He was responsible for the bodies of the 1989 Lancia Dedra (derived from the Fiat Tempra) and the 1993 Lancia Delta. He supervised the body engineering for the Fiat Bravo/Brava and Marea, Alfa Romeo 156, 166, and 147, and the Lancia Lybra and Thesis.

Dion, Wandonne de Malfiance de (Jules-Félix Philippe Albert) *see* de Dion, Wandonne de Malfiance (Jules-Félix Philippe Albert)

Di Virgilio, Francesco (1911–1995)

Design engineer of Lancia e C. from 1945 to 1975.

He was born on December 23, 1911, in Reggio Calabria as the son of a civil engineer. After local schooling and technical training, he went north to study mechanical engineering at the Turin Polytechnic. He began his career with Lancia e C. on February 1, 1939, and became an engine test engineer in 1940. He worked on the development of experimental engines during World War II and then became a draftsman. He was the principal designer of the V-6 engine for the Lancia Aurelia and its D-20, D-23 and D-24 derivatives. He was also active in chassis engineering, notably for the Flaminia. He worked on the Flavia and the Fulvia under Fessia's direction, and led the redesign of their engines with

bigger dimensions, fuel injection, and electronic ignition. He was an assistant to Sergio Camuffo in the development of the Lancia Beta including the "Montecarlo," and then retired. He remained a consultant to Lancia and Alfa Corse until 1981, and died on August 5, 1995.

Docker, Bernard (1897–1978)

Chairman of the board, the Daimler Group, Ltd., from 1941 to 1956.

He was born in London in 1897 as the son of F. Dudley Docker, and educated at Harrow School. He began his career with the Metropolitan Carriage, Wagon & Finance Company in Birmingham and worked on the assembly of battle-tanks from 1915 to 1918. He spent most of the year 1920 on company business in South Africa, and was elected chairman of the Metropolitan Carriage Company, Ltd., upon his return. His father put him on the Daimler board of directors in July 1940, and he was elected chairman of the Daimler board on December 22, 1941. In 1944 he succeeded his father on the board of the Birmingham Small Arms Company, Ltd., and was elected chairman. During the postwar years he also sat on the boards of the Midland Bank and the Guardian Assurance Company. His value to Daimler was mainly ambassadorial, maintaining ties with British royalty and foreign dignitaries as buyers of Daimler cars. His career was ruined by his notorious extravagance in having special cars built for his wife, the former Norah Collins, née Turner, whom he married in 1949 and put on the payroll as a director of Hooper & Co. (Coachbuilders) Ltd. He was ousted by Daimler and BSA directors led by Jack Sangster in May 1956. He retired to Rozel, Jersey, and died at Bournemouth on May 22, 1978, after a long illness.

Docker, Frank Dudley (1852–1944)

Deputy chairman of the Birmingham Small Arms Company, Ltd., (BSA) in 1909; managing director of the Daimler Company, Ltd., from 1910 to 1912; member of the board of directors, BSA, from 1911 to 1944.

He was born in 1852 at Smethwick as the son of a prominent Birmingham lawyer. He studied law but chose a career in business and finance. He pulled off his biggest coup in 1902 by merging five rolling-stock makers into one, which he named the Amalgamated Carriage & Wagon Company. In 1906 he was invited to take a seat on the BSA board of directors. The company was then reviving its bicycle production, and in 1907 purchased a disused munitions plant at Sparkbrook, Birmingham, to set up an automobile department under Colonel E.E. Baguley. The first BSA motorcycle was built in 1909 and production began at Small Heath in 1910. The automobile department fell deeper and deeper into deficit, and Docker had the idea of merging it with an established car manufacturer. His choice fell on Daimler in Coventry, and BSA purchased the Daimler company and the Motor Mills for £600,000 on September 27, 1910. Car production at Sparkbrook was halted, and from 1912 to 1914 a new BSA model, powered by a four-cylinder Daimler sleeve-valve engine, with a 3-speed gearbox enclosed in the rear axle, was assembled in the Motor Mills. In 1913–14 the same chassis was also supplied to Siddeley-Deasy, who marketed it as the Stoneleigh. In 1910 the Motor Mills began building KPL[1] gasoline-electric buses, and in 1912 began supplying sleeve-valve bus engines to AEC.[2] After 1912, Docker left the Daimler executives with a free hand, but he had a powerful voice in the appointment (and dismissal) of Daimler personnel. BSA resumed car production in 1920 with a 10-hp model powered by a Hotchkiss air-cooled V-twin, adding a four-cylinder 11-hp and a six-cylinder 12-hp model with Daimler sleeve-valve engines. The 10-hp model was discontinued in 1924, after W.R. Morris had bought the Hotchkiss engine plant and stopped making the V-twin. BSA then introduced a 14-hp car with a four-cylinder Daimler sleeve-valve engine, and BSA car production was transferred from Birmingham to Coventry, but stopped in 1926. The motorcycle subsidiary, BSA Cycles, Ltd., began building a light front-wheel-drive three-wheeler with its own V-twin engine at Small Heath in 1929. The V-twin was replaced by a water-cooled in-line BSA four for 1933–34, and then the three-wheelers were replaced by the four-wheeled Scout, assembled at the Daimler works. In 1931 Docker was one of the first to see the need for Daimler to offer a line of lower-priced cars, which led to the purchase of the Lanchester Motor Company, Ltd., for £26,000, and the transfer of Lanchester car production to the Daimler works. In 1940 he put his son on the Daimler board, and in spite of his advanced age, maintained his own seat on the BSA board of directors until his death in 1944.

1. Kriéger-Pieper-Lancheser. Louis Kriéger of Paris was a pioneer of electric vehicles, Henri Pieper had been building hybrid-drive gasoline-electric vehicles in Liège since 1901, and F.W. Lanchester designed a bus chassis.
2. Associated Equipment Company, Ltd., was a joint venture between the London General Omnibus Company and the Daimler Company.

Dombret, Emile (1876–1934)

Managing director of SA des Automobiles Motobloc from 1904 to 1987.

He was born on February 1, 1876, in St. Etienne and began his career with a local manufacturer of gas engines. In 1902 he, joined his brother-in-law, Charles Schaudel, in Bordeaux. Schaudel, a gunsmith by trade, had been producing small numbers of bicycles since 1895, and in 1897 built a prototype car with vis-à-vis seating for four, wire wheels, tiller and stub-axle steering. The 3 CV parallel-twin was mounted transversely below the back seat, with a three-speed gearbox and chain drive. Together they designed and built a car with a front-mounted slanted twin and chain drive to the rear axle. Dombret managed to raise capital to start making engines, and established G. Carde Fils & Cie in Bordeaux in 1903, which became SA des Automobiles Motobloc in 1904. Schaudel had invented an in-line four-cylinder engine with its flywheel sitting in the middle of the crankshaft. A compact, common housing for the clutch and gearbox, was bolted to the block ("bloc-moteur" giving rise to the Motobloc trademark). Production of four-cylinder Motobloc cars began in a large factory at Bordeaux–La Bastide in 1904. Shortly afterwards, Charles Schaudel left to become technical director of the Manufacture d'Armes et de Cycles at St. Etienne. By 1907 Motobloc was offering a range of modern-looking, powerful touring cars, the H-type 16/22 (built from 1906 to 1912), the K-type 24/30, the M-type 30/40, and the 12-liter P-type 70-hp model. From 1909 to 1914, the company concentrated on mid-range cars (10, 12, 16 and 20 hp) with F-head four-cylinder engines. Motobloc never abandoned the patented flywheel mounting between each pair of cylinders, though flywheel size, clutch and gearbox design evolved over the years. During World War I, Motobloc began by machining shells aid making aircraft parts, later making complete artillery shells, and in 1917–18, producing the 250-hp air-cooled radial nine-cylinder Salmson engine. The 12/22 was put back in production in 1920 to be joined by a 15/30 in 1922. Dombret resigned in 1927 and was engaged to design a 5 CV economy car for Sima-Standard, which went into production at Courbevoie in 1929. He watched from a distance as Motobloc made its last cars in 1930, and died in 1934.

Donnet, Joseph Albert Jerome (1885–1953)

Founder and president of SA des Automobiles Donnet-Zédel from 1924 to 1928; founder and president of SA des Automobiles Donnet from 1928 to the end of 1934.

He was born on February 10, 1885, at Trois-torrents in the Valais canton of Switzerland, as the son of a local farmer. His parents had ambitions for him in the tourist business and sent him to hotel school in Lausanne. He was in full agreement since the life of a farmer held no attraction for him. He went to Denmark as a hotel management trainee, married a local divorcée, and in 1907 secured the Renault franchise for the Nordic countries. He also set up a Renault dealership in Paris, and by 1910 he was maintaining homes in Copenhagen, Paris, Cabourg on the Channel coast, and Monte Carlo. In 1911 he launched himself into the aircraft industry in partnership with M.

Lévèque, and from 1915 to 1918 their Société FBA factory at Vernon on the Seine produced an average of five airplanes a day. One of the FBA models was a flying boat, powered by Clerget engines.

After the war, he began looking for a civilian-market activity, and in 1919 began negotiations with Automobiles Zédel of Pontarlier. He began investing in the Zédel enterprise and in 1924 rented the former Vinot & Deguingand factory at Nanterre, to which he transferred the production of Zédel cars, renamed Donnet-Zédel. In 1928 he purchased a property at Nanterre and erected a five-story factory to build cars wearing the Donnet label.

He tried to exploit the market for a very low-priced sports car, and he tried to cash in on the front-wheel-drive trend, but the economic climate defeated his cash reserves and resourcefulness. In 1935 he sold the Automobiles Donnet assets to H.T. Pigozzi (Simca) and purchased Société Cuttat (machine tools) from Emile Dumaine (maker of the Clerget engines).

He fled from France when World War II was declared in the autumn of 1939 and settled in Geneva, where he died in 1953.

Doorne, Hubertus Josephus "Hub" Van *see* Van Doorne, Hubertus Josephus "Hub"

Doriot, Auguste (18??–19??)

Works manager of Doriot, Flandrin & Cie from 1906 to 1908; managing director of Doriot, Flandrin et Parant from 1908 to 1988.

He began his career with Peugeot as a bicycle mechanic and became a foreman in 1890. He was associated with the design and construction of Peugeot cars from the very beginning, acquiring all-around experience of all aspects of making automobiles.

In 1906 he formed a partnership with Ludovic Flandrin, whom he had met when both were working for Peugeot. Flandrin also had engineering experience from Clément-Bayard. They moved into a small factory in rue Jules Ferry at Courbevoie and began building DF automobiles. The trademark was changed to DFP in 1908, when Jules-René Parant and Alexandre Parant became partners in the business. In 1912 they moved into a bigger factory in Boulevard St. Denis, Courbevoie, and set up their own engine plant. Earlier DF and DFP had been equipped with Chapuis-Dornier engines. All DFP engines were side-valve four-cylinder units, the E4 8/10 CV, the A 2000 10/14 CV and the DF 12/18. A six-cylinder Model G was developed, but produced only in limited numbers. The engine plant was converted to make aircraft engines during World War I, and idled in 1919. The A 2000 was put back in production with a 2-liter Altos engine and continued through 1923. In 1921 it was joined by the lighter EM, powered by a 1½-liter Sergeant engine. DFP got the engine plant working again in 1923, and introduced the V-type voiturette. The EM continued, and the ABM was added for 1924. DFP's biggest problem was high production cost, which meant that its cars were overpriced, and the sales curve plunged. The company fell into receivership in 1926 and continued to produce the VA 7/22 and ADM 11/30 CV models until 1928 when the factory was closed.

Douin, Georges (1945–)

Chief engine designer of Renault's passenger-car division from 1982 to 1985; director of mechanical design for Renault cars from 1985 to 1987; director of Renault vehicle engineering from 1988 to 1993; director of Renault's office of planning, projects and products from 1993 to 1997; director of Renault's international division from 1997 to 2000; joint managing director of Renault's car division and director of planning, product and international operations beginning in 2000.

He was born on July 5, 1945, at Saint-Florentin (Indre) and graduated from the Ecole Polytechnique at the precocious age of 19. In 1967 he joined Renault as a trainee engineer and was assigned to the research and development staff at Rueil. He worked in body engineering from 1969 to 1972 and in testing from 1972 to 1976. He worked for two years as as-

sistant to the director of the technical center at Rueil and was manager of Berex[1] at Dieppe, a Renault subsidiary for exploratory research and design from 1978 to 1982. He took charge of the Turbo-5 engine development and the EVE power train (low-emission, low fuel consumption). He upgraded Renault's basic engine program with the overhead camshaft F-series (1400 to 1900 cc) and E-series (1000 to 1400 cc), adding a 16 valve F-series unit and an F-series diesel. He directed the engineering of the Clio and Twingo, and the planning of the Safrane, Laguna, and Mégane.

1. Bureau d'Etudes et de Recherches Exploratoires.

Drew, Harold (1901–)

Assistant chief engineer of Vauxhall Motors, Ltd., from 1930 to 1950; chief engineer from 1950 to 1953.

He was born on December 1, 1901, in London, and educated at Bradfield College and the City & Guilds Engineering College. He held an engineering degree from the University of London.

He began his career as a draftsman with Sunbeam in Wolverhampton, but left in 1925 to join General Motors in London. GM sent him as a trainee to Oldsmobile in the spring of 1926, and when he returned 18 months later, he became an engine designer with Vauxhall. From May to August 1928, he was working in Detroit on the design of a new Vauxhall engine, a 2-liter six that went into the 1931 Cadet. He worked on the engines for the Big Six and the Light Six, and worked on both design and development for the 1935 14 and 20. He was involved with every new car project. During World War II, he led the design and development of the Churchill tank.

After the war, he was a member of the team that planned the 1951 Wyvern and Velox as well as the 1954 Cresta. In February 1954 he moved to Detroit as assistant chief engineer of GM Overseas Operations, preparing new models for Holden, Opel and Vauxhall, and was promoted to chief engineer in 1956.

Dreyfus, Pierre (1907–1994)

Chairman of the Régie Nationale des Usines Renault from 1955 to 1975.

He was born on November 18, 1907, in Paris, as the son of a banker. He studied finance, political science and law, ending up with a doctor's degree in Law. He began his career in the office of the budget director in the Finance Ministry in 1935, transferring a year later to the Ministry of Commerce and Industry as a technical advisor. In 1945 he was head of a government committee for "orphan" enterprises, such as Renault, Berliet and Lavalette, and was named vice president of Renault in 1948. In 1949 the government placed him in charge of the Lorraine Coal Mines[1] which he ran until Renault needed a new chairman to succeed P. Lefaucheux. He immediately began to decentralize Renault's production facilities, securing a site at Cléon near Rouen for an engine-and-gearbox plant in 1956. In 1958 he signed an agreement with Alfa Romeo for assembly of the Dauphine in Milan and production of a Renault truck-diesel engine at Naples. In 1961 he authorized the construction of a new assembly plant at Sandouville near Le Havre. In 1966 he signed an agreement with Peugeot for cooperation at several levels, including a joint engine-production venture at Douvrin.[2] He created RIET[3] and RMO[4] for manufacturing technology and automation.

He bought subsidiaries that Renault had no need for, such as Bernard Moteurs (small industrial engines) and Motobécane (mopeds and light motorcycles). He approved plans for another assembly plant in northern France, at Douai, which came on stream in 1971. He secured full control of Alpine in 1973 and Berliet in 1974. He began preparing the final closing of the Billancourt works before his contract ran out at the end of 1975. He served as Minister of Industry for about a year in the 1981–83 Mauroy government, then went into retirement and died in the last week of 1994.

1. Houillères de Lorraine.
2. La Française de Mécanique.
3. Renault Industrie Equipement et Techniques.
4. Renault Machines-Outils.

Dubonnet, André (1897–1980)

Head of Etablissements André Dubonnet from 1928 to 1938.

He was born in 1897 as the son of Marius Dubonnet, maker of the eponymous apéritif wine. He was privately educated and served as a fighter pilot in the Escadrille des Cigognes from 1916 to 1918, with five kills to his credit. After the war, he turned to motor racing and was a member of the Duesenberg team, taking fourth place in the Grand Prix of the Automobile Club de France at Le Mans in 1921. In 1924 he bought a Hispano-Suiza chassis which he equipped with a remarkable tulip-wood body by the Nieuport aeroplane company. He took a keen interest in chassis engineering and independently arrived at a rationale for very stiff frames and soft springs. He had formed a friendship with Gustave Chedru, who had been on the Richard-Brasier design team in 1903–04 and designed the 1924 Françon car, and also had new ideas about suspension systems. Dubonnet rented a workshop and drawing office at Courbevoie. They developed an intricate suspension system with oil-damped coil springs enclosed in cylinders, and wheel hubs carried on leading or trailing arms, for which French patents were issued in 1931. They built several demonstration cars with all-independent suspension and Hispano-Suiza engines, and took one on a tour of Detroit, which resulted in a contract with General Motors. That led to the adoption of "knee-action" front suspension on the 1934 Chevrolet and Pontiac, followed by Opel and Vauxhall in 1936. Fiat and Alfa Romeo also bought licenses for Dubonnet suspension, designed their own variations, and used them on both racing cars and production models. Dubonnet discovered streamlining about 1931. His starting point was an airplane fuselage on four wheels and the engine in an inboard/rear position. The "Narval" prototype was ready in 1936, with a Ford V-8 engine, Dubonnet all-independent suspension, side doors for the back seat, and access to the front seat by an opening front panel. In 1938 he designed a streamlined body for the 1931 Hispano chassis and had it built by Saoutchik, with such innovations as curved glass and sliding doors. He remained an active rally driver up to 1953 and was still driving a Chevrolet Corvette in everyday use until he was partly disabled in a traffic accident when touring in Spain in 1977. He died in Paris in January 1980.

Duckworth, David Keith (1933–)

Chairman and technical director, Cosworth Engineering, Ltd., from 1958 to 1998.

He was born in 1933 at Blackburn, Lancashire, as the son of a cotton-mill owner and was educated at Giggleswick School, Settle, from 1942 to 1951. He built model aircraft complete with radio control systems as a boy of eight and bought his first motorcycle at the age of 14. He joined the RAF and went in for pilot training, but failed his night-flying test and served his term as an aircrew member. He studied engineering at London University's Imperial College, South Kensington, graduating in 1957. He joined Lotus Cars, Ltd., as a gearbox development engineer, where he met Mike Costin. They agreed to go into business together and in 1958 established Cosworth Engineering, Ltd., with a capital of £100, working in the back yard of the Station Hotel, Friern Barnet Lane, New Southgate, London. The company made its income on tuning Coventry Climax engines and maintenance work on racing cars for private teams. He designed a Formula Junior car in 1959 and made the first Cosworth engine, using the 997-cc block from the Ford Anglia's "Kent" engine. Colin Chapman adopted it for the Lotus 18. In 1960 Cosworth moved into the racing car shops of Lotus Cars, Ltd., at Edmonton, London, as soon as Lotus Cars, Ltd., had gone to Cheshunt. Ford Motor Co. Ltd. gave Cosworth a contract for a single-overhead-camshaft Formula 2 engine, and the SCA dominated Formula 2 racing in 1964–65. In 1966 he won a Ford contract for a twin-cam Formula 2 engine, the Cortina-block FVA, and a Formula 1 3-liter V-8, the DFV with four valves per cylinder. The FVA powered cars of several makes to a total of 155 Grand Prix victories. Cosworth Engineering, Ltd.,

moved to a bigger plant on St. James Mill Road, Northampton in 1964–65. He designed a twin-cam head for the Chevrolet Vega 2.3-liter engine and began production in 1972. In 1978 he introduced the KAA engine for the Opel Ascona Rallye and opened an aluminum foundry at Worcester, specializing in cylinder heads. In 1985 he won a contract for 5,000 turbocharged 2-liter engines for the Ford Sierra Cosworth and set up an engine factory at Wellingborough. By 1990 Cosworth had created 55 different engines. To raise capital for these expansions, Cosworth issued new stock and majority control was ceded to Carlton Communications, Ltd., who sold its stake in Cosworth to Vickers, Ltd., for $278 million in 1990. Cosworth Engineering, Ltd., later became a group with its own subsidiaries: Cosworth Technology, Cosworth Castings, and Cosworth Racing. The Wellingborough plant was expanded and became the sole source of 2.3-liter 16-valve engines for the Ford Scorpio and Galaxy. In July 1998 Cosworth Technology was sold to Audi AG for $191 million, and Cosworth Racing was taken over by its biggest client, Ford Motor Co., Ltd.

Dufresne, Henri (1884–1950)

Assistant technical director of Citroën from 1930 to 1940; director of quality control for Peugeot from 1945 to 1950.

His career began in 1906 with an engineering post in the Brouhot factory at Vierzon, then producing a line of high-powered top-quality touring cars in modest numbers. In 1909, he moved to Paris as a tool engineer with Automobiles Mors, but left in 1911 when offered the title of chief engineer of Martini's automobile department in Neuchatel, Switzerland, where he directed the design of three new models from 12/16 to 25/35 hp. He returned to Paris in 1913 when André Citroën offered him the office of chief engineer at Automobiles Mors, whose car production was suspended during World War I. He worked in Citroën's munitions factory from 1915 to 1918, and organized a methods office for automobile production at the Quai de Javel in 1919–20.

During his years with Citroën, he made major contributions in the areas of quality and productivity, and Citroën sent him on missions to Detroit no less than six times. He went to Peugeot in 1940, spent the war years on a variety of assignments, and supervised the postwar return to car production at Sochaux. The high quality of the first new postwar car, the 203, won him great respect in the industry. He met an untimely death on November 26, 1950, following a surgical operation.

Dufresne, Louis (1890–1958)

Head of advanced design for Peugeot from 1927 to 1945; director of Peugeot's engineering center from 1950 to 1958.

He was born on June 20, 1890, at Bourgen-Bresse and graduated from the Ecole Centrale des Arts et Manufactures in 1913. He began his career as an engine designer with Panhard, where he worked on sleeve-valve engines for cars and overhead-camshaft aircraft-engine projects. In 1917 he was engaged, along with Ernest Artaut, by André Citroën to design a high-grade touring car. They created a big car with a six-cylinder sleeve-valve engine. Citroën shelved it in order to start production of a medium-size car designed by Jules Salomon. Dufresne took the 18 CV six-cylinder project to Voisin, who engaged him in 1920, and for a year he worked in the same office with Marius Bernard and André Lefebvre. He was signed up by Peugeot to design two sleeve-valve engines, 14 CV and 18 CV, for the Economy Run[1] at Tours in 1922. For the next ten years he planned, designed, tested and developed all Peugeot competition cars, including the 174 S that won the Coppa Florio in 1925 and the 24-hour race at Spa the same year. After 1932 he made an effort to get the industry as a whole to adopt government-approved standards, founded an association to put automobile engineering into the curriculum of technical schools, and authored a guide book for designers that was still in everyday use at Peugeot in 1966.

Dunn, Michael Donald David (1935–)

Director of product development, Ford Motor Company, Ltd., from 1980 to 1983; engineering director of Rolls-Royce Motor Cars, Ltd., from 1990 to 1993.

He was born on October 2, 1935, in Coventry, as the son of William M. Dunn, chief designer of Alvis cars. He was educated at Bablake School, Coventry, and studied engineering at Birmingham University and Sheffield University, graduating with a master's degree. He began his career by joining Ford Motor Co., Ltd., in 1955, but left in 1962 to become chief engineer of Alvis, Ltd. He put power steering and automatic transmission on the 3 Litre, and developed the TF-21 engine to an output of 150 hp. Production of Alvis cars stopped in 1967, and he signed up with Leyland Motors, Ltd. In 1970 he became chief engineer of Leyland's truck and bus division, only to return to Ford in 1973 as chief engineer of research. A year later, he was placed in charge of Ford's heavy truck drawing office, and in 1977 was transferred to Merkenich near Cologne as head of Ford Advanced Technology. He returned to Dagenham in 1980, took charge of the Sierra program, and the next-generation Escort/Orion. In February 1983 he joined Rolls-Royce Motors and developed automatic ride control for the Silver Spirit and Silver Spur. He became the company's engineering director in 1990 but resigned in 1993.

Durin, Michael (1928–)

Group technology director, Peugeot SA, from 1981 to 1996.

He was born on January 25, 1928, in Clermont-Ferrand as the son of a Michelin engineer. He was given a university education and graduated with a doctor's degree in science. He began his career by joining Citroën in the research department on October 1, 1955, followed by career stages in the methods department, in the laboratory, and in design engineering. He was in charge of tooling and methods for the DS-21 engine which went into production as a 1965 model. He was promoted to chief engineer of engine testing in 1964 and director of car testing in 1966. In 1968 he was named director of "Methods" (manufacturing, processes, assembly) and led the teams that laid out the transmission plant at Metz, the foundries at Charleville-Mézières, the engine factory at Trémery, and the final-assembly plants at Rennes, Aulnay-sous-Bois, and Vigo in Spain. He also devised the methods and tooling for Comotor (Wankel) engine production at Altforweiler in the Saarland. He was transferred to Peugeot, SA, at group level as director of research in 1980 and technical director in 1981. His biggest accomplishment was to streamline Peugeot's way of creating new models. Up to 1990, new Peugeot cars were designed and finished prototypes made at La Garenne, then handed over to the production engineers at Sochaux, who designed the same car all over again so that they could built it most efficiently. This was costly and time-consuming. Durin applied Citroën's way of involving the production engineers at the inception of new vehicle projects, beginning with the Peugeot 406, which went into production at Sochaux at mid-year, 1995.

Dürkopp, Ferdinand Robert Nikolaus (1842–1918)

Chairman of Dürkopp-Werke GmbH from 1899 to 1913; chairman of Dürkopp-Werke AG from 1913 to 1918.

He was born in 1842 at Bielefeld in Westfalia as the son of a mechanic. In 1867 he took over his father's shop and reorganized it as Bielefelder Maschinenfabrik AG vorm. Dürkopp & Co. He started production of sewing machines, added ball bearings and bicycles to his product line in 1881, and in 1886 began to build gas engines. In 1894 he reached a license agreement with Panhard & Levassor for their car design (but not the Daimler engine). Dürkopp made test cars with 1-, 2- and 3-cylinder engines of German design. In 1897 he set up a separate automobile department under the direction of George Hartmann, and in 1899 introduced two models with front-mounted parallel-twin engines, a 6/7-hp and a 9/10-hp car. That year Dürkopp also made a smaller car with a rear-mounted horizontal flat-twin and two-speed gearbox. In 1901 the factory made its first four-cylinder car and truck models, and the first six in 1902. In 1904 he took over Karosseriefabrik Wiemann & Co. in Magdeburg and in 1906 introduced the Knipperdolling light car designed by Nikolaus Henzel, with a choice of 6/12 or 8/15 power unit. In 1910 he took over the Oryx-Werke of Berlin-Reinickendorf, which was reorganized as Oryx Motorenwerke. During World War I the factory made important numbers of 4-ton military trucks.

Dusio, Piero (1899–1975)

Chairman of Cisitalia Automobili S.p.A. from 1946 to 1949.

He was born on October 13, 1899, at Scurzolengo d'Asti in Piedmont. With only a grade-school education, he went to Turin to seek his fortune. He became a salesman for a textile company, and by 1926 he was the owner of an oilcloth factory at Racconigi. He bought a Maserati and started racing, with some success. He began to make sports clothes and military uniforms, and started a side line in machine tools and garage equipment. In 1943 he combined his holdings into the Consorzio Industriale Sportiva Italia (CIS-Italia) and in 1944 began making plans to build racing cars in shops at Corso Peschiera, Turin. The first model was an 1100-cc single-seater designed by Dante Giacosa, with a tuned Fiat engine and lots of Fiat parts. It jump-started the revival of motor racing in postwar Italy. He wanted a touring car, and Giovanni Savunuzzi designed the Cisitalia 202 chassis in 1946, using a maximum of Fiat components. Dusio nurtured ambitions to compete in Grand Prix racing, and bought a design from Porsche, with a supercharged 1½-liter flat-12 and four-wheel-drive. But his cost estimates were far below reality, and Cisitalia Automobili S.p.A. went into liquidation in January 1949. Dusio went to Argentina with the Grand Prix test cars and plans, which President Juan Domingo Peron agreed to support, to start an auto industry in that country. He became an executive of Autoar SA Industrial, Comercial, Financiera y Mobiliara, which had a plant at El Tigre, outside Buenos Aires, and late in 1950 began building cars and off-road vehicles in cooperation with Willys-Overland. In 1951 he set up a private company, Cisitalia Argentina SA, as machine-tool importers. It expanded in 1956, with a big factory at Matedero, to assemble farm tractors and cars from parts kits arriving from Italy. Operations remained modest, however, until 1960, when the government stepped in, formed a joint venture with Fiat Argentina SA, and launched a new range of Cisitalia cars. But the business closed in March 1965. Dusio remained in Argentina and died there in November 1975.

Dyckhoff, Otto Eduard Maria (1884–1947)

Works manager of Hanomag from 1921 to 1930; technical director of Wanderer's automobile department from 1930 to 1935; production manager of Adam Opel AG from 1935 to 1937; production expert with Dr. Ing. F. Porsche Konstruktions-GmbH from 1937 to 1942.

He was born in 1884 and had a technical education. During World War I he was a production engineer with Hansa-Lloyd Werke in Bremen, and in 1921 he was engaged by Hanomag to organize tractor production. In 1924 he laid out the assembly hall for the Hanomag 2/10 PS economy car, with a moving assembly line. He modernized all of Hanomag's production facilities and became widely known as one of Germany's foremost automotive production engineers. In 1930 he joined Wanderer at the Chemnitz works, which he retooled for new models. Beginning in 1932, he coped with Wanderer's integration into Auto Union AG's production plans. In 1935 he was engaged by Adam Opel AG, with a big increase in salary, and he reworked the Rüsselsheim plants so well that they turned out over 125,000 cars in 1937. The Porsche organization tempted him with their plans for mass-producing the KdF-Wagen, and he went on the Porsche payroll, attached to Gezuvor.[1] In the summer of 1937 he was a member of a Gezuvor delegation to Detroit, along with Dr. Ing. F. Porsche and his son "Ferry," Dr. Anton

Piëch and Bodo Lafferentz, to order machinery and hire German-speaking production men with experience in America's auto industry. Dyckhoff was given responsibility for planning the shop layouts, materials flow, sequence of operations, purchasing and installation at Wolfsburg. He also had to organize the conversion of the plant to war production, but using his age as a pretext, he managed to disengage himself from Wolfsburg in the middle of World War II, and retired to private life.

1. Gesellschaft zur Vorbereitung des deutschen Volkswagens.

Eberan von Eberhorst, Robert Emmerich Manfred (1902–1982)

Auto Union design engineer from 1937 to 1940; chief engineer of ERA Ltd. in 1949–50; chief engineer of Aston Martin Ltd. from 1950 to 1953; technical director of Auto Union GmbH in 1956.

He was born on April 4, 1902, in Vienna and graduated with a degree in mechanical engineering from the Vienna Technical University. He continued his studies at the Dresden Technical University, graduating with a doctor's degree and then became an assistant to Professor Wawrziniok in his automobile research institute in Dresden in 1927. From 1928 to 1933 he worked as a scientist in the Saxony State Test Establishment for Motor Traffic[1] and then joined Horch in Zwickau as an engine development and test engineer. He led the design of the Auto Union D-Type racing car with its V-12 supercharged 3-liter engine and de Dion rear suspension. He also worked on the development of a 1500-cc racing car engine in 1939–40. During World War II he lectured on automotive fuels and lubricants, engine and chassis design at the Dresden Technical University. In 1947 he joined the Porsche organization and worked on calculations for the four-wheel-drive 12-cylinder Cisitalia racing car project. He was engaged by Leslie Johnson, managing director of ERA Ltd. at Dunstable in 1949 and designed a roadster with a tubular steel frame, a design which was first known as the ERA Jupiter, and then became the Jowett Jupiter production car. David Brown brought him to Aston Martin in 1950, where he was responsible for the design of the DB 3 sports/racing car. He also designed a Lagonda sedan, Project 117, with all-independent suspension, only to find that David Brown was not ready to put it in production. He went back to Germany late in 1953 and joined Auto Union GmbH in Düsseldorf, working as a technical assistant to Fritz Zerbst. When Zerbst retired in 1956, Eberan was named technical director. But he resigned in October 1956 to take up a position as head of the mechanical engineering department at the Battelle Institute in Frankfurt am Main, where he remained for 18 months. He then returned to Austria and became head of the Institute for Internal Combustion Engines[2] at the Vienna Technical University. He spent the final years of his career searching for cleaner combustion processes and other approaches to environmental protection.

He retired in 1972 and died in Vienna in February 1982.

1. Amtliche Prüfstelle für das Kraftverkehr Sachsen.
2. Institut für Verbrennungskraftmaschinen.

Edge, Selwyn Francis (1868–1940)

Board member of D. Napier & Son from 1900 to 1912; chief executive of AC Cars, Ltd., from 1922 to 1929.

He was born in Sydney, New South Wales, Australia, and came to England with his parents at the age of three. His formal schooling was skimpy and he had no technical education at all. Tall and athletic, he became an expert bicycle racer and a leading member of the Bath Road Cycling Club in London, where fellow members put him in contact with Harvey Du Cros, owner of the Dunlop Pneumatic Tyre Company, Ltd. He became London manager for Dunlop and learned to know pioneer motorists and racing drivers. After Ferdinand Charron had given him a demonstration ride on a Panhard-Levassor in 1895, his enthusiasm switched from two wheels to four. In 1898 he bought his own Panhard-Levassor secondhand and very soon decided to make major modifi-

cations to it. He wanted a radiator (it only had a water tank for cooling), a steering wheel (it had tiller steering) and of course, pneumatic tires. Again, his friends at the cycling club were helpful in suggesting where he could get this work done: at Napier's in Lambeth, London. The conversion was done with first-class engineering and workmanship. The idea grew upon him that the Napier works might have the ability to produce complete cars.

In 1899 he became a partner in the de Dion–Bouton British & Colonial Syndicate, to import and distribute the motor tricycles made by de Dion–Bouton. He also met often with Montague S. Napier, outlining his ideas on car design and urging Napier to make automobiles.

On the strength of one two-cylinder prototype, he won the backing of Harvey Du Cros for a sales company having exclusive rights to Napier's cars. In October 1899, Edge and Napier signed a contract for delivery of 396 cars from March 31, 1900, to the end of 1904. To promote the Napier name, Edge began driving them in international races in 1900. Unsuccessful at first, he drove a 30-hp Napier to win the Gordon Bennett Cup in 1902. He also raced Napier-powered speedboats, visited the United States to look at their automobile works, and lectured on automobiles in South Africa. When the Brooklands race track opened, he announced that he would drive a six-cylinder Napier for 24 hours at an average speed of 60 miles per hour. He proved as good as his word, bettering his claim by averaging over 65 miles per hour on June 28–29, 1907.

He reached a new agreement with Napier, agreeing to purchase cars for an amount of $480,000 per year until the end of 1921. His sales organization became a joint stock corporation, S.F. Edge, Ltd., but at the peak of its success, he began giving less than full-time attention to it. In 1906 he developed an interest in farming, and by 1910 he owned ten farms in Sussex, covering a total of 2000 acres. A year earlier, he experienced difficulty in selling enough Napiers to meet the contract. The situation worsened in 1910–11, and pressure grew from Napier's side to open a sales department of their own.

The rupture occurred in 1912. Edge sold his shares in S.F. Edge, Ltd., to Montague S. Napier for $640,000, resigned from the Napier board and agreed not to re-enter the auto industry or trade for a period of seven years. Instead he became the world's foremost breeder of pedigree hogs. During World War I he volunteered for government service and was named comptroller for agricultural machinery, in which capacity he organized imports and distribution of farm tractors. He also came into contact with a small engineering firm at Thames Ditton, Auto Carriers, Ltd., and in 1919 began to invest in it. In 1921 he was invited to join the board of directors, and a year later he acquired full control, reorganizing it as AC Cars, Ltd. He also bought control of Cubitt's Engineering Company, Ltd., at Aylesbury, Buckinghamshire, who were manufacturing Anzani-designed four-cylinder engines for some AC models. AC made its own six-cylinder engines.

But Edge was slow to modernize the products and failed to control mounting costs. He reorganized the business in 1927 as AC (Acedes) Cars, Ltd., but by 1929 it was bankrupt. Edge went back to farming and never worked in the motor industry again. He died in February 1940.

Edge, Stanley Howard (1903–1990)

Designer of the 1923 Austin Seven.

He was born on August 5, 1903, at Old Hill, Staffordshire, attended local schools and at the age of 14 went to work for Austin where he was assigned to the drawing office, making blueprint copies and running errands. By the time he was 16, he was a junior draftsman. His big opportunity knocked two years later, when the Austin Motor Company was in receivership, and Herbert Austin went home to undertake another car project, independently of his company. Edge was invited to join him, and for the next 18 months they worked in a makeshift drawing office centered around the billiards table in Austin's private residence at Lickey Grange. The plan was to make an ex-

tremely low-priced four-passenger economy car, and Austin borrowed an 8-hp Rover, with its aircooled flat-twin, to use as a basis. Edge talked him out of it, arguing that a small, simple four-cylinder engine could be manufactured at no higher cost than a twin. He also disliked air-cooling for car engines and eventually won Austin's agreement to accept the cost of a radiator. The lightweight, tapered steel frame was Edge's idea, and the front and rear suspension systems were inspired by the Rover. Overall dimensions were set by Austin's rule that it must fit in any shed that could house a motorcycle-and-sidecar combination. The prototype was built in a corner of a Longbridge machine shop, and the Austin Seven was officially presented in July 1922 as a product of the Austin Motor Company. From modest sales of 2,600 units in 1923, the demand soared and over 29,000 were sold in 1929. Bodies for the Austin Seven were modernized from time to time, but the basic chassis remained unchanged up to 1938. Copies of the Austin Seven were produced under license by Dixi of Eisenach in 1927–28 and by Rosengart in Paris from 1928 to 1939.

Edge left Austin in 1925 and joined Triumph, who offered a higher salary, and worked as a production engineer for the Super Seven (designed by Frank G. Parnell). In 1928 he joined Clayton-Dewandre, brake systems manufacturers, where he made his way through the ranks of the engineering staff to the position of chief engineer. When he resigned in 1956, he was assistant general manager of Clayton-Dewandre.

He died on July 24, 1990.

Edwardes, Michael (1930–)

Chairman and chief executive officer of British Leyland from 1977 to 1982.

He was born in 1930 in South Africa and educated at St. Andrew's College and Rhodes University, Grahamstown, graduating with a degree in Law. At the age of 20 he went into business for himself, forming enterprises that made lawns from rough terrain, traded in timber, and transported timber. He was interviewed by an executive from Chloride and was sent to London in 1951 as a trainee, returning to South Africa after two years as assistant sales manager for the Cape Battery Company. He worked for Chloride in African markets from 1963 to 1965, and then transferred to British operations. He restructured the nickel-cadmium battery activity at Redditch and was given a seat on the Chloride board in 1969. He was appointed chief executive of the Chloride Group in 1972 and its chairman in 1974.

He became a member of the National Enterprise Board early in 1975 at the request of Sir Don Ryder, industrial advisor to the Prime Minister, Harold Wilson. The government had legated its 95 percent ownership of British Leyland to the National Enterprise Board, which in October 1977 asked Edwardes to take over as chief executive of BL. He signed a three-year contract and won a three-year leave of absence from Chloride. BL was deep in debt and losing more money every month. The model lineup was irrational and the production apparatus an illogical amalgamation of engine, body, parts and assembly plants.

When he took command at BL on November 1, 1977, no one could tell him which models they were losing money on, or how much. He stopped production of the Austin Allegro and Maxi, Morris Marina and Triumph Toledo, and closed marginal plants at Abingdon (MG), Coventry (ex-Morris engine plant), Speke Road (Rover-Triumph) and London (Vanden Plas). The former Standard works at Canley were closed in 1980. Jaguar-Rover-Triumph was made into a separate unit from Austin-Morris in 1978, and Land Rover, Ltd., was established as a group of its own. The last Triumph Spitfire and Dolomite were made in 1980. For the surviving models, chaotic moves were made to new assembly points. Production of the Rover SD 1 (3500) was moved from Speke Road to Solihull in 1980 and to Cowley in 1982. The TR7 was moved from Canley to Solihull in 1980 and discontinued in 1981. He canceled the proposed Mini replacement ADO-88 after £300 million had been spent on it, in favor of the LC 8 (Metro) planned for production in 1980. He is credited with forging links with Honda for access to

modern technology and joint production, starting in 1981 with the Triumph Acclaim (similar to the Honda Ballade) at Cowley. His contract was extended for another two years. In 1981 he presented a new corporate organization plan, with just four groups (Cars, Unipart, Land Rover and Leyland). He killed a proposal to use Renault components for a new model based on the R-18, and was equally swift to stifle the idea of putting Leyland Vehicles into IVECO (though it eventually became part of DAF). From 1975 to 1981 the British government had poured £1771 million of taxpayers' money into keeping BL afloat, yet in 1980 Edwardes won approval for a further £415 million in state aid, conditional upon closing another 13 plants and eliminating 25,000 jobs. He disposed of Alvis, Ltd., Coventry Climax, Prestcold and Aveling-Barford, and began preparing the LC 10 introduced as the 1983 Maestro and the LM 11 introduced as the 1984 Montego. When he left BL on October 31, 1982, it was still losing money, but he had worked wonders to cut costs and raise productivity, reduced capacity from 850,000 to 450,000 cars a year and trimmed the payroll from 200,000 to 125,000. He returned to Chloride but resigned in March 1984 to become president of ICL (International Computers, Limited).

Edwards, "Joe"

Director of manufacturing, British Motor Corporation's Longbridge Works, Birmingham, from 1954 to 1956; managing director of British Motor Corporation from 1966 to 1968.

He held an engineering degree and joined Austin Motor Company, Ltd., in 1928 as a production engineer on C.R.F. Engelbach's staff. Except for two years' absence (with the Hercules Cycle Company), his career was made at Longbridge. He held major responsibilities for the production of military vehicles during World War II and was able to resume production of prewar passenger-car models with astonishing rapidity at the end of the war. He coped with the Longbridge end of all the production problems stemming from the merger with the Nuffield Organization in

1952, and the subsequent coordination of programs. BMC chairman L. P. Lord regarded him as a potential rival and arranged to have him sidelined as director of corporate labor relations, which he refused, preferring to resign in 1956. Almost immediately, he joined Pressed Steel Ltd., supplier of body shells to Morris and others, as managing director. He reorganized the company's operations and in 1965 sold it to British Motor Corporation. In 1966 George Harriman invited him back to BMC, where he began a major effort to cut costs, with plant closings and layoffs, which led to serious labor strife. When the merger with Leyland was planned, he was offered a seat on the British Leyland board of directors, but foresaw an inevitable conflict with Sir Donald Stokes, and resigned in April 1968.

Egan, John Leopold (1939–)

Chairman of Jaguar Cars, Ltd., from 1980 to 1990.

He was born on November 7, 1939, and educated at Bablake School, Coventry. He obtained a degree in petroleum engineering from London University and a master's degree in business studies from London Business School. He began his career with General Motors and ran the UK parts division of AC-Delco from 1971 to 1976. He joined British Leyland as head of Unipart, and later served as parts director of Massey-Ferguson. In 1978 he turned down an offer to run Jaguar-Rover-Triumph, but changed his mind when Margaret Thatcher was elected prime minister in 1979.

He made big waves at Jaguar when he cut the payroll from 10,500 to 7,200 but still raised production from less than 15,000 cars in 1981 to 22,000 in 1982. He made heavy investments, such as a highly automated transfer line for engine production at the Radford plant, a new body-in-white facility at Castle-Bromwich, and took over the ex–Chrysler technical center at Whitley. In 1984 he organized the partial-privatization of Jaguar, and the company had four profitable years. But Jaguar lost $84 million in 1989 and the government sold its remaining Jaguar shares to

Ford, which then proceeded to get full ownership of Jaguar. Egan resigned and served as managing director of British Airports Authority until 1999. He left to become chairman of MEPC, a property giant, and in 2000 was named chairman of Inchcape, worldwide traders in automobiles.

Ehrhardt, Heinrich (1840–1928)

Chairman of Fahrzeugfabrik Eisenach AG from 1896 to 1904; chairman of Heinrich Ehrhardt Automobilwerke AG from 1904 to 1922; chairman of Automobilfabrik Zella-Mehlis GmbH from 1924 to 1927.

He was born on November 17, 1840, at Zella-St. Blasii in Thuringia and had a technical and commercial education. He invented a recoil-absorbing cannon which became the basis for the Ehrhardt industrial group. He also held patents on new methods and machines for pressing sheet steel. He sold manufacturing rights to arms manufacturers in several nations and collected substantial license fees. He was given a seat on the supervisory board of a major client firm[1] in Düsseldorf in 1889. When he became interested in automobiles, he bought a license from Decauville, which served as the basis of Wartburg cars from 1897 to 1903. In 1904 the Wartburg evolved into the Dixi, and he sold his shares in the Eisenach-based company. He maintained the Decauville license, set up a factory in Düsseldorf to be run by his son, Gustav, and introduced a line of Ehrhardt cars. By 1906 Decauville's designs had evolved considerably, and Ehrhardt's production concentrated on medium and powerful touring-cars. The company also built a small economy car with the Fidelio label, and began truck production in 1909. Both the Dusseldorf and Zella-St. Blasii factories built military vehicles in World War I, and both found great difficulty in reviving production for civilian markets. The 1919–22 models were prewar designs. The Düsseldorf plant was closed in 1921 and sold to Szabo & Wechselmann, who built the six-cylinder Ehrhardt-Szawe luxury car up to 1924. The Zella–St. Blasii factory became home to a new Ehrhardt-owned company which produced the Pluto 4/20 and 5/30 light cars under Amilcar license from 1924 to 1927.

1. Rheinische Metallwaren und Maschinenfabrik AG, Düsseldorf (later Rheinmetall).

Elizalde y Rouvier, Arturo (ca. 1865–1925)

Technical director of Biada, Elizalde y Cia from 1910 to 1915; chairman and managing director of Fabrica Española de Automoviles Elizalde from 1915 to 1925.

He was born about 1865 in Barcelona to a Catalan father and French mother, and was educated in Paris. He went to work in a Barcelona auto repair shop which he later took over and expanded into manufacturing a variety of precision-engineered auto parts. In partnership with his two brothers-in-law, he acquired the Delahaye importers, José Maria Vallet y Cia, in the summer of 1910. Plans to make cars of his own began taking shape and a prototype 15/20 four-cylinder model was ready in 1913, built to the designs of two ex–Delahaye engineers. It went into production in 1914, received a bigger engine in 1917, and continued through 1924. In 1918 he put together a new team of engineers to design the ultimate luxury car. The 1921 Tipo 48 had a straight-eight aluminum-block 8-liter engine with four valves and two spark plugs per cylinder, and four-wheel brakes. But only two or three prototypes were built. The company survived mainly on the strength of the 15/20 and 18/30 four-cylinder models and the six-cylinder 20/32. Following the market, he added a low-priced Tipo 51 6/8 model in 1924. He died on December 4, 1925.

Enever, Sydney (1906–1993)

Chief planning engineer of the MG Car Company, Ltd., from 1938 to 1939; chief engineer of the MG Car Company, Ltd., from 1954 to 1971.

He was born on March 25, 1906, at Colden Common in Hampshire and grew up in Oxford. His formal schooling ended at the age of 14 when he joined Morris Garages as an errand boy. He evolved into a jack-of-all-trades and occasional troubleshooter for years before get-ting really close to the cars. In 1930 he was brought into the experimental department under Cecil Cousins, and for the next four years he worked on prototype construction and racing-car preparation. When the MG design staff was transferred to Cowley in 1936, he stayed on at Abingdon as a liaison engineer. He became head of the experimental department, where he and Reg Jackson built the EX 135 speed-record car, and in 1938 was placed in charge of new-model planning. During World War II he was engaged in the Nuffield Organization's production programs for the Ministry of Supply, and returned to building MG cars in 1946. He was the senior designer for the MG TD, which was made up from a shortened Y-series sedan frame and suspension, with a TC-like body. Assisted by Cecil Cousins, Reg Jackson, and Alec Hounslow, he had the MGA design ready in 1952, but its production was delayed until 1955 because L.P. Lord gave priority to the Austin-Healey. The MGA had a new frame, Austin B-series engine, gearbox and rear axle from the MG Magnette, and a very modern roadster body. A total of 13,410 MGAs were made at Abingdon up to 1962. He also led the design team for the 1962 MGB, with a bigger Austin B-series engine and a chassis evolved from the MGA. The MGB GT coupé went into production in 1965, and the MGB GT V8 arrived in 1973. Over half a million MGB units were produced over a ten-year period.

He retired in 1971 and died in 1993.

Engelbach, Carl R.F. (1876–1943)

Works manager of Wilson-Pilcher from 1904 to 1907; director of Armstrong-Whitworth's motor car department from 1908 to 1918; works director of Austin Motor Co. from 1918 to 1938.

He was born in London in 1876 as the son of a clerk in the War Office. He attended school at Southport, leaving at the age of 16 to go to work in the Royal Arsenal at Woolwich. He joined Armstrong, Mitchel & Co. at Newcastle-on-Tyne as an engineering trainee and ended up as manager of the Elswick Works which produced the Roots & Venables motor cars from 1902 to 1904.

After holding responsibility for the production of Wilson-Pilcher cars for nearly four years, he came to Armstrong-Whitworth when that giant organization took over the Wilson-Pilcher business. During World War I he organized and directed howitzer production at the Coventry Ordnance Works. His appointment with Austin was assured by Austin's creditors who wanted a common-sense man to counterbalance Austin's sometimes fanciful initiatives.

In 1924 he began a complete restructuring of the Longbridge works, with new layouts, a moving final-assembly line, new methods and equipment. Alongside Ernest Payton, the finance director, Engelbach was Herbert Austin's closest associate for nearly two decades. He retired in 1938 and died in 1943.

Engelhardt, Paul (1877–1948)

Head of the Turcat-Méry drawing office from 1903 to 1914; chief engineer of Rochet-Schneider from 1914 to 1919; managing director of Rochet-Schneider from 1919 to 1935.

He was born in 1877 and held an engineering degree from the Institut du Nord. His designs for Turcat-Méry were heavy cars with long-stroke four-cylinder engines and chain drive. From 1909 onwards, all new models had shaft drive. They provided the basis for generations of Lorraine-Dietrich cars, as produced under the direction of Léon Turcat and Simon Méry. He designed a powerful six-cylinder car in 1907, but only a few chassis were built. From 1908 the entire range consisted of four-cylinder models from 14 CV and up, the biggest having been a 25 CV in 1910, and a 35 CV in 1913. He left Marseilles when Turcat and Méry returned, and moved to Lyon. The factory built trucks during World War I, and one big hall was converted to make Renault aircraft engines. During the early postwar years, he became known for promoting standardization, but being reluctant to spend heavily on new-product development. The 1919 range included four models with four-cylinder engines from 12 to 25 CV, four-speed gearboxes and shaft drive. A 6.2-liter six was added in 1920 and produced up to 1929. Rochet-Schneider cars were not particularly fast, but great hill-climbers. In 1923 all engines were redesigned with overhead valves, and some models could reach 120 km/h. But all were overpriced, and demand fell. In 1924 the company built only 54 cars with 20 CV specification and less than 100 units of the 14 CV model, after several years of stable production at around 1000 units. The cars were modernized, some cost was taken out, prices were cut, but the rate of industrial investment was stagnant. Failure to modernize the plant prevented any major change in the cars. Engelhardt ordered all existing models phased out in 1928–29, to be replaced by a new lineup with only two models on the same chassis, a 3.8-liter 20 CV six and a 4.7-liter 26 CV six. But car production was pushed in the background as truck sales were increasing. The last Rochet-Schneider cars were built in 1932. He resigned early in 1935 and joined Société Lorraine, Automotrices et Chars, makers of railcars and military tanks at Lunéville, as chief engineer.

Engellau, Gunnar (1908–1988)

Managing director of AB Volvo from 1956 to 1971; president of AB Volvo from 1971 to 1978.

He was born in 1908 in Stockholm as the son of a lawyer and held an engineering degree from Sweden's University of Technology. He served as a flight engineer in the Swedish air force and upon discharge, took a job as a roundhouse mechanic with the Swedish State Railways to learn the business from the bottom up. He began his career with a company that built bridges and produced railway locomotives, where he was continuously moved from one position to the next, learning every aspect of production, sales and management. In 1939 he joined Electrolux (household appliances) as sales director. Soon afterwards he was approached by Assar Gabrielsson, president of AB Volvo, about taking over as technical director of Volvo's aircraft-engine division.[1] He turned it down, since he was making more money with Electrolux than Volvo could offer. But two years later, in the middle of World War II, when business was slow, he accepted a renewed offer, this time as managing director of Volvo's aircraft engine division, even though it meant a 66 percent cut in salary. He led the division through a spectacular postwar expansion as turbojet engine production was taken up. In 1956 he was invited to AB Volvo as managing director of a company that was building about 30,000 cars a year, 7500 trucks and buses, plus farm tractors[2] and Penta marine engines.

He replaced Gabrielsson's autocratic rule with a modern management team of 20 young executives, invested heavily on the car side and quickly doubled production, while energetically developing new export markets, notably the United States. In 1958 he bought a tract of land west of the main Volvo plant and began construction of a big passenger-car production center at Torslanda. He broadened Volvo's model range, gradually moving upmarket, and adding a six-cylinder model in 1968. Volvo produced 145,000 cars in 1967, with profits of $17 million on sales of $659 million. In a reversal of policy, he began buying into DAF in 1976, obtaining small-car technology and a production base in the Netherlands. He retired in 1978 and died ten years later.

1. AB Svenska Flygmotor, Trollhättan.
2. Bolinder-Munktell AB.

England, Frank Raymond Wilton "Lofty" (1911–1995)

Deputy managing director and deputy chairman of Jaguar Cars, Ltd., from 1961 to 1972; managing director and chairman from 1972 to 1974.

He was born on August 24, 1911, at Finchley, London, and educated at the local Christ's College. He served an apprenticeship with Daimler's service center at Hendon, London, from 1927 to 1932. He bought his first motorcycle, a 1921 Douglas, while still at college, and spent his spare time in the pits at Brooklands race track, where he grew into a first-class mechanic and worked on Sir Henry Birkin's Bentley, Whitney Straight's Maserati, Dick Seaman's Delage and Prince Bira's ERA. He pursued a parallel career racing motorcycles and won second place in the 1936 TT on

a Cotton. He joined Alvis in Coventry as a service engineer, joined the RAF in 1941, went to Texas for flight training and became an instructor to the U.S. Army Air Force before returning to Britain, where he flew on daylight bombing missions over Germany. In 1945 he returned to Alvis and was named deputy service manager but was engaged by Jaguar as service manager a year later. He acquired the additional title of competitions manager in 1951 when the company got serious about exploiting the racing potential of the XK-120. He led the teams that won at Le Mans in 1953 and 1954 and was promoted to service director in 1956 when Jaguar withdrew its racing team.

From 1961 onwards he was Sir William Lyons's right-hand man and ran the company to the founder's greatest satisfaction, eventually succeeding him as chairman, though only for a troubled two-year period. Jaguar had come under British Leyland control, and Lord Stokes put Geoffrey Robinson in his office to be groomed for taking it over. England retired to Austria in 1974, moved to Scottsdale, Arizona, six years later, and died there at the age of 83.

Ennos, Clive (1932–)

Chief chassis engineer, Ford of Britain, from 1983 to 1986; Ford of Britain director of body, chassis and electrical engineering in 1987; executive director of Ford of Britain product development from 1988 to 1990; director of product engineering operations, Jaguar Cars, Ltd., from 1990 to 1997.

He was born on April 15, 1932, in London and was a student apprentice with the Electricity Board from 1948 to 1956, which allowed him to take classes in electrical and mechanical engineering at East Ham Technical College. He was a surveyor in the Royal Army and served as a test engineer with Westland Aircraft from 1956 to 1966. He joined Ford of Britain in February 1966 and ten years later was promoted to executive engineer. From 1980 to 1983 he was Ford's chief electrical engineer. He made preliminary specifications for the Mondeo, and was transferred to Jaguar in

September 1990. He held overall engineering responsibility for the 1995 XJ6 and the subsequent XJR and XJ8, and retired in February 1997.

Enrico, Giovanni (1851–1909)

Technical director of F.I.A.T. from 1901 to 1906.

He was born in 1851 at Casale Monferrato and received his engineering diploma in 1876. He started a workshop with a drawing office to design and build steam engines, steam turbines, boilers, and electrical equipment, but was constantly underfinanced and had to liquidate his business in 1884. From that time until 1897 he held technical positions all over Italy with industrial companies in the power-generation business, finally settling in Turin as an engineering consultant. F.I.A.T. engaged him to modernize its automobile concepts and create a full model range. He designed F.I.A.T.'s first four-cylinder engine, a 12-hp side-valve unit produced from 1901 to 1903. The 16/24 and 24/32 followed in 1903 and 1904, respectively.

Like many others, Enrico was strongly influenced by the Mercedes type of vehicle, and many F.I.A.T. cars created under his direction were perceived as bearing a too-obvious resemblance to certain Mercedes models. He designed a 75-hp racing car with a 14-liter four-cylinder engine in 1904 and the 100-hp Gordon Bennett model with a 16.3-liter engine in 1905. The latter was beaten by a Richard-Brasier but took second and third places, beginning to build F.I.A.T.'s reputation for power and speed. In 1906 Enrico left F.I.A.T. and went to Fides, who was making cars under a Brasier license. But he had to retire to nurse his failing health and died at Pinerolo in 1909.

Epron, Luc (1940–)

Product planning and program manager of Automobiles Citroën from 1992 to 1996; joint managing director of Automobiles Citroën from 1996 to 1998; Peugeot SA group director of product planning from 1998 to 2000.

He was born on December 17, 1940, in Paris

and held engineering degrees from CESTI[1] and ISMCM.[2] In 1967 he joined Peugeot at the Sochaux plant but was soon transferred to the technical center at La Garenne. By 1970 he was working on the coordination between aerodynamics and body structures. Three years later he led a study on new test equipment, and in 1976 was promoted to manager of assembly plants and testing. In 1982 he was named manager of Peugeot body production with responsibility for three centers (Sochaux, Mulhouse, and Poissy). In 1984, he was given charge of factory equipment in all Peugeot plants. He was transferred to Citroën in 1987 to work in marketing and planning. He directed the planning of the Citroën ZX, Xantia and Saxo. He had nothing to do with the planning of the XM but the problems of changing it to attract more buyers landed on his desk. Yet the XM continued for several years with sales at 10 percent of planned volume. After 1996 he helped establish a new identity for Citroën, as distinct from mechanically related Peugeot models, and was promoted to a group-level office in 1998. He retired in 2000.

1. Centre d'Etudes Supérieures des Techniques Industrielles.
2. Institut Supérieur des Matériaux et de Construction Mécanique.

Erle, Fritz (1875–1957)

Manager of the Benz drawing office from 1903 to 1920.

He was born on November 12, 1875, as the son of a hotel owner in Mannheim. He was still in grade school when he became familiar with Benz gas engines, and in 1894 began a four-year apprenticeship in the Benz factory on Waldhofstrasse in Mannheim. He studied mechanical engineering at the Technikum Ilmenau in Thuringia, graduating in 1902, and promptly returned to Benz & Cie. He headed one drawing office while Marius Barbarou led another, rival office.

Erle directed the design of the 16/20 model with its flat-twin engine, and then the four-cylinder 16/20 which soon replaced the former. Barbarou left in 1903, Carl Benz in 1904, but Erle was not named chief engineer. Fritz

Hammesfahr was named technical director, and Hans Nibel arrived. Erle was a proficient rally driver and made important contributions, along with Georg Diehl, to the evolution of Benz cars. He stayed on during World War I, but resigned in 1920. He lived on until 1957 but never had any further connection with the automobile industry.

Eugen, Charles Marie Van *see* Van Eugen, Charles Marie

Evernden, Harold Ivan Frederick (1896–1980)

Chief designer of Rolls-Royce, Ltd., from 1939 to 1949; chief project engineer of Rolls-Royce, Ltd., from 1949 to 1961.

He was born on August 11, 1896, at Bromley, Kent, and educated at Sutton County School and King's College, London. He began his career by joining Rolls-Royce, Ltd., at Derby in 1916, where his first task was to tool up for producing the Danish-patented Madsen rifle. In 1917 he was transferred to the Airship Design Office, and in November 1918 joined the newly established Motor Car Design Office. In 1921 he was selected to Henry Royce's personal team of designers at West Wittering, Sussex, where the 20-hp model was just going on the drawing boards. He designed the four-door sedan and the "sports saloon" both produced by Barker & Co., Ltd. He stayed with Royce at West Wittering until 1933. The 1927 experimental Phantom I Continental was his creation, leading to the production-model Phantom II Continental in 1930, with coachwork by Barker, Park Ward, and Thrupp & Maberly. He designed the Park Ward body for the Bentley 3½-Litre while still at West Wittering, but upon Royce's death he was transferred to A. G. Elliott's staff at Derby. In 1934 he became head of a new department for body hardware, and in 1935 was transferred to Bernard Day's department as project designer, where he was associated with the Bentley Mk V and the Corniche streamliner. In 1940 he was part of a team formed to convert an unsupercharged version of the Merlin V-12 aircraft engine into a suitable prime

mover for battle tanks, then spent two years as a member of the Lucas tank mission to the United States. In 1945 he resumed his normal duties and turned John Blatchley's styling designs into fully engineered bodies for the Silver Wraith and Bentley Mk VI. He also assured liaison with the coachbuilders and Pressed Steel, Ltd., who produced the Standard Steel Saloon body shells for the Mk VI and Silver Dawn. After his promotion to chief project engineer in 1949 he created the Bentley R-Type Continental, whose spectacular fastback coupe body was produced by H. J. Mulliner. He updated it as the S-Type Continental in 1956. Several of Britain's coachbuilders went out of business during this period, but Rolls-Royce, Ltd., had owned Park Ward since 1937 and now took control of H. J. Mulliner. He retired in August 1961 and settled at West Wittering in 1964.

He survived a heart attack in 1971 but died in 1980.

Faccioli, Aristide (1848–1919)

Technical director of F.I.A.T. from 1899 to 1901.

He was born on December 23, 1848, in Bologna and studied at the Royal School of Applied Engineering[1] in Turin, graduating in 1873. He obtained a patent for a double-acting gas engine in 1883 and a separate patent for a petroleum-fueled engine in 1891. In 1895 he was granted a further patent, relating to improvements in four-stroke engines. In 1896 Giovanni Ceirano engaged him to design an automobile. Appearing in 1898 as the Welleyes, it was not a coy of any other design. It had Benz influence, and touches of Peugeot and Darracq. Production was carried out on a small scale in the Ceirano bicycle shops in Turin for almost a year when the newly founded F.I.A.T. bought the Ceirano company, Faccioli's patents, and the Welleyes car. It became the basis for the first F.I.A.T. of 1899, a four-passenger Vis-à-Vis with a 3½-hp 697-cc horizontal parallel-twin engine mounted under the driver's seat, chain drive and tiller steering. He was named technical director of F.I.A.T. but seems to have had great difficulty in making the transition to more advanced concepts of automobile design, and left the company in 1901. He joined the Crosio brothers, Ambrogio and Angelo, who produced the Faccioli car until their enterprise failed in 1905. Faccioli's attention, by that time, had switched to aviation. He wrote a pamphlet "Theory of Flight and Aerial Navigation" and in 1906 established a new company, Società Faccioli Ferro Rampone, to make radiators for the motor industry, but after two years, it was out of business. Matteo Ceirano engaged him in 1909 to design aircraft engines for SPA.[2] The SPA engines were highly successful and before World War I he decided to try his hand at the design of a complete airplane. He designed two, but they may not have been built, for he retired from engineering and spent his last years meditating on philosophical and religious matters.

1. Rego Scuola di Applicazioni per Ingegneri.
2. Società Piemontese Automobili Ansaldi-Ceirano, Turin.

Falchetto, Battista Giuseppe (b. 1896)

Design engineer of Lancia e C. Fabbrica di Automobili from 1927 to 1938 and 1950 to 1960.

He was born on April 1, 1896, in Piedmont and began his career with Audodli & Bertola, makers of hydraulic pumps in Turin. Called up for military service in World War I, he was assigned to the Air Corps, working first in the drawing office and later as a test engineer. He was discharged in December 1919 and was signed up by Vincenzo Lancia in January 1920 as a junior draftsman. He soon became Lancia's most valuable technical assistant. He designed the front suspension for the Lancia Lambda, as well as a number of alternatives which were discarded. It was Falchetto who proposed borrowing from naval-engineering practice for the body structure of the Lambda, eliminating the conventional frame. He was also instrumental in getting Vincenzo Lancia to adopt four-wheel brakes. He suffered a nervous breakdown in 1925, went to Fobello for a rest cure, and resumed his duties with Lancia in 1927. He developed improved versions

of the sliding-pillar independent front suspension with integral hydraulic damping for the Dilambda, Artena, Augusta, and the Astura. He never stopped experimenting with unit-construction bodies, and obtained a patent (in Lancia's name) for the Aprilia body shell. He also designed the Aprilia's unique rear-suspension system with trailing arms and torsion bars. A transverse leaf spring was added later to prevent excessive cambar angles in the rear wheels. After the founder's death in 1937, he began withdrawing from Lancia, and spent World War II as a designer of agricultural equipment. He also designed the Motom 48 moped and Motom light motorcycles, put in production by De Angeli Frua in Milan in 1947. In 1950 he returned to Lancia and was responsible for the Ardea body and chassis. He retired in 1960 after assisting Antonio Fessia in the development of the Flavia.

Falkenhausen, Alex von (1907–1989)

Chief engineer of engine design and development for BMW from 1959 to 1975.

He was born on May 22, 1907, in the Schwabing district of Munich and attended elementary school in Munich, spending his vacations on the family estate near Gunzenhausen. He studied mechanical and aeronautical engineering at the Munich Technical University, where Willy Messerschmitt was a lecturer. Even before graduation, he was given an offer to join Flugzeugbau Messerschmitt at Bamberg. At the same time he received an invitation from BMW, which he accepted, and on May 1, 1934, he began design work on his first BMW motorcycle engine, under the direction of Rudolf Schleicher. He had raced motorcycles since he was 19, and Schleicher made him an official member of the BMW racing team. When BMW introduced the 315 roadster, he began to race cars as well, and assisted in the development of the 328 roadster. In 1940 he was named head of the motorcycle test department, and in 1943 became chief development engineer for motorcycles. He withdrew in the chaotic circumstances of 1945. In 1947 he entered the AFM[1] car in a race at Hockenheim and won. He also built some AFM sports cars with Fiat-based engines, for which he developed cylinder heads with splayed overhead valves, one big inlet valve and two small exhaust valves per cylinder. He also made a Fiat-based supercharged 750-cc engine and sports car for Teddy Vorster, a leader in the textile industry. In 1949 he made a new AFM chassis for a 135-hp rebuilt BMW 328 engine for Formula 2 racing, and built and sold eight such cars over a two-year period. But he had reached the limit for the amount of power that could be taken out of the BMW engine. Richard Küchen, former chief designer with the design of a four-overhead-camshaft V-8 with fuel injection, and a new AFM F 2 car took to the race tracks in 1951. He also built a prototype sports coupé with a 2½-liter Opel Kapitän engine but failed to find backing for putting it into production. In 1954 he returned to BMW and led the development of the BMW 280 motorcycle/sidecar combination. He worked on the development of engine and chassis for the BMW 507 sports car, and in 1957 took charge of all BMW engine development. It did not take him long to dramatically raise the power output of the air-cooled flat-twin for the BMW 700 economy car, and in 1959 he also assumed responsibility for engine design. He designed the engines for the BMW 1500 and its derivatives, and the sixes for the 2500/2800 models. Inevitably he was also associated with BMW's first steps to get back into racing. In 1972 he designed, assisted by R. Henning, a new family of V-8 engines from 3.6 to 5-liter size, but the program was shelved. He retired in 1975 and died on May 28, 1989.

1. Alex (von) Falkenhausen Munich.

Falkenhayn, Fritz Von (1890–1973)

Sales director of NSU-Werke AG from 1930 to 1933; member of the NSU management board from 1933 to 1936; chairman of the NSU management board from 1936 to 1945.

He was born on September 27, 1890, in Berlin as the son of General Erich von Falkenhayn, Prussian minister of war and chief of the Imperial general staff in 1914. He was destined for a military career and served in the Kaiser's air force during World War I. From 1919 to 1930 he occupied a number of administrative positions in German industry and developed good contacts in financial circles. He was placed at NSU by Deutsche Bank, which had consolidated its debts in 1928. At his instigation, NSU took over Opel's motorcycle activity in 1932, the motorcycle division of Wanderer, and the D-Rad motorcycle from Deutsche Werke in Spandau. NSU prospered under his direction. He led the company through World War II and in 1945 fell into the hands of the U.S. military authorities. When he returned to civilian life, he became a director of Auto-Staiger, a retail and wholesale dealer group in Stuttgart. In 1954 he was invited back to NSU as a member of the supervisory board, on which he held a seat until 1959.

He died in 1973.

Farina, Pinin (1893–1966)

Founder and chairman of Pininfarina, major coachbuilder and designer.

Gianbattista Farina was born November 2, 1893, in Turin. He was the tenth of eleven children in a poor family from Piedmont and was given the nickname Pinin, meaning "baby" in the Piedmontese dialect. In 1905 he was apprenticed to one of his elder brothers, a carriage and car repairer who later turned to building bodies. When Pinin was only 17 years old, his design for the Fiat Zero was chosen ahead of those of many men much older. He was offered a job by Henry Ford in 1924, but chose to stay in Turin. In 1930 he struck out on his own as a coachbuilder, under the name Carrozzeria Pinin Farina. The new company began with around 90 employees, making one-off prototypes and very small runs of 5 to 10 special models for direct sale. The first cars were mainly for Italian companies (Lancia, Fiat, Alfa Romeo, Isotta Fraschini). By the early 1930s, a Cadillac and a Mercedes Benz were produced in their workshops. Around the same time an interest in aerodynamics began to emerge and it did not take long for this to influence the work of Pinin, notably with the 1933 Lancia Astura. By 1939, he was regarded as one of the leading coachbuilders in Europe, with a factory of 500 workers making two cars a day, many of them glamorized bodies on standard chassis such as the Fiat 1100. After the war, he was responsible for numerous ground-breaking designs, such as the 1947 Cisitalia coupe. In 1952, he styled the American Nash range. In 1958, he started an association with the British Motor Corporation, which resulted in all Austin, Morris, Wolseley and Riley cars having Farina styling for several years. He also styled the Peugeot 404. In 1961, he changed his surname and that of his company to Pininfarina, which the company still retains today. Upon his death in Lausanne on April 3, 1966, at the age of 73, he was succeeded as chairman by his son Sergio Pininfarina.

Farman, Henry (1874–1958)

Director of Société H. & M. Farman from 1912 to 1937.

He was born on May 16, 1874, at Cambrai as the second son of Thomas Farman, Paris correspondent for *The Standard*, a daily newspaper published in London. The three boys, Dick Farman (1872–1940), Henry, and Maurice (1877–1963) were British citizens at birth, but opted for French nationality when coming of age. They were privately educated and became racing cyclists in their teenage years. Henry and Maurice began motor racing in 1896 on Panhard-Levassor cars, later also driving Renault and Darracq cars. They were also associated with two car manufacturing ventures, in partnership with Micot in 1898 and with Bonissou in 1899. Their interest in aviation started with hot-air balloons, Maurice making his first ascent in 1894 and setting a distance record in 1905. Henry made his first flight in 1907 after buying a Voisin biplane, and on January 3, 1908, he won the Deutsch-Archdeacon Cup for the first officially controlled flight of one kilometer's distance, at Issy-les-Moulineaux. He came up with several ideas for improvements that Voisin adopted. Maurice stopped auto racing in 1903, after the experience of seeing the wreck of Marcel Renault's car in the ditch at Couhé-Vérac in the Paris–Madrid road race. He broke off his race and went to get a doctor, but it was too late. Henry continued to drive in major events until 1908. He was then busy drawing up plans for his own airplane. In 1909 he started building it in a big tent on the Camp de Chalons-de-Marne. Independently, Maurice also designed an airplane and built it in rented premises near Satory. In 1911 the two brothers joined forces, started a pilot's school, and erected a factory in rue de Silly at Boulogne-Billancourt. In 1912 they demonstrated to the French military authorities the Farman biplane with its 80-hp Gnome air-cooled radial engine. Dick Farman took a financial interest in their company and kept their books. In 1917 Farman began producing its own aircraft engines. The plant was fully committed to the war effort, and at the end of 1918 the government canceled all orders. Some of the capacity was retooled for production of Alda cars for Fernand Charron, which gave them the idea of making Farman cars. The 40 CV six-cylinder Farman luxury car made its debut in 1921, and for some years the bulk of the company's income came from cars. In 1925 the demand for aircraft engines revived for civil aviation, mainly for flying overseas mail, and later for passenger aircraft. By 1929 the cars were a sideline, and the last ones were built in 1933. In 1937 the company was nationalized and later merged with Blériot and ANF at Les Mureaux. The Farman brothers started an airline company, Société Générale des Transports Aériens, whose activity ended abruptly when the Wehrmacht occupied Paris in June 1940.

Henry Farman died in Paris in July 1958, and Maurice survived until 1964.

Farmer, Lovedin George Thomas (1908–1996)

Secretary of the Rover Company, Ltd., from 1945 to 1957; joint managing director of the Rover Company, Ltd., from 1957 to 1963; chairman of the Rover Company, Ltd., from 1963 to 1973.

He was born on May 13, 1908, at Bridnorth in Shropshire and educated at Oxford High School. He made postgraduate studies in accounting and qualified as a chartered accountant. In 1940 he joined the Rover Company, Ltd., as assistant secretary, taking charge of finances and accounting. During the war years he was works manager of the Solihull shadow factory and had the foresight to secure title to most of the surrounding land with a view to future expansion. He became company secretary in 1945 and was given a seat on the board of directors in 1952. He became joint managing director in 1957 and provided a budget for developing the P6 (1963 Rover 2000). As chairman, he negotiated Rover's merger with Leyland in 1966 (it was announced in March 1967). In March 1972, Triumph was merged with Rover under his direction.

He retired to the Isle of Man in 1973 and died in November 1996.

Farnham, Joe (1926–)

Technical director of Chrysler UK at Whitley from 1968 to 1971; technical director of Chrysler-France from 1971 to 1978; product engineering director of Austin-Rover from 1981 to 1985.

He was born in 1926 at Springfield, Massachusetts, studied at Yale University and served in the U.S. Air Force. He joined Chrysler in 1949 and was a classmate of Burt Bouwkamp and John Z. DeLorean at the Chrysler Institute. He developed Chrysler's first power-steering system for the 1950 Imperial and led the studies on noise, vibration and harshness for the full-size unit-construction body shell introduced for the 1960 range of Chrysler Corporation cars. When he arrived at Whitley, the Hillman Avenger was in an advanced development stage, but he improved the product in many ways, particularly in terms of quality. It went into production early in 1970. He was transferred to Paris and directed the Simca 1307-1308 program, which brought forth two front-wheel-drive mid-size cars at the end of 1975. He also supervised the European version of the Horizon, which went into production at the beginning of 1977. Chrysler brought him back to Detroit for two years,

1978–80, after which he was signed up by Michael Edwardes to create a new line of front-wheel-drive family cars. He directed the design and development of the Montego which went into production in 1984, and went back to America.

Fend, Fritz M. (1920–)

Technical director of the Regensburg Steel and Metal Construction Company[1] from 1953 to 1956; managing director of the FMR[2] vehicle and machinery company from 1957 to 1964.

He was born in 1920 in Rosenheim, Bavaria, as the son of a grocer, and educated as an aircraft engineer. He spent World War II in a Luftwaffe technical office and returned to civilian life in 1945. He repaired farm machinery and worked in a foundry before taking over his father's business. In transport-starved postwar Germany he saw a great need for a bicycle with weather-protection, and made scores of drawings. A prototype pedal car, the Fend Flitzer, appeared in 1946, but he could not get the materials to start production. In April 1947 he obtained a permit to open an engineering workshop and six months later, a new three-wheeler with wood-and-plastic-wrap body was shown. There was strong demand from disabled war veterans who needed basic transport. Many were legless, and Fend designed a hand-crank propelled version in 1948. Production took place in a disused shed of an industrial park. Customers then wanted auxiliary power for going uphill, and he began fitting 38-cc Victoria moped engines on the single rear wheel. Those who wanted more power were offered a 98-cc Riedel or Sachs engine. He displayed a three-wheeled bubble-car with wheelbarrow-wheels at the Frankfurt Fair in 1949, leading to an influx of orders, and also developed a cargo scooter with a moped rear section and a tray between the front wheels. Operating as the Fend Motor Vehicle Co.[3] he produced some 250 units from 1950 to 1952, but lost money on every one. Looking for industrial backing, he went to see Willy Messerschmitt who was then banned from making aircraft and built steel structures, wood-working machines and repaired railway carriages in his Regensburg factory. He liked Fend's cabin scooter, but said it must have two seats. Fend went home and redesigned as a two-seater, with seats in tandem. In the summer of 1952 Messerschmitt signed a contract for producing it under license, and set up a subsidiary at the Regensburg plant for that purpose. The Messerschmitt KR 175 came on the market at mid-year 1953, and 10,000 were built in the first 18 months. In 1956, when the ban on making aircraft was lifted, Willy Messerschmitt began bidding for air-ministry contracts and turned his back on the loss-making cabin rollers. With a partner (and supplier), Valentin Knott, Fend took over "his" section of the Regensburg plant, renaming it FMR[2] which also became the make of the vehicle. He introduced the four-wheel Tiger model in 1957, with a two-cylinder 493cc Sachs engine, but only a few hundred were produced. The market for such vehicles collapsed, and production was discontinued in 1962. Fend eased his way out of FMR and set up a consulting office in Regensburg in 1964.

1. Regensburg Stahl und Metallbau GmbH.
2. Fahrzeug und Maschinenbau GmbH, Regensburg.
3. Fend Kraftfahrzeug-GmbH, Munich.

Fenn, George Richard (1928–)

Director and general manager of Rolls-Royce Motors, Ltd., from 1976 to 1982; chief executive from 1982 to 1990.

He was born on March 25, 1928, and educated at Casterton School and Stamford Technical School. He began his career with Blackstone & Co. of Stamford, prominent machine-tool makers, in 1944. From 1950 to 1956 he held an executive position with Allis-Chalmers in the United Kingdom, and then joined the Oil Engine Division of Rolls-Royce, Ltd. In 1959 he was named purchasing manager for the Oil Engine Division, rising to materials director in 1961. In 1973 he was transferred to Rolls-Royce Motors, Ltd., as materials director, with a direct impact on the production side and evolution of the Silver Shadow and the T-Series Bentley. In 1976

he authorized an upgrading and upscaling the Rolls-Royce product, resulting in the 1980 launch of the Silver Spirit and the Silver Spur. He revived the Bentley Continental and in 1985 introduced the Bentley Turbo R, followed by a new Rolls-Royce Corniche in 1986.

He retired in 1990.

Ferrari, Enzo (1898–1988)

Owner and general manager of the Scuderia Ferrari from 1929 to 1940; managing director of Auto Avio Costruzioni Ferrari S.p.A. from 1940 to 1960; chairman of SEFAC[1] Ferrari Automobili S.p.A. from 1960 to 1969; chairman of Ferrari S.p.A. SEFAC from 1969 to 1977.

He was born on February 18, 1898, as the son of a metal-working artisan in Modena. He began studying at the Modena Technical Institute but had to get a job when his father died in 1915, and became an instructor for the recruits of the Modena fire brigade. He served with a mountain artillery regiment on the Austrian front in 1917–18 but was discharged because of illness. He went to Turin in 1919 to seek employment with Fiat, and when that failed, he worked in a Turin factory that produced frames for automobiles and trucks before moving to Bologna to work for a company that rebuilt used trucks with roadster or torpedo bodies. During 1919–20 he was a test driver with SA Costruzioni Meccaniche Nazionali in Via Fatebenefratelli, Milan, makers of the CMN car. He arrived at Alfa Romeo as a test driver in 1920, full of racing ambition. He became a test engineer for the racing team, and competed in 40-odd races with mixed results up to 1931. He left Alfa Romeo at the end of 1928 and early in 1929 founded Scuderia Ferrari in Viale Trento e Trieste, Modena, as a racing stable with Alfa Romeo cars, and distributors of Alfa Romeo cars for a big territory: Emilia-Romagna and the Marche provinces.

Scuderia Ferrari replaced the Alfa Romeo factory team and for eight years was given the pick of Alfa Romeo racing cars, such as the Tipo B in 1932, the 8C-35, and the 12C-37. At the beginning of 1938, Alfa Romeo bought an 80 percent stake in Scuderia Ferrari and revived its own team, Alfa Corse, naming Ferrari as its director. But he had bitter quarrels with Wilfredo Ricart, and resigned in 1939. He formed a new company in 1940 and built two sports cars, Auto Avio Tipo 815, for the Mille Miglia. Soon afterwards Italy joined World War II on Hitler's side, and Ferrari began making machine tools and aircraft parts for the Compagnia Nazionale Aeronautica in Rome, Piaggio SA of Pontedera, and RIV, Fiat's ball-bearing subsidiary. The Ferrari family owned some farmland and buildings at Maranello, south of Modena, where he built a factory in 1943. It was 50 percent damaged by Allied aerial bombing on November 4, 1944, and rebuilt in 1946. The Ferrari car was born in 1947 with a 2-liter V-12 engine. Next came a 1½-liter supercharged V-12 Grand Prix car, and then a succession of progressively bigger V-12 sports and grand touring cars. Production climbed slowly from 26 cars in 1950 to 81 cars in 1956, to reach 113 in 1957 and 183 in 1958. The company was reorganized as a joint stock corporation on May 23, 1960. That summer he was given an honorary degree in industrial engineering by Bologna University, and in 1963 he financed the establishment of the IPIA[2] in Modena. Production increased from 306 cars in 1960 to 598 in 1963 and 740 in 1965. In June 1969 he went to Turin for a meeting with Gianni Agnelli, and on June 21, 1969, it was announced that Fiat S.p.A. had taken a 50 percent stake in SEFAC, which then underwent a change of corporate structure. Fiat invested heavily at Maranello, and production rose to 1246 cars in 1971 and 1844 cars in 19872. Ferrari S.p.A. SEFAC took control of Carrozzeria Scaglietti of Modena in 1975 and Enzo Ferrari resigned as chairman on March 19, 1977, but continued as director of racing car construction and the racing program. On January 28, 1988, Modena University gave him an honorary doctorate in physics, and he died in his apartment in Modena on Sunday morning, August 14, 1988.

1. Società Esercizio Fabbriche Automobili e Corse.
2. Istituto Professionale per l'Industria e l'Artigianato.

Fessia, Antonio (1901–1968)

Vice director of Fiat's cars and truck design offices from 1928 to 1932; director of Fiat's aircraft-engine design office from 1932 to 1936; director of Fiat's central engineering office from 1936 to 1946; director of the CEMSA[1] car design office from 1946 to 1950; technical director of Lancia from 1955 to 1968.

He was born on November 27, 1901, in Turin, studied at the Turin Polytechnic and graduated in July 1923 with degrees in mechanical and industrial engineering. He began his career as a project engineer with Fiat in February 1925 and rose rapidly through the technical ranks. But years went by before he could put his creative imprint on a new car. His chance came in 1935 when Giovanni Agnelli asked him to make a proposal for a new type of very small and low-priced economy car. He proposed a two-passenger coupe with a four-cylinder engine carried in the front overhang, and front-wheel drive. It was altered to rear-axle drive and developed for production, which began in the spring of 1936, with the model name "500." After that, he was given design and engineering responsibility for all new Fiat cars and trucks. In 1945 Fiat received a front-wheel-drive test car from Simca for approval. Fiat's findings and decision were negative. Fessia alone felt it was time to make front-wheel-drive cars, not only Simca but also Fiat. That led to his resignation in March 1946.

Gianni Caproni engaged him to design a front-wheel-drive car, to be produced by one of the Caproni aircraft group's subsidiaries, CEMSA.[1] The prototype had a water-cooled 1100-cc flat-four mounted in the front overhang, permitting a short hood and a low, flat floor. But Caproni kept postponing the production plans, which were finally canceled. During his years with Caproni, Fessia frequently gave lectures at the University of Bologna, and from 1950 to 1952 he also served as technical consultant to Ducati motorcycles in Bologna. In 1953 Pirelli (tires and cables) gave him a contract as a consulting engineer, so he moved to Milan, and began lecturing at the Milan Polytechnic. At the same time, he renewed his ties to Fiat and made frequent trips to Heilbronn to give advice on the experimental cars of Deutsche Fiat. At the beginning of 1955 he was hired as a consultant to Lancia, becoming their technical director of April 2, 1955. He spent five years modernizing the CEMSA-Caproni, and developing it for production. It emerged in 1960 as the Lancia Flavia, with front wheel drive and a 1500-cc flat-four engine in the front overhang. He enlarged the aging Aurelia into a stylish Flaminia, and created the little Fulvia, with a V-4 engine and front wheel drive, which went into production at Lancia's new Chivasso plant in 1963. He never ran short of ideas, and he never retired. He died in his home at Borgomasino near Turin on August 19, 1968.

1. Costruzioni Elettro-Meccaniche Saronno.

Fiala, Ernst (1928–)

Head of Volkswagenwerk's research and development department from 1970 to 1973; director of VW research and development from 1973 to 1976; technical director of Volkswagenwerk from 1976 to 1988.

He was born on September 2, 1928, in Vienna as the son of a baker. He studied mechanical engineering and got his diploma in 1952 from the Vienna Technical University. He joined Daimler-Benz AG in 1954 and spent nine years in the body-structure test section at Sindelfingen. In 1963 he moved to Berlin as a lecturer at the Institute of Motor Vehicles at the Berlin Technical University. His research work for Volkswagen did not show up in the first-generation Scirocco, Golf or Passat, but he had some influence on the Polo, Type 181, and the LT vans. He bore overall responsibility for the second generation Golf in 1984 and the second generation Passat of 1988.

He retired at the end of August 1988.

Fiedler, Fritz (1899–1972)

Chief engineer of Horch from 1930 to 1932; chief engineer of BMW's car division from 1932 to 1945; consultant to AFN Ltd. and the Bristol Aeroplane Co., Ltd., from 1945 to 1950; technical director of BMW from 1951 to 1964.

He was born in Austria on January 9, 1899, attended local schools and studied engineering in Vienna and Munich. He worked for Aga[1] in Berlin from 1920 to 1923, joined Stoewer in 1926 and helped Bernhard Stoewer in the design of the 6/30 PS F-6 model car. He also designed the straight-eight engines for the Stoewer S 8 and S 10 of 1929–30. He formed links with Horch-Werke in 1928 and designed the V-12 Types 600 and 670, the straight-eight Type 850 of 1932 and the V-8 Type 830. When he arrived at BMW's Eisenach works in 1932, they were producing the ex–Dixi under Austin Seven license. He introduced independent front suspension and a bigger, overhead-valve engine in 1932, created the 303 in 1933 and the 309 in 1934. BMW's first six, the 315, came on the market in 1934, evolving into the 320, 321 and 326. He personally designed the chassis for the 327 and 328 sports models, and the bigger 335. He worked on the development of military vehicles from 1941 to 1944. Through connections established by the British War Reparations Board, H.J. Aldington invited him to join his company at Isleworth, Middlesex, and design a new Frazer Nash sports car.

The Frazer Nash was powered by a copy of the BMW 1971 cc six-cylinder engine made by Bristol Aeroplane Company, and when Bristol planned to produce cars, he was again consulted. In 1950 he returned to Germany and signed up with Opel as a chassis development engineer. But he was soon contacted by BMW's management and went to Munich as a member of the BMW board of directors. He was responsible for the BMW 503 and 507 sports models, derived from Alfred Böning's 501 chassis. He created the 600 and 700 models with rear-mounted motorcycle engines, and led the program that put the 1500 into production in 1963. The 1500 brought BMW into a new epoch, and he directed the engineering of its derivatives, the 1800, 1800 Ti, 2002, and 2000 Ti. He made his official retirement on January 1, 1964, but remained a consultant to BMW until 1968.

He died on July 8, 1972.

1. Aktien-Gesellschaft für Automobilbau, Berlin-Lichtenberg.

Flick, Friedrich (1883–1972)

First deputy chairman of the Daimler-Benz AG supervisory board from 1955 to 1968, honorary chairman of the supervisory board from 1968 to 1972.

He was born in 1883 in Austria and went into business. He became a force in brokering and investment banking and in 1937 established a trustee and investment office[1] in Düsseldorf. By 1940 his industrial holdings ranked second only to Krupp's. But he lost 70 percent of his fortune in 1945 when the Soviet powers seized all his property in East Germany. The Western powers put him on trial for war crimes in Nuremberg and gave him a seven-year sentence for using slave labor and giving financial support to Hitler's SS (Schutz-Staffel) "security" brigades. He was pardoned in 1950, but then had to dispose of his coal and steel holdings in the Ruhr district to comply with new de-cartelization laws. He managed to keep his most valuable investment, the Maximilian-Hütte steel works at Sulzbach-Rosenberg, which became his base for renewed growth. He sold his remaining coal mines for DM 250 million in 1952 and used the income to buy shares in Daimler-Benz AG, and to secure controlling equity positions in the Buderus and Krauss-Maffei engineering companies. By mid-year 1954, he held 15 percent of Daimler-Benz AG and on July 18, 1955, he sent a representative to the Daimler-Benz AG annual general meeting to announce that he held a blocking minority exceeding 25 percent of its capitalization.

A new supervisory board was elected, with Deutsche Bank's Hermann Abs as chairman (the bank held 15 percent of the capital) and Flick as his deputy. In 1956 Maximilian-Hutte secured 30 percent ownership of Auto Union GmbH, later acquiring full ownership, which he transferred to Daimler-Benz AG in a stock-swap deal in 1958. His share of Daimler-Benz AG grew to 39 percent. Flick also held controlling interests in Feldmühle paper, Dyna-

mit Nobel, and Gerling insurance. He held stock in Maybach Motorenbau GmbH which he sold to Daimler-Benz AG in 1960. He was behind Daimler-Benz' purchase of Hanomag-Henschel and Tempo-Werke, as well as the site and buildings of the ex–Borgward plant in Bremen. He died on July 20, 1972.

1. Friedrich Flick K-G und Verwaltungs-GmbH.

Folz, Jean-Martin (1947–)

General manager of Automobiles Peugeot from 1996 to 1997; chairman and chief executive officer of Peugeot SA beginning in 1997.

He was born in January 11, 1947, in Strasbourg, and graduated from the Ecole Polytechnique as a mining engineer in 1965. He began his career as a civil servant, holding a position as technical advisor for Craftsmanship and Commerce. In 1968 he joined private industry as plant manager with Rhone-Poulenc (petrochemicals). He was managing director and chief executive of Jeumont-Schneider from 1984 to 1987, then joined Pechiney (aluminum) as managing director and second-in-command to Jean Gandois. The two clashed over strategy, and in 1991 Folz left to join Eridania-Beghin-Say (sugar) as managing director. He was head-hunted by Peugeot and became head of Automobiles Peugeot in February 1996. On October 1, 1997, he succeeded Jacques Calvet as chairman of the parent organization. In 1998 he announced a new plan for sharing plant capacity: Citroën plants with excess capacity would assemble Peugeot models, and vice versa. In 1999 he negotiated a plan for joint diesel-engine production with Ford and in 2001 signed an agreement for joint production of a small car with Toyota.

Fornaca, Guido (1870–1928)

Technical director of Fiat from 1906 to 1916; general manager of Fiat from 1916 to 1920; managing director of Fiat from 1920 to 1928.

He was born on September 20, 1870, in Turin and educated at the Valentino School. He studied law and engineering at the Turin Polytechnic and held degrees in both disciplines. He began his career as a trainee engineer on a railway construction project in Rumania in 1895. Once back in Italy, he joined Fiat's locomotive and rolling-stock works at Savigliano. He was noted for his energy and for getting results, speeding up production and raising productivity. His work came to the attention of Giovanni Agnelli, who gave him the top technical office at the headquarters in Turin. He gave close personal attention to the big, powerful cars, such as the four-cylinder 5-liter 28/35 and 7.4-liter 35/40, and the six-cylinder 11-liter 40/50, all coming on the market in 1909. He organized a major expansion of Fiat's truck production in 1911, and prepared the war-production programs in 1915. He was given administrative responsibilities in 1916, coping with budgeting, personnel, and a host of legal matters. He did everything so well that he was promoted again in 1920, holding executive office with powers second only to Agnelli's. Fiat enjoyed tremendous growth under his leadership, which came to an unexpected end when he died on January 14, 1928.

Fornichon, Michel (1928–)

Technical director of Automobiles Peugeot from 1976 to 1981; director of product engineering, Peugeot SA from 1981 to 1993.

He was born in 1928 in Paris and graduated from the Ecole Centrale des Arts et Manufactures with a degree in mechanical engineering in 1954. He went to work for Peugeot and was assigned to production engineering, modernizing old plants and laying out new ones. In 1955 he was transferred to the technical center at La Garenne to work in research and development and in 1962 he was named head of the transmission sector. He became assistant technical director in 1969 and worked on the evolution of the 304. He led the 104 and 604 design programs, new steps into ever-smaller and ever-bigger cars. Neither was a sales success, but the 104 provided the basis for the Citroën Visa and the Talbot Samba. He had top responsibility for the 205 which won international acclaim and topped the sales charts for years. Production began in November 1982 in the Mulhouse plant and ended early in

1998, after more than five million units. He directed the 405 and 605 engineering teams and retired in 1993.

Frankenberger, Victor (1910–)

Technical director of NSU-Werke AG from 1951 to 1959; technical director of NSU Motorenwerke AG from 1959 to 1969; deputy chairman of the NSU management board from 1969 to 1972.

He was born on August 19, 1910, studied engineering in Ulm and Munich, and worked as a trainee engineer with a succession of motor-industry companies and supplier firms. Returning to school, he graduated in 1937 with a degree in mechanical engineering from the Munich Technical University. He began his career with NSU-Werke AG as a tool designer and subsequently held several production-engineering assignments. NSU was then strictly a motorcycle manufacturer, and big enough to compete with DKW, BMW, Horex and Zündapp. During World War II, the factory made an important number of "Kettenrad," a miniature half-track steered by a single front wheel, for the Wehrmacht. Bomb and artillery damage to the plant was extensive, and the useable facilities became truck repair shops for the U.S. Army in 1945.

Frankenberger was able to start building small industrial engines in 1946, and resumed production of the Kettenrad for use in forestry and agriculture. The first new two-wheeler was the NSU Quick moped, followed in 1947 by the 100-cc NSU Fox motorcycle. In 1955 NSU produced nearly 300,000 motorcycles. Frankenberger was one of the three directors who decided to start making cars. A prototype with a rear-mounted 20-hp 600-cc parallel-twin was ready by 1957, and Frankenberger laid out a new assembly hall that began turning out the NSU Prinz early in 1958. He provided a budget for the development of the Wankel rotating combustion engine, which paid off handsomely over a long period in the form of license fees. The Sport-Prinz with a Bertone-designed coups body was introduced in 1959 and the Prinz 4 sedan in 1961. Early in 1963 NSU began production of a four-cylinder car, the NSU 1000, and towards the end of the year, presented the Spider, powered by a single-rotor Wankel engine. The last NSU motorcycles were made in 1965, after the tooling, designs, and patents had been sold to Marshal Tito's government in Yugoslavia. That freed up factory space for another car, the up-market front-wheel-drive Ro-80, powered by a twin-rotor Wankel engine, which was introduced in 1967.

He lost his office when Volkswagenwerk bought 60 percent of NSU Motorenwerke AG in 1969, and had to work under the orders of Ludwig Kraus, technical director of Audi. He resigned in 1972.

Frazer Nash, Archibald Goodman (1899–1965)

Managing director of Godfrey & Nash (later GN, Ltd., and GN Motors, Ltd.) from 1910 to 1922; managing director of Frazer Nash Ltd. from 1923 to 1927.

He was born on June 30, 1899, at Hyderabad, Sindh province in India. After preparatory school and the William Ellis School in England, he studied engineering at the City & Guilds College at Finsbury, London, from 1905 to 1907. There he met H.R. Godfrey, his future business partner. He worked as a trainee engineer with Willans & Robinson, steam-engine manufacturers in Rugby, from 1909 to 1910, then returned to London, where he and Godfrey began building sports cars in a shop at Hendon. During World War I he was attached to the Air Armament Design Section, was commissioned and became a pilot with the Royal Flying Corps. Nash and Godfrey resigned from GN Motors, Ltd., in 1922 and went their separate ways. Nash founded a new company and introduced the Frazer Nash sports car with four-cylinder engines (Anzani, Meadows or Blackburne) and his aptly nicknamed "chain-gang" transmission. He fell ill in 1927 and while he was in hospital, his company was sold to H.J. Aldington. After a long period of convalescence he joined Parnall Aircraft in 1933 as technical director and developed the Frazer Nash hydraulic gun turret, which found wide application during World War II. He remained with Parnall Aircraft

throughout the war and in 1945 opened an office as a consulting engineer on Kingston Hill, Kingston-on-Thames. He died in 1965.

Freminville, Charles de (1856–1936)

Technical director of Panhard & Levassor from 1904 to 1911; joint managing director of Panhard & Levassor from 1911 to 1915.

He was born in 1856 as the son of a thermodynamics professor and graduated in 1978 from the Ecole Centrale des Arts et Manufactures. He began his career with the Chemins de Fer de Paris–Orleans, where he spent 20 years in charge of the locomotive fleet and rolling stock, renewal and maintenance of the equipment, and the development of the network. He joined Panhard & Levassor in 1899 as assistant to the technical director, and was promoted to technical director in charge of production in 1904. He was a keen proponent of Taylor's time-and-motion studies and theories of efficient production. He coped with the company's adoption of the Knight-patented sleeve-valve engine and a proliferating model range (11 different models with nine different engines (one six and eight four-cylinder units) in 1909.) After 4½ years as joint managing director he resigned in order to join Schneider as a consulting engineer at the Penhoët naval yards in Brittany. He never returned to the automobile industry but remained active in engineering associations and was a frequent lecturer on the organization and management of engineering departments in industry right up to the time of his death in 1936.

Frisinger, Haakan (1928–)

Director of product and product coordination, Volvo Car Division from 1966 to 1971; works director of AB Volvo's Köping plant from 1971 to 1973; production director of Volvo Car Division and a member of AB Volvo's board of directors from 1975 to 1977; director of AB Volvo, Industry Division, from 1977 to 1979; managing director of Volvo Car Corporation from 1979 to 1983; AB Volvo group managing director from 1983 to 1986.

He was born in 1928 at Skövde and graduated from Chalmers Institute of Technology in Gothenburg in 1951. He began his career by joining AB Volvo as manager of materials handling in the Skövde engine plant in 1952 and was soon appointed to engineering manager for engine production. He resigned from AB Volvo in 1959 to take a position as engineering manager of Bahco (tools) at Enköping, but returned to Volvo a year later as assistant plant manager at Skövde. In 1963 he was named chief engineer and production-engineering manager at Skövde, and transferred to the head office in Gothenburg in 1966. He supervised the completion of the Torslanda plant, which doubled Volvo's car-assembly capacity. At the Köping plant, he held top responsibility for gearbox production for cars, axles and transmissions for trucks and buses. In 1977 he became responsible for AB Volvo's industrial subsidiaries. He ran the car division during a period when the 760 and 740 replaced the 260 and 240 series, but in 1986 he requested an overseas posting and spent the next seven years in Volvo's Paris office. He spoke out against Gyllenhammar's planned merger with the Régie Nationale des Usines Renault, and returned to Gothenburg after Gyllenhammar's departure and played a senior-statesmanlike role in the ensuing corporate restructuring process. He steered the formulation of a post–Renault strategy, and retired in 1997.

Friswell, Charles (1871–1926)

Chairman of the Standard Motor Company, Ltd., from 1907 to 1912.

He was born in 1871 in London as the son of a prosperous businessman, and was privately educated. His father held wide investments in trade and industry and helped the young man set up a bicycle agency in London. In 1894 he became a Peugeot car dealer and drove a Peugeot in the London-to-Brighton "Emancipation Run" in November 1896. His dealership grew into the Automobile Palace, Ltd., in Albany Street, London. He first made contact with R.W. Maudslay at the Olympia motor show in 1905, and after a visit to the Standard factory in Coventry, he offered to take every car Standard could make, and signed a contract as sole distributor for the home and export markets. He provided fresh cash enabling Standard Motor Company, Ltd., to take over a bigger factory at Bishopgate Green, turn the old premises into a coachbuilding works, and set up a service and repairs facility in Aldbourne Road. In 1907 he obtained control of the company and changed the radiator badge to a Union Jack emblem. He began to dictate policy in areas including model range and product specifications which led to conflict with R.W. Maudslay and John Budge, the chief draftsman. While Friswell was on an extended marketing tour of India in 1911, Maudslay established Friswell's failure to fulfill his contract as sole distributor. With fresh capital provided by C.J. Band, a prominent lawyer, and S. Bettmann of the Triumph Cycle Company, Ltd., Maudslay was able to regain control of the Standard Motor Company, Ltd. Friswell's contract was torn up, and the Automobile Palace, Ltd., was liquidated. He tried to restart a dealership as Friswell's Limited in 1913, but that venture collapsed in 1915 due to the priorities of war. He returned to the management of the investments inherited from his father, including mining in South Africa and a metallurgical business in Sheffield.

Froede, Walter G. (1910–)

Director of research, NSU Motorenwerke AG from 1959 to 1975.

He was born on April 28, 1910, in Hamburg. After graduating from high school in 1928 he served a one-year apprenticeship in the Blohm & Voss shipbuilding yards in Hamburg. He studied engineering at the Technical Institute of Berlin, graduating in 1933, and stayed on in Berlin for post-graduate studies. He was noted for his research into the use of X-rays for visualization of the scavenging process in two-stroke engines, and later developed a fuel-injection system with magnetic control by condenser discharge, using compressed air to assure atomization. This work came to the attention of Curt Bükken, technical director of NSU-Werke AG in Neckarsulm. Froede was engaged as a development engineer for the NSU two-stroke motorcycle engines. He also became involved with experimental work on supercharged four-stroke engines. He survived World War II and continued his work at NSU. In 1951 he was approached by Felix Wankel who wanted NSU to use his rotary disc valve, an improved version of the disc valve used on Junkers torpedo engines during the war. That came to nothing, but two years later, Wankel came back with a small compressor, which Froede decided to test on NSU racing motorcycles. Froede knew well that any pumping mechanism has potential use as an engine, and over the next three-year period, Froede, Dr. Wankel, and Wankel's assistant, Ernst Hoeppner, turned the compressor into a rotary engine. The NSU-Wankel DKM was successfully tested on February 1, 1957. But it was not a practical automotive engine. Further rethinking led to the NSU-Wankel KKM, with a revolving eccentric shaft carrying an arcuate-triangular piston. It ran with minimal vibration and formed a compact installation package. Its existence was revealed in 1960, and NSU and Dr. Wankel got busy making licensing agreements all over the world. Froede developed the first production engine, KKM-502, for the NSU Spider in 1963. A two-rotor engine, the KKM-512, went into production for the Ro-80 sedan in 1967. The engine still required a lot of development work, which Froede headed along with other research projects until his retirement to Bad Wimpfen at the end of January 1975.

Fuhrmann, Ernst (1918–1995)

Technical director of Dr. Ing. h.c. F. Porsche K-G in 1971–72; chief executive of Porsche from 1972 to 1980.

He was born on October 21, 1918, in Vienna as the son of a high-ranking jurist and studied at the Vienna Technical University from 1936 to 1941. After Austria's "Anschluss" with Nazi Germany, he was called up for military service and was an officer in the Wehrmacht from 1939 to 1943. Due to his technical qualifications, he was taken out of the uni-

formed ranks and attached to the AEG[1] research institute in Berlin. In 1945 he escaped to Austria and for two years eked out a precarious living repairing watches, clocks, door locks, and other devices. In 1947 he joined the Porsche establishment at Gmünd in Carinthia, and worked on Robert Eberan von Eberhorst's team, notably on the Type 360 (Cisitalia) racing car. He accompanied the Porsche family to Stuttgart-Zuffenhausen in 1950 and completed a thesis on valve gear for high-speed spark-ignition engines which won him a doctor's degree in engineering from the Vienna Technical University. He went on to design the four-overhead-camshaft flat-four engine for the 1954 Porsche Carrera and was named head of engine testing. He resigned from Porsche in 1956 to join Goetze-Werke (piston rings) in Burscheid, and became technical director of Goetze-Werke in 1962.

He returned to Porsche as technical director in September 1971, a period when Porsche cars were winning the CanAm series at Le Mans, at Sebring, at Spa-Francorchamps. He did not become directly involved with the 930 Turbo or the 934/935. He put an end to the 914 series produced as a joint venture with Volkswagenwerk AG, but collaborated closely with Audi engineers on the EA 425 project which ripened into a front-engine sports coupe, Porsche 924, in 1976. The engine was made by Audi and the car was assembled in the former NSU plant at Neckarsulm. Beginning in 1973, he personally directed the Porsche 928 program a sports coupé with a front-mounted water-cooled V-8 engine. He wanted to put a total stop to the 911 rear-engine car and its derivatives and give the Porsche name a new identity as a maker of comfortable, safe, quiet luxury cars. The 928 came on the market in the beginning of 1977. But the 911 continued, with exciting new versions called 911 SC Targa and SC Cabriolet. The Porsche family, led by Ferry Porsche and Ferdinand Piëch, defied Fuhrmann bitterly and deposed him on November 28, 1980. It did not count to them that he more than tripled the company's turnover within nine years. It did count that he tried to do away with one of their family icons, the 911.

He went into retirement and died on February 6, 1995, in his home at Teufenbach in Steiermark, Austria.

1. Allgemeine Elektrizitäts-Gesellschaft.

Fuscaldo, Ottavio (1886–1952)

Technical director of OM from 1922 to 1927.

He was born on March 6, 1886, in Verona and educated as an engineer. He began his career in 1908 with Fides-Brasier, makers of Brasier cars under license in Via Monginevro, Turin. He worked for Fiat as a design engineer from 1909 to 1911, and then joined Felice Nazzaro and Arnold Zoller in starting Fabbrica Automobili Nazzaro in Foro Boario, Turin. He left Nazzaro in 1914 to become a design engineer with Fabbrica Automobili Zust, makers of Brixia-Zust cars in Brescia. He left when the company was taken over by OM[1] in 1917 and went to work for an aircraft-engine manufacturer.[2]

All his life he had a fixation on the lozenge-form chassis, with a central axle, and single wheels front and rear. His first patent for a car he called Rombo, featuring this layout, was granted in 1912, and a prototype was made in the Chiribiri works in 1913. In 1920 he founded Rombo, Societa Anonima Brevetti Fuscaldo, in Brescia, and tried to get Citroën interested in the lozenge configuration. In 1921 he built a new Rombo prototype which he sold to OM, along with manufacturing right to his patents. But the OM board of directors had it shelved. As technical director of OM he modernized their product line, cars and trucks. He changed from straight bevel to spiral bevel gears in the rear axles and in 1924 brought out the 1½-liter Type 469 S with a Roots-type supercharger, and the 2-liter Type 665 S "Superba" with an overhead-camshaft six. In 1926 he designed a V-12 1½-liter prototype racing car and a third version of the Rombo.

Leaving OM prior to its big reorganization of 1928, with the severance of the automobile branch in Brescia from the parent locomotive and rolling-stock works in Milan, he moved to Paris to try to exploit his patents. He may have influenced Gabriel Voisin, who built a lozenge-shape prototype in 1934, but never concluded a license contract. He wrote technical publications for Editions Dunod, including a book on high-speed two-stroke engines published in 1932. He returned to Italy and designed a military tractor with four wheels in lozenge formation, which he sold to Isotta Fraschini in 1935. From 1934 to 1943 he was head of an advanced experimental department in the CEMSA-Caproni[3] works at Saronno, where a light armored car and a scout car, both with lozenge-configuration chassis, were built as prototypes in 1941. There was no follow-up, however. In 1935 he obtained a patent for an aircraft-engine fuel-injection system which was adopted by Caproni, and also adapted for Moto-Guzzi racing motorcycles, and tested by Alfa Romeo in a car that ran in the 1940 Mille Miglia. He also designed a preselective gearbox which was extensively tested by Fiat in 1937. In 1946 he produced his final Rombo design, which was presented in the United States but without success. He died on August 31, 1952, in Milan.

1. Officine Meccaniche.
2. Società Bresciana per Motori di Aviazione.
3. Costruzioni Elettro-Meccaniche Saronno, subsidiary of Caproni aircraft group.

Ganz, Josef (1898–1967)

Independent engineering consultant; tireless proponent of small economy cars.

He was born in Hungary in 1898, and educated in Austrian and German schools. He settled in Germany, and became a German citizen in 1915. In 1923 he wrote an article in the magazine *Auto Kritik*, pointing out the advantages of a rear-mounted flat-twin engine in combination with swing-axle rear-suspension for a light car. He became editor of *Motor Kritik* and in 1929 drew up a complete small car for Dr. Wilhelm Wittig of Zündapp, motorcycle manufacturers in Nuremberg. But the timing was wrong, for Zündapp had no budget for a car project. Ardie Motorenwerk AG, rival motorcycle makers in Nuremberg, were better off and built a prototype to Ganz's designs. The Ardie-Ganz had a backbone frame and all-independent suspension, with a 175-cc Rinne engine and chain drive to the rear swing axles. In 1931 Ganz offered an improved design to his friend H.G. Röhr of Adler-Werke in Frankfurt, and a prototype "Mai-käfer" was built. But Röhr would not consider producing it as an Adler. One year later Ganz had developed it further and offered it to Wilhelm Gutbrod, motorcycle maker in Ludwigsburg. With a rear-mounted two-cylinder two-stroke 400-cc engine, it went into production in 1932 as the Standard Superior and advertised as the "Deutschen Volkswagen." In 1934 he fled to Switzerland to escape Nazi persecution for his Jewish origins.

In 1938 his name figured as the designer of the Erfiag prototype car, a modernized version of the Maikäfer. It returned in 1945 as the Rapid, made by a lawnmower manufacturer in Dietikon, with an air-cooled 350-cc MAG flat-twin. But only 34 cars were produced. Ganz emigrated to Australia where he joined the engineering staff of GM-Holden. He died in 1967 in Melbourne.

Garcea, Giampaolo (1912–)

Head of the experimental department of Alfa Romeo S.p.A. from 1941 to 1956; head of the Alfa Romeo design office and research department from 1956 to 1977.

He was born in June 1912 in Padua and graduated with a degree in mechanical engineering in Padua. He won an Alfa Romeo sponsorship to the Turin Polytechnic where he picked up a degree in aeronautical engineering. In 1935 he went to work for Alfa Romeo as a test engineer in the aircraft engine department. After 1945 he worked on the development of current models, notably the 6C 2500, and new projects, beginning with the 1900 series. He also played a part in the 430 and 900 truck projects, and was prominent in the development of the Giulietta. He laid the ground work for the company's first front-wheel-drive vehicle, the Romeo van, which went into production in 1958, with a choice of a Giulietta engine or a four-cylinder Perkins diesel. He shared the main responsibility for the Alfetta with Giuseppe Busso and was

involved with the planning of the Alfa Six. He retired in 1977 but remained a consultant to Alfa Romeo until 1982.

Gentil, Edmond (1874–1945)

Managing director of Edmond Gentil from 1895 to 1902; chairman of Société Gentil & Cie from 1902 to 1914; chairman of Société des Automobiles Alcyon from 1914 to 1929.

He was born on November 13, 1874, at Velentigney (Doubs) and went to Paris as a young man. He held positions in the bicycle industry and among its suppliers, becoming manager of Société des Cycles La Française and a director of Usines Thomann. He began building bicycles at Courbevoie in 1895 and introduced the first Alcyon motorcycle in 1904. He set up Société Gentil & Cie as a holding for his industrial interests in 1902. In 1906 he opened another factory in Neuilly for producing single- and four-cylinder engines, and assembling Alcyon cars. He also founded Société des Moteurs Zürcher at Courbevoie, producing engines under Zédel license. From 1909 to 1914 the Alcyon cars were built under Zédel license and powered by Zürcher engines. The Neuilly plant was closed in 1912 and auto assembly transferred to the motorcycle works at Courbevoie.

His enthusiasm for cars faded after World War I and the Alcyon cars made from 1923 to 1927 were pure SIMA–Violet. Motorcycles accounted for the bulk of Alcyon turnover, and he held office as president of the bicycle dealers' association from 1918 to 1927 and a syndicate of suppliers and traders.[1] In later years he served as president of Société Centrix, headquartered in Paris with a factory at Caluire outside Lyon, and president of Société des Etablissements Chavanet at Saint-Etienne.

He died unexpectedly in 1945.

1. Syndicat des Négociants pour Fournitures du Cycle, de l'Automobile, de l'Aviation et des Sports.

Georges, Yves (1914–)

Technical director of the Régie Nationale des Usines Renault from 1964 to 1976.

He was born in 1914 in Paris as the son of a general in the French Army. He started out aiming for a military career and was accepted at the Ecole Polytechnique. But he graduated in engineering, and joined Avions Mathis in 1945. He was responsible for the design of a radial air-cooled seven-cylinder engine, but left Mathis to became head of the Chenard-Walcker drawing office, then working almost exclusively for Chausson. He led the Chausson mechanical and metallurgical research laboratories until the end of 1955 and joined Renault on January 2, 1956, as assistant to the technical director, Fernand Picard. He was soon promoted to director of automobile research and development, and was at the origin of the front-wheel-drive Estafette van that went into production in 1958. He directed the R 3–R 4 economy car program, putting two front-wheel-drive models in production in 1961. The R 4 was built up to 1992, reaching a total of over eight million units. He directed the engineering of the R 16 front-wheel-drive family car and was named technical director in 1964. He planned to go front-wheel-drive across the board and had the R 8 and R 10 taken out of production in 1972. The R 12 came out of a joint program with Ford and the R 14 came out of collaboration with Peugeot. The R 30 stemmed from an experimental. safety car. In 1976 he was transferred from holding overall responsibility for research, engineering and development to take charge of advanced research. He retired in 1980 but remained a scientific consultant to Renault.

Ghidella, Vittorio (1931–)

Managing director of Fiat Auto S.p.A. from 1978 to 1988; president of Automobili Ferrari from 1984 to 1988; and chairman of Alfa Lancia Industriale S.p.A. in 1987–88.

He was born in 1931 in Vercelli and broke off his schooling to go to work in the sales department of an oil company. He attended evening classes in engineering and graduated from the Turin Polytechnic. He joined Fiat in 1956 and worked on a number of production-engineering assignments before becoming mechanical production manager at the Lingotto plant. In 1963 he was transferred to RIV (ball bearings), serving as plant manager and pro-

duction manager until he was named managing director of RIV in 1974. He proved himself a live-wire executive and in March 1978, Gianni Agnelli picked him to replace Nicola Tufarelli at the head of Fiat Auto. He made a big push for new models, and the Panda arrived in February 1980. The Uno replaced the 127 in March 1983, and the Regata replaced the 131 in October 1983. He got the Lancia Thema into production in 1984 followed by the Fiat Croma in 1985 and the Fiat Tipo in 1987. He led Fiat's side in the negotiations with Ford in 1985–86, aimed at merging the European passenger-car operations of the two groups, but the plans were torn up because both wanted to be "master." In 1987 Ghidella successfully organized Fiat's take-over of Alfa Romeo. A year later, he fell out with Cesare Romiti over policy and investment priorities, and resigned in November 1988. Ford wanted to hire him as a consultant, but his contract with Fiat barred him from working with any rival groups for six months, and that was too much for Ford. Due to stock options accruing from his title with Ferrari, he had accumulated a big block of Ferrari shares. He sold them back to IFI[1] and founded his own holding: GIG (Gruppo Industriale Ghidella), took over Graziani Trasmissioni and merged it with Oto Trasm, a former branch of Ernesto Breda. In January 1992 he sold GIG to Dr. Tito Tettamanti who controlled the Saurer Group Holding, and became an executive of Saurer. But in June 1994 he sold his parts in Saurer Group and retired.

1. Istituto Finanziaria Industriale, the Agnelli family holding.

Giacosa, Dante (1905–1996)

Head of Fiat's passenger-car drawing office[1] from 1935 to 1945; director of Fiat's motor-vehicle technical office[2] from 1946 to 1955; director of Fiat's top management for motor-vehicle technical affairs[3] from 1955 to 1966; Fiat's director of project management and motor-vehicle studies,[4] member of the board of directors from 1966 to 1970.

He was born on January 5, 1905, in Rome into a family from Alba in Piedmont. He studied mechanical engineering and graduated from the Turin Polytechnic in 1927. He began his career with Olivetti (office machines) but left within months to take up a position on the SPA[5] engineering staff. He was assigned to the military-vehicle design office and worked on the Pavesi P4 tractor, the CV-31 armored vehicle, the Fiat 611 truck chassis, and OM gun carrier. He worked in Fiat's passenger-car design office in 1931, but a year later was transferred to the aircraft-engine department as a design engineer. Late in 1932 he was promoted to section head and led the design of the Fiat A33 RC engine. Unofficially he became involved with a project for a two-passenger economy car, which at the outset had front-wheel-drive, with a 500-cc four-cylinder water-cooled engine mounted in the front overhang. He had to abandon front-wheel-drive because reliable low-cost constant-velocity universal joints could not be found, and the final version had a live rear axle.

He was transferred to the passenger-car section when his chassis drawings were approved. It was given a modern, attractive body designed by Rodolfo Schaeffer and went into production in the spring of 1936 as the Fiat 500. He led the development of the 508 C "Balilla" and headed the design team for the six-cylinder Fiat 2800 of 1938. On January 1, 1940, he was given additional duties as vice-director of the automobile division. He spent the war years in Turin, mainly occupied with aircraft-engine projects. In 1943 he began preliminary studies for postwar cars and trucks. Privately, in his spare time, he designed a small racing car chassis for Piero Dusio. It became the first Cisitalia. In 1947 he patented a power-train layout for a front-wheel-drive car, with a horizontal, transverse, four-cylinder in-line engine. He planned a total product renewal with a new numbering system. The numbers, starting at 100, were allocated chronologically, by date of the go-ahead order. Type 101 was the first to be put in production, starting in 1950 as the Fiat 1400. Type 103 followed in 1953 as the Fiat 1100. Type 100, after a long gestation period, appeared in 1955 as the Fiat 600, with a rear-

mounted 767-cc water-cooled in-line four in a tight four-passenger body measuring less than 3.3 meters in length. Type 110 followed in 1957, as the Nuova 500, with an air-cooled parallel-twin engine installed in the rear overhang. Type 102 was based on his front-wheel-drive patent, and led to Type 109, an early prototype for the Autobianchi Primula of 1964.

Type 104 was the 8V sports coupé of 1952 and Type 105 was the Campagnola off-road vehicle for civilian and military markets. Types 112 and 114 were six-cylinder sedans (1800 and 2100) launched in 1959, followed by the four-cylinder 115 and 116 (1500 and 1300) in 1961. The 1500 roadster of 1960 had number 118. Type 120 was a 500 station wagon with its engine laid underfloor. From 1961 onwards, he devoted himself to making a success with front-wheel drive and paid scant attention to the 125 and 124 projects. The front-wheel-drive revolution began with the Autobianchi Primula and continued with the Fiat 128 in 1969, and 127 in 1971, along with the Autobianchi A-112.

He retired in 1970 but remained a technical consultant to Fiat's board of directors. He died on March 31, 1996 in his home in Turin.

1. Capo Ufficio Tecnico Vetture.
2. Direttore degli Uffici Tecnici Autoveicoli.
3. Direttore della Direzione Superiore Tecnica Autoveicoli.
4. Direzione Progetti a Studi Autoveicoli.
5. Societa Ligure Piemontese di Automobili, acquired by Fiat in 1926.

Gieschen, Wilhelm H. (1908–)

Director of product planning, testing and engineering for BMW from 1961 to 1964; director of BMW production and quality control from 1964 to 1971.

He was born on November 2, 1908, in North Germany (Oldenburg) and received a technical education. He held a diploma in mechanical engineering. He joined Hansa-Lloyd as a trainee in 1924 and was a member of the Hansa-Lloyd engineering staff until 1931, when he joined Focke-Wulf Flugzeugbau. He became a glider pilot and aircraft designer, and in 1941 was named works manager of

Focke-Wulf, where he remained throughout the war years.

In 1945 he joined Borgward and played a central role in getting truck production restarted in the Bremen factories. He organized production of the Lloyd and Goliath cars, and the Hansa 1500, introduced in 1949. During the 1950s he supervised the reorganization of the Borgward group's factories with modern equipment, and in the final years of Borgward, he served as the group's head of planning and special developments.

He arrived in Munich on September 1, 1961, to take charge of new car programs. He suggested that the 1100-cc flat-four of the Lloyd Arabella, whose design and tooling could be picked up at a low price, would be a good basis for a replacement for the BMW 700. But the idea was turned down in favor of an all-new 1500-cc model slotted well up-market from the 700.

BMW's new management board decided to remove him from product planning in order to make the best use of his primary qualifications: production engineering. He was in charge of integrating Glas production at the Dingolfing plant with BMWs at Munich.

He retired from BMW in August 1971.

Girardot, Léonce (1864–1922)

Technical director of CGV (Charron, Girardot & Voigt) from 1901 to 1905; technical director of GEM (Société Générale d'Automobiles Electro-Mécaniques) from 1905 to 1910.

He was born on April 30, 1864, in Paris and was educated as an engineer. In 1895 he went into partnership with Emile Voigt and Fernand Charron as automobile dealers in Paris, and began an illustrious career as a racing driver. He won the Paris–Amsterdam–Paris road race of 1899 on a Panhard-Levassor and drove a CGV to second place in the Circuit des Ardennes in 1903. In 1902 he designed and built an eight-cylinder car (it was actually two four-cylinder engines mounted end-to-end). He erred in thinking that with torque of that magnitude, no gearbox was required, and the car was a failure. The touring cars had T-head four-cylinder engines and four-speed gear-

boxes mounted in unit with the differentials and jackshafts for the chain-drive sprockets. In spite of high prices, CGV produced and sold 196 cars in 1903 and 265 cars in 1905. He left CGV when the company was taken over by Charron, Ltd., in 1905. Emile Voigt went to America. Girardot began to develop ideas for a new type of electric propulsion. His "mixte" (hybrid-electric) system appeared on the 1907 GEM car which combined a 20-hp four-cylinder engine with an electric clutch, a generator, electric motors, and a pack of storage batteries.

After the liquidation of GEM, he became the French representative of the Daimler-Knight interests, selling licenses for the sleeve-valve engine.

Giusto, Nevio Di *see* Di Giusto, Nevio

Glas, Hans (1890–1969)

Chairman of Hans Glas GmbH Isaria Maschinenfabrik from 1949 to 1961.

He was born on June 12, 1890, in Dingolfing as the son of Andreas Glas, founder of Landmaschinenfabrik Glas in 1883, a manufacturer of farm machinery. At the age of 15 he was apprenticed to his father's shops, and in 1909 left the family to work for Massey-Harris in Berlin. A year later he was selected for training at the Toronto, Ontario, main plant, where he stayed until 1914. He had put his savings in a New York bank which crashed, leaving him stranded. He went to work for McCormick as a factory hand in Chicago, was an assembly worker in a Ford plant, became a foreman with American Bosch, and production manager of Hendee Manufacturing Company, Inc., makers of the Indian motorcycle at Springfield, Massachusetts. He returned to Germany in 1920, to find the family enterprise, now named Isaria Landmaschinen-GmbH, in the hands of strangers. He began buying Isaria shares and 12 years later emerged as the sole owner. He modernized the product line and made the factory one of Europe's biggest makers of seed spreaders. Production practically stopped during World War

II and the factory survived on repair work. The postwar market brought renewed growth, and he started to think about other product lines. He reorganized his private company as a family-owned GmbH (limited liability stock corporation) in 1949 and prepared a motor scooter which he called Goggo (nickname of his newborn grandson). Scooter production began in 1951, and some 60,000 were made by 1956. He soon decided to get out of the scooter market, not only because it had peaked, but because he wanted to offer his customers personal transport with full weather protection. He began dreaming of a car, but it must be a real car, with seating for four and adequate power. The Goggomobil was designed by Karl Dompert, who had served in the Luftwaffe with Andreas "Anderl" Glas (son of Hans) and appeared in October 1954 as a two-door sedan with an all-steel body, all-independent suspension and 10-inch wheels, and a rear-mounted air-cooled two-stroke parallel-twin, mounted transversely. Felix Dozekal had designed the engine when he was with Adler, and took it with him to Dingolfing. Production began in March 1955 and over a ten-year period, Glas made over 245,000 of them. Coupé and convertible versions were added in 1957. A front-wheel-drive model with an air-cooled flat-twin was developed, but was redesigned with rear wheel drive before production of the Glas Isar T600 and T700 began in 1958.

He retired in 1961, leaving the enterprise in the hands of "Anderl" who pursued a relentless march up-market but sold out to BMW in 1966. Hans Glas lived to see the demise of the Glas car and died on December 14, 1969, in Dingolfing.

Gobbato, Pier Ugo (1918–)

Assistant to Fiat's director of production from 1967 to 1969; deputy managing director of Lancia in 1969–70; managing director of Lancia from 1970 to 1975.

He was born on June 12, 1918, in Florence as the son of Ugo Gobbato. He attended school in Turin and college in Milan. In 1941 he graduated from the Milan Polytechnic with a degree in mechanical engineering. He was a fighter pilot in World War II, had four kills to his credit, and was shot down once. In January 1946, he joined Motori Marini G. Carraro in Milan and was named managing director of Carraro in 1950. He joined Fiat in 1955 and was detached to Grandi Motori Trieste, where the huge two-stroke diesels for big ships were made, as assistant to the director. In July 1958, he was transferred to SPA in Turin, Fiat's main truck-and-bus division, and was named vice-director of SPA in April 1959. Fiat's OM truck division in Brescia took over Motori Marini G. Carraro, and in 1962 he returned to Milan. He held a director's title with Ferrari in 1965–66 and then returned to Turin as a production engineer at the new Rivalta works. He was responsible for production-engineering the Dino V-6 engine and the Fiat Dino chassis, and supervised production and quality control for the 124 Sport coupe. When Fiat took over Lancia in 1969, he was sent to examine their plans for new models. There weren't any. He rushed the Beta into production to keep the plants going, and originated the concepts for the Beta HPE and Beta "Montecarlo." When he saw Bertone's Stratos prototype, he drove Lancia's engineers to develop a Group 4 racing car with the Dino V-6 in an inboard-rear transverse position. He cut the Fulvia range back to a GT coupe and a rally car and organized the production of the Gamma, planned jointly with Citroën. In 1975 he was replaced by Giovanni Sguazzini, but stayed on with Lancia as general manager until his retirement in 1977.

Gobbato, Ugo (1888–1945)

Managing director of Alfa Romeo S.p.A. from 1933 to 1945.

He was born on July 16, 1888, at Volpago del Montello into a family of small farmers and was educated at schools in Treviso and the Vicenza Professional Institute. He continued his studies in Germany and held diplomas in electrical and mechanical engineering from the Zwickau Polytechnikum in Saxony. He returned to Italy in 1909 and reported for military service. After discharge in 1911 he began his career with a manufacturer of hydraulic presses in Milan. From 1912 to 1915 he was a plant manager for Società Ercole Marelli in Sesto San Giovanni, producing industrial motors, ventilators and centrifugal pumps. He was back in uniform in World War I, and in 1917–18 was in charge of a depot at Taliedo for military vehicles and materiel. In 1918 he was reassigned to an aircraft factory[1] by the Commissioner of Aviation. He joined Fiat in March 1919, and Guido Fornaca gave him the task of reorganizing its passenger-car production. As early as 1916, Fiat had bought 400,000 square meters of open land between Via Nizza and the railroad tracks in south-central Turin, where construction work began in March 1917. The five-story building was designed by an industrial architect, Giacomo Matte-Trucco, and Gobbato laid out the machine shops, materials flow paths, assembly lines, and ordered the tooling. The Lingotto works came on stream in October 1923 with 6,900 workers. By mid-year 1927 Fiat had 14,000 workers at Lingotto. During this period he lectured on industrial engineering at the Turin Polytechnic, and his lectures from 1928 to 1931 were published in book form. Vittorio Valletta sent him to Heilbronn in 1929 after buying NSU's passenger-car factory, and in 1930 he was delegated to Spain to supervise the production of the Hispano car at Guadalajara assembled from Fiat 514 parts shipped from Turin. He spent the years 1931–33 in Moscow, setting up the world's biggest ball-bearing factory, using RIV/SKF technology. In 1933 Mussolini wanted him to run Alfa Romeo, and Fiat released him. Within months of arriving in Milan, he had given Alfa-Romeo a new corporate structure, making five separate divisions: automobiles, aircraft engines, foundries, forges and sheet-metal processing, and auxiliary activities. He secured a diesel-engine license from Deutz and an aircraft engine license from Daimler-Benz. Alfa Romeo had also been making air-cooled radial aircraft engines under Bristol license. Aircraft engines was the company's main line of activity, for in 1935 the plant turned out only some 500 cars and 350 trucks and buses. In 1938 Gobbato built a new air-craft-engine plant, on order from Mussolini, at Pomigliano d'Arco near Naples. In 1945 he was arrested and tried for collaboration with the enemy (Nazi Germany changed from an ally to an enemy on Mussolini's removal from power in July 1943). On April 27, 1945, Gobbato was cleared of all charges. The following day he was assassinated by a volley of gunfire from a passing car as he was walking home from the office.

1. Stabilimento di Costruzioni dell'Aviazione SCAF, in Florence.

Gobron, Gustave (1846–1911)

Chairman of the Société des Moteurs Gobron et Brillié from 1898 to 1911.

He was born in 1846 into an old Ardennaise family of landowner-industrialists. He went into politics and represented the Ardennes in the Senate. He first came to fame in 1871 when he escaped from Paris, which was besieged by Bismarck's troops, by balloon, to join the Armée de la Loire. He was addicted to technical inventions, and the opposed-piston engine of the Gobron-Brillié cars may have originated in his brain, though Eugène Brillié was the engineer who turned it from an idea into reality. They became business partners, financed by Senator Gobron, and opened a factory in rue Philippe de Girard in Paris in 1898. Two years later they moved to roomier premises in the rue de la Révolte, Levallois-Perret, and annual production climbed to 150 cars. He sold manufacturing rights to Maurice Nagant and the Société d'Automobiles Nancéiennes in 1900. The partnership with Brillié broke up in 1903, when the engineer left to join Schneider, but the Senator continued car production, specializing in heavy, high-performance touring cars, with the same type of opposed-piston engine. On July 17, 1904, Louis Rigolly drove a Gobron-Brillié powered by a 120-hp 16½-liter engine at a timed speed of 166.16 km/h on the Ostend-Nieuport highway. With a new chief engineer named Prod'-homme, the company branched out into aircraft engines in 1908, sticking strictly to the opposed-piston principles.

Godfrey, Henry Ronald (1887–1968)

Partner in GN Limited from 1910 to 1920 and GN Motors, Ltd., from 1920 to 1922; managing director of the HRG Engineering Company, Ltd., from 1935 to 1958.

He was born in 1887 in London and went to work as an office boy with the agents for Werner motorcycles. He studied engineering at Finsbury Technical School where he met A. Frazer Nash. Both went to work as trainees with Willans & Robinson, electrical engineers, at Rugby. Returning to London in 1910, they began production of the GN cyclecar at Hendon, Middlesex. After World War I they moved to Wandsworth, London, but left GN in 1922. Godfrey went into partnership with an aeronautical engineer, Stuart Proctor, and they began building a small sports car in 1928 with the powertrain from the Austin Seven, but no more than ten were built. He worked in general engineering positions from 1930 to 1935, but was active in the Junior Car Club, where he met Edward A. Halford, who had retired with the rank of major from the Army in 1930. They began making plans for a new sports car, and were joined by Guy H. Robins, a former production engineer of Leyland Motors, Ltd. The first HRG was completed in a garage of the Mid-Surrey Gear Company at Norbiton in 1935, with a Meadows 4Ed engine, Moss gearbox, ENV rear axle, and Rubery, Owen frame. They bought a site at Tolworth, off the Kingston Bypass, and erected a small factory in 1936. They made about three cars a week, and survived on subcontracting work, which was abundant during World War II. HRG had fitted both Triumph and Singer engines in 1939, but after 1945 used Singer power exclusively. With a production of three to five cars a week, the make achieved a fame out of all proportion to its commercial importance by consistent class wins in endurance races, Le Mans, Spa-Francorchamps, and Donington. A streamlined body by Fox & Nicholl was tried on the old chassis, but only a few were built. Godfrey's old friend Stuart Proctor joined HRG as technical director in 1950 (when Guy Robins retired) and designed a new model. The 1955 prototype featured a twin-cam head on the Singer engine, disc brakes and all-independent suspension. But the last HRG cars were made in 1956, and Godfrey retired in 1958, settling near Guildford where he died in 1968.

Good, Alan Paul (1907–1953)

Chairman of L.G. Motors (Staines) Ltd. from 1935 to 1937; chairman of Lagonda Motors, Ltd., from 1937 to 1947.

He was born in 1907 in Ireland. He was about 18 when his father gave him £1000 to make his start in the business world. He went to London, studied law and became a partner in a law firm. He practiced as a solicitor at Lincoln's Inn and made a fortune in the property market and trading in industrial stocks and bonds. In 1935 he learned that the Lagonda Motor Company, Ltd., was for sale. Its managing director since 1928, brigadier-general C.H.N. Metcalfe, had invested in a small sports car, the Rapier, and prepared to build 400 of them, which broke the company's solvency. Metcalfe died in 1935 and Basil Holden was named receiver. T.G. John tried to buy it, but the Alvis board of directors restricted his bid to £35,000. Sir Arthur Sidgreaves of Rolls-Royce, Ltd., was prepared to bid higher, but stalled when a Lagonda won the 24-hour race at Le Mans, boosting the make's fame and its price. That's when A.P. Good stepped in with a bid for £67,500, which Basil Holden accepted when Good agreed to pay an additional £4000 for the stores. Good was acting for a partnership including R. Gordon Watson, former sales director of Hillman and Humber. Watson became managing director of L.G. Motors (Staines) Ltd. and persuaded his friend W.O. Bentley to join them as technical director.

Good immediately sold the Rapier to its designer, Timothy Ashcroft, who founded Rapier Cars Ltd. and continued production of the little sports car with its twin-cam 1100-cc engine up to 1939. Watney told Frank Feeley to restyle the Lagonda for 1937, with more daring, sportier lines. The results were spectacular. The availability of a Lagonda V-12 added tremendous prestige, too. During World War II the Staines factory produced gears and rockets, flame throwers, aircraft seats, and trolleys for loading bombs on planes. W.O. Bentley and his engineering team had a new Lagonda model ready in 1946, but production was slow in getting under way. A.P. Good was devoting more time to a scheme for combining Britain's diesel-engine makers into a group centered on Petter's of Yeovil. It did not work out, though Petter's later merged with R.A. Lister of Manchester. In March 1947, A.P. Good sold Lagonda to David Brown — and the Staines factory to Petter's. He served for a period as managing director of the Brush Electrical Engineering Company, and became chairman of the Heenan & Froude Engineering Group, and chairman of the Associated British Engineering.

He fell ill and died on February 10, 1953.

Grade, Hans (1879–1946)

Technical director of Grade Motorenwerke GmbH from 1908 to 1945; managing director of Grade Automobilwerke und Aviatik from 1916 to 1930; chairman of Hans Grade-Werk GmbH from 1921 to 1945.

He was born in 1879 in Magdeburg, graduated from high school in Berlin-Charlottenburg, and had some further technical education. He built a motorcycle in 1903 and two years later designed a range of air-cooled two-stroke engines suitable for small road vehicles and motor boats. Engine production began in a roomy factory at Bork bei Brück in 1908. He made his own fuel-injection system for an aircraft version of his engine, built his first airplane in 1908, and set up Flieger-Werke H. Grade at Bork bei Brück in 1910. In 1915 he made a prototype car with two seats in tandem and a torpedo-shaped body, powered by a 500-cc Grade engine with a four-speed gearbox and belt drive to the rear wheels. He founded a company to make cars in 1916, based at Mulhouse-Burzweiler in Alsace, and took it back to Germany at the end of World War I, with headquarters at Bork and a branch

factory, set up for engine production in 1911, at Freiburg in Breisgau. In 1920 he also set up a consulting, test and development business[1] in Berlin. He redesigned his little tandem-car with a welded-up steel-tube frame and canvas body panels, patented the design, and began production in 1921. In 1923 he introduced a new model with a 1440-cc flat-twin and a unit-construction sheet-steel body shell, friction drive with chain to one rear wheel, the other free-wheeling. This machine was produced, with periodic additions to the equipment, up to 1930. Hans Grade-Werk GmbH was a holding company, founded on February 17, 1921, with his own capital, plus financial support from the Kolonialbank AG, Gefila[2] and a wholesale merchant in Berlin, Paul Herzberg. In 1931 he invented a new automatic transmission with a hydraulic torque converter, using water instead of oil. He removed the flywheel from a Ford Model A engine, put the torque converter on, added a two-speed gearbox, mounted it all in a chassis, and began testing. But it went nowhere.

After 1933 he was on the fringes of the aircraft industry. He survived the war but died in 1946.

1. Versuchs-GmbH zur Ausarbeitung und Verwertung technischer Konstruktionen.
2. Gesellschaft für Industriemotoren und Landwortschaftliche Maschinen mbH.

Grégoire, Jean-Albert (1899–1992)

Technical director of Automobiles Tracta from 1926 to 1992; technical director of Société Anonyme Française de l'Automobile from 1937 to 1941; technical director of Hotchkiss from 1950 to 1953.

He was born on July 7, 1899, in Paris and graduated from the Ecole Polytechnique in 1923 with degrees in law and engineering. He began his career as an engineer with a textile machinery firm, but went to Madagascar in 1924 on assignment for the petroleum-drilling company.[1] In 1925 he was back in Paris as co-founder of an auto agency and repair shop[2] in partnership with Pierre Fenaille (1888–1967). Fenaille was also the inventor of the Tracta constant-velocity universal joint and held the basic patents. They decided to demonstrate the fine qualities of the Tracta joint by building and racing sports cars, which Grégoire designed. The first Tracta car appeared in 1926 with a supercharged 1100-cc SCAP engine and front-wheel drive. Tracta cars were entered in the 24-hour race at Le Mans for several consecutive years, showing remarkable stamina and reliability. From 1928 to 1934 Grégoire also took out a number of patents for applications of Tracta joints, one of which brought him a fortune in World War II when the Jeep and a multitude of American and British all-wheel-drive trucks, scout cars and armored vehicles adopted the Tracta joint. Tracta car production ended in 1934, but the company continued as a holding for the patents and a base for new projects.

He entered a small-car design contest with a front-wheel-drive entry that attracted the interest of Jean Dupin (1891–1976) of Aluminium Français because of its extensive use of light alloys. The makers of the Amilcar agreed to make a production version, which became the Compound, with a cast-aluminum frame supplied by Aluminium Français. Under a contract with the Compagnie Générale d'Electricité he made a small battery-electric car in 1942. In 1943 Jean Dupin sponsored his next project, a front-wheel-drive economy-car with an air-cooled flat-twin, variable-rate coil-spring rear suspension, and a one-piece cast-aluminum frame and bulkhead. The AF-Grégoire was a brilliant concept, and the French government maneuvered to have it put into production by Simca. When that foundered, it was offered to Panhard — and accepted. It became the basis of the 1946 Dyna Panhard. Licenses for the AF-Grégoire were also sold to Denis Kendall in Britain, Laurence J. Hartnett in Australia, and Henry J. Kaiser in the United States. But none of these projects proved viable.

He designed a bigger car with a water-cooled flat-four, front wheel drive, and a wind-tunnel tested low-drag body, the Grégoire R, which was sold to Hotchkiss. From 1951 to 1953 Hotchkiss built 247 of them, at outrageous cost. He was also the designer of the SOCEMA-Grégoire[3] prototype of 1952. It did not have front wheel drive, but a Turboméca gas turbine driving the rear wheels, and an ultra-streamlined coupé body. After leaving Hotchkiss, he redesigned the car as the Grégoire Sport, with an enlarged flat-four, equipped with a Constantin supercharger, and signed up Henri Chapron to make coupé and cabriolet bodies. But very few were built in 1956–57. About 1963 he reactivated his thoughts about battery-electric cars and built a new electric city-car prototype in 1966. But it led to nothing.

He died on August 18, 1992, in Paris from heart failure.

1. Compagnie Minière des Pétroles de Madagascar.
2. Société des Garages des Chantiers de Versailles.
3. Société de Constructions et d'Equipements pour l'Aviation, a subsidiary of CEM (Companie Electro-Mécanique).

Griffin, Charles (1918–1999)

Chief experimental engineer for the Nuffield-origin makes of British Motor Corporation from 1952 to 1961; chief experimental engineer for all BMC cars from 1961 to 1963; BMC director of engineering from 1963 to 1974; technical director of British Leyland from 1974 to 1975.

He was born on August 13, 1918, and served an apprenticeship with BSA,[1] one of Britain's biggest motorcycle manufacturers. He joined Wolseley as a draftsman in 1940 and helped design the wings of the Horsa glider plane. In 1945 he became experimental engineer for Wolseley and designed the 1947 Nuffield taxicab. When the Wolseley factories closed in 1949 and Wolseley car production transferred to Cowley, Oxford, he moved to Cowley as deputy chief experimental engineer.

He directed the Wolseley 1500 and Riley 1.5, Austin A-99 and Wolseley 6/99 programs. He was the principal designer of the ADO 16,[2] sold as the Morris 1100, Austin 1100 and MG 1100. He held overall responsibility for the ADO 71 (Princess) range and supervised the Metro, Maestro and Montego. He retired in 1975 and died on October 31, 1999.

1. Birmingham Small Arms, Ltd.
2. Austin Design Office.

Grillot, Pierre Marcel Alphonse (1887–1967)

Head of final assembly, Renault passenger cars, from 1928 to 1941; director of Renault production from 1941 to 1945; joint managing director of Renault from 1945 to 1956.

He was born in 1887 in Burgundy as the son of a farmer but was more interested in engineering than in agriculture. He graduated with top marks from the Ecole Centrale des Arts et Manufactures at Chalons-sur-Marne in 1907. He began his career with Delaunay-Belleville at St. Denis, leaving in 1911. After a stint with Clément-Bayard in 1912 and Hanriot in 1912–13, he joined Renault's aviation department in 1913. By 1917 he was Renault's production manager for aircraft engines. In 1928 he was placed in charge of passenger car assembly, where he introduced an elementary type of "just-in-time" supply method. His colleagues regarded him as a human computer, for Grillot synchronized the output from the manufacturing branches so that frames, engines, transmissions, front suspension, steering, brakes, rear axle, and body were available when needed on the assembly line. In 1931 he told Louis Renault how uneconomical it was to produce so many different models of similar size when just one would satisfy the market. The Juvaquatre was supposed to be the answer, but it did not go into production until 1937, and did not replace the Primaquatre, Novaquatre or Vivaquatre until World War II put a stop to car production.

In 1944 he got the task of putting the 4 CV, a completely new type of car with rear-mounted engine and unit body construction, into production. In September 1945, he told the board members that "we could lighten the Juvaquatre and produce it at the same cost as the 4 CV." The Juvaquatre was in fact built as a sedan through 1948 and as a station wagon through 1960. After the death of Pierre Lefaucheux in 1956, Grillot was de facto chief executive of Renault for a period until Pierre Dreyfus took over. Grillot retired in 1956 and died in 1967.

Grinham, Edward G. (1892–1968)

Chief engineer of Humber, Ltd., from 1922 to 1930; chief engineer of Standard Motor Company from 1931 to 1936; technical director of Standard from 1936 to 1941 and 1945 to 1956.

After serving an apprenticeship with Vickers, Ltd., he began his career as chief draftsman of Arrol-Johnston in Dumfries in 1917, where he assisted G.W.A. Brown in designing the Victory model with its overhead-camshaft 2.8-liter four-cylinder engine. Leaving Dumfries for Coventry, he designed the 1923 Humber 8 hp, renamed 8/18 for 1925 and 8/20 for 1926. He created the 12/25 of 1925 and the 20/55 of 1927, which was upgraded to 20/65 for 1929. His last design for Humber was the 20/70 marketed as the Snipe. He directed the evolution of the Standard Nine, culminating with the Flying Standard Nine of 1937. He was also responsible for the 1933 Ten and Twelve, the 1936 Flying 16, and 1937 Flying 14. After the Flying 20 with its 2.7-liter V-8 engine in 1938, he concentrated on replacements for the older models. In 1941 his talents were called upon to organize the production of the Mosquito light bomber for De Havilland. He worked briefly for the Rootes Group in 1945 before returning to Standard, where he started the Vanguard program, an all-new postwar car that went into production in 1948. He also directed the teams that created the Triumph Renown, Mayflower, and Triumph TR2. He retired at the end of 1956, but remained a consultant to Standard-Triumph. He also acted as a consultant to Dunlop in connection with pneumatic suspension systems, until 1962. He died at his home in Coventry on March 15, 1968.

Grosseau, Albert (1929–)

Director of the Citroën drawing office from 1966 to 1976; technical director of the Peugeot SA group from 1976 to 1981; head of the PSA research laboratories and scientific affairs from 1981 to 1994.

He was born on June 7, 1929, in Paris and held an engineering diploma from the Ecole des Arts et Métiers in Paris. He began his career with Citroën in 1953 and was assigned to technical customer service, where he soon instituted a new program for the maintenance of hydraulic systems and the training of repair-work personnel. In 1958 he was transferred to the design office, where he worked on suspension systems, roadholding, and gearboxes. He also invented a new type of synchromesh. In 1962 he was placed in charge of small-car development and directed the evolution of the 2 CV, Dyane, Ami-6 and Ami-8. He was given overall responsibility for all product design and development in 1966, and created the DS-20 and DS-21, D Special and D Super. He headed the GS program, aimed to fill the gap between the two-cylinder cars and the D-series with a mid-size four-door sedan powered by an air-cooled flat-four and riding on Citroën's oleo-pneumatic suspension system. He directed the CX program, creating a more modern successor to the D-series, with transversely mounted four-cylinder engines, introduced in September 1974. From 1976 to 1981 he was head of product planning for the entire Peugeot group, including Talbot (ex-Simca) since 1979, and held responsibility for all production-car engineering.

He developed the Citroën Visa and the Talbot Samba from the Peugeot 104 in record time, and started a project that brought forth the Peugeot 205 in 1983. In 1981 he was given overall responsibility for technical research for the Peugeot group and held that position until his retirement in 1994.

Grylls, Shadwell Harry (1909–1984)

Chief engineer of Rolls-Royce's Car Division from 1951 to 1968; technical director of Rolls-Royce's Car Division from 1968 to 1969.

He was born in July 1909 at St. Neot in Cornwall and educated at Rugby and Trinity College, Cambridge. He joined Rolls-Royce in 1930 and worked in the motor car experimental department until 1939. He was associated with chassis development for the Phantom III, the Bentley Corniche, and the 4-B-50 straight-eight Bentley prototypes of 1939. During World War II he was personal secre-

tary to Ernest W. Hives. He remained at Derby from 1945 to 1949, overseeing the evolution of the Silver Wraith and Bentley Mark VI, and in 1949 was transferred to a new design office at Crewe. He directed the engineering of the Silver Cloud and S-Type Bentley which went into production in 1956, promoted the standardization of automatic transmissions, and won management approval for the V-8 engine program. The V-8 was designed by Jack Phillips and Ronald West and replaced the sixes in September 1959. He started design work on a car with unit-body construction and all-independent suspension, which went into production in 1965 as the Silver Shadow and T-Series Bentley. As technical director in his final year at work, he was freed from responsibility for current models to concentrate his attention on future projects. He retired in 1969 but continued his association with Rolls-Royce as a regular consultant and advisor on engineering policy and technical affairs.

Guédon, Philippe (1933–)

Chief engineer of Matra Automobiles from 1965 to 1969; engineering director of Matra Automobiles from 1969 to 1983; president and managing director of Matra Automobiles beginning in 1983.

He was born in 1933 in Paris as the son of a wealthy real-estate agent and was educated at the Lycée Charlemagne. He studied engineering at the Ecole des Arts et Métiers in Angers and graduated in 1956. He began his career with Simca as a trainee service engineer and in 1958 moved into service-department management. Two years later he arrived in the drawing office, and in 1962 he became an assistant to Vittorio Montanari, who was directing the Simca 1100 (front-wheel-drive) program. One day in 1964 he answered a small newspaper advertisement for a car designer in some company he had never heard of: Matra. He was called for an interview in a modest, temporary office at Vélizy, but he soon got an understanding of Matra's resources. Matra was then making the Djet, a small plastic-bodied Renault-powered sports coupé taken over from René Bonnet. Guédon designed the

Matra 530, powered by an inboard/rear-mounted Ford V-4 engine, that went into production at Romorantin in 1967. For the Matra Simca-Bagheera, he placed a 1.3-liter Simca engine transversely between the rear wheels of a steel chassis, with a plastic-paneled body seating three abreast in individual bucket seats. It went into production in April 1973 and had a seven-and-a-half-year production run.

His next creation, introduced in December 1980, was the Talbot Matra Murena with 1.6- and 2.2-liter engines, which inherited the layout of the Bagheera in a restyled body. He had an idea for a new type of van with a flexible seating arrangement and began preparing a proposal for Peugeot, with styling by Antoine Volanis. Two different prototypes were built for Peugeot before their collaboration came to an end. Guédon took his proposal to Renault, and Bernard Hanon ordered an engineering team to make up power trains and chassis components for a new vehicle that went into production in 1984 as the Renault Espace. It was a great success and opened up a whole new market segment. Matra was part of the production setup for three generations of the Espace before Renault took it over completely in 1998. Matra Automobiles was then given the final-assembly contract for the Renault Avantime.

Guy, Sydney Slater (1884–1971)

Works manager of the Sunbeam Motor Car Co., Ltd., from 1909 to 1914; chairman and managing director of Guy Motors, Ltd., from 1914 to 1957.

He was born on October 31, 1884, in Wolverhampton, and educated at Birmingham Technical College. He served an apprenticeship with Bellis & Morcom, Birmingham, and later joined the General Electric Co., Ltd., in Birmingham as a trainee. From 1906 to 1909 he held a position as repair shop manager with the Humber company in Coventry, and in 1909 accompanied Louis Coatalen to Sunbeam in Wolverhampton. Its Moorfield works on Upper Villiers Street were then among Britain's most important, with an excellent rep-

utation for quality, and S.S. Guy was the youngest man holding such high responsibilities in Britain's motor industry. He left Sunbeam at the end of May 1914 to start his own business, Guy Motors, Ltd., at Fallings Park, Wolverhampton. Guy began building one-and-a-half-ton trucks with overhead-valve engines and four-speed gearboxes with an overdrive top gear, but in August, truck production was stopped in order to make munitions. He was later awarded contracts for aircraft engines. Guy truck production resumed in 1919 and he also introduced a luxury car with a V-8 engine, which was made in small numbers up to 1925. The most successful Guy car was the 13/36 four-cylinder 2-liter, made from 1923 to 1925. From then on, Guy concentrated on heavy goods vehicles, and launched the Goliath with a 7.7-liter engine in 1931.

He had begun to take an interest in front-wheel drive, and applied for two front-wheel-drive patents that were granted in 1930. He achieved considerable commercial success with bus chassis named Arab, Vixen, and Wolf, and adopted Gardner diesel engines as an option in 1935. During World War II the Guy factory was kept busy making three- and four-axle military vehicles. In 1948 Guy Motors, Ltd., took over the Sunbeam Trolleybus Company, Ltd. He retired in 1957, and saw his company sold to Sir William Lyons in 1961. He died on September 21, 1971, at Albrighton in Staffordshire.

Gyllenhammar, Pehr Gustav (1935–)

Managing director of AB Volvo from 1971 to 1983; chairman of AB Volvo from 1983 to 1990; chairman and executive president of AB Volvo from 1990 to 1993.

He was born on April 28, 1935, in Gothenburg, Sweden, as the son of an insurance-company executive. He graduated from the University of Lund in 1959 with a degree in law and made postgraduate studies at the Centre d'Etudes Industrielles in Geneva, Switzerland, and later studied international law in Britain and the United States. He began his career with Amphion Insurance Company in 1961 and nine years later took over his father's position as president of the Skandia Insurance Company. He joined Volvo in May 1971, succeeding Gunnar Engellau as managing director, and set Volvo's expansion on a new course. Pre–1971, Volvo's growth stemmed from takeovers of domestic suppliers (Penta engines, Köping gears, Olofström steel pressings) and related industries (Bolinder-Munktell farming and forestry equipment). Gyllenhammar set out to internationalize Volvo's car production while turning the parent company into a world-scale diversified holding. He made Volvo buy a 33 percent stake in DAF Car BV in 1972, took majority control in 1975 and reorganized it as Volvo Car BV. In 1975 he made Volvo a partner in a three-way joint venture with Peugeot and Renault[1] for joint production of V-6 car engines. For six months in 1977 he held talks with Peter Wallenberg about a Saab-Volvo merger which came to nothing.

In the spring of 1978 he was negotiating with Norway's socialist government for part ownership of certain North Sea oilfields in return for a 40 percent stake in AB Volvo. The Norwegians signed an agreement on May 22, 1978, but Volvo shareholders and other interests fought it down, and the scheme was buried on January 26, 1979. He propelled AB Volvo into non-automotive ventures in partnership with Beijerinvest and Procordia and invested AB Volvo funds in Norway's Saga Petroleum and Canada's Hamilton Oil. All of these had to be undone later, at great cost. He authorized spending (with government subsidies) on experimental assembly plants (Kalmar since 1974, Uddevalla since 1985) that were doomed to low productivity and inapplicable to volume production. Volvo Car BV ran up heavy losses in 1980–81 and AB Volvo appealed to the Dutch government for state help. The state took majority control, leaving Volvo with a 30 percent stake. In 1990 he reached a deal with Raymond Lévy for cross-shareholding with Renault and a merger of their purchasing and quality-control organizations. In 1992 he brought Mitsubishi Motors into a new

corporation, Nedcar BV, to take over Volvo Car BV in a three-way joint venture with Volvo and the Dutch government. Many Renault and French government officials looked askance at the arrival of a Japanese partner, but in September 1993, Renault still went ahead and proposed a full merger with AB Volvo. It led to a complete split between the two groups, made official in February 1994. Gyllenhammr resigned on December 2, 1993.

He returned to the world of insurance, and was also active in international trade associations. In 2002 he was chairman of Britain's biggest life insurance company.

1. Franco-Suèdoise des Moteurs, Douvrin.

Habbel, Wolfgang R. (b. 19??)

Chairman of the management board of Audi NSU Auto Union AG from January 1979 to November 1988.

He was born in Dillenburg and studied law in Bonn and Cologne. He was drafted into military service in 1942 and was taken prisoner-of-war. He renewed his law studies in 1946 and began his career in personnel management with Auto Union GmbH in Ingolstadt in 1948. He was placed in charge of organizational projects for the Düsseldorf and Ingolstadt factories, then making DKW motorcycles and light vans, soon to resume production of the DKW car. He left in 1957 to join Ford-Werke AG of Cologne, in the personnel department. After the creation of Ford of Europe, Inc., he was given the task of coordinating pay scales, work hours, and other social questions between the German, British and French Ford companies. He left Ford in 1970 to become personnel director of C.H. Böhringer (pharmaceuticals) in Ingelheim, Toni Schmücker brought him back to Audi in October 1971 as deputy director for personnel, social programs, and labor relations. Later he also became head of a section in the legal department dealing with contracts and license matters. As soon as he became chairman of Audi, he became very assertive about the make's identity as something separate from Volkswagen. He introduced the quattro and 80 diesel in 1980, the five-cylinder Audi GT

5S coupé in 1981, the big, streamlined 100 and the 80 quattro in 1982, the 200 Turbo in 1983, the "new look" Audi 80 in 1984, the Audi 90 and 90 quattro in 1988.

In November 1988 he was relieved of his duties and given a seat on the supervisory board of Audi, with an agreement to act as their advisor when invited. They never invited.

Hahn, Carl Horst (1926–)

Chairman of Volkswagenwerk AG (Volkswagen AG since 1985) from 1981 to 1993.

He was born on July 1, 1926, in Chemnitz, Saxony, as the son of Carl Hahn, then sales manager for DKW motorcycles. He graduated from high school in 1946 and studied business administration at Cologne University in 1947 and at Zurich University in 1948. He studied social science in Bristol in 1949 and Paris in 1950. He delivered his doctoral thesis in 1952 at the University of Bern, Switzerland, discussing the Schumann plan for a European coal-and-steel union. He held an office post with the OECD[1] in Paris in 1953–54, and then joined Volkswagenwerk AG as export manager. He served as president of Volkswagen of America, Inc., at Englewood Cliffs, New Jersey, from 1959 to 1964 when he returned to Wolfsburg as a member of the management board and director of sales. He abruptly resigned in protest when he was passed over for promotion to chairman in 1972 and went to Continental Gummiwerke AG in Hannover as chief executive officer. He led the company through the acquisition of Uniroyal's European subsidiaries and a decade of renewed growth. He returned to Wolfsburg in November 1981 when Toni Schmücker resigned.

He arranged VW's purchase of a 51-percent stake in Seat for $282 million in October 1982 and secured majority control of Seat in 1986. In December 1990 he bought 30 percent of Skoda, right under the nose of Renault officials who thought it was "in the bag" and committed VW to heavy investments in the Czech Republic before gaining majority control of Skoda. He was sometimes blamed for blindly adding plant capacity and seeking to maximize the VW group's market share while not caring enough about mounting costs. On balance, however, his administration was highly successful. His contract expired in November 1991, but it was renewed for 18 months to give the board more time to pick his successor. He retired in April 1993.

1. Organization for Economic Development and Cooperation.

Hahnemann, Paul Gustav (1912–1997)

Sales director of BMW from 1961 to 1969; deputy managing director of BMW from 1969 to 1971.

He was born on September 21, 1912, in Strasbourg, a city then under German administration. He studied economics and mechanical engineering at the universities of Karlsruhe, Munich, Berlin and Heidelberg. He went to Detroit in 1937 and became a trainee with General Motors, where he could see corporate management from the inside and learn the latest advances in production technology and methods. He returned to Germany in 1939 and survived World War II. In 1948 he joined the sales department of the Opel branch in Frankfurt am Main, working simultaneously in management and at the retail level. He developed a way of sensing trends by direct contact with the public and minimal use of statistics. In 1957 he became marketing director of Auto Union GmbH, which was preparing to introduce the DKW F-12 Junior in the crowded economy-car market.

He joined BMW in 1961 as sales director and quickly disposed of its stock of BMW 700 models while the Munich factory was tooling up for production of the new 1500 family car. His functions expanded from market research into product planning, and he became known inside and outside BMW as "Nischen-Paul," as a tribute to his uncanny talent for identifying small but significant niche markets and suggesting the right kind of product to fill them. He was the guiding spirit behind the 1800 TI of 1964, 1800 TISA, 2000 CS of

1965, 2000 Ti-Lux and 1600-2 of 1966, 1600 TI of 1967 and 2002 in 1968, the six-cylinder 2500 and 2800 planned jointly with Helmut Werner Bönsch, and finally the 1600 Touring of 1971. No one contributed more towards turning BMW into a modern business organization, and he tripled sales volume within nine years. By 1970, BMW was producing an impressive range of models at a rate of 750 cars a day. But during 1971 he came into conflict with the new chairman, Eberhard von Kuenheim, and resigned abruptly in November 1971. He died on January 23, 1997, in Munich.

Hands, George William (b. ca. 1870)

Managing director, Calthorpe Motor Company, Ltd., from 1904 to 1912; managing director of Calthorpe Motor Company (1912) Ltd. from 1912 to 1919; chairman of the G.W. Hands Motor Company, Ltd., from 1920 to 1924; managing director of Calthorpe Motor Company (1912) Ltd. from 1924 to 1926.

He was born about 1870 in the Midlands and became a partner in the bicycle business of Hands & Cake in Birmingham. They started making their own bicycles in 1897, operating as the Bard Cycle Manufacturing Company, Ltd. The corporate title changed to Minstrel Cycle Company in 1901, Minstrel & Rea Cycle Company in 1905, and Calthorpe Motor Cycle Company, Ltd., in 1909. Separately, Hands had set up the Calthorpe Motor Company, Ltd., in 1902 to build cars, and a prototype was built in the Minstrel & Rea factory in 1904. In 1906 the Calthorpe Motor Company, Ltd., opened its works in Cherrywood Road, Bordesley Green, Birmingham, and began production of the 12/14 hp car, powered by a four-cylinder White & Poppe engine. In 1908 he switched to Alpha engines, made by Johnson, Hurley & Martin, Ltd., in Gosford Street, Coventry. The car company was reorganized, with fresh capital, in 1912, and began producing its own engines. The 9.5 hp Calthorpe Minor was introduced in 1913 but production was suspended in 1915 as the factory was converted to making

grenades and mines for the War Department. Bodies were built, practically next door, by Mulliner's, Ltd., which was taken over in 1917. In 1919 the Minor was upgraded to a 10.5 hp car. Hands left the company and founded the G.W. Hands Motor Company in Birmingham to make the Hands light car, using Dorman engines. This venture folded after turning out some 150 cars from 1922 to 1924, and he returned to take charge of Calthorpe which was on the brink of financial ruin. The Calthorpe cars had a reputation for quality but prices ranged from £299 to £430. He slashed prices, but had to stop production in 1924. In 1926 he sold the plant and went into retirement.

Hanks, Reginald F.

Production manager of Nuffield Industries, Ltd., in 1938–39; general manager of Nuffield Mechanizations, Ltd., from 1941 to 1945; vice chairman of the Nuffield Organization from 1947 to 1952.

After elementary schooling in Oxford, he began a premium apprenticeship at the Swindon works of the Great Western Railway. He was drafted into military service in 1915 and served in the RASC[1] until the end of World War I. In 1922 he joined Morris Motors, Ltd., in the service department, rising through the ranks to the title of assistant service manager in 1929. In 1936 he was promoted to chief inspector for the cars branch of Morris Motors, and won great respect for his insistence on quality control. Lord Nuffield named him production manager for the whole group in 1938, which meant responsibility for eight different Morris models, five Wolseley models, two MG models and two Riley models, plus the Morris Commercial vans and trucks. The start of World War II put an end to his efforts to streamline the manufacturing setup, for in September 1939 he was delegated to the RAF as manager of its civilian repair organization. He spent most of the war on production systems for the war-materiel industries, headquartered in Common Lane, Washwood Heath, Birmingham.

In 1945 the government spurred a major drive for export sales, and Hanks was given high responsibilities in a new affiliate, Nuffield Exports, Ltd.

In 1947 he was named vice chairman of the Nuffield Organization, with a seat on the board of directors. One of his first decisive actions was to arrange the retirement of nine of the oldest-serving directors and replace them with younger men. He was enthusiastic about the Morris Minor, scheduled for operation late in 1948, and strongly opposed to the merger talks with Austin, which began in 1950. He resigned in November 1951 after the first announcement of British Motor Corporation's coming into being.

1. Royal Army Service Corps.

Hanon, Bernard (1932–)

President of Renault, Inc., from 1972 to 1976; deputy managing director of Renault's car division from 1976 to 1981 and its managing director in 1981–82; president of the Régie Nationale des Usines Renault from 1982 to 1985.

He was born on January 7, 1932, at Bois-Colombes and graduated from the HEC[1] commercial college in 1955. He went to New York and obtained a master's degree in business administration from Columbia University in 1956. In 1962 Columbia University awarded him a doctor's degree in mathematical economics. He joined Renault, Inc., of Englewood Cliffs, New Jersey, as marketing director in 1959, but left in 1963 to lecture as assistant professor of management at New York University. He returned to Renault in 1966 as chief of economic studies and programming, and was promoted to computer and planning director in January 1970. He was given credit for pushing the R 5 through to production, but he never found an internationally acceptable replacement for the R 16. As head of the car division, he continued his forerunners' habit of running Renault as a social experiment and not a showcase for efficient management. He was blamed for overmanned factories with an aging work force, inferior products, and a disastrous North American venture (an association with American Motors that lasted from 1979 to 1987). He was widely praised for putting the

Espace into production but the Régie was losing a lot of money and in 1984 its accumulated debt soared to almost $9 billion. He was deposed in 1985 and became a private business consultant.

1. Hautes Etudes Commerciales.

Harriman, George William (1908–1973)

Deputy managing director of British Motor Corporation from 1952 to 1956; its managing director from 1956 to 1961, and its chairman and managing director from 1961 to 1966; chairman and managing director of British Motor Holdings from 1966 to 1968.

He was born on March 3, 1908, in Coventry and served an apprenticeship with the Hotchkiss factory, firearms and engine manufacturers, which was taken over by Morris Motors, Ltd., in 1923 and renamed Morris Motors (Engine Branch). He stayed with Morris and was transferred to the Cowley, Oxford, plant, where he came to know L.P. Lord. In 1938 he moved back to Coventry as assistant works superintendent of the Engines Branch. At L.P. Lord's invitation, he joined Austin Motor Co., Ltd., as production superintendent in 1940. He was promoted to works manager at Longbridge in 1944 and general works manager in 1945. Harriman and Issigonis laid down BMC product policy from 1956 to 1962, resulting in the Pinin Farina-styled 1959 Morris Oxford also appearing as an MG Magnette, Riley 4/68 and Wolseley 15/60. The Morris Mini was also marketed as an Austin Seven, Riley Elf and Wolseley Hornet, and the 1100 (ADO 16) was made with Austin, Morris, MG, Riley and Wolseley badges. This did not sit too well with those who like each make to have its own character. In 1965 he arranged the purchase of Pressed Steel, and a year later, the merger with Jaguar Cars, Ltd. During his chairmanship period, his biggest failings were seen as an inability to achieve proper financial control, to understand the effects of taxation on trends in the market, and to develop the foresight needed for forward planning. He took a strong stand against Lord Stokes regarding the merger of

British Motor Holdings with Leyland Motors, Ltd., but lost the battle. He resigned in September 1968 and died on June 23, 1973.

Hartwich, Günter (1935–)

Volkswagenwerk AG board member for production from 1972 to 1993.

He was born on April 23, 1935, in Berlin and broke off his formal schooling at the age of 16 to go to work as an apprentice mechanic, qualifying in 1953. He held technical positions in mechanical industry for a year and then decided to go back to school. He attended Bayreuth Engineering School in Berlin, graduating in 1957, and joined Volkswagenwerk as a trainee engineer. In 1959 he became a planning engineer for chassis production methods in the Wolfsburg plant, and a year later was transferred to the Kassel plant in a similar capacity. In 1968 he was named plant manager at Kassel and then returned to Wolfsburg in 1969 as head of a group in the production planning office. On July 1, 1971, he was promoted to works director of the entire Wolfsburg complex, and a year later, production director for all VW plants. He coped brilliantly with the phasing out of the "Beetle" and the introduction of the Scirocco, Golf, and Polo. He planned and supervised the modernization of the Salzgitter engine plant, and the Hanover, Brunswick, Kassek, Emden and Wolfsburg factories. In 1986 he was also made a director of Seat. By 1987 the Wolfsburg plant alone was turning out 2900 Golf cars a day, and by 1989 the Emden plant reached a daily output of 1000 Passat cars. In 1993, F. Piëch pressured him into resigning from Volkswagen AG.

Harvey-Bailey, R.W. (1876–1962)

Chief technical production engineer of Rolls-Royce, Ltd., from 1922 to 1937; chief engineer of the Rolls-Royce chassis division from 1937 to 1939.

He was born on February 22, 1876, and educated at Tiffins School in Kingston-on-Thames. He was accepted at King's College, London, and graduated in 1896 with a degree in civil engineering. He began his career as a drafts-man for Edward Joel Pennington at the Horseless Carriage Company, but left to design steam engines for Coulthard's of Preston, Lancashire. In 1903 he joined Critchley-Norris and redesigned the front-wheel-drive Brightmore steam struck for rear-axle propulsion. He also designed a bus chassis powered by a Crossley gas engine. In 1906 he became a consultant to the Pullcar Company at Preston and designed the Pullcar, a front-wheel-drive automobile with a transversely mounted White & Poppe four-cylinder engine and three-speed planetary gearbox. He was co-designer with Francis Leigh Martineau of the 1909 Little Pilgrim, another example of front-wheel drive, with a water-cooled flat-twin engine. Claude Johnson answered his application to Rolls-Royce in 1910, and he moved to Derby as section-chief designer. Throughout World War I he was engaged on aircraft-engine design, and did not get back to the motor cars until 1920. He was responsible for the drawing office when the 20/25 and the New Phantom were designed, and was personally in charge of Bentley from 1931 to 1936. He also led the team that prepared the Bentley Mark V (with its independent front suspension), the Corniche prototype, and the eight-cylinder Bentley that never got to the production stage. In 1940 he was transferred to the Aero Engine Division, where he stayed till his retirement at the end of 1945. He died in September 1962.

Haspel, Wilhelm (1898–1952)

Daimler-Benz works director at Sindefingen from 1927 to 1936; deputy member of the Daimler-Benz management board from 1936 to 1941; full board member in 1941; chairman of the management board from 1942 to 1945 and 1948 to 1952.

His talents were both commercial and technical, and he earned a diploma in mechanical engineering. He began his career with Daimler Motoren Gesellschaft in 1924 and was placed in charge of handling the economics of the merger with Benz & Cie. Wilhelm Kissel named him works director at the Sindelfingen assembly plant in 1927, where he cut the payroll by half in the next five years. Due to slack demand for cars, he installed a line for bus bodies. In 1928 he retooled the stamping plant with a row of Weingarten heavy presses, and in 1932, with Mercedes-Benz car sales at a low point, he secured contracts and tooled up for making bodies for BMW and Wanderer.

In 1936 he moved to the corporate head office at Untertürkheim, fighting for fresh funds to streamline production. From 1933 to 1938 Daimler-Benz made plant investments of over RM 100 million, not counting the new aircraft-engine factory at Genshagen.

As chairman since 1942, his responsibilities took on new dimensions. He was no Nazi, but he was surrounded by them, and Himmler wanted him removed from his position because Mrs. Haspel was half–Jewish. Twice Hitler's friend Jakob Werlin saved him and his family.

Haspel had no illusions about the outcome of World War II and in September 1944 he arranged for the transfer of all Daimler-Benz funds deposited in Berlin banks to banks in Stuttgart, so that in 1945 the company had cash reserves of RM 250 million. The factories were situated in the American and French occupation zones, but the buildings had been 80 percent destroyed in Allied air raids and only 47.5 percent of tooling and equipment was still useable. He had a few hundred Mercedes-Benz 170 V ambulances, vans and pickups built at Sindelfingen, 5-ton Mercedes-Benz trucks at Gaggenau and 3-ton Opel Blitz trucks at Mannheim before the U.S. Forces had him removed from office in October 1945. He was surreptitiously returned to Untertürkheim in September 1947 and officially reinstated on January 1, 1948. By February 1949 monthly car production had climbed to 1000 units.

He shepherded new models into production, the 170 S in 1949 and the 220 and 300 in 1951. It has been said that he worked himself to death. He succumbed to a cerebral hemorrhage on January 6, 1952.

Hassan, Walter "Wally" (1905–1996)

Development engineer with Jaguar from 1943 to 1950; chief engineer of Coventry Climax from 1950 to 1966; head of power unit engineering with Jaguar from 1966 to 1975.

He was born in 1905 in Upper Holloway, London, where his father had a clothing store. He was educated at Northern Polytechnic and the Hackney Institute of Engineering. He began his career as a trainee with Bentley in 1922, and stayed with Bentley as experimental and racing engineer until 1936. He had a role in the development of the six-cylinder models from 1925 to 1930, and worked on modifications of Rolls-Royce engines for use in Bentley cars after 1931. In 1936 he joined ERA[1] as superintendent of racing car assembly but left a year later to become service manager of Thomson & Taylor, sports car dealers on the Kingston By-Pass who also had race-preparation shops at Brooklands. He joined SS Cars, Ltd., in Coventry as a development engineer in 1938. The company stopped car production to fill War Office contracts, and in 1940 he transferred to Bristol Aero Engines, Ltd., as a development engineer, helping to make improvements in the 14-cylinder Hercules engine and speeding the 18-cylinder Centaurus engine into the production stage.

He was invited back to SS Cars, Ltd., in 1943, as a development engineer on experimental military vehicles and post-war Jaguar cars. He had a central role in the program that led to the XK engine, and became chief engineer of Coventry Climax in 1950.

Ably assisted by Harry Mundy, he designed the 1100-cc fire-pump engine that powered Lotus, Lola, Elva and some other British sports and racing cars from 1958 onward. They also created the V-8 and the flat-16 racing engines. He returned to Jaguar when Sir William Lyons bought control of Coventry Climax in 1966, and began design work on the V-12 Jaguar engine which went into production in 1972. He retired in 1975 and died in July 1996.

1. English Racing Automobiles, Ltd., Dunstable.

Hauk, Franz (1926–)

Chief engineer of Audi engines from 1974 to 1987; director of the VW Group Power Unit Development Center from 1987 to 1991.

He was born in 1926 in Neuburg on the Danube and educated as an engineer. He joined Henschel & Sohn in Kassel as a drafts-man in the diesel-engine design office in 1948, and later came to Daimler-Benz where he worked on the design of both spark-ignition and diesel engines. The company was then alone in producing diesel-powered passenger cars. Ludwig Kraus brought him to Auto Union in 1964, where he designed the Heron-head slant-four engine for the reborn Audi 60 of 1965. He also created the "827" engine family, built in sizes from 1300-cc to 2000-cc, first appearing in the Audi 80 in 1972, and then extended to the VW Passat, Scirocco and Golf.

He designed Audi's five-cylinder in-line of 1976 and initiated the programs that led to the Audi and VW V-6 and V-8 engines.

He went into retirement in July 1991.

Hayter, Don (1927–)

Chief design draftsman with MG from 1956 to 1958; head of the MG Projects Department from 1958 to 1973; chief engineer of MG from 1973 to 1980.

He was born in 1927 and ended his formal schooling at the age of 14, when he began an apprenticeship with the Pressed Steel Com-pany, Ltd., at Cowley, Oxford. He became a sheet-metal tool-and-die designer, but after 12 years at Cowley, he felt his career was in a rut, and he was looking for wider horizons. He joined Aston-Martin, Ltd., in 1954 and spent two years at Feltham, where he designed a new nose shape for the DB 3. He resigned when body production was transferred to Newport Pagnell and signed up with MG. He designed the coupé for the MGA, first seen at Le Mans in 1960, and started a project of Aston Mar-tin size. Sid Enever then told him about the plans for the MGB, and suggested downscal-ing his design. Hayter made the first pencil sketch for the MGB roadster in the summer of 1958 and designed the MGB GT coupé in 1961. He held technical responsibility for the MGC, the ADO 21 prototype with inboard/rear-mounted engine, and the ongoing evolu-tion of the MGB. He resigned when the Abingdon plant was closed in 1980 and went into retirement.

Healey, Donald Mitchell (1898–1988)

Experimental manager of Triumph in 1933–34; technical director of Triumph from 1934 to 1939; founder and director of the Donald Healey Motor Company, Ltd., in 1946; originator of the Austin-Healey, the Nash-Healey and the Jensen-Healey sports cars.

He was born on July 3, 1898, at Perran-porth, Cornwall, attended Newquay College and went to Cambridge but broke off his stud-ies when he was accepted for an apprenticeship at Sopwith Aviation Company in Kingston-on-Thames. Sopwith had its assembly shops at Brooklands, where he first saw high-speed racing cars in action. He volunteered for the Royal Flying Corps and became an "air me-chanic" in 1916. After getting his pilot's li-cense, he flew anti–Zeppelin raid patrols and night-bomber missions over France. He was wounded and discharged in 1917, returned to Perranporth and took a correspondence course in automobile engineering. He opened an auto-repair garage and started a chauffeur-dri-ven hire-car service. He started driving in tri-als and hill-climbs and became a Triumph dealer in 1924. His reputation as a driver won him a factory-sponsored Invicta in 1930.

He joined Riley early in 1933 to prepare the rally-team cars but left to join Triumph where he was given responsibility for product design and development. He built up an extremely competent engineering team and they made one model, the Dolomite, with Alfa Romeo's cooperation.

The Triumph plant made Claudel-Hobson carburetors for aircraft engines when the war broke out. Before the end of 1939 he joined Humber, Ltd., to develop military vehicles. He was also active in the RAF voluntary re-serve during the war. With fellow Humber engineers A.C. Sampietro and B.G. Bowden, he designed a sports car for postwar produc-tion. He leased factory space at The Cape, Warwick, and obtained engines from Riley. Coupé bodies were made by Elliott. He introduced the Healey Silverstone roadster in 1950. From a chance meeting with George W. Mason, president of Nash-Kelvinator Corporation, he invented a plan to have Austin build the Nash Metropolitan at Longbridge, and created the Nash-Healey sports car.

He displayed a new Healey with the engine and gearbox from the Austin A 90 Atlantic at Earls Court in 1952. Its look (styling by Gerry Coker) were so sensational that L.P. Lord of Austin took over the production at Longbridge, giving birth to the Austin-Healey. Donald Healey received a royalty for every car they built.

The collaboration expanded with the addition of the Austin-Healey Sprite in 1958 and the Austin-Healey 3000 in 1959. But in 1970 Lord Stokes of British Leyland ordered a stop to their production and the removal of the Healey name from the payroll.

Soon afterwards Healey became chairman of Jensen Motors, Ltd., in an international venture to revive the Jensen name, financed by W.M. Brandts Son & Co., Ltd., and produce the Jensen-Healey sports car with a Lotus engine. It ended up in bankruptcy court in 1973. He retired to Cornwall and died in 1988.

Healey, Geoffrey Carroll (1922–1994)

Technical director of Donald Healey Motor Company, Ltd., from 1947 to 1954; creator of the Austin-Healey Sprite.

He was born on December 14, 1922, and educated at Warwick School. Bent on an engineering career, he began studies at the Camborne School of Mines but left school when he was accepted for an apprenticeship at Cornercroft in Coventry in 1939. Called up for military duty in 1944, he served with the Royal Engineers, Mechanical/Electrical, being discharged with the rank of Captain in 1947.

He joined his father's company and was associated with all its products. With Barrie Bilbie, he was the chassis engineer for both the Austin-Healey 100 and the Sprite.

He died on April 30, 1994.

Heinrich, Jean (1939–)

Works manager of Comotor SA from 1972 to 1975; director of SMAE[1] from 1978 to 1989; project manager for the Citroën Saxo from 1990 to 1996.

He was born in 1939 and graduated from the Ecole Polytechnique. He joined Citroën in 1963 in the methods department and in 1967 was assigned to study production methods for the Wankel rotary engine. Citroën established Comotor SA as a joint venture with NSU Motorenwerke AG to produce Wankel engines in a new plant at Altforweiler in the Saarland. When Citroën gave up on the Wankel engine in 1975, he was sent to Romania, where Citroën International was a partner in a joint venture to build the Oltcit car, an offshoot of the Dyane. He also spent some time in the Paris office for international and industrial planning. In 1978 he was named director of the SMAE subsidiary, which operated a gearbox factory at Metz-Borny and a hyper-modern engine plant at Trémery in Lorraine. Citroën's AX, intended as a modern-world replacement for the 2 CV and developed from 1981 to 1986, was a big disappointment to the customers and the manufacturer alike. Heinrich was placed in charge of coming up with its successor. That was the Saxo, making its debut in 1996, assembled at the Aulnay-sous-Bois plant.

1. Société de Mécanique Automobile de l'Est.

Henze, Paul

Chief engineer of Automobiles Impéria from 1907 to 1910; technical director of Reichenberger Automobil-Fabrik from 1910 to 1913; technical director of Maschinenfabrik Walter Steiger & Cie from 1917 to 1921; technical director of Walter Steiger AG Automobilfabrik in 1921–22; chief engineer of Automobilbau Simson & Co. from 1922 to 1928; chief engineer of Selve Automobilwerke AG in 1928–29; consulting engineer with NAG[1] from 1929 to 1933.

He began his career as an engine draftsman for Max Cudell in Aachen (Aix-la-Chapelle) in 1903, working alongside Carl Slevogt on the design of new two- and four-cylinder engines to replace the power units Cudell was producing under license from de Dion–Bouton. Adrien G. Piedboeuf brought him to Nessonvaux near Liège in 1907, where he designed a fast 3-liter four-cylinder touring car and a giant 9.9-liter 50/60 model. A medium-priced 12 CV Imperia followed in 1909. In 1910 he moved to Reichenberg in Bohemia to join a company whose majority owner was Baron Theodor von Liebieg and had been building RAF cars since 1907. The company had just purchased a license for the Hansa 6/16 and 7/20. His task was to design an upmarket family of high-performance touring cars. He began carefully with a four-cylinder 14/30. He disagreed with von Liebieg's purchase of a license for the Knight sleeve-valve engine and departed in 1912. He returned to Nessonvaux where he designed four-cylinder 16-valve Abadal for Adrien G. Piedboeuf. In 1917 he joined Walter Steiger at Burgrieden bei Laupheim in Swabia and designed a 10/50 sports/touring car with an overhead-camshaft 2.6-liter four-cylinder engine which was produced from 1920 to 1926. But he deserted Swabia for Thuringia in 1922 and designed a succession of Simson cars, beginning with the Type S "Supra" of 1924, with a twin-cam 16-valve 2-liter four-cylinder engine. He toned it down for the Supra So, with two valves per cylinder, and created the six-cylinder Type R 3.1-liter model for 1926. In 1927 he joined Walther von Selve in Hamelin to develop the Selve Selecta luxury car, which was sadly underpowered. He also designed a front-wheel-drive Selve 12/50 prototype which was orphaned when the company went out of business. He went to NAG in Berlin and designed the Type 218 with a 4½-liter overhead-valve V-8 engine, produced from 1931 through 1933. He also redesigned the front-wheel-drive NAG-Voran models 212 and 220. But NAG had merged with Büssing in 1932 and stopped passenger-car production in 1934.

1. Nationale Automobil-Gesellschaft AG, Berlin-Oberschöneweide.

Hereil, George (1909–1980)

President of Automobiles Simca from 1963 to 1970; vice president of Chrysler-France in 1970–71; consultant to Chrysler International in Geneva until 1980.

He was trained as a lawyer and held a doctor's degree in law. From 1936 to 1946 he was employed as a judiciary liquidator, helping to put a legal gloss on some of the French government's takeovers of private industry. He carried out this role up to 1940 and from 1944 to 1946 disposing of the assets in bankruptcy cases, such as Lioré & Olivier and the Dewoitine aircraft companies. He made useful contacts in the French aircraft industry and was invited to join SNCASE[1] at the executive level. He realized that piston engines had no future in modern airplanes, and started the Caravelle twin-jet airliner project in 1953.

From 1957 to 1962 he served as chairman and managing director of Sud-Aviation. He was approached by the Chrysler Corporation, who had bought a 25 percent stake in Automobiles Simca in 1958, boosting it to majority control (64 percent) in 1963, about heading the French auto maker. He had no qualms about accepting.

The Simca works at Poissy were mass-producing the 1000 (derived from Fiat's 850) and had just introduced the 1300–1500 family sedans. He gave the green light to a front-wheel-drive car project, which went into production in 1967 as the 1100.

In 1964 he proposed reviving the Talbot name which had come into Simca's possession in 1958, but the American executives nixed that idea and decided that new models should carry the Chrysler name. His influence diminished when Automobiles Simca became Chrysler-France in June 1970, and he resigned in 1971, though still retaining the title of "honorary president."

He was taken ill in September 1980 and died in the Hopital Cantonal in Geneva, Switzerland, on December 1, 1980. He had lived long enough to see the French Chrysler models renamed Talbot after Chrysler's European affiliates had been sold to Peugeot SA.

1. Société Nationale de Constructions Aéronatiques du Sud-Est, which evolved into Sud-Aviation.

Heynes, William Munger (1904–1989)

Chief engineer of SS Cars, Ltd., from 1935 to 1945; chief engineer of Jaguar Cars, Ltd., from 1945 to 1961; Jaguar's vice chairman in charge of engineering from 1961 to 1969.

He was born on December 31, 1904, at Leamington Spa and attended Warwick School from 1914 to 1921. He was undecided about his choice of a career and considered studying medicine before settling on engineering. In 1922 he was accepted as a trainee with Humber Ltd. in Coventry, and was promoted to the drawing office three years later. From 1930 to 1935 he was head of Humber's drawing office. After modernizing the Humber Snipe and Pullman, he left in 1935 to join SS Cars, Ltd. as chief engineer, a newly created position in that company.

His first task was to design a frame for a new roadster which became the SS 90. He also revolutionized body engineering, for all earlier SS models were built up with a wooden structure. The SS Jaguar saloon which went into production in September 1937 had an all-steel body.

He directed the Mark V project, which introduced independent front suspension. As early as 1934 he had experimented with independent front suspension for the Hillman Minx, and in 1938 he tried two systems on SS Jaguars, one with pneumatic struts, the other with coil springs. For the Mark V he chose torsion bars.

In 1945 he also started an engine program for Jaguar whose engines were supplied by the Standard Motor Company. It evolved from a motorcycle engine he had designed in 1922, with twin overhead camshafts and hemispherical combustion chambers. A series of experimental units, beginning with the XA, XB, and so on, were tested and discarded. The 11th, designated XK, went into production in 1949 for the XK 120 roadster. It was installed in the Mark VII saloon and was adopted for the C-Type and D-Type racing cars. He pioneered the use of disc brakes on the racing cars in 1953 and on production cars as an option for the 1958 2.4 and 3.4 liter models.

He led the design and development of the Mark X, the E-Type, and the XJ saloon, retiring in 1969. He died in September 1989.

Hieronymus, Otto (1879–1922)

Chief engineer of Laurin & Klement from 1907 to 1911.

He was born in 1879 in Cologne and given a technical education before joining Benz and Cie in Mannheim as a trainee in 1895. After four years with Benz he moved to Vienna, joined Arnold Spitz and designed a range of single-, twin- and four-cylinder cars that were produced for Spitz by Gräf & Stift from 1902 to 1907.

He raced Spitz cars with some success, and won some important events at the wheel of a Mercedes. He arrived at Mlada Boleslav in 1907, stopped Laurin & Klement's production of little cars with V-twin engines, and created a range of new cars, trucks and buses. He designed a racing car, Type FCR with 100-hp 5.5-liter four-cylinder engine, and the F-series touring-car chassis with a choice of four overhead-valve four-cylinder engines from 1994 cc and 53 hp to 3486 cc and 96 hp. He also made up a rally car with two F-Type engine blocks combined to form a straight-eight.

The 1908 G-Type replaced the previous two-cylinder models, a simple chassis with a side-valve four-cylinder 1555-cc engine having a one-piece cast-iron block, three-speed gearbox and shaft drive. The 5.7-liter 50-hp model EN introduced in 1909 evolved into the ENS in 1910, with the engine enlarged to 7.5 liters.

In 1911 he rationalized car production to a two-model lineup, the S-Type light car with a 1771-cc side-valve four-cylinder engine and the medium-size K-Type with a 4253-cc side-valve four-cylinder engine. The K-Type was produced up to 1915 and the S-Type until 1916.

Hieronymus left the company in 1911, traveled to France where he learned to fly, and began to design aircraft engines, his first effort being a modest 50-hp water-cooled in-line four. Returning to Mlada Boleslav, he reached an agreement with Laurin & Klement to put it in production. In 1914 he designed a new family of aircraft engines for production by the Warschawski Austrian Industry Works, which were also built under license by Werner & Pfleiderer in Stuttgart.

After the war, he served briefly on the Austro-Daimler engineering staff (under Porsche) and then settled at Steyr, working on engine design, car development, race-preparation, and as a racing driver. He was killed in a 100-km/h accident during practice for a local race on May 8, 1922.

Hill, Claude (1907–1983)

Chief designer of the Aston Martin from 1936 to 1940; technical manager from 1940 to 1942; technical director and chief engineer from 1942 to 1949; chief engineer and director, Harry Ferguson Research, Ltd., from 1950 to 1971.

He was born on February 16, 1907, in Birmingham and educated at King's Heath School and the Sparkhill Institute. At the age of 16 he went to work for the Spencer Sprinkler Company at Handsworth, but soon found employment with Best & Lloyd, makers of auto accessories. In 1924 he answered a newspaper advertisement for a junior draftsman and was accepted by Renwick & Bertelli, engine manufacturers at Tyseley. Their overhead-camshaft 1½-liter four-cylinder unit was installed in Aston Martin cars as soon as W.S. Renwick purchased the bankrupt Bamford & Martin business at the end of 1925. Hill was transferred to the Aston Martin works at Feltham but resigned to take a position in the drawing office of Morris Motors (Engine Branch) Ltd. in Coventry. A.C. Bertelli lured him back to Feltham in 1930 by making him chief craftsman and paying him a higher salary. He resigned again in 1934 and spent 18 months with Vauxhall Motors as a chassis draftsman, returning to Aston Martin in 1936. In 1939–40 he designed the Atom prototype, a concept car that broke with Aston Martin's traditions. After the war he designed the new 2-litre model, using elements of the Atom chassis, with a new four-cylinder engine.

He resigned in 1949 because David Brown chose a six-cylinder engine designed by W.O. Bentley over Hill's four-cylinder unit for the DB 2 model. A.P.R. Rolt engaged him to work on four-wheel-drive projects for Harry Ferguson Research, Ltd., at Toll Bar End, Coventry, where he spent the rest of his career. He resigned in 1971 and died in 1983.

Hiller, Karl Gustav (1863–1917)

Chairman and owner of the Phänomen-Werke from 1898 to 1917.

He was born on March 30, 1863, in Zittau in Saxony and served an apprenticeship in a local machine shop. In 1888 he joined Müller & Preussger and set up a factory for textile-making machinery. He had a side line selling bicycles, and in 1891 came to an agreement with James Starley to assemble Rover bicycles from British-made parts at Zittau. He invented a new type of ball-bearing, and soon was making most of the parts for the bicycles. In 1898 he took over Müller & Preussger and reorganized the business as Phänomen bicycle works.[1] In 1900 he started production of Phänomen motorcycles with single-cylinder engines and belt drive. Within three years, Phänomen was making more motorcycles than bicycles. In 1907 he formed an alliance with Cyklon Werke in Berlin and obtained the rights to built three-wheeled cars and utility vehicles based on the Cyklonette. The Phänomobil introduced in 1908 resembled the Cyklonette in having front-wheel-drive and a single front wheel with tiller steering. The 7-hp 880-cc air-cooled V-twin was carried on the steering fork and had chain drive to planetary gearing in the hub. A new model for 1912 had a four-cylinder 6/12-hp 1½-liter engine mounted transversely on the fork. The first four-wheeled Phänomen was designed in 1911. It was purely conventional in concept and engineering and was built up to 1919.

1. Phänomen Fahrradwerke Gustav Hiller.

Hillman, William (1848–1921)

Chairman and managing director of the Hillman Motor Car Company, Ltd., from 1910 to 1921.

He was born on December 13, 1848, at Stratford, Essex, as the son of a shoemaker. He attended local schools and received a technical education. He began his career with Penn of Greenwich as a marine engineer, and later worked at the Cheylesmore factory of the Coventry Machinists' Company. Financed by William Herbert, older brother of Alfred Herbert, Britain's biggest machine-tool manufacturer and trader, he established Automachinery in Coventry in 1875 to make parts for the bicycle industry. They also made sewing machines and roller skates. Separately, they set up the Hillman & Herbert Cycle Company, which evolved into the New Premier Cycle Company in 1896. He prospered in the bicycle industry and was a millionaire before his fifty-second birthday.

His interest in cars was awakened several years later, when he felt a desire to see his name on a Tourist Trophy winner. In 1906 he engaged Louis Coatalen to design the car, founded the Hillman-Coatalen Motor Car Company, Ltd., and moved into a factory at Pinley, Coventry. The car looked like a winner, with its 25-hp 6.4-liter four-cylinder engine, but only one was built in time for the 1907 Tourist Trophy, and though it proved fast, it did not finish. Coatalen went to Sunbeam in 1909, and in 1910 Hillman reorganized the firm as the Hillman Motor Car Company, Ltd. Production concentrated on the 12/15 medium-size touring car, and a 10-hp van powered by a water-cooled V-twin was added in 1912. Yet the company was not profitable until the Hillman Nine, designed by A.J. Dawson, came on the market in 1913. He aged considerably during World War I, when the factory was working under defense contracts, and in 1919 he entrusted the management of the company to his two sons-in-law, Captain John Black and Spencer Wilks. He died in 1921.

Hirst, Ivan (1916–2000)

Major in the REME (Royal Engineers, Mechanical-Electrical); de facto chief executive officer of the Volkswagen works from 1945 to 1949.

He was born on March 1, 1916, at Saddleworth and educated at a nearby grammar school. He studied engineering at the University of Manchester, and upon graduation began his career in the Hirst family's optical-instrument business. He volunteered for service with the Territorial Army and served at the Huddersfield drill hall until 1939 when he was enrolled in the Duke of Wellington's regiment with the rank of Major. He was transferred to the REME in 1942 and in the summer of 1944 followed the Allied armies in Normandy, repairing tanks and other motorized equipment. His unit advanced to Brussels where he became head of a tank repair depot. Arriving at Wolfsburg in 1945, he never had precise orders. At his own initiative and the help of his commanding officer, Colonel Michael McEvoy, he made up a prototype Volkswagen from salvaged parts, demonstrated it to the Allied commanders, and received an order for 20,000 cars.

The plant was in ruins, as 66 percent of the buildings had been destroyed by aerial bombardment, but most of the machine tools were intact. Production began while the factory was being rebuilt. In 1946 VW had 8000 workers and turned out 1000 cars a month. That was nearly doubled in 1948 and quadrupled in 1949.

He left Wolfsburg in August 1949, reassigned to the Foreign Office's German section. He later joined the OECD[1] staff in Paris, retiring to his house in the Pennine Hills in 1975. He died on March 10, 2000.

1. Organization for Economic Cooperation and Development.

Hitzinger, Walter (1908–1975)

Chairman of the Daimler-Benz AG management board from 1961 to 1966.

He was born on April 8, 1908, in Linz on the Danube, Austria. After graduating from the Technical Institute of Vienna with a degree in mechanical engineering in 1934, he began his career as an engineer in the special-vehicles department of the Steyr plant of Steyr-Daimler-Puch AG. He evolved into a drawing-office administrator rather than a designer. In 1943 he was named technical director of the Steyr-Daimler-Puch AG aircraft-engine works in Wiener-Neudorf, and in 1945 he was appointed general manager and director of the Austrian Saurer (truck) company.

In 1952 he was named president of the Voest-Alpine United Austrian Iron and Steel Works in Linz, where he raised steel production tenfold within an eight-year span. In that position, he came into contact with Friedrich Flick, majority shareholder of Daimler-Benz AG. In 1961 Flick invited him to come to Stuttgart as chairman of the Daimler-Benz AG board of directors. He accepted and led the company through five years of steady growth, but retired in 1966 and died on July 26, 1975.

Hives, Ernest Walter (1886–1965)

General works manager of Rolls-Royce, Ltd., from 1937 to 1946; managing director of Rolls-Royce, Ltd., from 1946 to 1950; chairman of Rolls-Royce, Ltd., from 1950 to 1957.

He was born on April 21, 1886, and left school at 14 to start an apprenticeship in a garage at Reading, Berkshire. In 1903 he went to London and joined C.S. Rolls & Co. in the service department at Fulham. From 1904 to 1908 he worked in D. Napier & Son's automobile department at Acton Vale, London, and then joined Rolls-Royce at Derby as a chassis tester. In 1911 Rolls-Royce opened its own test track, with a high-speed oval and banked curves at each end. He was an excellent test driver and his reports on experimental cars were given great weight in the drawing office. Soon he became involved in non-flight testing of aircraft engines, and in 1914 he became head of the aircraft engine experimental department. In 1923 the motor car experimental department also came under his authority. He pushed the New Phantom through its development phase so that production could begin in 1925. He directed the development of the "R" V-12 engine for seaplanes competing for the Schneider Trophy and its subsequent evolution into the Merlin V-12 which powered the Supermarine Spitfire, Avro (A.V. Roe, Ltd.) Lancaster, and De Havilland Mosquito. When Arthur Wormald retired in 1936, he applied for the title of works manager and was promoted. He directed all Rolls-Royce production programs and facilities throughout World War II, and led the aircraft engine department into the jet age. When he found out how far Rover had got in development of the Whittle gas turbine, sometimes complementing but usually overlapping with work being done at Derby, he invited Spencer Wilks to dinner. "You give us Barnoldswick (gas turbine laboratories) and we'll give you the Meteor tank engine," he said. It was a done deal. The Rover W2B became Rolls-Royce B-23 and went into production as the Welland turbojet. In 1945 he determined that there was a lot more money to be made in civil aviation than in automobiles, and built up the aircraft-engine division to a size where it could compete in world markets. He also reinforced the motor car division with an Oil Engine Division in 1950, but took a hands-off supervisory attitude to the production and evolution of Rolls-Royce and Bentley cars. He retired in January 1957, suffered a stroke in March 1963 and died in a London hospital on April 24, 1965.

Hobbs, Dudley Erwin (1911–)

Chief designer of Bristol Cars, Ltd., from 1949 to 1975.

He was born on October 11, 1911, in Bristol, and was educated at Queen Elizabeth's Hospital, Bristol. He studied engineering at Merchant Venturers Technical College in Bristol and began his career with the Bristol Aeroplane Company as a trainee in 1928, graduating to draftsman in 1930. He worked on wing design for a succession of Bristol aircraft until 1941. Bristol had to boost aircraft production to meet orders from Lord Beaverbrook, Churchill's man in charge of keeping the RAF in the air. This forced Bristol to procure a lot of components from outside sources, and Hobbs was transferred to a new office for liaison with suppliers. He was named sub-contracts manager in 1943 and chief planning engineer for the aircraft division in 1945. He was associated with Bristol's car projects from the outset, and designed the body for the Bristol 400, a striking and tasteful extrapolation of prewar BMW styling trends. He was transferred to Bristol

Cars, Ltd., in 1946, and in 1949 was named chief designer for engine, chassis and body design. He was responsible for the 403, based on a Carrozzeria Touring prototype of 1948, the 404 and the 405. When T.A.D. Crook acquired a 50 percent stake in Bristol Cars, Ltd., in 1960 and asked Hobbs to stay on, he accepted and determined the shape of Bristol cars including the 407 and 411.

Hodkin, David (1924–1980)

Managing director of ERA, Ltd., from 1951 to 1959; chief experimental engineer of the Rootes Group from 1959 to 1963; chief design engineer of the Rootes Group from 1963 to 1964; director of business and product planning for Austin-Morris in 1979; director of strategic and component planning, Austin-Morris and Jaguar-Rover-Triumph from 1979 to 1980.

He was born in 1924 and held an engineering degree from Cambridge University. He began his career with De Havilland Aircraft in 1949, but left to join ERA, Ltd.,[1] in 1951 where he designed the G-Type racing car. The prototype was sold to Bristol Cars, Ltd., and served as the basis for the Bristol 450 of 1953–55. That ended ERA's existence as a racing car constructor, and Hodkin successfully transformed the company into an automotive engineering consultancy. The Rootes Group became a prominent client, and after declining several invitations, he left ERA and joined Rootes in 1959. He was associated with the development of the Sunbeam Rapier, Singer Gazelle and Hillman Super Minx. In 1964 he joined Ford to work on new vehicle concepts. Remaining with Ford, he privately organized Fenair Air Services in 1966 as a business aviation carrier. He resigned from Ford in 1968 and joined Perkins Engines, Ltd., as a member of the board. He left Perkins in 1976 to work full-time for Fenair, but accepted a contract with British Leyland to act as a design consultant for the Austin Metro. He joined British Leyland on a full-time basis in 1979 and died unexpectedly in 1980.

1. English Racing Automobiles, Ltd., Dunstable, Lincolnshire.

Hofbauer, Peter (1941–)

Head of Volkswagenwerk's research program on unconventional power systems from 1970 to 1977; department head for power systems from 1977 to 1979; chief engineer of VW–Audi power unit development from 1979 to 1987.

He was born on February 19, 1941, in Znaim, Czechoslovakia, and graduated from the Vienna Technical University with a degree in mechanical engineering in 1966. The following year he joined Volkswagenwerk at Wolfsburg as an engine designer. He worked on VW's first experimental water-cooled engines and developed the first generation VW diesel engine in 1972–75. In the 1980–83 period he led the development of a new generation of single overhead camshaft 1.6- and 1.8-liter spark-ignition engines.

He left Wolfsburg in 1987 to become vice president of product development and sales of the Pierburg Group.[1] In 1994 he left Pierburg to serve as management board member for research and development of the Viessmann-Werke, heating-systems manufacturers.

1. Formerly Deutsche Vergaser GmbH (German carburetors).

Holbein, Hermann (1910–)

Technical director of Champion Automobilwerk GmbH at Paderborn from 1951 to 1952.

He was born in 1910 and had a technical education. He became head of chassis testing for BMW in 1936, stayed with BMW during World War II, and settled at Herrlingen near Ulm in 1945. He found a wrecked BMW 328 by the roadside and rebuilt it as a racing single-seater, which he drove himself in competition. In 1947 he made contact with his old friend Dr. Albert Maier of ZF,[1] who had designed and built a miniature roadster with a flat-twin air-cooled engine in the tail. He bought the drawings and began developing the car. He secured a supply of 6-hp Triumph[2] motorcycle engines and began building a few Champion 250 cars in a garage at Herrlingen. The deal was legitimized with a ZF contract in 1949, and he produced about 400 cars from 1948 to 1951. He gave Helmut Benteler, son of an industrialist in Paderborn with a steel stampings and tube factory employing 4000 men, a test drive in the Champion, and Benteler wanted to start mass-production, but Holbein insisted that the car was not yet fully developed. They came to a compromise and started a joint venture, Champion Automobilwerk, with Holbein providing the product and Benteler the capital and the industrial facilities. Drauz of Heilbronn was contracted to supply the bodies. Over 2000 Champion cars were made at Paderborn in less than two years, but at the end of 1952, Champion Automobilwerk was bankrupt. The car was still in considerable demand, and Holbein was able to arrange for its production under license by no less than three manufacturers: Hennhöfer in Ludwigshafen, Thorndal in Ludwigshafen, and Maico in Pfäffingen. Hennhöfer stopped in 1954, after building nearly 2000 cars. Thorndal made less than 300, up to 1954. Only Maico continued, broadening the model range with a coupé, but was forced to stop in 1958 when the market evaporated.

1. Zahnradfabrik Friedrichshafen, gear and gearbox manufacturers, founded by Count Zeppelin.
2. Triumph-Werke Nürnberg, motorcycle manufacturers, later merged with Adler.

Holbrook, Claude Vivian (1886–1979)

Works manager of Triumph Cycle Company, Ltd., from 1919 to 1930; works manager of Triumph Motor Company, Ltd., from 1930 to 1933; managing director of Triumph Motor Company, Ltd., from 1933 to 1936; chairman of Triumph Motor Company, Ltd., from 1936 to 1938.

He was born in 1886 in London as the son of newspaper publisher Sir Arthur Holbrook and was privately educated. He was not destined for a military career but volunteered as soon as the war began in 1914, and served as an officer in a unit that later took the name Royal Army Service Corps. He was discharged with the rank of colonel in 1917 and placed at the head of a military procurement office in Whitehall. He handled major orders for transport equipment of all kinds, including motorcycles, and became acquainted with Siegfried Bettman of Triumph Cycle Company. In 1919 he accepted an offer to join Triumph in Coventry. He believed that low-priced economy cars, not cyclecars, would soon eat big chunks of the motorcycle market, and urged Bettmann to compete on both sides. In 1921 Triumph purchased the former Dawson Car Company, Ltd., factory in Clay Lane, Stoke, Coventry, and retooled it to produce the Triumph Ten, which came on the market in 1923. The 13/35, a bigger companion model, was added for 1925, and a year later, both were replaced by the 15/50 Type TPC. The motorcycle factory was still profitable, but Holbrook began arguing for selling it off in order to broaden the range of cars and increase production. Bettmann resisted, but in 1933 Holbrook was able to sell it to John Y. Sangster. The Super Seven, introduced in 1928, and its derivatives, Super Eight and Super Nine, were very competitively priced, and quite popular. But they were unprofitable, and Triumph's accounting practices prevented its discovery until 1933. Nothing daunted, Holbrook brought out a fleet of new models. The Southern Cross had been launched in 1932 and the Gloria Ten in 1933. The Gloria 12 and Gloria Six (six cylinders 1476 cc) arrived in 1934 with fresh new styling. But Holbrook also made mistakes. In April 1935 he had purchased White & Poppe's engine plant in Coventry, only to find out too late that it was totally unsuited for making Triumph cars. The board of directors replaced him in 1936 with Maurice Newnham, the leading Triumph dealer in London, as managing director, while piously electing Holbrook as chairman of the board. But the company's financial situation continued to deteriorate, and it was declared bankrupt in 1938, and sold to Thomas W. Ward, a steel-industry captain, in 1939. Holbrook retired in 1938 and died in 1979, having witnessed the postwar rise and fall of Standard-Triumph.

Holste, Werner (1927–)

Director of VW research and development programs from November 1968 to March 1972.

He was born on September 19, 1927, in

Westfalia and studied physics, chemistry and mathematics at the University of Münster. He took engineering courses at Aachen Technical University, specializing in thermodynamics. He became a management board member of Demag (construction equipment) and sat on the supervisory boards of Mannesmann and Simag (general-engineering companies). He became widely known as an outstanding futurologist and theoretician. Kurt Lotz brought him to Wolfsburg in 1968 and charged him with the burden of creating a successor model for the "Beetle." With no auto-industry background, he hardly knew where to begin. But he quickly grasped the situation. The "Beetle" was a mass-produced car and its replacement must have mass appeal, a potentially long production life, and make profits for its makers. Porsche had been given a VW development contract for the EA-266, a two-door four-passenger car with an in-line horizontal water-cooled four-cylinder engine under the back seat. He did not reject it out of hand, but let it face evaluation against rival projects. Lotz's failures with the VW 411 (a four-door car with an air-cooled engine in the rear overhang) and the front-wheel-drive K-70 inherited from NSU told him that his brief was not just a "Beetle" replacement, but a new-generation range of cars.

Reading the industry-wide trend correctly, he opted for front-wheel drive, and started a project identified as EA-272. When he got wind of Audi's EA-838 (which went into production as the Audi 80), he replaced EA-272 with EA-400, making use of Audi 80 components to a great extent. It went into production in 1973 as the VW Passat. He started a smaller project, EA-337, with a transverse four-cylinder engine, which after his departure from Wolfsburg, provided the basis for the 1974 VW Scirocco and Golf.

He resigned from VW to become a professor at the Aachen Technical University in 1972, and stayed there until 1977, when he took over the position as director of the Wuppertal Technical Academy.

Hooven, John L. (1919–)

Product engineering director, Ford-Werke AG, from 1964 to 1967; vice president and director of product development, Ford of Europe, Inc., from 1967 to 1973; vice president for special product and manufacturing programs, Ford of Europe, Inc., from 1973 to 1976.

He was born on April 11, 1919, in Toledo, Ohio, and attended the Henry Ford Apprentice School. He went on to study engineering at Wayne State University and the Lawrence Institute of Technology. He joined Ford Motor Company as an hourly employee at the River Rouge plant in 1939. He became a junior designer, but was drafted into the U.S. Navy in 1944. He returned to Ford and in 1952 led the MUTT team, designing a small military vehicle with all-independent suspension that Ford proposed as a replacement for the Jeep. He supervised the design and development of the Ford Falcon and the Econoline van, developed the Mercury Comet, and was assistant chief engineer in charge of final design and development for the 1961–1964 Ford Fairlane and Mercury Meteor. Arriving in Cologne, he first tackled quality problems with the Taunus 12M and slowly improved the existing models. He had oversight authority over the new Escort and Capri, and directed the Taunus/Cortina program for 1970 introduction. From 1973 onwards he was mainly occupied with long-term planning and the coordination of design with manufacturing. He returned to Dearborn in 1976 and retired shortly afterwards.

Horch, August (1858–1951)

Chairman and managing director of A. Horch & Co. from 1899 to 1909; managing director of Audi Automobilwerke GmbH from 1910 to 1914; chairman of Audi-Werke AG from 1914 to 1920.

He was born on October 12, 1868, into a family of blacksmiths at Winningen on the lower Moselle River. He began an apprenticeship at the age of 13 and went off on a journeyman's career in 1884, gaining experience in steam mills and machine works, railway-building and bridge construction, all over

Hungary and Serbia, where he caught malaria, forcing his return to Winningen in 1887. He went back to school and received his engineering diploma after three years at Mittweida Technical College. In 1891 he joined Maschinenfabrik Spierling in Rostock, where he designed a manure spreader, cranes, and steam engines. He wanted to learn marine engineering and went to work for the Neptun Werft in Rostock, only to be laid off when the company fell into financial straits. His boss at Neptun found him a place, however, with Grob & Co. of Leipzig, who was then making an 800-hp petroleum-fueled engine for motor torpedo boats. Suddenly Horch realized that his future was in engine design. In 1896 he saw the Hildebrand & Wolfmüller motorcycle, which prompted him to seek employment with Benz & Cie in Mannheim. Carl Benz made him assistant works manager for engine construction, and eventually put him in charge of the motor vehicle assembly shop.

Leaving Benz & Cie in order to start his own business, he established A. Horch & Co. on November 14, 1899, in Cologne-Ehrenfeld and began to produce Horch automobiles. Within months he took over a disused spinning works at Reichenberg in Vogtland and retooled it for car production. He thought the plant had ample capacity, yet by 1903 it had proved too small. In 1904 he secured financial backing from the Saxony-Thuringian Automobile Club and moved to a bigger factory at Zwickau. He personally tested the cars and drove them in rallies, his successes stimulating the order flow. But he was spending too much money, and his chief accountant, Jakob Holler, plotted for his ouster. On June 19, 1909, the board voted for Horch's dismissal. Nothing daunted, he moved into a former woodworking plant in Zwickau and founded August Horch Automobilwerke on July 16, 1909. But Holler's management team brought suit to prevent him from trading under the Horch name. He had to find a new name for his cars. His loyal business partner Fritz Fikentscher had a small son, still in school, who amused himself by translating the word "horch"[1] into Latin, quoting "Audi et Altera."

Horch took the name Audi for his car and changed the corporate title to Audi Autombilwerke GmbH. He maintained a busy rallying schedule and won quite a reputation for the Audi name. During World War I, the Audi factory produced trucks, and August Horch was called to Berlin to work in the Kaiser's war-supply administration. In 1917 he was appointed director of the German Motor Vehicle Manufacturers' Association[2] and in 1920 gave up his office at Audi to remain in Berlin and lobby for the industry. He kept a seat o Audi's supervisory board, however, until 1933. He held office in the State Traffic Ministry and became chairman of a committee for the automobile accessory industries in the government's motor vehicle bureau. In 1924 he was elected chairman of a committee for standardization in the automotive industries. Also in 1924, the Audi-Werke fell into receivership. At Horch's initiative, it was recapitalized by Carl Leonhardt, owner of a paper mill, Franz Fikentscher and Hans Zimmerman, a Leipzig banker. They took over Audi's debts and trimmed the payroll from 800 to 200 men, and then began talks with the Saxony State Ministry and banking groups with a view to arranging a merger with Horch-Werke. The plan was rebuffed by Dr. Moritz Straus (1882–1959), managing director of Horch-Werke and its biggest shareholder. August Horch then approached Saxony's most successful engine and motorcycle manufacturer, J.S. Rasmussen of DKW, and met with great interest, which resulted in Rasmussen's purchase of Audi-Werke and all its assets in October 1928. In the next few years, August Horch served as an expert in many cases of patent litigation. He was also invited to examine Dr. Porsche's drawings for the proposed P-Wagen racing car by the management of Auto Union AG. On May 22, 1933, Horch was elected to the supervisory board of Auto Union AG.

He became an honorary citizen of Zwickau on February 15, 1939, and spent the war years in retirement. He died on February 3, 1951, at Münchberg in Oberfranken.

1. Horch = listen (second person, singular, imperative mood).
2. Verein Deutscher Motorfahrzeugindustrieller.

Hörnig, Rudolf (1930–)

Daimler-Benz AG management board member for research and development from 1984 to 1990.

He was born on April 26, 1930, at Spaichingen in Württemberg and attended high school at Rottweil. He studied mechanical engineering at the Stuttgart Technical University, graduating in 1955. He joined Daimler-Benz AG as a test engineer for large engines in 1956 and was promoted to senior engineer in 1968. In May 1971 he was named manager of the technical reliability department, with responsibility for all parts and components in every combination, including units from outside suppliers. He was promoted to senior department manager, passenger-car organs development, in August 1974. The "organs" included engines, transmissions, steering gear, final drive units, etc. His responsibilities included every version of new "organs" and their mutual compatibility. He served as deputy chief engineer of passenger-car research from 1980 to 1984, and moved into the seat vacated by Werner Breitschwerdt in May 1984. He had expected to continue his work of preparing long-term plans for future products, but beginning in 1986 his time was taken up by organizing cooperation and division-of-labor among MTU,[1] Dornier, M-B-B,[2] AEG[3] and making plans for specific synergy projects. He opted for early retirement and left office on April 25, 1990.

1. Motoren- und Turbinen-Union.
2. Messerschmitt-Bölkow-Blohm.
3. Allgemeine Elektrizitäts-Gesellschaft.

Horrocks, Raymond (1930–)

Chairman and managing director of Austin-Morris from 1978 to 1979; managing director for passenger cars, BL Cars from 1979 to 1981; group chief executive for BL Cars from 1982 to 1986.

He was born on January 9, 1930, and educated at Wigan and District Miners and Technical College. He later attended Liverpool University. He began his career with Ford in 1953 and became head of Ford Advanced Vehicles before leaving to join Eaton in 1962. He rose in the ranks of Eaton's European management, and was selected for promotion to the head office in Cleveland, Ohio. He wanted to remain in Britain, and let it be known that he was available for a high-level post in industrial management. Michael Edwardes contacted him and brought him to British Leyland, Ltd., in November 1977 as assistant managing director under Derek Whittaker. He was a central negotiator in the talks with Honda Motor Company which led to an agreement for joint production of a medium-size passenger car to be marketed in two versions: Honda Ballade and Triumph Acclaim. At the same time, BL needed a replacement for the high-volume Morris Marina, but the Montego was not ready until 1984. He canceled plans to replace the Triumph Dolomite with a car combining the TR7 platform and a scaled-down Rover SD1, and to replace the Triumph Stag with a stretched TR7 platform and the Rover 3500 V-8, tentatively called Lynx. In April 1983, he signed a new contract with Honda Motor Company for joint production of an executive-class car, which went into production in 1986 as the Rover 800 Sterling with a Honda V-6 engine. He resigned from BL Cars in 1986 to become chairman of Overbell Limited, and in 1988 was appointed deputy managing director of Applied Chemicals, a privately owned Australian company.

Hounsfield, Leslie Hayward (1877–1957)

Managing director of Trojan Commercial Vehicles, Ltd., from 1914 to 1922; chief engineer and managing director of the Kingston branch factory, Leyland Motors, Ltd., from 1922 to 1928; chief engineer and managing director of Trojan, Ltd., from 1928 to 1930.

He was born on July 20, 1877, in Watford, Hertfordshire, and educated at Brighton Grammar School, Battersea Polytechnic, and the Royal College of Science. He served an apprenticeship with James Simpson & Sons, Ltd., makers of hydraulic pumps, and held engineering positions with the Crompton Electrical Company and Ransomes, Sims and Jefferies, Ltd., before going to work at the

Royal Arsenal, Woolwich. He did his military service with the Electrical Reserve Volunteers during the Boer War and in 1902 assisted R.E. Crompton in the design of a high-speed military tractor. In 1904 he set up his own shop, the Polygon Engineering Works in Clapham, London, and taught evening classes in engineering at City & Guilds College, London. About 1912 he invented an extreme economy car with a water-cooled horizontal parallel-twin two-stroke engine placed under the passenger's seat, planetary gearbox, chain drive, no differential, and solid rubber tires on disc wheels. He aimed to start production in 1914 but the Trojan existed only as a prototype until 1921, when he built a series of six vehicles in his shops at Vicarage Road, Croydon.

Any large-scale production was beyond his means, but in 1922 he sold the design to Leyland Motors, Ltd., who had a recently idled plant at Ham Common, Kingston-on-Thames. The Trojan's technical specifications evolved over the years, though the basic concept was never violated. Nearly 17,000 units, cars and vans, were built until 1928, when Leyland canceled the contract. Hounsfield transferred production to a new plant on Purley Way, Croydon, but resigned from his company in November 1930, to promote other inventions he had patented, such as a tensiometer and the Housfield Safari camp bed. He retired from business in 1953 and died in September 1957 in Croydon.

Hruska, Rudolf (1915–1994)

Technical director of Alfa Romeo S.p.A. from 1954 to 1956; managing director of Alfa Romeo S.p.A. from 1956 to 1959; technical advisor to the chairman of Simca from 1959 to 1964; technical director of Alfasud S.p.A. from 1968 to 1972; engineering supervisor of Alfa Romeo S.p.A. from 1972 to 1975; technical advisor to the chairman of Alfa Romeo S.p.A. from 1975 to 1980.

He was born on July 2, 1915, in Vienna and graduated from the University of Vienna with a degree in mechanical engineering. He began his career by joining Magirus in Ulm in 1937. During his student days, he had made the ac-quaintance of Karl Rabe, who invited him to come and work for Porsche in 1938. After being briefed on the KdF-Wagen project, he was sent to Wolfsburg as a production engineer and stayed there till 1942. He spent a year as a liaison engineering for Tiger tank production, traveling between Krupp, Siemens, and the Semmeringer Waggonfabrik. In 1943 he was put in charge of the production of a small Porsche-designed farm tractor. His research for suppliers and assembly facilities led him to OM in Brescia, where he stayed till 1945. Rejoining the Porsche organization at Gmünd, Katschberg in Austria, he completed the tractor project, which was sold to Allgaier. In 1947 he was sent to Turin as Porsche's man-on-the-spot to supervise the construction of the 12-cylinder supercharged four-wheel-drive Cisitalia (Porsche 360) racing car. In 1950 Porsche released him on loan to Alfa Romeo, where he directed the installation of their first moving assembly line for production of the 1900-series at the Portello works in Milan. He made a great effort to help Italy's coach-builders convert their operations to an industrial level and signed contracts for series-production of body shells by Bertone (Giulietta Sprint coupé), Pinin Farina (Giulietta Spider) and Touring (1900 Sprint and Super Sprint). He resigned from Alfa Romeo in 1959 during a dispute with the government-appointed officials of IRI[1] and the Finmeccanica group which controlled Alfa Romeo S.p.A., and signed up with Fiat. He was delegated to Paris as technical advisor to H.T. Pigozzi, chairman of Simca. He organized production of the Simca 1000 and the 1300/1500 models, and returned to Turin in 1964. He revised the 850 chassis for the Sport Spider, and made the 124 chassis acceptable for the Coupé and Spider. He also made modifications to the Fiat Dino chassis.

In November 1967 Giuseppe Luraghi telephoned him to invite him back to Alfa Romeo. The government, eager to create industrial jobs in southern Italy, had decided to finance an Alfa Romeo plant at Pomigliano d'Arco near Naples to produce 1000 cars a day with 14,000 workers. Hruska was asked to design the car and lay out the factory. He was still under contract with Fiat, but Gaudenzio Bono agreed to release him with immediate effect. Hruska then asked Luraghi for four months to conduct a feasibility study. With a small staff of 27 men, mostly former Fiat and Simca engineers pulled out of retirement, he had the factory plans, full specifications and plastic models of the Alfasud car ready on January 1, 1968. On February 16, 1968, the Alfa Romeo board of directors gave their approval, and Hruska committed himself to production startup in exactly four years. Only then was his name put on the Alfa Romeo payroll, and only then were the investment funds unblocked. In spite of strikes on the construction site and among suppliers, the Alfasud went into production in the summer of 1972, only three months behind schedule, the whole program completed well within the budget.

He held high technical offices with Alfa Romeo until his retirement in 1980. He had a busy retirement, working closely with I.DE.A. Institute and consulting for General Motors in Detroit, and for Japanese clients. He cut back his schedule after 1990, and died without warning from a cerebral hemorrhage on December 9, 1994.

Hübbert, Jürgen (1939–)

Director of passenger-car engineering, Daimler-Benz AG, from 1987 to 1990; director of the passenger-car division of Daimler-Benz AG beginning in 1990.

He was born on July 24, 1939, at Hagen in Westfalia, attended local schools and studied engineering at Stuttgart Technical University. He joined Daimler-Benz AG at Sindelfingen in 1965 and for the next eight years he was involved with a variety of technical planning tasks in the process-engineering sector. In 1973 he was named head of production engineering and took charge of the expansion of the Bremen plant. He became director of the production planning office in 1984 and head of corporate planning a year later. With his next promotion in 1987, he was appointed deputy member of the management board, becoming a full member in 1989. In 1990 he

was given responsibility for the entire passenger-car division, and pushed through such programs as the M-Series sport-utility vehicle, the SLK sports car and the Viano van. He authorized the Formula 1 engine for the McLaren racing team and the experimental CLK-GTR road-racing prototype.

Huber, Guntram (1935–)

Director of passenger-car body development, Daimler-Benz AG, from 1977 to 1997.

He was born in 1935 and graduated from the Munich Technical University in 1959 with a degree in mechanical engineering. He began his career as a test engineer for body structures and safety research with Daimler-Benz AG and in 1962 was promoted to group leader for body development, with responsibility for structural stiffness, vibration, acoustics, aerodynamics, and safety. He led the W-123 midsize car body through to production startup in 1975, introducing the company's first-ever station wagon. He was also responsible for the W-126 (S-Class), W-201 (C-Class), the 300 SL and 500 SL bodies. He later led the W-140 (S-Class) and the W-124 (E-Class) body programs. During his career he amassed 38 patents, mainly relating to body construction and safety but also including inventions in the areas of transmission shift control and front suspension design. From 1981 onwards, he was also a regular lecturer at the Darmstadt Technical University. He retired in December 1997.

Hunt, Gilbert A. (1915–)

Managing director of Rootes (Motors) Ltd. from 1967 to 1969, managing director of Chrysler United Kingdom from 1969 to 1976.

He was born in 1915 in Wolverhampton and educated at Malvern College, Worcestershire. He served as a trainee with a steel company and joined the Hawker-Siddeley Group in 1940, working on the management side of aircraft and aircraft-engine production. In 1944 he was named director and general manager of High Duty Alloys Ltd., a Hawker-Siddeley subsidiary. In 1960 he became managing director of Massey-Ferguson (UK) Ltd., joint managing director of Massey-Ferguson-Perkins Ltd., and a director of Massey-Ferguson Holdings Ltd. On March 30, 1967, he replaced Geoffrey Rootes (who became chairman) as managing director and chief executive of Rootes (Motors) Ltd. The company had lost money almost every year since 1960, and the Chrysler Corporation had increased its ownership from an initial 30 percent in 1964 to 100 percent in 1967. In 1966 Rootes paid $25 million for a stamping plant at Linwood near Glasgow; in 1967 Hunt arranged the sale of its Dodge truck plant at Kew, Surrey, however, and the Thrupp & Maberly coachworks at Cricklewood, London. The Hillman Arrow was upgraded into a Hillman Hunter, but still, Rootes (Motors) Ltd. declared a $30 million loss in 1967. It was reorganized as Chrysler UK in 1969 and told to develop tighter co-operation with Chrysler France. Dodge truck production was relocated to Chrysler Espana.[1] Hunt was still optimistic and in preparation for the new Hillman Avenger, a small family car of conventional specifications, authorized investments in modernization totaling $472 million at the Ryton-on-Dunsmore plant, a new engine plant at Stoke, Coventry, plus retooling at Linwood. Hunter production was transferred from Ryton to Linwood. In 1976 the Ryton plant began assembly of the French-designed Chrysler 1307-1308 (marketed as the Alpine in Britain), after the Avenger had replaced the Imp on the Linwood assembly lines. Hunt retired and on August 10, 1978, John J. Riccardo of Chrysler Corporation announced the sale of Chrysler UK, Chrysler France and Chrysler Espana to Peugeot SA of Paris.

1. Formerly Barreiros Diesel SA, Villaverde, Madrid.

Hurlock, William Albert Edward (1886–1965)

Chairman and managing director of AC Cars, Ltd., from 1930 to 1965.

He was born on June 8, 1886, in London as the son of a tailor, and educated at the British School, Great Yarmouth. He began his career with Alford & Alder, an engineering company and supplier to the automotive industry, where he was promoted to manager of the motor department in 1909. In 1910 he set up his own business, William Hurlock, Jr., on Denmark Hill in southeast London, as motor dealers and road haulers. During World War I he served as an inspector in H.M. Admiralty's Mines Inspection Department. In 1919 he began buying up war-surplus trucks and rebuilding them for the civilian market, keeping many of them to expand his haulage fleet. He opened local dealer branches and set up service stations. In 1929 he learned that AC (Acedes) Cars, Ltd., of High Street, Thames Ditton, Surrey, was in liquidation. He went to inspect the premises, accompanied by his brother Charles Fleetwood Hurlock (1901–1969), and bought the property with the intention of using the works as a truck repair shop and to build truck bodies. They also discovered many semi-finished AC chassis, lots of body parts, and the complete tooling for the 2-liter overhead-camshaft AC six. They rehired some of the idled workers and built a few AC cars, first for their own use, and then for sale. What began as an interesting sideline blossomed into regular production of the AC 16/56 Magna in 1931, along with the more sedate AC 16/40 Royal and the three-carburetor AC 16/60 Sports. The 16/56 AC Ace roadster was added in 1933, with styling by the Earl of March[1] who provided great-looking body designs for a succession of AC models. In 1936 the Ace was offered with an optional Arnott supercharger. Yet the total output of AC Cars, Ltd., from 1930 to 1939 hardly exceeded 600 cars. In 1940 the company acquired an auxiliary plant on Taggs Island in the Thames for making Smith six-pound guns. The Thames Ditton works made parts for the Fairey Swordfish bomber and landing gear for the Hamilcar glider, fire-pump trailer and flame-throwers. In 1946 the Taggs Island plant was retooled to make invalid carriers powered by a small BSA engine. Car production at Thames Ditton resumed in 1947 with new models. In 1950 Buckland Body Works Ltd. began offering a four-passenger tourer on the AC chassis. Most of the paintwork was done in another shop, where John Tojeiro was making sports cars. Ernest Bailey, owner of Buckland, arranged for Tojeiro to demonstrate one of his cars to AC's sales manager and later to the Hurlock brothers. They had been looking for new directions in product engineering, and felt sure they had now found one. Alan Turner and Desmond Stratton developed the Tojeiro prototype into an AC roadster with a narrow twin-tube chassis, all-independent suspension with transverse leaf springs, and a Ferrari-inspired body. The AC Ace went into production in May 1954 with the aging AC engine, followed by a coupé version, the Aceca, a year later. At the same time, the Taggs Island plant was building the AC Petite three-wheeler (single front wheel) with a single-cylinder Villiers engine, and made 2000 of them up to 1958. In 1956 the Bristol 2-liter engine became available in the Ace and Aceca, which ended when Bristol took that engine out of production in 1961. AC distributor Ken Rudd showed the Hurlock brothers his installation of a 2.6-liter Ford Zephyr engine in the Ace, and AC built a small number of Ford-powered cars. In 1961 AC Cars Ltd. contracted with Carroll Shelby to supply AC Ace cars without engine and gearbox to his shop in Venice, California, where Phil Remington installed a Ford V-8 with a Ford gearbox. That became the AC Cobra, later Shelby Cobra and Ford Cobra.

1. March Special Body Designing Consultancy, led by Freddy March, later Duke of Richmond and Gordon.

Hüttel, Franz Louis (1862–1919)

Chief engineer of Cyklon Maschinenfabrik GmbH from 1899 to 1906, technical director of Neue Kraftfahrzeug GmbH from 1907 to 1919.

He was born on May 8, 1862, at Erlau in Saxony, held an engineering degree and worked in local industries up to 1899, when he went to Berlin. He designed the first Cyklon motorcycle in 1900, and two years later invented a three-wheeler with a 450 cc single-cylinder air-cooled engine carried atop the single front wheel, with belt drive to the wheel hub. It was covered by three separate patents and series production of the Cyklonette began in 1904. He developed a bigger version with an 880 cc V-twin and chain drive, which he sold to Phänomen-Werke Gustav Hiller AG of Zittau in Saxony. In 1907 he established a non-manufacturing company in Berlin, Neue Kraftfahrzeug GmbH, to handle distribution of the Phänomobil and exploit his patents. He died in 1919.

Iden, George Walter (ca. 1867–)

Works manager of the Motor Manufacturing Company, Ltd. from 1898 to 1903, managing director of the Iden Motor Car Company Ltd. from 1904 to 1906, chief engineer of Crossley Brothers, Ltd. from 1907 to 1910.

He was born about 1867 and began his career with the London, Brighton and South Coast Railway, rising in the ranks from mechanic to foreman. He went to Coventry in 1898 and invested in the newly-founded Motor Manufacturing Company, Ltd. occupying a section of the Motor Mills. He designed new cars, free of Daimler heritage, with rear-mounted horizontal engines. Unfortunately his ideas of automobile design were vague at the time and he wasted much time and money on fruitless experiments. It was typical of his methods to draw up nine different engines when only one was needed. He resigned in December 1903 and set up his own company at Parkside, Coventry, to build Iden cars with front-mounted, vertical four-cylinder engines with a shaft drive to the rear axle. He sold the company to H.H.P. Deasy in March 1906.

In 1907 he designed and built a front-wheel-drive car with a V-twin engine, but no production ensued.

Iden joined Crossley Brothers Ltd. at Openshaw, Manchester, where he adopted four-speed gearboxes and shaft drive for Crossley cars. He resigned when the Crossley motor car department was spun off as a separate company and became a consulting engineer to the London General Omnibus Company. Iden went to work for AEC[1] at Southall works, Middlesex, in 1912. He returned to Crossley Brothers Ltd. in 1917 and spent the rest of his career as a designer of gas engines.

1. Associated Equipment Company, Ltd.

Indra, Fritz (1940–)

Chief engineer of Alpina (Burkard Boven-siepen K-G) from 1970 to 1978, senior departmental engineer, Audi AG research and development staff from 1978 to 1985, chief engineer of engine pre-development with Adam Opel AG from 1985 to 1989, becoming chief product development engineer, Adam Opel AG from 1989 to 1996, director of Advanced Engineering, Adam Opel AG, since 1996.

He was born in Vienna on March 22, 1940, and studied mechanical engineering and marketing at the Technical University of Vienna. In 1965 he joined Burkard Bovensiepen at Buchloe, whose Alpina garage specialized in making high-performance road cars from standard products. He began as an engine tuner and his conversions for BMW power units were extremely successful. He left Buchloe in December 1978 to join Audi at Ingolstadt, where he worked on engine development under Franz Hauk's direction from 1979 to 1983. He spent the next two years as head of an Audi engine design section, and joined Opel at Rüsselsheim in 1985. He developed the 2-liter Calibra 16V engine for 1990 and the Calibra Turbo engine for 1992. Indra was promoted to director of Advanced Engineering in 1996.

Innocenti, Ferdinando (1891–1966)

Chairman of the board, Innocenti Società Generale per l'Industria Metallurgica e Meccanica from 1933 to 1966.

He was born on September 1, 1891, at Pescia near Pistoia. He worked his way through high school but dropped out at the age of 18 to start his own business. He moved to Rome in 1922 and opened a machine shop. A few years later he was able to expand into making steel tubes. In 1931 he moved to Milan and set up a factory at Lambrate to make steel tube scaffolding. He spent the World War II years in Rome. His factory was destroyed in 1944. A year later he came back to Milan, rebuilt the factory, and began making machine tools and heavy presses for sheet metal. In 1947 he began production of the Lambretta motor scooter, designed by Pierluigi Torre, a direct rival for the Vespa from Piaggio.

Innocenti's next ambition was to build cars, and he secured a license for the Austin A-40 in 1960. Production began a year later. He also made the 950 Spider, based on the Austin-Healey Sprite, with an OSI[1] body designed by Carrozzeria Ghia. In 1963 he added the IM3 based on the Morris 1100 and two years later, the Mini. BMC[2] began buying Innocenti stock in 1964, leading to full ownership in 1972. But the founder had died from a heart attack in June 1966.

1. Officine Stampaggi Industriali, Turin.
2. British Motor Corporation, Birmingham.

Irat, Georges (1892–1971)

Managing director of Automobiles Georges Irat SA from 1921 to 1934; chairman of Irat & Cie from 1935 to 1953.

He was born in 1892 at Arcachon and went to Paris at the age of 20 to work in the motor trade. By 1914 he ran a car dealership on Boulevard Pereire in the east end of Paris, selling Majola and other makes. At the end of World War I he took control of the Majola factory in rue Nay at Saint-Denis. His sales organization, SA d'Exploitation des Automobiles Georges Irat, had shops in Boulevard de la République, Chatou, where he began production of the Georges Irat car with a 2-liter four-cylinder engine and four-wheel brakes in 1921. The little factory completed 150 chassis in 1922 and 200 in 1924. No other models were offered until 1927, when a 3-liter six was added. In 1928 he closed the Majola plant at Saint-Denis and prepared to move Georges Irat car production to Boulevard de Levallois in Neuilly-sur-Seine. In 1929 he took over the defunct Chaigneau-Brasier company at Ivry-Port and reorganized it in his son's name: Automobiles Michel Irat SA, Avenue de Villiers in Paris. Only one model Michel Irat was ever built, the CB2 with an 1100-cc engine. A bigger Georges Irat with a 4-liter Lycoming engine appeared in 1929, and a 5-liter Lycoming-powered car was offered in 1932. It led to ruin and he closed the Neuilly plant in 1934. He obtained financial backing from

Etablissements Godefroy & Leveque, makers of Ruby engines at Levallois-Perret, and organized production of a new sports car in an idle hall of the Ruby engine works. It was a front-wheel-drive car with a very low-slung chassis. The engines were Ruby, 950-cc and 1100-cc. Some 1500 were produced over a four-year span. In 1938 he brought out a new model with rubber springs and the Citroën 11 CV powertrain. Nearly 200 were made before car production stopped in 1940. He spent World War II in the Bordeaux area, where he set up two subsidiary companies, Irat Moteurs and Irat-Diesel, with a factory at Bègles-Tartifume. He made engines for farm tractors, trucks, and Isobloc buses. He went back to Paris in 1945 and proposed a new sports car with very modern styling and the Citroën 11 CV powertrain. In 1949 he displayed a smaller prototype roadster with an 1100-cc flat-four engine, front wheel drive, and a curb weight of 600 kg. He made plans for its production at Bègles-Tartifume, but it did not work out. He bought a basic-transportation car design from Emile Petit, and set up the Société Chérifienne d'Etudes des Automobiles Georges Irat on Boulevard Ballande in Casablanca, Morocco, where the market was supposed to be. It had a flat-twin air-cooled engine in the rear overhang and a utility-type four-passenger body. But production never got under way. He retired to Arcachon where he died on January 20, 1971.

Irving, John S. (1880–1953)

Development engineer with the Sunbeam Motor Car Company, Ltd., from 1919 to 1927; technical advisor to Rootes, Ltd., from 1928 to 1931.

He was born in 1880 in Manchester and held a degree in mechanical engineering. He joined Daimler Motor Co., Ltd., in Coventry in 1903 and worked on the design of engines, gearboxes, axles, suspension and brake systems until 1915. He served in the Royal Flying Corps in World War I and was discharged in 1919 with the rank of captain. He joined Sunbeam Motor Car Company, Ltd., at Wolverhampton and came under the orders of Louis

Coatalen. He was attached to the racing car department and also prepared special engines for motorboat racing. In 1926 he was named manager of Sunbeam's competition services and designed a speed-record car driven by two Sunbeam Matabele aircraft engines which became the first car in the world to top 200 mph. He resigned in July 1927 to become chief engineer of the company which held the Humfrey-Sandberg free-wheel and clutch patents. He was privately engaged by C.C. Wakefield[1] to design a speed-record car with a Napier Lion 24-liter W-12 engine, which was named Golden Arrow and set the world's record at 231 mph in 1929. In 1928 he accepted an invitation from William Rootes to supervise the engineering of Hillman and Humber cars. He even proposed a ready-made Hillman Segrave[2] car which was announced in 1928 but never reached production. He led the design team for the 1931 Hillman Wizard chassis and gave guidance for the engineering of the original Hillman Minx chassis. He left Rootes, Ltd., in 1931 to join Bendix, Ltd., and later became technical director of Girling, Ltd.[3] In 1945 he served on the British Intelligence Objectives Sub-Committee investigating technical developments in the German motor industry from 1934 to 1945.

1. Charles Cheers Wakefield (1859–1941) founded a world-wide petroleum products sales organization in 1899, placed the Castrol brand name on the market in 1909, and was a prominent sponsor of motor racing projects between 1920 and 1940.
2. Henry O'Neil DeHane Segrave (1896–1930) was a member of the Sunbeam Grand Prix racing team since 1921 and the driver of the 1000-hp Sunbeam that was timed at 203.79 mph at Daytona Beach in 1927. He was killed during a speedboat record attempt in 1930.
3. Albert Henry Godfrey Girling (1881–1971), inventor of Girling brake mechanisms, and head of a company that took over the Bendix brake business in Britain.

Issigonis, Alexander Arnold Constantine "Alec" (1906–1988)

Chief engineer of Morris Motors, Ltd., from 1947 to 1952; chief engineer of British Motor Corporation from 1957 to 1961; technical director of British Motor Corporation from 1961 to 1968.

He was born on November 18, 1906, in Smyrna (now Izmir, Turkey) as the son of a Greek-born, British-citizen father and a German mother from a Bavarian family of brewers. His father owned foundries and an engineering works in Smyrna, but sent the son to England to complete his schooling. He attended the Battersea Polytechnic in London, but never graduated, repeatedly failing in the mathematics tests. In 1928 he went to work for a firm of consulting engineers in London who were developing a semi-automatic transmission, but left when Norman Wishart gave him the chance of joining Humber in 1930. He spent the next six years in the drawing office for Hillman and Humber cars. From 1933 to 1937 he spent a lot of his spare time building a race car in partnership with J.M.P. "George" Dowson. they called it the Lightweight Special for its unit-construction body. Powered by a 750-cc supercharged Austin engine, it also featured all-independent suspension with rubber springs and small-diameter cast-elektron wheels. At the same time, he was involved with Joseph G. Fry, David Fry, Richard D. Caesar and Robin Jackson in the Freikaiserwagen hill-climb racing car team. His career changed when he left Humber in 1936 and joined Morris Motors, Ltd., as a suspension engineer. He had an idea for a new small-car concept in 1939 and made a scale model in 1942. It became a Morris project under the code name Mosquito in 1943, a four-passenger sedan with a front-mounted flat-four engine and rear-axle drive. By 1945 it was slated to replace the E-Series Morris Eight when ready, and it went into production in the fall of 1948 as the Morris Minor, inheriting the E-series engine in place of the flat-four. It was a tremendous sales success, in export markets as well as in Britain. When Morris Motors, Ltd., merged with Austin Motor Company to form BMC late in 1952, R.F. Hanks advised Issigonis to "get out" and he signed up as chief engineer of Alvis, Ltd. He designed a powerful V-8 sports car which never reached production. L.P. Lord lured him back to BMC in 1955, with a brief to unify the Austin and Morris model ranges, starting at the bottom.

That led to the Mini, introduced in 1959. Issigonis had been toying with transverse engine and front-wheel-drive layouts since 1950. Now his ideas matured, with assistants named Jack Daniels, Chris Kingham and John Shepherd, the latter two recruited from Alvis. The Mini (ADO 15) was followed by the 1100 (ADO 16) in 1962 and the 1800 (ADO 17) in 1964. They were the same basic design in three sizes, but most successfully applied to the smallest car. Due to this single-concept strategy, BMC lost the fleet-market to Ford (mainly) and Vauxhall. Issigonis's last BMC design was the Maxi, launched in 1969. In the spring of 1968, after the BMC merger with Leyland, he was given the title of Advanced Design Consultant, free of direct product responsibilities. He retired in 1971 and died on October 3, 1988.

Jacoponi, Stefano (1941–)

Head of the Alfa Romeo engineering department from 1987 to 1991; technical director of Fiat Auto S.p.A. from 1991 to 1999.

He was born in 1941 and studied engineering. He joined Ferrari in the drawing office in 1964 and worked on the 330 GTC and 275 GTB before being assigned to the racing car office. He led the development of the 330/P4 and during 1969 replaced Mauro Forghieri as chief engineer of the Grand Prix racing team. But a few months later, he left Maranello to join Abarth in Turin as chief engineer of engine development. After the Fiat Abarth 131 Rally engine, he was transferred to Fiat's car design office as an engine specialist and developed the supercharged 131 engine and the 124 Turbo engine. He led the team that designed the FIRE[1] 1000 which went into production at Termoli in 1984. Three years later he was given the task of correlating Alfa Romeo's engineering programs with Fiat's and getting the 164 (based on the Lancia Thema) ready for production. In 1991 he was chosen to succeed Paolo Scolari as technical director of Fiat. He reorganized the way new models were created for Fiat, Lancia and Alfa Romeo, putting them under central control, with shared production facilities. In September

1999, he was named managing director of Fiat's Research Center at Orbassano.

1. FIRE = Fully Integrated Robotized Engine.

Janecek, Frantisek (1878–1941)

Chairman of Zbrojovka Ing. F. Janecek AS from 1928 to 1941.

He was born in 1878 in Bohemia, then part of the Austro-Hungarian (Habsburg) empire, and educated as an engineer. He held technical positions in a succession of enterprises and in 1915 set up his own company, Prager Waffenfabrik, makers of artillery shells, hand-grenades, and typewriters. After World War I the company survived by subcontracting work, making steel stampings, pots and pans, and machining semi-finished parts for other enterprises.

In 1928 he set up a subsidiary to exploit a newly acquired license for the Wanderer 500 cc overhead-valve shaft-driven motorcycle with its pressed-steel frame, opening a branch factory at Prague-Nusle. His motorcycles were sold with the Jawa trade mark (Ja [necek] Wa [nderer]). In 1933 he negotiated with Auto Union AG for manufacturing rights to the front-wheel-drive DKW car with its two-cylinder two-stroke engine. His first car went into production in May 1934 as the Jawa 700, mechanically very faithful to DKW specifications, but with bodies designed and built at Kvasiny, engines made in Prague, and final assembly at Tynec nad Sazavou.

In 1936 he engaged an American engineer, George W. Patchett, who designed new two-stroke engines not covered by the DKW/Schnürle (reverse-flow scavenging) patents and brought in an ex–Praga engineer, Rudolf Vykoukal, who had radical ideas about chassis design. They brought forth the Jawa Minor in 1937, with a 616 cc two-cylinder two-stroke engine, backbone frame, and all-independent suspension. The company built nearly 2000 of them in less than three years. Then the plants were converted to war production, under German management. He died on June 4, 1941.

Jano, Vittorio (1891–1965)

Chief designer of Alfa Romeo racing cars from 1923 to 1937, director of Alfa Romeo's production car design office from 1926 to 1937, chief engineer of product development, Lancia e C. from 1937 to 1942, technical director of Lancia e C. from 1942 to 1956.

He was born in Turin on April 22, 1891, the son of the director of the Turin Arsenal and grew up in San Giorgio Canavese. He had some technical schooling and practical experience of the manufacture of firearms in the Arsenal. He began his career in 1909 as a draftsman with Società Torinese Automobili Rapid, then offering a range of five well-made touring cars from 12/14 to 50/70 hp. He came to Fiat in 1911 and became a car designer on the staff of Carlo Cavalli, who made him head of the technical racing service in 1914. He worked on aircraft engines during World War I and was involved in racing car design for Fiat in 1921–23. At that time Nicola Romeo looked over the dismal racing record of his cars and decided to find a new racing-car designer. He sent one of his directors to Turin to interview Jano, who moved to Milan in September 1923. His first masterpiece was the P2 racing car with a straight-eight two-liter engine, made up of two blocks of four with integral twin-cam heads and carrying a gear-drive Roots-type blower. It went on to become a distinguished racing car from 1924 through 1930. His first production model was the exquisite 6C 1500 of 1927. After enlargement to 6C 1750 in 1929, it became available with a twin-cam cylinder head and supercharger.

He designed the Tipo B 8C 2300, with twin diagonal propeller shafts, one to each rear wheel, for the 1932 racing season. His 6C 1900 production model gave way to the 6C 2300, replacing the former gear train with a timing chain, and introducing a factory-built steel body in 1933. The 6C 2300/B of 1935 became the first Alfa Romeo with all-independent suspension (under Porsche license, but Alfa execution). He designed the 12C-312 racing and the 8C 2900 sports car, both mechanical marvels built regardless of cost.

He left Alfa Romeo in October 1937 to join Lancia e C. in Turin, where he created the 1939 Ardea four-door sedan with a 903 cc V-4 engine, a tiny jewel with perhaps the highest cost-per-kilogram of all European production cars. The 1948 Ardea was the first modern car with a five-speed gearbox in its standard specifications. Jano coordinated a design team to create the Aurelia compact sedan with a V-6 engine, going into production in 1950. A smaller model with a V-4 engine, the Appia, replaced the Ardea in 1953. Jano designed the D20 sports/racing prototype in 1952/53 with a four-overhead-camshaft V-6, and its D23 and D24 derivatives. In 1954 they won both the Mille Miglia and the Carrera Panamericana Mexicana. Jano also designed the D50 Formula One car with a V-8 engine. In 1955/56 Lancia pulled out of racing and gave its D50 cars and engines to Scuderia Ferrari. Jano accompanied them to Maranello as a consulting engineer.

He helped the young and fatally ill Alfred (Dino) Ferrari design a 65° V-6 that was produced in a variety of sizes by Ferrari from 1965 to 1975 and by Fiat from 1966 to 1972.

Jano had lost his brother to cancer and feared that he had it too. He committed suicide on March 12, 1965.

Jaray, Paul (1889–1974)

Pioneered aerodynamism in the first part of the twentieth century.

He was born in Vienna and became the chief engineer of the Zeppelin airship company. In 1921, the young Austrian engineer spent a great deal of time and effort researching the air-resistance of vehicles and calculating the aerodynamic components and variables for chassis of all kinds, testing scale models of streamlined cars in a wind tunnel. Jaray's chassis design concepts were laid out in a 1922 patent. By building an experimental vehicle which had a teardrop saloon shape (instead of the traditional long narrow cylinder) on a Ley chassis, he proved his thesis that the engine capacity required to overcome the aerodynamical resistance of ground vehicles is proportionate to the third power (cube) of the speed of the vehicle. Following the discovery, his thesis gradually drew worldwide attention. Later he designed special bodies for Benz, Adler, Hanomag and Audi, but the designs were too radical to be accepted by the public.

The Tatras 77 and 87 were his only commercial successes. In the late 1920s he left for Switzerland to create his own consulting company. He continued with aerodynamic cars into the 1930s, on Chrysler and Mercedes-Benz chassis among others. His innovations earned him royalties in addition to recognition as Chrysler (Airflow) and Peugeot (402) used his patents for their designs. By the end of the 1930s almost every automobile manufacturer applied his concepts in the design and development phase of their manufacturing. Jaray's thesis and experiments continue to influence the automobile industry today.

Jellinek, Emil (1853–1918)

Founder of Mercédès, Société Française d'Automobiles.

He was born in Leipzig on April 6, 1853, as the son of a Czech father and Hungarian mother. He was privately educated, then sent to boarding school at Sonderhausen, returning to the family home in Vienna at the age of 16. Within a year he "flew the coop" and some time later he was found, making his way as a minor railway official at Rotkosteletz in Moravia. Once family communication was reestablished, his father, a man with some influence in the administration, persuaded Emil to accept a position in the Austro-Hungarian consulate in Tangiers, Morocco. He was promoted to commercial secretary of the consular service, and in 1872 was delegated as vice-consul to Tetuàn.

Two years later he left the consular service, and moved to Oran in Algeria, where he entered private business as a trader in tobacco. He made a visit to Vienna in 1880 to sound out the business climate, and settled in Vienna in 1884 as the agent of a French insurance company. A keen cyclist, he developed a strong enthusiasm for motor cars. His agency prospered, and he began spending winters in Monte Carlo. About 1895 he bought a residence on the Promenade des Anglais in Nice, and moved there with his family as Consul-General of Austria-Hungary, later serving also as Consul-General of Mexico. Privately he took over the Hotel Scribe and Château

Robert in Nice, Hotel Astoria and Hotel Mercédès in Paris.

He bought his first motor vehicle, a de Dion–Bouton tricycle, he owned a succession of Daimler cars which he found unsuitable for racing, preferring a Mors. He initiated a busy correspondence with Gottlieb Daimler and Wilhelm Maybach, making lots of suggestions for improving their products. He held out the promise of setting up an agency for Daimler cars and engine in Monaco, while racing the latest Daimler models in events such as the Nice–La Turbie hill-climb. He told Wilhelm Maybach he needed a four-cylinder engine, longer wheelbase, and a lower frame.

A basic contract for the car which became the Mercedes was signed on April 2, 1900, in Nice by Wilhelm Maybach, Gustav Vischer and Emil Jellinek. On seeing the prototype, Jellinek signed an order for 36 cars, on condition of having exclusivity rights to that model. The first Mercedes car was road-tested on November 2, 1900, and gave a sensationally good performance in the Nice Speed Week of 1901.

In 1903 the authorities in Vienna granted him the right to sign his name as Jellinek-Mercedes. His relationship with the Daimler Motoren Gesellschaft, itself torn by strife among the directors, worsened, and his letters to the factory, frank and uninhibited to the point of offensiveness, took on an openly insulting tone since 1905, accusing its technical personnel of "stupidity and inefficiency."

In 1908 he broke off all relations with Cannstatt. During the years that followed, his health began to deteriorate and in June 1914 he went to Bad Kissingen for a cure. He was resting in Baden, near Vienna, when Germany declared war on Russia and France. The French authorities confiscated his Villa Mercédès in Nice and all his hotels as enemy property. He spent the winter of 1914-15 in Merano, South Tyrol, and settled in Geneva in April 1915. He died from a stroke in his room at Hotel National, Geneva, on January 21, 1918.

Jenschke, Karl F. (1899–1969)

Chief engineer of Steyr-Werke AG from 1930 to 1934; chief engineer of Adler-Werke AG from 1935 to 1945.

He was born in 1899 in Austria and held a university degree in mechanical engineering. He joined the makers of Steyr cars in 1923 and three years later became assistant to the chief engineer, Anton Honsig. He had a hand in the design of the Steyr XII with its overhead-camshaft 1½-liter six and rear swing axles, and its heavier derivative, the Steyr XX of 1928. In 1929 he worked under Ferdinand Porsche on the Steyr XXX and Austria, and was appointed chief engineer on Porsche's departure. He designed the 430 and its successor, the 530, and then prepared Type 100, the first of a new generation of Steyr cars with all-independent suspension, powered by a low-cost side-valve 1½-liter four-cylinder engine. The same chassis was used for Type 120, which had a smaller-bore version of the six from the 530. Steyr's car production had fallen from 8700 cars in 1929 to 1200 in 1933, and talks began for a three-way merger with Puch-Werke in Graz and Austro-Daimler. Steyr-Daimler-Puch AG was registered on January 1, 1934, and the Steyr drawing office became subordinate to the group technical director, Oscar H. Hacker. Jenschke preferred to resign. He had nearly completed design work on Type 50, a four-passenger economy sedan with a water-cooled 980-cc flat-four in the front overhang and rear-wheel drive, under a dome-shaped unit-construction body, but it did not get into production until 1936.

In 1935 Ernst Hagemeier engaged him as chief engineer of Adler-Werke AG in Frankfurt, where he began to upscale the Steyr 50 concept for a 2½-liter six, with a longer, better streamlined body. It went into production in 1937. He remained with Adler throughout World War II, when the company produced military patrol cars, half-tracks, Maybach tank engines and Maybach multi-speed transmissions, but resigned in 1945.

Jensen, Frank Alan (1906–1994)

Joint managing director of Jensen Motors, Ltd., from 1936 to 1966.

He was born on February 9, 1906, in Moseley, Birmingham, and educated at Woodrough School and King Edward's School, Birmingham. After working as a trainee with Serck Radiators in Coventry, he joined New Avon Body Co. Ltd., at Warwick as a designer and engineer. Privately he and his younger brother Richard rebuilt an Austin Seven, giving it a striking sports-car look. It came to the attention of A.H. Wilde, chief engineer of Standard Motor Company, Ltd., who asked them to create a sports body on the Standard Nine chassis. It went into production at the Avon factory in 1929. He left Avon and went into partnership with Joseph Patrick, Wolseley dealer in Birmingham, to build sports-car bodies on the Wolseley Hornet chassis. They quarreled and broke off the partnership. The Jensen brothers bought stock in West Bromwich Motor & Carriage Works from the founders, W.J. Smith & Sons, and designed sports car bodies for Wolseley Hornet, Singer, Standard, and the Morris Eight. They designed racing bodies for the MG Midget and MG Magnette, and sports bodies on Ford chassis for a prominent Ford dealer. In 1934 they created a sports body on the Ford V-8 chassis for Edsel Ford. They took full ownership of the factory at Carters Green, West Bromwich, in 1934 and reorganized it as Jensen Motors, Ltd., in 1936. They produced spectacular cars, styled in good taste, with Ford V-8 and Nash engines up to 1940.

A completely new car powered by a 130-hp straight-eight Meadows engine was announced in 1946. Jensen was also active in commercial vehicles and began building a truck with an aluminum frame in 1948. For the 1949 Interceptor, Jensen chose a six-cylinder Austin engine. For the 1953-model 541 Jensen adopted fiberglass-reinforced plastic bodies of very modern design. In 1960 Jensen Motors signed a contract with Volvo to assemble and paint steel bodies for the P-1800 coupé, and assemble the final product. In 1962 the 541 was replaced by the CV8, abandoning Austin power in favor of a Chrysler V-8. In 1966 the Jensen brothers sold the company to Norcros, Ltd.

Jensen, Richard Arthur (1909–1977)

Joint managing director of Jensen Motors, Ltd., from 1936 to 1966.

He was born on April 13, 1909, in Moseley, Birmingham, and educated at Woodrough School and King Edward's School in Birmingham. Upon graduation he began an apprenticeship with Wolseley Motors (1927) Ltd. and four years later, he joined Joseph Lucas, Ltd., as an engineering trainee.

In 1931 he joined his older brother Alan as designer and engineer with the West Bromwich Motor & Carriage Works and his subsequent career took place at his brother's side.

John, Thomas George (1880–1946)

Chairman of Alvis Car & Engineering Company, Ltd., from 1921 to 1924; managing director of Alvis Car & Engineering Company, Ltd., from 1921 to 1935; chairman and managing director of Alvis, Ltd., from 1935 to 1944.

He was born in 1880 at Pembroke Dock, South Wales, as the son of a longshoreman. After serving an apprenticeship in Her Majesty's Dockyards at Pembroke, he was enabled to study engineering at the Royal College of Science in London and the Royal Naval College at Greenwich. In 1911 he joined Armstrong, Whitworth & Company at Barrow-in-Furness as manager of the shipbuilding department, and in 1915 he went to Coventry as works manager of Siddeley-Deasy, then in the process of tooling up for aircraft-engine production. In 1920 he purchased the Holley Brothers factory which had been set up during the war to make carburetors, and founded his own company, T.G. John, Ltd. He also engaged G.P.H. de Freville to design a car which became the first Alvis.

Perpetually underfinanced, Alvis Car & Engineering Co., Ltd., fell into receivership in 1924, and he had to give up the title of chairman to one of the main creditors. With new Alvis models, John enjoyed strong business growth up until 1929 when his financial resources were again stretched to the limit. He was saved by Charles Follett, head of a dealer organizations, who agreed to buy one-third of Alvis's production.

In 1935 he proposed to start aircraft-engine production, secured a license for the Gnome-Rhône air-cooled rotary, and changed the company title to Alvis, Ltd. But the Ministry of Supply did not order any engines from Alvis, and John took up subcontracting work for Rolls-Royce to keep the plant busy. He was also interested in building military vehicles and in 1936 formed a joint venture with Nicholas Straussler, a Hungarian engineer and specialist in all-wheel-drive vehicles. But the first orders from the MoS were not received until 1939. His health began failing during the war and he resigned in 1944. He died on August 9, 1946.

Johnson, Claude Goodman (1863–1926)

Managing director of Rolls-Royce, Ltd., from 1906 to 1926.

He was born in 1863 in London as the son of an official at the Imperial Institute, South Kensington. As a schoolboy he devoted many hours to arranging exhibits in the museum, and from 1883 to 1886 he was a staff member in the South Kensington Fisheries, Healtheries, Inventions, Colonial and Indian Exhibitions. That was followed by ten years as chief clerk at the Imperial Institute. He came to know F.R. Simms and helped him organize the Automobile Club of Great Britain and Ireland (later the Royal Automobile Club), serving as the club's first secretary. In 1901 he joined Paris E. Singer in starting the City & Suburban Electric Vehicle Co., which failed within two years, and in August 1903 went to work for a car dealer in London, C.S. Rolls & Co. He had a personal friendship with C.S. Rolls, dating back to the 1000 Miles Trial of 1900, which Johnson organized and Rolls had won. The agency added the Royce car to their franchises in 1904. Two years later, he organized the complete merger of the Royce and Rolls companies. He masterminded the promotion and advertising that build the Rolls-Royce reputation and took part in motor sports events, driving Rolls-Royce cars himself. When Royce fell ill in 1910, Johnson took

it upon himself to nurse the great engineer back to health and accompanied him on leisurely travels to France, Italy and Egypt. He also arranged Royce's purchase of a winter residence at Le Canadel on the French Riviera.

In the face of the overwhelming military demand for aircraft engines in World War I, he won the use of the National Shell Factory at Derby, and parts of the National Projectile Factory at Dudley in 1915, adding to Rolls-Royce's production capacity. In 1917–19 he held office in Barlby Road, North Kensington, for the works of Clement Talbot Ltd. had been requisitioned as repair shops for Rolls-Royce aircraft engines. After the war he took charge at the Derby headquarters, hard hit by the cancellation of all military orders. He turned the company around, giving priority to the cars, and pushed for a smaller, lower-priced model. The "Twenty" came on the market in 1923 and did a lot to stabilize the company's finances. A replacement for the 40/50 was long overdue, and the New Phantom went into production in 1925. Still solicitous over Royce's health, he neglected his own for years, and was eventually overcome by fatigue. He died on April 11, 1926.

Johnston, George (1855–1945)

Chief engineer of the Mo-Car Syndicate, Ltd., from 1895 to 1905; co-founder of the New Arrol-Johnston Car Company, Ltd., in 1906.

He was born in 1855 and became a steam-locomotive engineer. He was fascinated by road locomotion but never built a steam car, embracing the internal-combustion engine long before steam power lost popularity. In 1895 he bought a Daimler car, imported from Germany, but did not copy it when designing his own. Instead, his vehicle had large-diameter wheels and high ground clearance, with an underfloor engine, chain drive to the gearbox, and final drive by chain to the rear axle. His engine was quite unusual: a flat-twin with four opposed pistons, cranks, rods and gearing on both sides, with a common output shaft. To get production started, he inked up with his cousin, Norman Fulton, and another engineer, T. Blackwood Murray. Together, they won the financial backing of Sir William Arrol, civil engineer and architect of the Fourth railway bridge, in forming the Mo-Car Syndicate, Ltd., at Bluevale, Camachie, in Glasgow in 1895. The first vehicle was road-tested in March 1896, and their products were marketed either as Mo-Car or Arrol-Johnston cars. They also made commercial vehicles. Production was moved to Paisley in 1901, after Fulton and Murray had withdrawn to found another enterprise.

By 1903 the Arrol-Johnston was the most archaic-looking car on the Scottish market, for Johnston never altered the basic makeup or disposition of mechanical elements. Two years later, it was no longer in demand. Sir William Beardmore stepped in to refinance the enterprise, which was reorganized as the New Arrol-Johnston Car Company, Ltd., at Paisley. Sir William also brought in a new engineering team from England, and Johnston resigned.

In 1906 he allied himself with the All-British Car Company at Bridgeton, Glasgow, and brought out another high-wheeler, far more powerful, with an eight-cylinder version of the horizontal twin. A simpler but bigger and heavier four-cylinder version was installed in the All-British bus chassis. All-British production lasted barely two years. Johnston, now aged 53, withdrew from the automobile industry but remained active in Scottish engineering for another decade. He died in 1945.

Jones, David Brynmor (1910–)

Styling manager of Vauxhall from 1945 to 1971.

He was born on February 25, 1910, in Birmingham and after basic schooling until the age of 15, he began to study design and sculpture at the Birmingham School of Art. In 1928 he moved to London to take classes at the Royal College of Art, became an apprentice of the sculptor Barry and came under the influence of Eric Kennington. In 1933 he joined Vauxhall as a body designer and set to work on futuristic shapes. When Harley Earl, GM styling vice president, saw his work, he brought him to Detroit for 18 months' training in the Art & Color Section. Back at Luton,

his first complete car design was the 1938 Vauxhall Ten coupé. He next styled the new 14 and then spent the war years leading a camouflage team at Leamington for the Ministry of Home Security. He adopted American trends for the 1948 Wyvern and Velox bodies, and came up with the unremarkable Victor in 1957 and Cresta in 1958. Until 1961, the Vauxhall styling studio was subordinate to the engineering director, but then it became a separate department reporting to the managing director. He gave a balanced, rational look to the 1962 Victor. His boxy Viva from 1963 was replaced by a restyled, curvaceous Viva in 1966. He retired on December 31, 1971.

Jong, Sylvain de *see* de Jong, Sylvain

Jordan, Maurice (1899–1977)

Joint managing director of Peugeot from 1933 to 1937; managing director from 1937 to 1947; vice president of Peugeot from 1947 to 1964; president and chief executive from 1964 to 1972.

He was born on August 24, 1899, in Orleans as the son of a general of the French army, though they belonged to a family of bankers. He volunteered for military service at 17, in the middle of World War I. He attended the Ecole Nationale Supérieure from 1922 to 1924 and joined Peugeot as a machine tool engineer on August 1, 1924. The following year he became a body engineer at the Sochaux plant and in 1927 was named director for all the Peugeot factories in the Doubs area. A year later he was called to Paris as the company's secretary-general. Within five years he had risen to executive rank. After World War II he held responsibilities in finance, plant reconstruction, organization and planning.

He approved the adoption of front-wheel drive for the smallest cars (204 and 304), and authorized the 504 program with production startup in 1969. In 1966 he broke off a three-year-old agreement with Citroën in order to start cooperation with Renault at several levels including a joint venture for engine production. He started aluminum foundries and a car assembly plant at Mulhouse. He was made honorary president after his retirement in 1972 and kept a seat on the supervisory board. He died in the first week of 1977.

Joseph, Paul (1884–1967)

Chief engineer of Automobiles Cottin & Desgouttes from 1921 to 1930.

He was born in 1884 and held an engineering degree from the Ecole des Arts et Métiers. He went to work for Marius Berliet as a production engineer in 1910. During World War I he organized the production of the Berliet CBA ordnance truck which proved so vital in running supplies to besieged Verdun and in the battles along the Marne River. He was also responsible for the assembly of Renault tanks in the Berliet works at Venissieux, turning out nearly 1000 units in 1917–18. He left Berliet in 1921 to design a new range of touring cars for Cyrille Cottin. The four-cylinder 14/16 and 23/25 went into production in 1922, followed by the six-cylinder 18/20 a year later. The 18 CV 4-liter M6 appeared in 1924 along with the 16 CV 3-liter four-cylinder M3L, which led to high-performance models sold as Grand Prix Tourisme and Grand Prix Sport. He became widely known for a new chassis with all-independent suspension advertised as "Sans Secousse" (no-shock), first shown at the Paris Salon in October 1925. Front and rear wheels were mounted on sliding pillars and connected by lower transverse leaf springs. The rear brake drums were carried on the inner ends of the drive shafts, and each front wheel had its own steering-gear and linkage (to avoid shimmy). The Sans Secousse was made in a variety of wheelbases, with four-cylinder engines from 1.7 to 3 liter size and six-cylinder engines from 1.7 to 3½ liters. There was even a 1931 prototype with an eight-cylinder Lycoming engine. But he realized that the company was headed for ruin, and at the end of 1931 he joined Citroën Transports as managing director, where he organized bus lines and fleet maintenance until 1940. He was named chairman of Citroën Transports in 1945 and retired in 1950, but held the title of non-executive chairman of Citroën Transports up to 1964.

Julien, Maurice François Alexandre (1902–1979)

Project engineer with SA André Citroën from 1928 to 1935; director of Ateliers M.A. Julien from 1941 to 1949.

He was born on May 11, 1902, in Tananarive, Madagascar, and educated as a mining engineer. Citroën engaged him in 1928 to work on noise and vibration problems. With the help of Pierre Lemaire, lecturer at the University of Lyon, he developed a balanced engine mounting system which had some similarities with Chrysler's patented "Floating Power" for which André Citroën had prudently taken out a license. The Julien-Lemaire system was introduced on all Citroën cars beginning in 1933. Julien also designed the torsion-bar front and rear suspension systems for Citroën 7 CV and 11 CV "Traction" models. He resigned from Citroën when the company came under Michelin ownership, joined Jean-Félix Poulsen and helped him set up the Société Paulstra (industrial rubber products). He left Paris for Toulouse in the spring of 1940 and began to design a microcar. The Julien MM5 was displayed at the Paris Salon in 1946, a tiny four-wheeled two-seater with the rear wheels very close together, fully enclosed, powered by a rear-mounted single-cylinder 25-cc motorcycle engine. He also built the VUP two-seater with equal tracks front and rear, but only the MM5 was produced in any quantity. He returned to Paris and assisted J.-A. Grégoire in the development of variable-rate coil-spring applications and acted as a consultant to RATP,[1] the Paris Metro underground railway system, for the adaptation of pneumatic tires to their carriages in 1952–55. He remained a consultant to Paulstra until 1975 and died at his home in Toulouse on May 25, 1979.

1. Régie Autonome des Transports Parisiens.

Kales, Josef (1901–)

Chief engineer of Volkswagenwerk from 1945 to 1958.

He was born on March 21, 1901, at Lewin in Sudetenland and held a degree in mechanical engineering. He began his career as a design engineer with Oesterreischische Waffenfabriks-AG, makers of Steyr cars and trucks. After a year's experience with Steyr he joined Avis Flugzeugwerke where he worked mainly on aircraft engines. He came to Austro-Daimler in 1927, under Karl Rabe, and developed the six- and eight-cylinder overhead-camshaft engines for the ADR models of 1929–34. He returned to Steyr in 1929 and worked under Ferdinand Porsche's orders on the Steyr XXX and Austria models. In 1930 he accompanied F. Porsche to Stuttgart, became head of the Porsche engine design office, and in 1932–33 designed engines for Wanderer and Röhr. He was the principal designer of the V-16 engine for the P-Wagen (later Auto Union) racing car and the air-cooled flat-four of the KdF-Wagen. In 1941 he was delegated to Wolfsburg as the chief liaison engineer between the Porsche organization and Volkswagenwerk, where he spent the war years on military-vehicle projects. In 1945 he opted to stay in Wolfsburg, and was named chief engineer. He directed the development of the "Beetle" and in 1955 began planning the specifications of Type 3 (VW 1500, launched in 1961).

Kalkert, Werner (1934–)

Director of product development, Ford-Werke AG, from 1985 to 1997.

He was born on June 22, 1934, at Wissen in Siegerland, attended the Rhineland-Westphalian Technical University at Aachen, and graduated in 1962 with a doctor's degree in mechanical engineering. Being attracted to research and teaching, he stayed on at the university as a lecturer until he joined Ford-Werke AG in January 1966 as a manufacturing engineer. He was promoted to production manager of the Cologne engine plant. He served briefly as general manager of the Düren axle plant, and later was delegated to the Blanqueville plant near Bordeaux as coordinator of production of Ford's automatic transmissions. He was given overall responsibility for Ford of Europe's engine production in 1976. The lineup then included different V-6 engines of similar size, made in Britain and Ger-

many, and V-4s made in Germany. He tried to streamline the situation, but was transferred to chassis-and-transmission design and development as chief engineer. In his product-development office, he overhauled the mid-range Escort and Orion for 1991, with simultaneous production at Halewood in Britain, Almusafes in Spain, and Saarlouis in Germany. He also supervised the Mondeo program, with production startup in 1993, replacing the Sierra. He retired in 1997.

Karcher, Xavier (1932–)

Managing director of Automobiles Peugeot from 1977 to 1979; managing director of Automobiles Citroën from 1979 to 1990; vice chairman and managing director of Automobiles Citroën from 1990 to 1997.

He was born in 1932 and held an engineering degree from the Ecole Centrale des Arts et Métiers in Paris. He joined Automobiles Peugeot in 1956 and worked in production engineering at the Sochaux plant for two years. In 1958 he was transferred to the technical center at La Garenne and assigned to experimental and advance projects, divorced from current new-model programs. From 1968 to 1972 he was head of the new-model committee and set out the master plan for the Peugeot 104. He also laid down the basic definition for the Peugeot 604, originally with a V-8 engine. In 1972 he was sent to Buenos Aires with a brief to revitalize Peugeot's operations in Argentina, serving as president of SAFRAR[1] until 1975. On his return to Paris, he was given a seat on the Automobiles Peugeot board of directors and held chief executive office for the car division from 1977 to 1979. He was then transferred to Citroën in the same capacity. He directed the gradual integration with Peugeot SA and ran Citroën's day-to-day operations, got the BX into production in 1982 and the AX in 1986. He authorized production of the XM in 1989 and prepared the ZX for 1991. He was named vice-chairman of Automobiles Citroën in 1990 and held that office until his retirement in 1997.

1. Sociedad Anonima Franco-Argentina de Automotores Ci y F.

Karen, Tom (1926–)

Managing director and chief designer of David Ogle, Ltd., from 1962 to 1963; managing director and chief designer of Ogle Design, Ltd., from 1963 to 1992; design director and group chairman, Ogle Design, Ltd., beginning in 1992.

He was born on March 20, 1926, in Vienna, as the son of a brickworks owner, and attended schools in Brno. The family fled to Brussels in 1939, and had to flee again a year later. He arrived at Bristol in 1942 and enrolled at Loughborough College, graduating in 1946 with a degree in aeronautical engineering. He began his career in the aircraft industry, but disliked mathematics. His mind was more attuned to concepts. He took a course at the Central School of Arts & Crafts, London, graduating in industrial design, and worked in Ford's styling department from 1955 to 1959. He worked for David Ogle, Ltd., for a brief period in 1959, and then joined the design studio of Hotspot home appliances, later holding a similar position with Philips. When David Ogle was killed in a traffic accident in 1962, the new owner, John Ogier, asked him to take charge of the small company based on Birds Hill, Letchworth, Hertfordshire. He completed David Ogle's design for the Daimler SX 250 coupé which became the basis for the 1964 Reliant Scimitar. After producing 70 to 80 Ogle coupés on the BMC Mini platform, he restricted the company's activity to original design and prototype construction. He entered a non-exclusive, long-term contract with Reliant Engineering Company and designed the Reliant Robin, Scimitar GTE, and Kitten. He also designed the body for the Bond Bug three-wheeler and the Triplex Glassback prototypes of 1978 and 1979. By then he was lecturing on automotive design at the Royal College of Art in London and also teaching at Manchester College of Art & Design. He was responsible for the 1982 Lucas hybrid-electric experimental car body, and the 1984 Project 2000 aerodynamic one-box, followed by the P. 2001 four-passenger compact-car. He handed over car design responsibilities to Ron Saunders in 1985, and was named group chairman in 1992.

Karmann, Wilhelm (1871–1952)

Founder and chairman of Wilhelm Karmann GmbH; automotive designer known for his work on convertible sports cars.

From an early age, Wilhelm Karmann worked with his father in his business, learning the craft of cartwright after school. In 1901, Wilhelm Karmann purchased Klases, a coach-building firm established in 1874, and immediately renamed it after himself. Car body building began the next year, and soon production was converted entirely to motor bodies. Unlike some rivals, he made bodies in quantity instead of bespoke designs, so was able to supply Adler with 100 bodies per month as early as 1910. In 1924 he delivered 1000 taxicab bodies to Aga in three months. Starting a close association with Volkswagen in 1949, he supplied a 4-seater cabriolet for the Beetle. He became well known for the Volkswagen/Porsche–developed Karmann-Ghia, which debuted in 1955. In his 64-year automotive career, Karmann also worked on the designs of the VW Scirocco Coupe and Golf Cabriolet, and did design work for BMW, Ford, Opel and Mercedes-Benz. Wilhelm Karmann died in 1952 and was succeeded by his son, Wilhelm Jr. The company continues its history as the makers of fine automobile bodies for a number of European manufacturers, and of tools, dies, and parts for many others. Since the early 1940s, the company has built more than 3 million vehicles.

Kieffer, Joseph-Nicolas (1877–1968)

Director of Société Générale des Automobiles Porthos from 1906 to 1914.

He was born in 1877 at Hollerich in Luxembourg as the son of a cabinetmaker. The family moved to Paris in 1881 and he attended local schools. He had his first automotive experience at the age of 18, when he assisted in the construction of a car with a two-stroke engine for André Hartman, an Alsatian textile-industry magnate. He became a journeyman mechanic, working in factories in Moscow, Yokohama, Oran and San Francisco. Returning to Paris in 1903, he went to work for Clément Ader's telephone company which then had a sideline in making cars with a V-4 engine. In 1906 he formed a partnership with Armand Farkas, a Hungarian-born engineer, and opened a factory at Boulogne-Billancourt. Their first Porthos car appeared in 1907 with a 4.6-liter four-cylinder T-head engine, and in 1908 they offered 50-hp and 60-hp six-cylinder models. Their technical imagination far outstripped their financial means, however, and the company was liquidated in 1909. A year later they formed a new company under the same name, and introduced new models in 1911, the 16/20 and 24/30 with four-cylinder engines and the 20/25 and 30/40 with six-cylinder power units. They were continually dogged by slow sales, and tried moving the Porthos down-market with a 14 CV in 1913 and 10 CV in 1914. But that year the factor was closed for good.

Kieffer spent World War I working as a mechanic in a garage in Chartres, but in 1919 moved to Frankfurt am Main, where he helped Otto Schulz make his start as a manufacturer of automotive instruments. After a number of years in German supplier industries, he returned to Luxembourg and settled at Beggen, operating a precision-engineering shop in partnership with his son Jean.

Kimber, Cecil (1888–1945)

Managing director of the MG Car Company, Ltd., from 1930 to 1935.

He was born on April 12, 1888, in Dulwich, London, as the son of a printing engineer who owned a printer's ink plant in Manchester. He attended Stockport Grammar School but went to work as an ink salesman at the age of 16. He took evening classes in accounting and engineering at Manchester Technical School. In 1915 he joined Sheffield-Simplex as a purchasing agent but also helped design airships during World War I. He went to Autocarriers (1911) Ltd. as a purchasing agent and became a designer with Martinsyde Aircraft. In 1919

he joined E.G. Wrigley & Company, Ltd., makers of gearboxes and axles in Birmingham, where among other tasks, he worked on the production of frames and drive-line units for Aston Martin. In 1921 he became sales manager of Morris Garages in Longwall Street, Oxford, and was promoted to general manager in 1922. In 1923 he made a sports car on the Morris Cowley chassis, and started production in a small way, with coachwork by Carbodies in Coventry. In 1924 he began putting the MG badge on his cars and W.R. Morris gave him factory space in a wing of a radiator plant on Bainton Road. In 1927 MG assembly was moved again, to Edmund Road, Oxford, and two years later to Abingdon-on-Thames. He resigned from Morris Garages and devoted all his time to MG.

He had the financial backing of W.R. Morris who also let him have his pick of Morris and Wolseley components for a proliferating range of MG roadsters. But in 1935 W.R. Morris sold his personally held shares in MG Car Co. Ltd. to Morris Motors, Ltd., and Kimber no longer held any authority at Abingdon. He continued to work at MG, with unspecified duties, until November 1941, when Miles Thomas asked him to leave. He joined Charlesworth, then busy making aircraft components, and later became works director of Specialloyd Limited, piston manufacturers. He was killed in a railway accident at Euston Station, London, on February 4, 1945.

Kimberley, Michael John "Mike" (1938–)

Chief engineer of Lotus Cars, Ltd., from 1974 to 1976; managing director of the Lotus Car Companies from 1976 to 1983; chief executive officer of Group Lotus from 1983 to 1990.

He was born August 24, 1938, and held a degree in mechanical engineering from the Lanchester Polytechnic in Coventry. He later collected a diploma from the Gosta Green College of Advanced Technology. After four years as an apprentice with Jaguar Cars, Ltd., he became a design and development engineer for Jaguar. He joined Lotus in 1969 as a project engineer and became engineering manager in 1972, chief engineer in 1974 and a director in 1975. He had a major input in the 1974 Elite and its derivative, the Eclat, as well as the Esprit. After Colin Chapman's death in December 1982, the hollow nature of Lotus's finances was revealed, and ruin threatened. In July 1983 the Group Lotus was rescued by British Car Auctions which put in £3.5 million against a 20.4 percent equity holding. Shortly afterwards, Toyota Motor Company offered £10 million for Lotus, lock, stock and barrel, but the sale was vetoed by Margaret Thatcher's government. Still, Toyota acquired 22 percent of Lotus, and in January 1986, General Motors bought a 58 percent stake in Lotus for £22.7 million. Kimberley became a non-executive vice president of GM Overseas Corporation. He left GM in March 1994, at the invitation of the new Indonesian owners of Automobili Lamborghini to come to Sant'Agata as chairman of the company. He accepted, but in November 1996 was downranked to a position as vehicle projects manager for production in Indonesia, where he was given the task of developing the Timor car with technical assistance from Kia Motors of Korea. The scheme collapsed and he returned to London. In 2000 he was named chairman of MVI, former importers of the Lada, and made agreements to import the Tata Safari and Tata Indica from Bombay, India.

King, Charles Spencer (1925–)

Chief engineer of Rover Advanced Vehicles from 1952 to 1959; chief engineer of Rover from 1959 to 1968; director of Standard-Triumph engineering from 1968 to 1974; technical director of British Leyland from 1974 to 1979; vice chairman of BL Technology from 1979 to 1985.

He was born in 1925 as the son of a solicitor and a nephew of Spencer B. Wilks. He was educated at Haileybury School and began an apprenticeship with Rolls-Royce in 1942. In 1945 he joined the engineering staff of Rover and worked as an assistant to F.R. Bell on the JET 1 experimental gas-turbine car which was timed at 213.3 km/h (151.2 mph) on the Jabbeke highway in Belgium in June 1952. He held a high position in Rover's ex-

perimental laboratories and was one of the engineers behind the four-wheel-drive T.3 gas turbine prototype of 1956 and began preparing the front-wheel-drive T.4 test car, also powered by one of Noel Penny's gas turbines. The T.4 was a variant of the planned P 6 production model, and he led the P 6 program from its inception. It went into production in 1963 as the Rover 2000 with a four-cylinder single-overhead-camshaft engine and rear-wheel-drive with a de Dion type suspension. He developed it into a more powerful 2000 TC and masterminded its conversion to V-8 power, with a 3½-liter all-aluminum ex–Buick V-8 engine.

In 1965 King and Gordon Bashford began design work on a moderately priced sports car with the same V-8 engine, mounted in an inboard/rear position. It was exhibited in New York in 1968 but when British Leyland began considering it as a potential production model, Sir William Lyons protested to Lord Stokes that it would ruin the market for his E-Types—and it was shelved. He was the chief designer of the Range Rover, an up-market four-wheel-drive on- and off-highway vehicle that went into production in 1970. He was in overall charge of the SD 1 program, a big six-cylinder sedan, assisted by Michael Lewis and Gordon Bashford. The SD 1 went into production as the Rover 3500 in 1976. He took a supervisory role in the preparation of the Triumph TR7 and Dolomite Spring. After 1979 he took charge of BL's research work in automotive safety and led the design of the ECV-1, ECV-2 and ECV-3 experimental cars. He retired at the end of March 1985.

Kissel, Wilhelm (1885–1942)

Benz & Cie board member in charge of finance from 1925 to 1926; management board member of Daimler-Benz AG from 1926 to 1937; chairman of the board from 1937 to 1942.

He was born on December 22, 1885, at Hasloch in the Palatinate as the son of a railway engineer. He attended commercial college and held trainee positions in the steel and machinery industries. He joined Benz and Cie in Mannheim in 1904 as a correspondent. He did his work so well and carried out so much that his superiors took notice, and a series of promotions followed. He held high office in the purchasing department, and in 1917 became secretary for the management board. After the merger with Daimler Motoren Gesellschaft, he was instrumental in preventing Jacob Schapiro from gaining control and in forcing the dismissal of Ferdinand Porsche. He joined Hitler's NSDAP party, became an SS man, and was active in the NSKK motorcycle dispatch riders' brigade. Such political connections earned Daimler-Benz vast contracts for military trucks, aircraft engines, and other warfare equipment when rearmament began in earnest. He died from a heart attack at the age of 57.

Klement, Vaclav (1868–1938)

General manager of Laurin & Klement from 1896 to 1907; managing director of Laurin & Klement AG Automobil-Fabrik from 1907 to 1925.

He was born on October 16, 1868, at Velvary in Bohemia as the son of a laborer who alternated between road construction and farm work, and was apprenticed to a bookseller in Slany. He graduated from county school in Slany and joined a major bookseller in Prague. In 1886 he was engaged as manager of a book shop in Mlada Boleslav, which he bought in 1891. Soon afterwards he set up a sideline as a bicycle agent, selling Brennabor, Humber, Premier and Puch bicycles. He knew of Vaclav Laurin's bicycle operations and approached him with a merger plan, which was accepted. In 1898 he visited Paris and came back with a Werner motorcycle which he wanted to copy. On the Werner, the front wheel was belt-driven from an engine mounted above it. Laurin and his assistants redesigned it with the engine carried on the frame, and belt drive to the rear wheel. He made a sales tour of Great Britain in 1900 and in the following years set up dealer organizations in Germany and France. He wanted to make cars but did not have a suitable design until late in 1905. By 1907 cars were their main line of business, and the partners sold their company

to a group of industrial investors who named him managing director. With strong financial backing, he multiplied his initiatives. In 1911 he started a joint venture with Ruston-Hornsby to make steam rollers and in 1912 he purchased RAF[1] and a license for the Minerva-Knight engine. In August 1914 he rushed to Vienna, returning with lucrative orders for ammunition, assuring growth throughout the war years. He also secured orders for "blimp" winches. Foreseeing the trend to mechanized farming, he had his engineers design the Excelsior motor plow, which became L&K's main product line, and for which he sold licenses to Johann Puch and Austro-Fiat. The company's weakness on the car side prompted merger talks with Skoda-Werke. He took part in the negotiations and voted in its favor, even though he full well realized that it meant the end of his career.

1. Reichenberger Automobil Fabrik, Liberec. L&K sold the RAF factory to the textile industry in 1916.

Kleyer, Heinrich (1853–1932)

Managing director of Adler Bicycle Works[1] from 1888 to 1906; managing director of the Adler-Werke AG from 1906 to 1923; chairman of the Adler-Werke AG from 1923 to 1932.

He was born in Darmstadt on December 13, 1853, as the son of a machinery builder. He practically grew up in the machine shop and became familiar with tools, materials and processes while still a mere boy. After completing grade school he served a three-year apprenticeship with an iron-works and machinery company in Frankfurt. He enrolled at Darmstadt Technical University to study engineering but dropped out in order to take paid employment as a sales engineer for a steel-maker in Siegen. In 1875 he joined Biernatzky & Co. of Hamburg, prominent importers of machinery, mainly from the United States. Biernatzky opened the way for him to go to America, and he spent a year with the Sturtevant Mill Company of Boston, Massachusetts. He returned to Darmstadt shortly before his father's death in 1879 but lost out on his inheritance claim to the parental factory. With next to no capital, he went to Frankfurt and in March 1880 set up a machine and bicycle shop, acting as sales agent and repair shop. In 1888 he began making Alpha wire wheels and Adler bicycles. Big orders for wire wheels began arriving from Benz & Cie in Mannheim in 1893, which spurred him on to thinking about making Adler motor vehicles. The bicycle market was still booming, bringing good profits to Kleyer, who helped finance the setting up of a Dunlop branch factory at Hanau in 1893, making tires and rubber products. In 1895 he began the construction of the first Adler motor tricycle, powered by a single-cylinder de Dion–Bouton engine. A year later, he began manufacturing Adler typewriters (under American patents). The Adler motor tricycle came on the market in 1899, the four-wheeled Adler car in 1900, and the Adler motorcycle in 1901. All were powered by single-cylinder de Dion–Bouton engines, but in 1903 Adler began making its own two-cylinder engines. As demand for bicycles diminished, the car market blossomed, and in 1904 Kleyer set up a separate department for automobiles. The word "bicycle" was omitted from the corporate title in 1906, and in 1903 Kleyer discontinued the production of motorcycles and motor tricycles. By 1914 he had a work force of 7000 and approximately 20 percent of all cars registered in Germany carried Adler's stylized eagle trademark.

During World War I, with the Adler plants busy filling government orders, he made the mistake of granting an interest-free RM 4-million loan to the government. It was never repaid, and Kleyer went heavily into debt to restart civilian production in 1919. The Adler-Werke came under the control of Deutsche Bank in 1920.

Kleyer stayed at the head of the management until 1923 when he handed over the day-to-day management responsibilities to his son Erwin Kleyer (1888–1975), taking the title of chairman. He was still the boss when it came to planning future products — not always with the happiest results. During 1929 Adler was once again in financial straits. The Danat Bank granted extensive credit, but only on condition of having its own representative, Dr. Ernst Hagenmeyer (1888–1966) named chief executive of Adler. On Kleyer's death in 1932, Hagenmeyer was installed as chairman.

1. Adler Fahrradwerke vormals Heinrich Kleyer which became an Aktiengessellschaft (joint stock corporation) in 1895.

Knight, Robert J. (1920–)

Engineering director of Jaguar Cars, Ltd., from 1969 to 1978; managing director of the Jaguar-Daimler part of British Leyland's Jaguar-Rover-Triumph group from 1978 to 1980.

He was born in 1920 and held a Bachelor of Science degree in engineering. He joined the aircraft-engine section of Armstrong-Siddeley as a draftsman and later worked as a power-unit design engineer with Bristol Aeroplane Company, Ltd. He joined SS Cars, Ltd., in Coventry in 1944 as a chassis engineer and worked on independent front suspension designs. The torsion-bar front suspension used on the Mark V, XK-120 and the Mark VII was mainly his work. In 1951 he became a member of W.M. Heynes's personal staff and was involved with the C-Type and D-Type racing models. He led the chassis design for the Mk II and the Mark X, and was named chief vehicle engineer of Jaguar in 1961. He started work on the project that led to the XJ as early as 1962, but the development took six years. Introduced in September 1968, the XJ6 had an all-new body shell, derivative front and rear suspension systems, and an engine dating back to 1948. It replaced the 420 and the Mk II lines, and went on to a production life of 18 years. W.M. Heynes had begun planning a V-12 in 1955 for future endurance-racing cars, but shelved it when Sir William Lyons canceled the racing program. Wally Hassan designed a new "milder" V-12 that went into the 1971 Series-3 E-Type, and early in 1972, Knight managed to install a V-12 in the XJ. That led almost immediately to production of the XJ 12 and the Daimler Double Six, and the creation of the XJS in 1975. In 1978 he was offered the position of engineering director for the whole Jaguar-Rover-Triumph group, but he maintained that they should be kept separate. He served two years running the Jaguar and Daimler operations, but resigned in 1980.

Koch, Hans C. (1930–)

BMW management board member for production from 1972 to 1990.

He was born on June 23, 1930, at Nümbrecht, Kreis Oberberg, and graduated from the Aachen Technical University in 1954. He began his career with Ford-Werke AG in Cologne in 1955 but left within a year to join the Bergischen Achsenfabrik Kotz & Söhne (axle factory) at Wiehl, where he became head of quality control. In 1960 he returned to Ford-Werke AG as coordinator for a new engine plant, and was named head of engine-manufacturing technology in 1964. In 1968 he joined Boge (hydraulic dampers and suspension elements) at Eitorf, as a member of the Boge management board. He went to BMW in 1970, and was named deputy board member for production in 1971, becoming a full member in 1972. That year BMW completed 185,188 cars in two plants, the main works at Munich-Milbertshofen, and the Dingolfing factory taken over from Hans Glas K-G in 1967. Construction had begun in 1970, setting up a bigger plant at Dingolfing, which came on stream in 1973, assembling the 5-series cars. Under Koch's direction, BMW car production doubled in 12 years, without any deterioration in the quality. An engine factory at Steyr, opened as a joint venture with Steyr-Daimler-Puch AG, began producing the small spark-ignition sixes in 1981, adding diesel versions a year later, after which it came under 100 percent BMW ownership. A new final-assembly plant at Regensburg came on stream in 1986, for the 3-series models. The 7- and 8-series cars were made at Dingolfing. In 1990, car production was only fractionally short of 500,000 units.

Komenda, Erwin (1904–1966)

Chief of the Porsche body designing department (1931–1966); developed the car body construction of the VW Beetle.

He was born April 6, 1904, in Weyer, a little village in Upper Austria. From 1926 to 1929, he worked as a car-body designer in the Steyr factories, where he met Ferdinand Porsche when Porsche joined Steyr as technical director. Komenda's reputation at Steyr brought him the position of Chief Engineer and leader of the Porsche car-body construction department for Daimler-Benz. During his tenure (1929–1931), he was often able to reduce the weight of Mercedes cars through better design; a streamlined car with monocoque construction was developed. Komenda may be best known for his work on the car body construction of the VW Beetle, the most built car body of the last century. With his co-worker Josef Mickl, he also designed the famous Auto Union Grand Prix car and the Cisitalia Grand Prix car. After World War II, Komenda and Ferry Porsche designed and then improved the Porsche type 356. Komenda also designed the Porsche 550 Spyder. Due to his early death in 1966, his last Porsche project was the development of the Porsche 911.

Komnick, Karl Franz (1857–1938)

Chairman of Automobilfabrik Komnick from 1907 to 1922; chairman of Automobil-fabrik Komnick AG from 1922 to 1930.

He was born on November 27, 1857, at Marienburger Werder and attended schools in various parts of Germany. He learned engineering by working in machine factories in his home town and throughout the province of Posen. In 1898 he bought a machine factory at Elbing on the lower Vistula, close to Danzig. The area had long traditions in steel, sawmills and brewing industries. In 1907 he bought a steel-mill and metalworking factory which he retooled for automobile manufacturing. His son Otto Komnick (1887–1945) was works manager of Automobilfabrik Komnick, while the elder Bruno Komnick (1885–1945) became general manager of the Maschinenfabrik Komnick. The first Komnick cars appeared in 1908 with three sizes of four-cylinder engines. They had "coal-scuttle" hoods and radiators mounted behind the engines, three-speed gearboxes and shaft drive. He became a keen rally driver and in 1910 announced a 22/60 model with a 5½-liter four-cylinder engine which was produced up to 1914 alongside the lighter 10/30 and 17/50 from 1911. The factory made trucks and military tractors during World War I, along with other war equipment. Civilian car production resumed in 1921 with new designs by Joseph Vollmer. The 8/45 model C2 had a two-liter four-cylinder engine with an overhead camshaft driven by a vertical shaft rising between the two pairs of cylinders. Vollmer also designed a line of Komnick trucks and farm tractors. The company began losing money in 1925 and never recovered. Car production was discontinued in 1927, to concentrate on the current 2½-ton truck and a new 5-ton truck with a 75-hp six-cylinder engine. But Automobilfabrik Komnick was declared bankrupt on April 2, 1930. The municipality of Elbing took over its RM 2.25 million debts and the factory was sold to Büssing of Brunswick.

Könecke, Fritz (1899–1979)

Chairman of the management board of Daimler-Benz AG from 1953 to 1960.

He was born on January 16, 1899, and educated for a career in business. He held a degree in business administration. He began his career in the rubber and tire industry in 1919, and held an executive position with Phoenix Gummiwerke AG of Hamburg-Harburg in the mid–1920s. Next he went to Hanover in a management capacity with Continental Gummiwerke AG, and was elected to the Continental management board in 1934. He served as managing director of Continental Gummiwerke AG from 1938 to 1952. He was invited to join Daimler-Benz AG following the death of its chief executive, Wilhelm Haspel, and went to Untertürkheim as deputy chairman of the management board on June 1, 1952. He became chairman in January 1953 and set the corporation on a course of global expansion. Haspel had started a branch factory in Argentina in 1950. Könecke organized a subsidiary in Brazil in 1953 and started a joint venture for assembling trucks in India with Tata Engineering and Locomotive Company in 1954. He arranged the acquisition of Auto

Union GmbH from Friedrich Flick in April 1958 and secured majority control of Maybach Motorenbau GmbH in August 1960. He resigned in November 1960 after the tragic death of his only son and died in Stuttgart on March 26, 1979.

Kraus, Ludwig (1911–1997)

Chief engine designer of Daimler-Benz from 1956 to 1962; chief engineer of product development for Auto Union GmbH from 1963 to 1967; technical director of Audi from 1967 to 1972; vice chairman of Audi NSU Auto Union AG from 1972 to 1974.

He was born on December 26, 1911, at Hetteshausen near Ingolstadt in Bavaria, and studied mechanical engineering at the Technical Universities of Munich, Stuttgart, and Hanover before joining Daimler-Benz as an engine designer in 1937. He was in military service for practically the entire duration of World War II, returned to Daimler-Benz, and became occupied with theoretical and practical problems of engine applications to airships and speedboats. Late in 1945 he was assigned to the truck–diesel engine department.

In 1952 he was placed in charge of designing the M-196 straight-eight fuel-injected racing engine for the Grand Prix and 300 SLR in two sizes (2496 and 2998 cc). The M-196 brought the world championship to Stuttgart in 1954 and 1955. The racing department was then dissolved and he went on to design a new generation of production-car engines.

The Daimler-Benz management delegated him to Auto Union to propose a replacement for the DKW car and its two-stroke engine, and he led the teams that designed the F-12 and F-102. He also directed the complete design of the 1972 Audi 80 and the 1974 Audi 50 (later renamed VW Polo).

He retired in August 1974 and died on September 19, 1997.

Krebs, Arthur Constantin (1850–1935)

Head of technical services for the Société Anonyme des Anciens Etablissements Panhard &

Levassor from 1897 to 1916; managing director of the company from 1904 to 1916.

He was born in 1850 at Vesoul into a Rhinelandish family. After getting an engineering diploma from the Ecole Centrale des Arts et Manufactures, he embarked on a military career, studying naval engineering at the Ecole Polytechnique and attending the St. Cyr military academy from 1870 to 1873. He gained practical experience in the Arsenal at Brest and the Arsenal at Nantes, rising to the rank of lieutenant by 1875. The following year he joined the aerial communications team of Captain Charles Renard and in 1878 designed an electrically propelled dirigible airship. In 1884 he was named captain of the general staff of the Paris Fire Brigade,[1] an autonomous military unit, with responsibility for the fire-fighting equipment.

Taking a keen interest in automobiles, he designed one and showed his drawings to Emile Levassor in September 1896. In August 1897, he moved into Levassor's former office in Avenue d'Ivry. He created the Centaure engine to replace Daimler's, put the cars on pneumatic tires, replaced the tiller with a steering wheel, and developed the gilled-tube radiator.

In 1889 he had invented and patented a compensating carburetor with an automatic diaphragm, which was later adapted to the Panhard engines. Privately he designed a car for Adolphe Clément and arranged for its production in the Panhard factory. He began using shaft drive on new models in 1904, and in 1909 Réné de Knyff, racing driver and a member of the board of Panhard-Levassor, persuaded him to adopt the Knight-patented sleeve-valve engine.

In 1917 he designed an overhead-camshaft V-12 aircraft engine for Panhard and resigned shortly afterwards. He retired to Quimperlé in Brittany where he died in March 1935.

1. Pompiers de Paris.

Krieger, Louis (1868–1951)

An electrical engineer who built the best-known and most widely sold French electric cars in the early years of the twentieth century.

His first car was an avant-train attachment for horse-drawn vehicles with a separate electric motor in each wheel. This not only gave power steering but also 4-wheel braking, since there were already brakes on the rear wheels. His cab won First Prize for 4-seaters in the 1897 Paris Motor Cab Trials. In 1898 Krieger started making complete electric vehicles, and later sold licenses to auto manufacturers in Britain, Italy and Germany. Krieger made petrol electric cars in 1907, but in the following years gave up cars to concentrate on general electrical engineering. His company was associated with Milde during World War II, making electric conversions of petrol cars which were produced under the name Milde-Krieger.

Kuenheim, Eberhard von (1928–)

Chairman of BMW AG management board from 1970 to 1999.

He was born on October 2, 1928, in Königsberg, East Prussia, as the son of a horse breeder and a member of the Prussian aristocracy. His father died in 1935. At the age of 15 he helped man an anti-aircraft gun in his home town, and made his way to the West in 1945, without cash or resources. He completed his high-school education and in 1948 went to work for Robert Bosch GmbH as a trainee, taking evening classes. From 1950 to 1954 he was a full-time student at Stuttgart Technical University, graduating with a Master's degree in mechanical engineering. He then joined Max Müller, makers of machine tools in Hanover, as a production and sales engineer. He steered the company into the development of automated factory equipment and became Max Müller's right-hand man. In 1965 Harald Quandt hired him as a member of his personal staff, to work as a technical consultant and handle the coordination of the family holdings. In 1968 he was given additional duties as deputy board member of IWKA[1] and in 1970 he was chosen to take the helm of BMW AG. He led the company through nearly two decades of spectacular growth, and raised BMW's reputation to new heights. He began discussions about taking over Rolls-Royce Motor Cars, Ltd., with Sir Colin Chandler, chairman of Vickers, Ltd., which ended with BMW's acquisition of the Rolls-Royce name, though Volkswagen AG purchased the Rolls-Royce factory and the Bentley make. He also approved BMW's purchase of the Rover Group which proved a very costly mistake and ended with BMW AG actually paying a British consortium to take Rover off its hands in 1999. Land-Rover had been sold to Ford Motor Company, and what BMW chose to keep was the new Mini and the Cowley plant. He resigned from BMW AG in May 1999.

1. Industrie-Werke Karlsruhe-Augsburg, makers of production systems and equipment.

Lagaay, Harm (1946–)

Director of Style Porsche from 1972 to 1978; head of the advanced studio, Ford-Werke AG from 1978 to 1985; chief designer of BMW Technik AG from 1985 to 1989; design director of Dr. Ing. h.c. F. Porsche KG since 1989.

He was born on December 28, 1946, in The Hague as the son of a Shell Oil Company cartographer. He grew up in far-flung locations like Ecuador and Venezuela, and attended school in Brunei from 1952 to 1958. He graduated from high school in Arnhem in 1965 and studied commercial and technical subjects at the Institute for Automobil-Technik at Driebergen for two years. He worked as a technical illustrator for Olyslager publishing company and then joined the Simca service and parts organization in the Netherlands as an illustrator. He spent his vacations in the Ghia shops in Turin, working without pay, which led to an offer from Porsche. He went to Zuffenhausen in 1970. His sketches were chosen as the basis for the 924, and he directed the design of the 928. He left Porsche in 1978 to go to work for Ford-Werke AG at Merkenich, where he was associated with the design of the Probe and Sierra. During his years with Ford he also learned design management under Uwe Bahnsen. He styled the Z1 roadster for BMW but returned to Porsche on April 1, 1989, succeeding Tony Lapine. He led the body design for the 996 (marketed as a 911).

Lagardère, Jean-Luc (1928–2003)

Managing director of SA Engins Matra from 1963 to 1977; chairman of SA Engins Matra since 1977; chairman of Matra Automobiles beginning in 1965.

He was born in 1928 at Aubiet (Gers) as the son of the finance director of the National Aerospace and Development office. He attended the Lycée d'Auch and the Lycée Saint-Louis, and graduated with a diploma from the Ecole Supérieure d'Electricité. He began his career with Dassault Aviation in 1956 and joined Matra[1] in 1963. The founder, Marcel Chassagny, had decided to diversify the activities of Matra to avoid over-dependence on military orders. Lagardère arranged the purchase of René Bonnet's garage at Champigny-sur-marne with full rights to the Bonnet Djet sports car, and Brissonneau & Lotz's factory at Romorantin. The car was renamed Matra Djet in 1964 and was replaced in 1967 by the Ford-powered Matra 530. He started an ambitious racing program with a V-12 engine powering cars that were successful both at Le Mans and in Formula One. In 1979 Peugeot took a 45-percent stake in Matra Automobiles, after six years of producing the Matra-Simca Bagheera at Romorantin. It was replaced by the Matra-Talbot Murena in 1980. In 1983 Lagardère had a quarrel with Jean-Paul Parayre of Peugeot, and Peugeot disposed of its interest in Matra. Matra was then preparing its M-25 project, intended for Peugeot, but he offered it to Renault instead. It went into production in 1984 as the Renault Espace.

He led Matra's diversification into publishing, public-transit systems, machine tools and audio-visual media. He founded Lagardère SA as a holding for his private trading and industrial investments and became one of France's best-known businessmen. He died March 14, 2003.

1. Société Anonyme de Materiél Aéronautique et de Traction, established in 1941.

Lago, Antonio "Tony" (1893–1960)

Chairman and managing director of SA des Automobiles Talbot from 1935 to 1958.

He was born in 1893 in Venice and held an

engineering degree from the Milan Polytechnic. He moved to England in 1923 and worked for the London importers of Isotta-Fraschini cars. He became technical director of LAP Engineering, a small business specializing in making high-performance cylinder heads for low-powered cars such as the Morris Cowley and Austin Twelve. He held a position with Sunbeam from 1925 to 1927, and then joined Self-Changing Gears, Ltd., which held the rights to the Wilson preselector gearbox, and became its general manager. In 1933 he returned to Sunbeam as deputy general manager, and when the Sunbeam-Talbot-Darracq combine collapsed in 1934, he acquired the French Talbot company and its factory at Suresnes. Before the collapse, he had argued that S-T-D could have been saved by reorganization and reorientation of product policy. He had sold a license for the Wilson gearbox to Pont-à-Mousson in Lorraine, and began fitting their transmissions in the new Talbot cars. Pont-à-Mousson also made the four-speed manual gearboxes for Talbot.

He introduced Talbot models with bigger and more technically advanced engines, and conducted an active racing program. The Talbot name became one of France's most prestigious. After World War II, the prewar models were revived and updated, and the racing team reactivated. Its best year ever was 1950, when practically identical T-26 cars won both the Dutch Grand Prix and the 24-hour race at Le Mans. But Automobiles Talbot was living beyond its means and declared bankruptcy on March 6, 1951. The creditors voted unanimously for a committee to run the company and repay its debts over an eight-year period. Talbot disbanded the racing team, curtailed car production and went massively over to filling military contracts. New and modern-looking sports cars were introduced with BMW and Ford V-8 engines, but none ever proved profitable. The company and its assets, including the Talbot name, were sold to Simca in 1958.

Lago retired and died at his home in Paris near the end of December 1960.

Lamborghini, Ferruccio (1916–1993)

Chairman of Automobili Lamborghini from 1962 to 1971.

He was born on April 28, 1916, at Renazzo di Cento, as the son of a farmer. He was more interested in farm machinery than in agriculture, and was still a schoolboy when he rigged up a primitive machine shop on the family farm. He studied engineering at an industrial school in Bologna and returned to the farm in 1937, making plans for his future. During World War II he served in the Italian Air Force, and was discharged in 1946. His father needed a tractor and he built one for him, assembled from parts salvaged from a junkyard. Then he decided to start production of farm tractors with small diesel engines and founded Lamborghini Trattrici at Cento in 1949. The company prospered, and he began thinking about other product lines. He started Lamborghini Bruciatori in 1959 to make oil burners for heating and also began production of air conditioners in 1960. He was now an industrialist of some wealth and owned the most expensive cars (Maserati and Ferrari). Then he decided to create the Lamborghini car, purchased a factory site at Sant'Agata Bolognese, erected and equipped a spacious, modern plant. Production of the V-12 350 GT began in the spring of 1963. The Miura was an overnight sensation when presented in 1966, with a transverse V-12 mounted in an inboard/rear position. The Marzal four-passenger prototype (by Bertone) evolved into the Espada production model, a marketing success since 1217 of them were built and sold from 1968 to 1978.

He had once wanted to build helicopters and made a prototype in 1962, but later abandoned the project. Driven to broaden his industrial empire, he founded Oleodinamica Lamborghini to make high-precision tools and instruments for the petroleum and petrochemical industries. The car company began building the P 250 Urraco in 1970 and the LP 50 Countach a year later. Troubles began for the tractor company in 1972 when the Bolivian agents canceled a big order. He sold it to SAME[1] of Treviglio. He had sold 51 percent of the car company to Henri-Georges Rossetti in 1971, and three years later disposed of the balance to René Leimer.

He retired, sharing his time between his villa in the hills overlooking Modena, and his estate on Lake Trasimeno. He died on February 20, 1993, in Perugia after suffering a heart attack.

1. Società Applicazioni Motori Endotermici.

Lampredi, Aurelio (1917–1989)

Chief engineer of Auto Avio Costruzioni Ferrari S.p.A. from 1950 to 1955; chief engineer of Fiat's engine design office from 1955 to 1980; technical director of Abarth & C. from 1972 to 1982.

He was born on June 16, 1917, in Livorno and held an engineering degree from the Technical Industrial Institute in Fribourg, Switzerland. He began his career with a shipbuilding yard, Cantieri Navali Odero, in his hometown, as a draftsman. He was an aircraft-engine designer with SA Piaggio at Pontedera from 1937 to 1940, and with SA Reggiane of Reggio Emilia from 1940 to 1946. He returned to Livorno and worked for a few months with the Bassoli steel mills, and spent the winter of 1946-47 as assistant to Giuseppe Busso in the Ferrari drawing office. Antonio Alessio invited him to join Isotta Fraschini in Milan, where he worked on aircraft-engine design throughout the summer of 1947. He returned to Ferrari in November 1947 as an engine designer, and became chief engineer in 1950. He designed the 3.3-liter, 4.1 and 4½-liter V-12 engines, the twin-cam four for the 500 Mondial, an experimental two-cylinder racing engine, and the 4.4-liter twin-cam six of 1955. In all, he created 40 different Ferrari engines. Arriving at Mirafiori in 1955, he began design work on a new six, built in sizes from 1.8 to 2.3 liter, and its four-cylinder derivatives for the 1963 1300/1500 models. He directed the design of the four-cylinder engines for the 124 and 125, including the 124 AC/125 which was the first high-volume production engine in the world with twin overhead camshafts. From that time, Fiat standardized belt-driven camshafts. He also directed the 127 and 128 engine programs and designed the V-6 for the 130. He led Abarth's development of competition-car engines, notably for the 131 Rally, which won the European championship three years running. He retired when reaching the age limit in June 1982 and settled in Tuscany. He remained a consultant to Fiat's research center and also advised Cagiva and Ducati on motorcycle engines. He died on June 6, 1989.

Lanchester, Frederick William (1868–1946)

Co-founder, general manager and technical director of the Lanchester Gas Engine Company Ltd. from 1899 to 1904; consulting engineer to the Lanchester Motor Car Company Ltd. from 1905 to 1913; technical advisor to the Daimler Company from 1909 and B.S.A. from 1910 until 1929; consulting engineer to Beardmore's diesel-engine department from 1928 to 1930.

He was born on October 23, 1868, at Lewisham in southeast London, as the son of an architect, Henry J. Lanchester of Hove, Sussex. He was educated at Hartley College, the Royal College of Science, and Finsbury Technical College. He began his career as assistant works manager with T.B. Barker & Company, makers of the Forward gas engine, in 1889. Realizing the potential advantages of making engines run faster, he built an experimental single-cylinder unit in 1893, reaching 800 rpm. It put out 1½ hp and was installed in a stern-wheel launch, plying on the Thames River, in 1895. That year he also invented a balanced engine with two flywheels, turning in opposite directions.

He made his first prototype automobile in 1896, featuring an air-cooled 5-hp single-cylinder engine, planetary transmission, worm drive, steel tube frame, tiller steering, pneumatic tires, and four seats facing forward. A second prototype made in 1897 was powered by a flat-twin engine equipped with Lanchester's patented wick carburetor mounted in the furl tank.

In partnership with his brothers Frank and George, he founded the Lanchester Gas Engine Company, Ltd., on December 13, 1899, with a factory at Sparkbrook, Birmingham. Car production began in 1900, with a flat-twin underfloor central engine. A year later, he adopted preselective gear change, and in 1902 obtained a patent for a caliper-type disc brake. In 1904 Lanchester began installing a flat-four underfloor engine.

He had become interested in the science of flight before the Wright brothers made their demonstration at Kitty Hawk in 1903, and began writing a two-volume textbook on aerial flight. The first volume *Aerodynamics* was published in 1907 and the second, *Aerodonetics*, in 1908. In 1909 he applied for a patent for a new invention, the harmonic torsional vibration damper, consisting of a small, loose flywheel mounted on the end of the crankshaft and coupled to it by a number of thin discs immersed in oil. It was soon licensed to Vauxhall, Lanchester, and Willys-Overland.

Well before the outbreak of World War I, he was appointed to the British Government's Advisory Committee on Aeronautics, and in 1915 he authored a book entitled *Aircraft in Warfare* which turned out to be prophetic as well as technically accurate.

After World War I the Daimler Company assigned him to the development of the Renard road train, which made him feel that his talents were being wasted. In 1925 he was able to get Daimler's support as a 50/50 financial partner in the Lanchester Laboratories, Ltd., founded to take on consulting work for outside clients. He soon lost Daimler's support, however, and at the end of 1927, Daimler cut his fee by one-third. His basic contract with Daimler was canceled in 1928, effective November 1929.

He won the Guggenheim Gold Medal in 1931 for his contributions to the theory of aerodynamics and became a Fellow of the Royal Society. He wrote numerous technical papers on subjects from steering and suspension of automobiles to high-altitude flight and skin friction in aeronautics. He wrote a treatise on airships in 1937 and presented four papers on jet propulsion in 1939. He was awarded the

Ewing Medal in 1942 and the James Watt Medal in 1945. He died on March 8, 1946.

Lanchester, George Herbert (1874–1969)

Chief designer of the Lanchester Motor Car Company, Ltd., from 1909 to 1912; chief engineer of Lanchester from 1912 to 1925; technical director of Lanchester from 1925 to 1931; chief engineer of Lanchester from 1931 to 1936.

He was born on December 11, 1874, in Brighton, Sussex, as a younger brother of F.W. Lanchester. At the age of 15 he was apprenticed to the Forward Engine Company at Saltley and was promoted to works manager in 1893. He left Forward in 1897 to work full-time as an assistant to his brother, F.W. Lanchester, and their brother Frank joined them in the formation of the Lanchester Gas Engine Co., Ltd., in 1899. George became the tooling specialist, head of factory organization and methods, and personnel training.

He served as superintendent of the Lanchester Motor Car Company, Ltd., from 1904, redesigning F.W.'s drawings for simpler and more economical production. F.W. and George had equal parts in making the 38-hp six and the 25-hp four-cylinder models of 1910–11. The 38-hp model served as the basis for an armored car produced during World War I. The Lanchester factory also produced searchlight tenders, field kitchens, mine-sweeping paravanes and balloon-winch wagons. From 1916 to 1918 Lanchester also manufactured a Renault-based V-8 engine.

In 1915 he revealed the all-new 40-hp car with its 6.2-liter 90-hp overhead-camshaft engine and three-speed planetary transmission, which became the mainstay of Lanchester production in 1919.

He designed a six-wheeled armored vehicle in 1922 and saw it chosen by the Army in 1924. His next passenger car was the six-cylinder 21-hp model introduced in 1923, a lively luxury car that sold for less than a Rolls-Royce 20/25 yet went ten miles per hour faster. About 1927 he proposed a 7-liter six and a V-12 model, but both were rejected by the board of directors. Forecasting market trends downwards led him to design a six-cylinder 16-hp car, which the board of directors vetoed in 1929. The directors had, by way of contrast, voted in favor of the 30-hp straight-eight, which went into production towards the end of 1928.

After the Lanchester Motor Car Co., Ltd., was taken over by the Daimler Co. in 1930, he stayed on and took part in the design of Daimler overhead-valve engines to replace the sleeve-valve engines favored by Daimler since 1909. He left Daimler in 1936 and joined Alvis where he proposed replacement models for the Sixteen and the 12/70 before being transferred to the Mechanized Warfare Department. In 1940 he signed up with Sterling Armaments Company and organized the mass production of the Sterling machine gun and the Lanchester 9-mm submachine gun. At the end of the war, he became a consulting engineer to one of Sterling's associate companies, Russell Newberry Ltd., and in 1952 he was named a director of the Russell Newberry Diesel Engine Company of Dagenham, Essex, where he remained until his retirement in 1961. In his final years, he drove a DAF for his personal transport.

He died on February 13, 1969.

Lancia, Vincenzo (1881–1937)

Co-founder of Lancia e C., Fabbrica di Automobili, and its head and majority owner from 1906 to 1937.

He was born on August 24, 1881, at Fobello in the Valle Sesia as the son of Giuseppe Lancia who had made his fortune selling frozen meat to the Western Forces during the Crimean War, settled in Turin, and ran a big business in the food-packaging industry. His education began at the Niccolo Tommasco elementary school, and the family wanted him to seek a career as a lawyer. He enrolled at the Scuola Tecnica Giuseppe Lagrange, but failed because his mind continually strayed from the studies as a trainee bookkeeper. In 1898 his father arranged for him to work in the bicycle factory of Giovanni Ceirano, next door to the Lancia family home in Turin.

When Ceirano began making the Welleyes automobile, young Lancia spent more time in the shop than in the accounting office, working without regard for the hours, and got a thorough understanding of gearing, steering, and mechanics in general. He never went back to school.

In 1899 the Ceirano factory was sold to Fiat who retained the personnel, making Lancia a test driver. He also served as a driver on Fiat's official racing team, visiting other European countries and the United States. By 1903 he was chief inspector in the Fiat plant, making a decent salary. What he now wanted most of all was independence. He resigned from Fiat in 1906 and on November 27, 1906, established his own company in partnership with Claudio Fogolin.

Their small factory burned down in February 1907, and production of the first Lancia car was delayed until 1908 when they moved into bigger premises on Via Ormea in Turin. Lancia continued as a member of the Fiat racing team, though his engagements grew less frequent after 1908. He ran his last race for Fiat in June 1913. In 1911 Lancia moved into a former Fides bicycle factory on Via Monginevro in Turin. During World War I passenger-car production was pushed into the background to concentrate on building Iota medium-duty trucks and a line of armored vehicles.

With a superbly competent engineering staff, Lancia began advanced technical projects, obtaining a patent for a 45° V-8 engine in 1917 and a 60° V-12 in 1919. The 35-hp Kappa of 1919 was a four-cylinder model, however, of no great distinction other than being well made. The Lambda, launched in 1923, was the first of Lancia's many revolutionary designs, introducing unit body construction, independent front suspension, and a narrow-angle staggered V-4 engine.

The majestic Dilambda with its 100-hp staggered V-8 engine, arrived at the end of 1929. It had been created for production in the United States but Lancia was the victim of a stock-promotion swindle, and the Dilambda was built in Turin instead. It set Lancia back financially, but he was still rich enough to provide backing for Pinin Farina to set up his own coachbuilding works in 1930.

All subsequent Lancia cars were packed with features of technical interest, from the light Augusta to the Artena and the eight-cylinder Astura. The final masterpiece that Vincenzo Lancia personally brought to reality was the Aprilia, backed by a number of patents covering its body construction, all-independent suspension, and ultra-short V-4 overhead-camshaft engine. For a small car, it was expensive but won wide international acclaim.

He died of a heart attack during the night of February 18, 1937.

Lange, Karlheinz (1939–)

Head of engine research and development for BMW from 1975 to 1989; director of BMW vehicle development from 1989 to 1992; BMW management board member in charge of quality from 1992 to 1999.

He was born in Bavaria in 1939 and graduated from Munich Technical University. He presented a postgraduate paper on the diesel-engine combustion process and joined BMW as an engine test engineer in 1965. He joined Porsche in 1971 as head of engine pre-development and was soon named chief engineer of engine testing in the Porsche technical center at Weissach. He returned to BMW in 1975 and developed BMW's first turbodiesel engine, the 524 td, and the "Eta" spark-ignition engine for exceptional fuel economy. His work also provided a sound basis for a new generation of engine programs, from four-cylinder to V-12 units. He directed research into exhaust-treatment technology, electronic engine control systems, and improved test and measurement equipment and methods. He had a great deal of influence on the 1994 7-Series.

He retired from BMW in 1999.

Langheck, Wilhelm (1907–)

Works director, Sindelfingen plant, Daimler-Benz AG from 1943 to 1954; management board member in charge of passenger-car production from 1954 to 1976.

He was born on November 29, 1907, in Esslingen and educated as a production engineer. He joined Daimler-Benz AG in 1933 as plant manager at the Sindelfingen works, sole assembly point for Mercedes-Benz passenger cars. He was promoted to works director in 1943. The Sindelfingen works were 85 percent destroyed by repeated U.S. Army Air Force bombing raids in September 1943, and occupied by the U.S. Army in 1945. He was placed in charge of planning its reconstruction, and laying out the various facilities including new body assembly and welding equipment for unit-construction bodies. He became deputy board member in 1952 and a full member of the management board in 1954. He coped with a proliferation of models, from the low-volume 300 SL to the high-volume 180, 190 and 200 sedans, boosting the output from Sindelfingen from a total of 63,683 cars in 1955 to 174,007 cars in 1965 — and 350,098 cars in 1975.

He retired in 1976.

Lanstad, Hans (1879–1956)

Chief draftsman of WRM Motors, Ltd., from 1914 to 1919; chief engineer of Morris Motors, Ltd., from 1919 to 1926; works director of Morris Motors (1926), Ltd., from 1926 to 1936; technical director of Morris Motors, Ltd., from 1936 to 1948.

He was born on May 20, 1879, at Haa, Jaeren, Norway, as the son of a protestant bishop famous for editing a book of hymns. He was educated at Hamar and attended Christiania Technical College. He studied naval engineering at Hortens Technical College and took evening classes at Treiders Commercial College in Christiania (later Oslo). He went to sea and worked as a steamship engineer, landed in England in 1903 and secured employment with the Linotype company. Through family connections, he met P.A. Poppe who hired him as an engine draftsman for White & Poppe, who became engine suppliers to W.R. Morris in 1912. In 1914 he took an (unpaid) leave of absence to accompany W.R. Morris to America and spent six months with continental Motors, while also designing a smaller car, the Morris Cowley. He designed new Oxford and Cowley models after World War I and played a role in Morris's purchase of the Hotchkiss engine plant in Coventry in 1923. In 1925 he was stationed in France, designing the Morris-Léon Bollée car. He held top responsibility for all Morris production for ten years, and was placed in charge of the MG and Riley factories as well in 1938. He also organized the production of Bofors guns in a Morris plant and directed all war-production programs for the Nuffield Organization. In the postwar reorganization, he was one of the first to be pushed out. He remained in Oxford after his retirement in 1948 and died there on December 9, 1956.

Latil, Auguste Georges (1878–1961)

Co-founder and partner in Korn & Latil from 1898 to 1901; head of Latil et Cie from 1901 to 1914; chairman of the Société Industrielle des Automobiles Latil from 1914 to 1950.

He was born in 1878 in Marseille as the son of a notary. Both father and son loved to tinker with mechanical things and Georges acquired a thorough knowledge of tools and mechanical skills at an early age. In 1897 he built a motorized fore-carriage for the conversion of horse-drawn carriages to automobiles. Unlike the fore-carriages of Gauthier-Wehrlé, Gevin, Pretôt, Riancey and Vollmer, which steered by turning as a unit, Latil's had stub-axle steering with constant-velocity universal joints in the wheel hubs. His brother Lazare helped him draft the patent application in 1897; it was issued under No. 297,700 on March 1, 1900. Latil had begun production of fore-carriages with transversely mounted single-cylinder de Dion-Bouton and Aster engines in shops on Boulevard Rabatau in Marseille but left for Paris in 1898 where he went into partnership with Alois Korn, who had a shop at Levallois where complete vehicles were assembled. Latil had his office and a small workshop on Boulevard Voltaire in Paris. Several two-passenger voiturettes were built in 1898–1900, along with a number of light commercial vehicles, all with front wheel drive and Latil's patented universal joint. Technically it was a precursor of the Tracta joint and bore no resemblance to the Weiss (patented 1925) or Rzeppa (patented 1935) joints. In 1901 Korn & Latil attracted a major investment from Charles Blum, a graduate of the Ecole Polytechnique who later became mayor of St. Cloud. Ch. Blum offered factory space in one of his plants in Suresnes for vehicle assembly, and financed the erection of a new Latil factory in the same neighborhood, where production began in 1906. Latil was busy inventing four-wheel-drive and four-wheel steering systems, which were combined on the TAR, designed in 1913 and produced throughout World War I as an artillery tractor and ordnance truck. Powered by a 7250-cc four-cylinder Latil engine, it was also used as a road-train locomotive, hauling two drawbar trailers.

After the war, Alois Korn left to start his own road-haulage business, and Latil purchased Ch. Blum's shares. He was now in sole charge of the company, produced a range of conventional trucks, and later added a line of light-duty commercial vehicles. In 1935 he introduced the JTL forestry tractor with four-wheel-drive and four-wheel steering, and the M2 TZ three-axle all-wheel-drive military truck (6 x 6). But he had overspent on development costs and a consortium of investment bankers acquired majority control of the firm. He held on to his title and continued in charge of engineering until his retirement in 1950. He settled at Cagnes-sur-Mer on the French Riviera, where he died in June 1961.

Laurin, Vaclav (1867–1930)

Works manager of Laurin & Klement from 1896 to 1907; technical director of Laurin & Klement AG Automobil-Fabrik from 1907 to 1925.

He was born on September 27, 1867, in Kameni, Bohemia, as the son of a farm laborer. He was apprenticed to a mechanic in Turnov and later attended the Artisan's School in Mlada Boleslav for three years. From 1883 to 1893 he was employed in mechanical workshops and machine shops in Hus, Ceska Lipa, and Jablonec. In 1893 he received his certificate as a steam-engine service engineer from the authorities in Dresden. He returned to Turnov and went into partnership with Josef Kraus to repair bicycles. They also imported parts from British manufacturers and assembled new bicycles. When he fell out with Kraus, he met Vaclav Klement, a bicycle dealer in Mlada Boleslav. They teamed up to start bicycle production, using British-made parts at first, but gradually finding local suppliers and making critical parts in their own shop. They prospered and in 1898 bought a large factory site on Kosmonosy Road where the first building was erected within six months. They began making motorcycles and motor tricycles in 1899 and built a prototype four-wheeled single-seater car with front engine and chain drive in 1901. They introduced the Slavia motorcycles in 1902, some models with an air-cooled V-twin, others with a water-cooled V-twin. Bicycle production ended in 1905, and in 1906 they began building a voiturette car with a front-mounted water-cooled V-twin and shaft drive to the rear axle. Motorcycle production was phased out in 1909 to make room for a range of four-cylinder cars. The partners sold out to a group of ten main shareholders in 1907, and Laurin was named technical director. His responsibility was to run the plants and modernize them as required. Most of the factory capacity was converted to make artillery shells in World War I, and he also had to find ways to make his own machine tools. After the war, the car program was cut back to just two models, and the main product was a line of motor plows powered by compression-ignition engines made under Brons license. Parts of the factory were badly damaged by fire in June 1924, but he maintained output by going to three-shift operation. He took no part in the merger talks with Skoda and retired in 1925.

Lawson, Geoffrey (1945–1999)

Chief designer of the Jaguar styling studio from 1984 to 1999.

He was born in 1945 in Leicester and studied industrial design at the Leicester College of Art. He pursued his studies at the Royal College of Art in London, graduating with a degree in furniture design. He began his career with Vauxhall Motors, Ltd., at Luton and spent some time with GM Styling at Warren, Michigan. Returning to Britain, he joined Jaguar Cars, Ltd., in 1984. He designed the new XJ sedan which went into production in 1995 and the XJ8, announced in 1997. On the 1996 XK8 coupe and roadster he was assisted by Fergus Pollock. He then directed the design teams for the 1998 S-Type and 2001 X-Type. He died suddenly from a stroke in June 1999.

Lawson, Harry John (1852–1925)

Chairman of the British Motor Syndicate, Ltd., from 1895 to 1903; chairman of Daimler Motor Company, Ltd., from 1896 to 1897; chairman of the Great Horseless Carriage Company, Ltd., from 1896 to 1898; chairman of the Motor Manufacturing Company, Ltd., from 1898 to 1899.

He was born on February 23, 1852, in London, as the son of a Brighton-based model maker, brass turner and Puritan preacher. He worked as an apprentice with an engineering shop in London in 1868–69. He set up his own shop in Essex Road, Islington, London, but moved to Brighton in 1872, where he set up a shop to make bicycle wheels. He patented the "Crocodile" bicycle which was put in production and distributed by Singer & Co. in Coventry. He sold his wheels business in 1877 and moved to Coventry, becoming manager of the Tangent & Coventry Tricycle Company. In 1879 he obtained a patent for a chain-driven "safety" bicycle and joined Haynes & Jeffries, which was reorganized as the Rudge Cycle Company. In 1880 he was sales superintendent for Rudge bicycles. A few years later he floated Cattle Foods Company, which failed miserably. He left Coventry in 1889, returned to London, and established the London & Scottish Trustee and Investment Company and began promoting all sorts of stock issues, from breweries, bottlers of spring water, to bicycles. It was renamed Financial Trust & Agency in 1890. He became associated with Martin D. Rucker and Ernest T. Hooley in a private investment consortium. Rucker held interests in the bicycle industry and Hooley was a big-time financier. Lawson approached F.R. Simms about buying the Daimler patents, and paid £35,000 for the Daimler Motor Syndicate which he reorganized as the British Motor Syndicate, Ltd., on November 21, 1895. A new subsidiary, Daimler Motor Company, Ltd., was formed on January 14, 1896, with a payment of £40,000 made to Lawson for the patents. Lacking an industrial base, the new subsidiary imported Cannstatt-built Daimler cars for the British market. This trade continued for some time even after Lawson had purchased the derelict Cotton Mills in Coventry in May 1896, where the first British Daimler cars were made in 1897. In the summer of 1896 Lawson also established the Great Horseless Carriage Company, Ltd., as a holding for his non–Daimler patents and industrial interests. He paid £100,000 for E.J. Pennington's automotive patents, and a mere £20,000 to Léon Bollée for manufacturing rights to his "motor tandem" tricycle. Lawson, Rucker and Hooley were partners in the Humber Bicycle Company, Ltd., the Humber (Extension) Company, Ltd., to handle financial affairs and peripheral activities, and the New Beeston Cycle & Motor Company, Ltd. They also moved into the former Cotton Mills, renamed Motor Mills. The Léon Bollée machine, renamed Coventry Motette, was built as a joint venture between the Beeston Motor Company, Ltd., and Humber & Company, Ltd. As for the exploitation of the Pennington patents, Lawson brazenly told the shareholders of the British Motor Syndicate in April 1895 that Pennington-type motor cars were under construction by Humber & Co., Ltd., Coulthard of Preston, Lancashire, Robey & Sons of Lincoln, and the Fowler Brothers of Leeds. In January 1898, the Great Horseless Carriage Company, Ltd., was in liquidation, and Lawson reorganized it as the Motor Manufacturing Company, Ltd. A number of MMC cars were made at the Motor Mills up to 1907. In 1898 Lawson floated the London Steam Omnibus Company, Ltd., which was renamed the Motor Traction Company, Ltd., a year later, marketing vehicles made by Bayley, Ltd., of Northampton, powered by four-cylinder Daimler engines. His fortune shrank severely in 1900–01 due to failed lawsuits over patents (Pennington's and others) acquired at excessive

cost. The British Motor Syndicate, Ltd., collapsed in 1904 when Lawson was indicted for conspiracy to defraud and sentenced to 12 months of "hard labor." Its "master" patents were disposed of to D. Napier & Son, for a pittance, in 1907.

Ledwinka, Erich (1904–1992)

Chief engineer of Tatra-Werke from 1940 to 1948; chief engineer of Steyr-Daimler-Puch from 1951 to 1955; technical director of Steyr-Daimler-Puch from 1955 to 1976.

He was born on July 15, 1904, in Klosterneuburg as the son of Hans Ledwinka, attended local schools and graduated from the Technical University of Vienna with a degree in mechanical engineering in 1929. In 1930 he moved to Nesselsdorf and worked under his father's guidance on the air-cooled flat-four engine, the air-cooled V-8 and the body-engineering of the first streamliners (Tatra 77 and 87). He left Tatra in 1936 to work on aircraft-engine development for Büker Flugzeugbau in Berlin, returning to Tatra in 1940. During World War II he designed special-purpose vehicles with all-wheel-drive and air-cooled V-12 diesel engines. Like his father, he also came under indictment for collaboration with the Nazi occupants, but obtained an emigration permit in 1950 and joined Steyr-Daimler-Puch at Graz as a test engineer for engines and chassis. He designed the parallel-twin air-cooled engine for the Steyr-Puch 500 in 1958, a number of Puch motorcycle engines, and the Steyr Haflinger and Pinzgauer four-wheel-drive off-road vehicles. He remained in Graz until after his retirement in 1973 and remained a consulting engineer to Steyr-Daimler-Puch until his death.

Ledwinka, Hans (1878–1967)

Chief engineer of Tatra (Ringhoffer Tatra Werke AS) from 1921 to 1945.

He was born on February 14, 1878, at Klosterneuburg near Vienna, and educated at the monastic elementary boys' school[1] from the age of 6 to 11. He attended high school[2] in a Vienna suburb, and in 1892 was apprenticed to his uncle, Johann Zwiauer, a local mechanic and toolmaker. From 1895 to 1897 he studied at the shop master's school of the State Artisanry School[3] in Vienna. He began his career with a railway rolling-stock manufacturing company, Nesseldorfer Waggonfabrik, as a draftsman in 1897. Some of the company's engineers proposed making automobiles, and the owners gave their consent. The Nesselsdorfer prototypes made up to 1900 were not approved for production, but Ledwinka submitted his ideas, resulting in Type A, and a production of 22 cars. He then designed the more powerful Type B, which was built from 1901 to 1904. Ledwinka left Nesselsdorfer, however, in 1901, returning to Vienna where he designed steam-powered cars for Professor Richard Knoller and Alexander Friedmann. He soon saw that the future for steam cars was dim, and in 1905 returned to Nesselsdorfer, and was named director of the automobile department. For the next ten years, he designed conventional cars with four- and six-cylinder engines. In 1916 the automobile department was closed, and he again departed for Vienna, where he joined Steyr with a brief to design a big touring car. The first Steyr resembled the Nesselsdorf Type U and went into production in 1920.

At the end of World War I, the Nesselsdorf enterprise came under new management, which sent agents to talk to Ledwinka about coming back. He bargained long and hard to have the contract written on his terms, and it was not signed until 1921. When he left Vienna, he also had a consultancy contract with Steyr, and sent Steyr the drawings for the Steyr IV, V and VI models in the following years. The next-generation cars from Nesselsdorf wore the Tatra label, and showed Ledwinka's ability to create totally novel yet rational designs. The Tatra 11 and 12 were built up around a central tube "backbone" frame, with a flat-twin engine in the front overhang, driving the rear wheels, which were independently suspended in a swing-axle arrangement. He retained the backbone frame and swing-axle combination for cars with six-cylinder and V-12 engines, plus a number of truck and bus chassis. About 1930 he embraced the ideas of rear-mounted engines and streamlined bodies. The Tatra V 570 prototype of 1933 had a rear-mounted flat-twin engine and a rounded body shape. Its main significance rests on the fact that it was the basis of a lawsuit against Volkswagen, which was not settled until 1963 when VW had to pay damages of DM 3 million to the heirs of Franz Ringhoffer, former owner of Tatra-Werke.

When Ledwinka presented his first streamliner, it was a stunningly radical design, the Tatra 77 of 1934, with an air cooled V-8 engine in the rear overhang. It evolved into the more practical Tatra 87 of 1936, yet futuristic in its shape. His Tatra 57 of 1933 looked like an ordinary midsize family car, but had a backbone tube chassis with a front-mounted air-cooled flat-four engine and rear drive by swing-axles. An enlarged version, the Tatra 75, was built from 1933 to 1942.

The Nazi occupation of Czechoslovakia in 1938 put his office under military rule, and for six years his work consisted of supervising the engineering and production of military vehicles. Though it had been against his will, it resulted in a six-year internment in Communist-ruled postwar Czechoslovakia. Upon his release in 1951, he returned to Vienna. He became a consultant to Steyr, where his son was chief engineer, and also did some consulting work for Robert Bosch GmbH. He never retired completely and died in Munich on March 2, 1967.

1. Volksschule der Schulbrüder, at Wien-Fünfhaus.

2. Bürgerschule.

3. Werkmeisterschule der Staatsgewerbeschule, Vienna.

Leek, George Harold

General manager of Riley (Coventry) Ltd. from 1932 to 1934; chairman of Lea-Francis Engineering, Ltd., from 1937 to 1958.

He joined Humber in Coventry in 1901 and worked in a number of positions, from production to administration. He joined Riley in 1911 in the sales department, organized wartime production for the Ministry of Munitions from 1914 to 1918, went to work in the purchasing department in 1919 and a few years later was named purchasing director of Riley. In 1926 he planned and organized the production of the Riley Nine, and then returned to the sales department. He took over as general manager of Riley in March 1932 but quarreled with Victor Riley and resigned in 1934.

In 1936 he became a director of Leaf Engineering Company, established to take over the production of Lea-Francis cars, but dissolved within a year. In 1937 he purchased the tools, jigs, dies and fixtures, goodwill and patents, of Lea & Francis, Ltd., but not their factory on Lower Ford Street in Coventry. He was successful in making a bid for Triumph's Gloria works on Much Park Street, where production of Lea-Francis cars resumed. He also took over the Midland Light Body Company. During World War II, the plant was converted to subcontracting work for Short Brothers (aircraft), BSA[1] and A.V. Roe (aircraft). It was one of the few plants in the area to escape relatively unscathed, and the resumption of car production went smoothly, the first postwar model being completed in January 1946.

Lea-Francis was making 12 cars a week in 1950–51, which was barely enough to break even. But he kept the company afloat until 1958, when he took ill and retired.

1. Birmingham Small Arms Company, Ltd.

Lefaucheux, Pierre (1898–1955)

Chief executive of the Régie Nationale des Usines Renault from 1945 to 1955.

He was born on June 30, 1898, at Triel (Seine-et-Oise) and attended local schools. He served as a teenage soldier in World War I and was accepted at the Ecole Centrale des Arts et Manufactures, graduating in 1922 with a degree in civil engineering and a doctor's title in law. He began his career with a railway company[1] but left after two years to go to Africa with a contactor.[2] In 1925 he returned to France and joined a big organization[3] which was erecting gas works all over Europe. Mobilized in September 1939, he was placed in charge of the Arsenal at Le Lans, but transferred in 1940 to the Arsenal at Brive-la-Gail-larde. Throughout World War II he was a leader in the underground resistance movement, operating under the code name "Commandant Gildas."

He was chosen to be the provisional manger of the Renault factories on October 4, 1944, when they were still the property of the Renault family. On January 16, Général de Gaulle pushed through a new law authorizing the state's confiscation of the Renault enterprise, and Lefaucheux was named chief executive of the Régie. Car production resumed with just one prewar model, the Juvaquatre, and he rushed the small 4 CV with its outboard/rear-mounted engine into production at Billancourt in 1946. He authorized new plant construction at Flins in the Seine valley, but construction work did not get underway until the end of August 1950.

In 1948 he approved Project 108 for a big family car. The concept changed several times, finally resulting in the Frégate sedan, with production startup in 1951. He sold a 4 CV license to Hino Motors in Japan and began collaboration with Willys-Overland do Brasil. He reached an agreement on joint engine and small truck production with Alfa Romeo. His career was cut short by an accident on February 11, 1955, when he was driving a Renault Frégate towards Strasbourg, where he was to make a speech. The roads were icy, the car went out of control, and he died on the spot.

1. Compagnie des Chemins de Fer du Nord.
2. Société des Travaux Dyle & Bacalan.
3. Compagnie Générale de Construction de Fours.

Lefebvre, André (1894–1964)

Experimental engineer of Avions Voisin from 1915 to 1924; chief engineer of Avions Voisin from 1924 to 1931; untitled but undisputed director of design engineering, S.A. André Citroën from 1933 to 1958.

He was born on August 19, 1894, at Louvres and graduated from the Ecole Supérieure de l'Air with a diploma in aeronautical engineering. He joined Avions Voisin in 1915 and worked on the development of transport and bomber planes. He did not work on Voisin's earliest production cars, but was the principal creator of the ultra-low 2-liter six-cylinder racing car of 1923 and raced one himself in the Grand Prix de Vitesse at Tours, taking fifth place. In 1924 Gabriel Voisin placed him in charge of design for the big cars, the smaller ones being assigned to Marius Bernard. He was experimenting with chassis having rear-mounted engines in 1927, and two years later he began exploring front-wheel-drive. In 1931 Gabriel Voisin outlined for him the bleak prospects of his company, and urged him to join one of the big car companies. He worked for Renault in 1932 and joined Citroën in March 1933, where he was given a separate office in 48 rue du Théatre and a special mission. André Citroën wanted a front-wheel-drive car with unit-construction body. The inspiration came from a Budd Company prototype with a V-8 engine. The future Citroën would have a four-cylinder engine, torsion-bar suspension, and Budd-patented body construction. He bypassed the technical director's hierarchy and formed his own team of Citroën engineers and outside experts. The 7 CV Traction Avant went into production in March 1934, still unready from lack of adequate testing but magnificent in its promise. By 1937 all was in order, and in 1939 the six-cylinder model was in production. He drew up a small car with a transversely mounted four-cylinder engine in 1935, but Pierre Michelin decided against going ahead with it. Pierre Boulanger instead gave him the task of inventing an economical farmer's do-all car, and prototypes were running in 1937, with a flat-twin engine and interconnected front/rear suspension. A test series of 200 cars was built in 1939. In September 1939 he was mobilized as an officer in the service of the Ministère des Armées, and discharged on the collapse of France in 1940. He spent the war years in Paris, developing the farmer's car, which went into production as the 2 CV late in 1948. In 1950 he had a big streamlined sedan on his drawing board as a planned replacement for the "Traction" with a horizontally opposed six in the front overhang. But that engine was not ready in time, and the concept was adapted, as an after-

thought, for the old four-cylinder vertical power unit. mounted inside the wheel-base. Appearing in October 1955, the DS-19 was 20 years ahead of the competition, but still not the car that Lefebvre had intended. He retired, due to illness, in July 1958 and died in May 1964.

Legros, Augustin (1877–1953)

Production manager of Delage from 1905 to 1907; general manager of Delage from 1907 to 1935.

He was born in 1877 in Paris as the son of a railway employee. He graduated from the Ecole des Arts et Métiers of Cluny with an engineering degree in 1899, and worked in a succession of machine shops in the Paris suburbs before joining Peugeot in 1904. There he met and befriended the test engineer and chief draftsman, Louis Delage, and in 1905 threw in his lot with Delage rather than stay at Peugeot. He organized car production in Delage's small plant at Levallois, and moved production into bigger premises at Courbevoie in 1912. He introduced methods into the work processes and became known as a good administrator. Delage was never a mass producer, but in 1913 the factory was turning out 150 chassis per month. During World War I Legros converted the facilities to make artillery shells and reached a peak daily output of 5000 units. He reworked the plant with more logical materials flow for civilian work in 1919 and Delage's payroll swelled to 3000 workers. The engine plant produced a number of four-, six- and 8-cylinder units, culminating with the D8 in 1928. He retired in 1935, when Delage was in bankruptcy and the plant was closed, settled at Triel (Seine-et-Oise), and lived there until his death on March 12, 1953.

Lehideux, François (1904–1998)

Managing director of Usines Renault from 1933 to 1940; managing director of Ford SAF from 1950 to 1953.

He was born in 1904 in Paris as the son of a banker and graduated from the Ecole des Sciences Politiques (School of Political Science). In 1929 he married into the Renault family and went to work at Billancourt in June 1930, holding positions in manufacturing, purchasing, and personnel administration. In 1935 he was placed in charge of production management, social policy, real-estate investment, and future programs. In addition he was named president of Renault Aviation. Faced with a serious strike threat in 1938, he ordered a lockout. The labor unions caved in, troublemakers were not rehired, and production resumed. When he volunteered for military service in 1939, he was assigned to the Renault plant at Billancourt as government liaison officer. Louis Renault regarded his acceptance of this post as an act of treason and never spoke to him again. In November 1940 with northern France under German occupation, he became head of a newly created government office, the Comité d'Organisation de l'Automobile (Automobile Organization Committee) to oversee wartime motor vehicle production. In February 1941, he was appointed delegate-general for state-owned industries and public works in the Darlan government and held the title of Secretary of State for Automobile Production. But the return of Pierre Laval as head of the government in 1942 put an end to his ministerial career. After the liberation of Paris in 1944, he was indicted for collaboration with the Germans, but the court cleared him of all charges. In 1948 the 60-year-old Maurice Dollfus recommended Lehideux to be his successor as managing director of Ford SAF, and the appointment became effective in 1950. He disdained the economy-car market and adopted a one-model policy centered on the V-8 powered Ford Vedette, which proved too high-priced to lift Ford's sales curve in a rapidly expanding market. His management got Ford's executive committee in Dearborn so worried that in October 1952, they discussed whether to inject fresh capital in the French affiliate or simply liquidate Ford SAF. Ford Motor Company chose the former, advanced FRF 1.5 billion to Paris, and dispatched Francis C. "Jack" Reith to keep an eye on things, with plenipotentiary powers. Lehideux's authority was in shreds and he resigned in April 1953. He never worked in the

automobile industry again, and died on June 21, 1998.

Lehmann, Ernst (1870–1924)

Technical director of L'Auto Métallurgique from 1903 to 1907, and its managing director from 1907 to 1910; technical director of NSU in Neckarsulm from 1910 to 1912; chief engineer of Selve Automobilwerk in Hameln from 1919 to 1924.

He was born on September 2, 1870, in Leislau near Camburg on the Saale River. He spent two years as an engineering trainee and four years in college, graduating with a degree in civil engineering. He began his career in steam-engine design and later worked in bridge building. After brief spells with a sewing-machine company and a bicycle manufacturer, he joined Daimler Motoren Gesellschaft as an engineer in 1898. He witnessed the creation of the Mercedes and went to work for L'Auto Métallurgique at Marchienne-au-Pont in Belgium in 1903, where he designed large powerful touring cars with four-cylinder 26/32, 40/45 and 50/60 hp engines. Lehmann's cars were known for advanced engineering and high build quality. Resembling the contemporary Mercedes automobiles, his Metallurgiques featured pressed-steel chassis, high-tension ignition, swing axles and an electric lighting dynamo, and later models were successful on European racetracks. He departed in 1910 to join NSU, best known for its well-made small cars, and designed a 10/30 and a 13/35, both with four-cylinder F-head engines, introduced in 1911. He was occupied with military engineering projects during World War I and signed up with Selve in 1919. Though his background was in luxury automobiles, Lehmann realized that the market demanded economy cars, and he designed the 1½-liter 6/24 and 2-liter 8/32, produced from 1921 to 1925. When he died in an accident in 1924, Walther von Selve replaced him with Karl Slevogt.

Leiding, Rudolf (1914–)

Chairman of the management board of Volkswagenwerk AG from 1971 to 1974.

He was born on September 4, 1914, at

Busch in Kreis Osterburg, Altmark, as the son of a machinery salesman. He began a four-year apprenticeship as an auto mechanic in a local garage in 1930 and studied mechanical engineering at Magdeburg from 1934 to 1939, working intermittently in metalworking factories to finance his studies while simultaneously getting practical experience with machines and factory equipment. He was drafted into military service in 1939 and sent to the front as a forward scout in the Wehrmacht infantry. He spent several years as a prisoner of war. Freed in 1945, he joined Volkswagen as a production engineer, and then went through a succession of appointments in sales and service up to 1958.

He was plant director of the new gearbox factory at Baunatal near Kassel from 1958 to 1965, managing director of Auto Union GmbH (Audi) from 1965 to 1968, and president of VW do Brasil from 1968 to 1971. His work in Brazil produced innovative Volkswagens, including designs for more performance-oriented cars, and led to his assumption of leadership over the company in 1971. As soon as he held the highest authority at Wolfsburg, he shocked the board by demanding approval for a $3.25 billion budget to design, develop and tool up for an entirely new product range over a five-year period. His predecessors had been grappling for years with the problem of replacing the "Beetle." He argued that no more time and money must be wasted. The directors agreed with the plan, but gave him only half the amount of money he wanted.

His plan called for a front engine, water-cooled replacement for the "Beetle," resulting in the introduction of the Golf in 1974. Known as the Rabbit in America, the design was a strong seller, but was initially plagued by problems with reliability and high production costs. Nevertheless, it became VW's most successful automobile, accomplished the problem of replacing the "Beetle," and earned Leiding the title "father of the Golf."

He had complete confidence in Ludwig Kraus to provide the concepts and engineering for the new model range, and axed the Porsche-designed EA-266 (reported to have cost $55 million) with its engine below the rear seat.

He also put an end to the production of the NSU Prinz and 1200, but did not suspend the Ro-80 or the K-70. In 1974 he turned the Wolfsburg industrial complex completely around by transferring "Beetle" production to Hanover and Emden, making room for the Scirocco and Golf at Wolfsburg. He changed the Audi 50 into a VW Polo, giving not one but two successors to the "Beetle": Polo at the lower end, and the Golf at the upper end.

Leiding championed VW's globalization, and pushed for increased production outside of Germany to offset the rising wages and decreasing productivity at Wolfsburg. This caused a contentious internal struggle at Volkswagen, and several leaders threatened to resign over the supervisory board's reluctance to build a U.S. plant.

The board's failure to provide an adequate budget for this upheaval caught up with him, for the corporation ran up heavy losses in 1974 and he resigned under fire. Toni Schmuecker, who advanced Leiding's push toward a global strategy, succeeded him. Some of the directors wanted to keep Leiding on as head of product planning, but he retired to his home in Baunatal.

Le Quement, Patrick (1945–)

Design director, Ford of Germany, from 1981 to 1985; director of advanced design and strategy with Volkswagen AG from 1985 to 1987; director of industrial design from 1987 to 1994; director of industrial design and quality since 1994.

He was born on February 4, 1945, in England and held a degree in industrial design from Birmingham Polytechnic. He went to London and won a master's degree in business administration. He began his career as a designer in the Simca styling studio in 1966, but went to Ford of Britain in 1968. He became director of truck design for Ford and worked on new models of the Transit and the Cargo, as well as the commercially successful Ford Sierra. He made a guest appearance in Dearborn as head of Specialty Car Programs and then went to Wolfsburg where he never found a good reception for his ideas. He went to Renault in October 1987 to take charge of passenger-car, truck and bus design. Though not known for personal creativity, he has shown remarkable receptivity to new ideas, and his styling policy for Renault has evolved from exploratory to adventurous.

With an emphasis on unconventional designs, Le Quement's Renaults have been received by many critics as ugly and unappealing. Nevertheless, they have sold well, particularly in Western Europe, where the Mégane now rivals the Volkswagen Golf in sales. During Le Quement's tenure, Renault has steadily assumed a more powerful role in the global automobile market, and continues to introduce provocative new concept cars. Le Quement has adapted and pushed into production such radical concepts as the Clio by Michel Jardin, Scenic by Anne Asensio, Avantime by Thierry Metroz and Vel Satis by Anthony Grade.

Letts, William Malesbury (1873–1957)

Managing director of Crossley Motors, Ltd., from 1910 to 1944.

He was born on February 26, 1873, near Reading, Berkshire, as the son of a farmer. His parents wanted him to stay on the land, but he went to London and in 1897 went to work for the British Motor Syndicate, Ltd., driving German-built Daimler cabs. He spent a year's time, 1898–99, in the United States and brought back the Locomobile franchise. In July 1902, he was a co-founder of the Society of Motor Manufacturers and Traders, with F.R. Simms, Charles Jarrott, Dick Farman, S.F. Edge, H.G. Burford, and J.S. Critchley. In 1903 he went into partnership with Charles Jarrott as London agents for De Diétrich, Napier, and Oldsmobile cars. They engaged J.S. Critchley to design a powerful touring car and persuaded William J. Crossley to undertake its production. They became sole distributors for Crossley cars, and in 1910 he was a co-founder with Charles J. Crossley and his son, Kenneth Irwin Crossley, of Crossley Motors, Ltd. In 1914–15 he was responsible for building the No. 2 National Aircraft Factory at Heaton Chapel near Manchester, and sold important numbers of Crossley

20/25 chassis to the government for use as Royal Flying Corps tenders, ambulances, and mobile workshops. He directed Crossley Motors, Ltd., setting policy with great sensitivity to market trends. He also became managing director of aircraft-builder A.V. Roe, Ltd., when Crossley bought a controlling interest in it, and in 1919 opened negotiations with the Willys-Overland Company of Toledo, Ohio, to market its products in the British Isles. This led also to setting up an assembly plant for Willys-Overland cars at Heaton Chapel, and a joint venture for the development of a light car to be made in Britain, mainly for export in the United States. After 1931, he understood that Crossley could no longer compete in terms of cost with the major car makers, but kept trying to sell in niche markets until 1937. The factory continued to produce trucks and buses, and built special-purpose military vehicles during World War II. He retired in 1944 to Llandudno, where he died on February 27, 1957.

Levassor, Emile (1843–1987)

Partner in Société Panhard et Levassor and its technical director from 1886 to 1897.

He was born on January 21, 1843, at Marolles, a village in the flat Hurepoix farmland, and educated at the Collège Chaptal. He began studies at the Saint-Etienne mining school[1] but found it mentally and physically trying because the class had to spend long hours in the underground pits, and left. In 1860 he was accepted at the Ecole Centrale des Arts and Manufactures at Chalons-sur-Marne, where he became friendly with a fellow pupil, René Panhard. Both graduated in 1864. He began his career with Cockerill at Seraing, Liège, in Belgium, as a machine-shop engineer, where he came to know Edouard Sarazin. He also made improvements on some of Cockerill's machines, notably the multiple drills. He spent some weeks as a sales engineer for Cockerill at a trade fair in Paris in 1867, and decided to move back, to be closer to his parents. He spent another two years with Cockerill, and then joined Etablissements Durenne of Courbevoie as works manager.

On the other side of Paris, his friend René Panhard needed a chief engineer, and invited Levassor to join the firm[2] which he did in 1872. His efforts led to a major expansion and he organized the removal of operations from the Faubourg St. Antoine (near La Bastille) to a big factory in Avenue d'Ivry on the outskirts of Paris. In 1875 the factory began producing low-speed Deutz gas engines under license from Otto & Langen of Cologne and over the next ten years learned enough about engines to be receptive when his friend Edouard Sarazin approached him about building a high-speed (say 600 rpm) Daimler engine, and produced two single-cylinder units for Sarazin in 1886. That year, René Panhard made him a partner in the business, and in 1888 they purchased manufacturing rights to a V-twin Daimler engine. In October 1888, Levassor accompanied Sarazin's widow to Cannstatt for talks with Gottlieb Daimler. He established a good friendship with Daimler, but was (silently at first) most critical of the Daimler cars he was shown. On his return to Paris, he began to design a car according to his own ideas. It took about 18 months before a prototype was ready for road testing.

The first car had a dos-à-dos body, with the V-twin engine mounted between the front and rear seats. On the second car, built in 1891, the engine was moved backwards into an inboard/rear position. He was not pleased with either layout, and drew up a third, with the V-twin engine at the front end of the frame, with a separate gearbox and jackshaft with chain drive to the rear wheels. With that, he had invented what became the industry norm, and six such cars were built and sold before the end of 1891. He had won marine-engine contracts by demonstrating two small boats with V-twin engines on the Seine in 1889, and also contracted with Armand Peugeot to supply V-twin engines for the Peugeot cars then being designed.

He maintained a regular correspondence with Gottlieb Daimler, ceaselessly asking for more powerful engines, and discussing all kinds of problems. In 1892 he obtained drawings for the new Phénix parallel-twin engine, in sizes of two, three and four hp, and half the weight of the V-twins of similar ratings. Panhard et Levassor made 15 cars in 1892, 37 in 1893 and 39 in 1894. Series production began in 1895, when he introduced the first four-cylinder model (8 hp) and added a 10-hp twin. He was a splendid test driver, coping brilliantly with the tiller steering, two gear levers, handbrake, and multiple small hand controls. He also took part in road races, and was hours ahead of all rivals in the 1895 Paris-Bordeaux-Paris event, covering 1175 km in 48 hours 47 minutes (24.6 km/h average speed). Late in 1896 he had an accident on the approach to Avignon in the Paris-Marseille-Paris race, his car rolling over after hitting a stray dog. Unaware of the true extent of his injuries, he went back to work, and it was not until months later, in his office, that he felt pain and went home. He died on April 14, 1897.

1. Ecole des Mines.
2. Périn, Panhard et Compagnie.

Levy, Raymond Haim (1927–)

Chairman of the board, Régie Nationale des Usines Renault, from 1986 to 1992.

He was born on June 28, 1927, and given a broad education. He graduated as a mining engineer from the Ecole Polytechnique in 1946, won a diploma from the Ecole Nationale des Mines in 1949, and collected a Master of Science degree in nuclear physics from the Massachusetts Institute of Technology in 1950. He held positions in the coal-mining industry in 1951–53 and was stationed at the motor vehicle certification office in Toulouse from 1953 to 1957. After ten years in the steel industry and a succession of executive positions in the petroleum industry, he was named chairman of Elf-France in 1975 and executive chairman of Elf-Aquitaine in 1976. From 1981 to 1984 he was chairman of a special steels unit[1] and held the title of chairman with Usinor from 1982 to 1984. He held office as chairman of Cockerill-Sambre in 1985–86 and joined Renault at the end of 1986, succeeding Georges Besse. He started programs to raise product quality at Renault, and met with success. He was less successful in raising the productivity of the plants. He pushed the merger plan with Volvo in 1990 and began preparations for taking Renault out of state ownership.

He retired in 1992, having reached the statutory age limit.

Lewis, Edmund Woodward (18??–19??)

Chief engineer of Daimler Motor Company, Ltd., in 1902–03; chief engineer of the Rover Cycle Company, Ltd., from 1903 to 1905; chief engineer of the Deasy Motor Car Manufacturing Company, Ltd., from 1906 to 1909.

He joined Daimler Motor Company as chief engineer in March 1902 but in September 1903, Harry Smith, general manager of the Rover Cycle Company, Ltd., hired him away to design the first Rover cars. His 8-hp single-cylinder 1904 Rover was a car of striking technical originality, built up on a tubular "backbone" frame enclosing the three-speed gearbox. Its rear portion was integral with the rear axle casing. After cracks and fractures of some backbones, it was redesigned as a two-part structure, jointed aft of the gearbox. The road springs, transverse leaf on the front axle and parallel semi-elliptics at the back, were attached to an aluminum framework that provided a base for the bulkhead and body. The first cars had wire-and-bobbin steering, quickly replaced by rack-and-pinion. The gear change was mounted on the raked steering column, and the main brake acted on rear-wheel drums. In addition, the car had Rover's patented "engine brake" which automatically closed the inlet valve on the overrun. To alleviate the roughness of a single cylinder, Lewis gave it two heavy flywheels. It was water-cooled, mounted vertically, and 1327-cc in size, giving the car a top speed of 48 km/h. With a two-passenger body, the car weighed 484 kg and was priced at £210. Rover built over 3,000 of them in an eight-year span.

His 1906 6-hp Rover had a conventional ladder-type frame and a 780-cc single-cylinder engine, with a selling price of only £101. Lewis also designed a four-cylinder rotary valve engine that remained purely experimental. Assisted by Bernard Wright, he created two four-cylinder models that were introduced in 1905. The 10/12 was notable for its one-piece cylinder block and the 16/20 is best remembered for winning the Tourist Trophy in 1907. In 1905 Lewis met Captain H.H.P. Deasy who offered him a five-year contract. He left Rover before the end of the year and took office at Parkside, Coventry, as chief engineer, works manager and head of procurement. He designed the 24-hp 4½-liter four-cylinder 1906 Deasy which had two big defects. The cast-aluminum rear axle casing proved fragile, and the transverse-leaf-spring front suspension lacked resistance to shocks in the horizontal plane. Captain Deasy wanted to fire him, but the board of directors still had faith in Lewis and overruled Deasy. On the 8.6-liter four-cylinder 35-hp model introduced in 1908, the front axle was carried on parallel semi-elliptics. In 1909 Lewis was pushed out, before his contract expired, by the new general manager, John D. Siddeley. Lewis returned to Rover, but conflicts with the management led to his dismissal in March 1910.

Lincke, Wolfgang (1945–)

Director of technical research at Volkswagenwerk AG from 1973 to 1978; manager of VW passenger-car development from 1979 to 1989; chief engineer of Volkswagen AG from 1989 to 1993.

He was born in 1945 in Frankfurt am Main and studied engineering at the Darmstadt Technical University, graduating with a degree in aeronautical engineering. In 1964 he presented a thesis on "The influence of airframe elasticity on directional stability and control of airplanes." He went to work in Germany's aircraft industry and moved to Wolfsburg in 1967 as a research engineer. For ten years he was active in the areas of material substitution, alternative power units, and vehicle safety, active and passive. He led the development of the second-generation VW Golf, which went into production in 1983, and the 1988 VW Passat. He reworked the Polo range in 1990 and led the engineering team for the third-generation Polo which made its debut in 1994. He resigned in 1993.

Lindsay, William Henry (1911–)

Deputy chief engineer of Armstrong-Siddeley Motors, Ltd., from 1945 to 1953; chief engineer of Armstrong-Siddeley Cars from 1953 to 1956; technical director of the motor car division of Armstrong-Siddeley Motors, Ltd., from 1956 to 1959.

He was born on November 19, 1911, in Liverpool and educated at the Liverpool Institute High School and Selwyn College, Cambridge. He began his career as an engineering trainee with Armstrong-Siddeley Motors, Ltd., in 1933, and served as a technical assistant from 1934 to 1940. He was the company's chief research engineer from 1940 to 1945, while postwar cars were being engineered by Sidney Thornett, with elegant bodies styled by Percy Riman. They were powered by the same 2-liter six, which was bored out to 2.3-liter size for 1949. The sedan was called Lancaster and the convertible Hurricane. A Typhoon sports saloon was added in 1946 and the Whitley sedan arrived in 1949. He had a major role in the creation of the Sapphire, making its debut in October 1952, and was responsible for the Sapphire 234 and 236 introduced in 1955. He stayed on with Bristol Siddeley Engines, Ltd., when production of Armstrong-Siddeley cars stopped in 1959, and in 1963 was named technical director of Bristol Siddeley Engines, Ltd. He retired in 1971.

Lisle, Edward, Sr. (1852–1921)

Managing director of the Star Cycle Company, Ltd., from 1896 to 1909; managing director of Star Motor Company from 1898 to 1902; managing director of Star Engineering Company from 1902 to 1909 and Star Engineering Company, Ltd., from 1909 to 1921.

He was born in 1852 in Wolverhampton and went to work in a bicycle factory in 1869. In 1876 he went into partnership with Ernest John Sharratt to make Star bicycles in Stewart Street, Wolverhampton. Sharratt & Lisle became Star Cycle Company, Ltd., in 1896. He bought a Benz Vélo which he stripped down and copied, purchased a Benz license, and set up the Star Motor Company with

works in Frederick Street, Wolverhampton. By 1900 he realized that trends had bypassed the Benz, and a new Star appeared, with a de Dion–Bouton engine and a general resemblance to the de Dion–Bouton Vis-à-Vis. Subsequent Star models borrowed from Panhard-Levassor and Mercedes. Star Engineering Company had its own foundry on Ablow Street and coachbuilding works on Dobbs Street, with machine shops and final assembly in the Frederick Street works. A junior car line named Starling, powered by 6-hp de Dion–Bouton engines, went into production at the Star Cycle Company, Ltd., in 1905. By 1907 the Star Engineering Company was making two big four-cylinder models and a six. In 1909 he reorganized the Star Cycle Company, Ltd., as the Briton Motor Company, Ltd., and introduced the Briton car, whose production was moved to Walsall Street in 1913. In 1910 he established the Star Aeroplane Company which began building monoplanes powered by a 40-hp four-cylinder engine made by Star Engineering Company to a design by F. Granville Bradshaw. During World War I the Star companies subcontracted to make wings for Avror (A.V. Roe & Co.) aircraft and Renault 80-hp V-8 aircraft engines. Both Star and Briton cars were back on the market in 1920, but it was a new and different market. He committed suicide, drowning in a canal, on February 14, 1921.

Loasby, Kenneth Michael "Mike" (1938–)

Executive engineer of Triumph from 1969 to 1975; chief engineer of Aston Martin Lagonda from 1975 to 1978; director of engineering, DeLorean Motor Company from 1978 to 1982.

He was born on November 20, 1938, and served an engineering apprenticeship with Alvis in Coventry. He stayed on with Alvis until 1966, spent less than a year with Coventry Climax, and then joined Aston Martin Lagonda as a development engineer. By 1969 he was manager of the experimental department at Newport Pagnell. That year he joined Triumph as executive engineer for engine design, later being promoted to manager of engine design and development coordination. He worked on the Stag V-8 and the 16-valve Dolomite Sprint, returning to Newport Pagnell in 1975. John DeLorean hired him in 1978 to redesign his sports coupé, moving the engine into the rear overhang to make room for a back seat, in consultation with Lotus Engineering. When the DeLorean Motor Co. was declared bankrupt in 1982, he set up his own design office, Midland Design Partnership. He did some consulting work for FF Developments, Ltd., in 1983 and also became technical director of Gozalro, Ltd., in 1985.

Loeb, Ludwig (1873–1938)

Chairman of Ludwig Loeb & Co. GmbH from 1903 to 1917.

He was born on September 26, 1873, in Berlin and received a commercial education. He was one of the first citizens of Berlin to become a car owner, buying his first car in August 1897. He organized Maschinenfabrik Loeb to manufacture automobile accessories, and also made components for motor boats, airships, and machine tools. In 1906 he founded Kraftfahrzeug AG as an automobile retail outlet in Berlin, and began production of the LUC car in 1909. LUC production was suspended in 1915 as the factory was converted to war production, and in 1917 the Loeb & Co. GmbH assets were taken over by Dinos Automobilwerke. His days as a car manufacturer were over, but he remained active in business circles in Berlin until his death in 1938.

Lohner, Ludwig (1858–1925)

Chairman of Jacob Lohner & Co. from 1892 to 1925.

He was born on July 15, 1858, at Liesing near Vienna, and attended the Vienna Technical Institute from 1873 to 1878. He spent the next nine years as a journeyman, holding engineering posts in Europe, America, and the Orient. In 1887 he returned to Vienna, where the Lohner family had been wagon builders since the 1830s, and took over the management of the family business. In 1896 he engaged Ferdinand Porsche to design battery-

electric and hybrid-electric vehicles. The Lohner-Porsche patents covered a drive system with electric motors carried in the wheel hubs, supplied with power from either a battery pack or an electric generator driven by an onboard internal combustion engine. This system lent itself well also to four-wheel drive, which was first adopted for a battery-electric vehicle in 1898. Lohner-Porsche touring cars were produced up to 1906, when he sold the patents to Austro-Daimler. The Lohner enterprise maintained production of carriages and other horse-drawn vehicles up to 1910, when the main factory in the Floridsdorf district of Vienna was converted to aircraft-engine production. He led the company through the difficult postwar years and died on July 14, 1925.

Lohr, Fritz (1926–)

Technical director of Adam Opel AG from 1908 to 1991.

He was born on August 31, 1926, and joined Opel on April 1, 1940, as a toolmaker's apprentice. After being transferred to the drawing office in 1941, he was sent at Opel's expense to engineering schools in Friedberg and Darmstadt. He graduated in 1950 and spent nine years in experimental and pre-development with Opel, working on the evolution of the Rekord and Kapitän, and the creation of the new Kadett. He served as chief chassis engineer from 1966 to 1969, when he put de Dion rear suspension on the Kapitän, Admiral and Diplomat. From 1969 to 1974 he was head of a drawing-office department, and from 1974 to 1978, chief engineer of pre-development, the stage where all elements of a car are combined and compromises made towards a practical result. He started the Omega, Vectra and Calibra programs, and then served as chief body engineer from 1978 to 1980. In later years he directed the Astra program, and retired in 1991.

Lombard, Jacques (1923–)

Member of the board of directors, Automobiles Citroën, from 1975 to 1979; president of Automobiles Citroën from 1979 to 1982.

He was born on July 10, 1923, at Auxerre and graduated as a mining engineer in 1945. He began his career with a state-owned coal-mining group[1] where he spent the next eight years. In 1953 he joined Peugeot's export department, and served with Peugeot, Inc., in New York from 1957 to 1959. He became sales director of Peugeot in 1960 and presided over a steady expansion as Peugeot broadened its model range down-market (204) and up-market (604). He was transferred to Citroën as a member of the board in 1975. Citroën was deep in debt and running at a loss, producing an expensive sedan (CX), a compact sedan (GS) with an air-cooled flat-four, and the air-cooled flat-twin economy cars (2 CV, Dyane and Méhari). He became president of Citroën in 1979, and pushed the mid-range BX through to production late in 1982, but was replaced by Jean Baratte.

1. Houillères du Bassin du Nord et Pas-de-Calais.

Loof, Ernst (1907–1956)

Technical director of Veritas GmbH from 1948 to 1950; managing director of Veritas Automobilwerke Ernst Loof GmbH from 1951 to 1953.

He was born in 1907 in Bad Godesberg and educated as an engineer. He raced motorcycles and became German motorcycle champion. For a number of years he worked on motorcycle design for Imperia-Werke in Bad Godesberg and was noted for finding clever solutions to technical problems. He came to Auto Union AG at Chemnitz in 1932 and was associated with the development of the P-Wagen into a Grand Prix racing car. In 1938 he joined BMW in Eisenach as head of the sports-car department. He worked as a procurement staff officer in the Luftwaffe throughout World War II. In 1945 he arrived in Ulm where Hans Holbein helped him find lodgings and storage space for a cache of BMW engines and parts. He linked up with Lorenz Dietrich, former head of Bramo aircraft engines and Georg Meier, former BMW motorcycle champion, to make "new" cars from BMW parts. Meier was plant security chief at BMW's Allach plant, and searched all sources

of supply. They built seven cars in a former barracks at Hausen in 1947. In 1948 they founded Veritas GmbH at Messkirch, and built the Veritas RS racing roadster, with body by Hebmüller. Most of their business consisted of taking in customers' BMW 328s and rebuilding them as Veritas cars. The Veritas Meteor appeared in 1948 with a new engine designed by Ernst Zipprich, ex-technical director of BMW, now running an engineering establishment at Ponte Tresa on Switzerland's border with Italy. Engine production was subcontracted to Heinkel in Stuttgart-Zuffenhausen. In April 1949 the Veritas Comet prototype arrived from Zipprich, the first model without BMW parts. It went into production in part of an ex–Mauser factory at Oberndorf, with bodies by Spohn. It was followed by the Saturn coupé and Scorpion cabriolet. But Veritas could not build enough cars to make money. Lorenz Dietrich offered to sell the company to Daimler-Benz AG but was rebuffed by Wilhelm Haspel. In February 1950, Veritas GmbH was refinanced by Auto Handels Gesellschaft of Baden-Baden, which took a 50 percent equity in the company, and offered a factory site at Muggensturm in Rastatt. But by October 1950, Veritas GmbH was bankrupt. Only 62 cars had been built over a three-year period. Ernst Loof founded a new Veritas company, financed by Erwin Bonn, a washing-machine manufacturer, who held 51 percent of the capital (and Loof 49 percent). They leased the former Auto Union garage in the paddock of Nürburgring racing circuit, and introduced new models with Heinkel-built engines and Spohn bodies. In 1953 Heinkel cut off the engine supply, and a few Veritas cars were made with Opel and Ford power before Loof shut the plant in August 1953. He returned to BMW as a development engineer and sold the Veritas assets to BMW in 1955.

He died from a brain tumor on March 3, 1956.

Lord, Leonard P. (1896–1967)

Works manager of Morris Motors, Ltd., from 1928 to 1932; managing director of Morris Motors, Ltd., from 1932 to 1936; works director of the Austin Motor Company, Ltd., from 1938 to 1941; joint managing director of the Austin Motor Company, Ltd., from 1941 to 1945; chairman of the Austin Motor Company, Ltd., from 1945 to 1952; managing director and deputy chairman of the British Motor Corporation in 1952–53; chairman and managing director of BMC from 1953 to 1961.

He was born in 1896 in Coventry, attended Bablake School and served an apprenticeship with Courtauld, Ltd. After holding engineering positions with the Coventry Ordnance Company and the Daimler Company, Ltd., he joined the Hotchkiss Engine Company in Coventry in 1921. When W.R. Morris bought the Hotchkiss factory in 1923, he came into the Morris web of business expansion. In 1926 Morris bought Wolseley Motors, Ltd., and put Lord in charge of its reorganization and integration with the production of Morris cars. He installed the moving assembly line in the Cowley plant and organized the production of the Morris Minor. He streamlined the Morris model range to six basic types (Minor, 10/4, Cowley, Oxford, Isis and 25) for 1933, and in 1934–35 moved Wolseley engine production to Morris Motors (Engines Branch) Ltd. in Coventry. In 1936 he quarreled with Lord Nuffield over a profit-sharing scheme and departed in anger. Sir Herbert Austin brought him to Longbridge in February 1938 to take charge of production. He quickly assumed executive power and began preparing new cars for the postwar market. The Austin 16 (a new 2.2-liter overhead-valve engine in the prewar 12) was announced in September 1944. The big Sheerline and Princess limousines were introduced in 1947, and production of the A 40 began in October 1947. The A 70, powered by the 16 engine, arrived in 1948 and the A 90 Atlantic in 1949. He masterminded the merger with the Nuffield Organization to form British Motor Corporation in 1952. He gambled on putting the front-wheel-drive Mini with its rubber springs into mass production in 1959, and began replacing all earlier models with scaled-up Mini derivatives. He retired in November 1961 as Lord Lambury, and died in September 1967.

Lotz, Kurt (1912–)

Deputy chairman of Volkswagenwerk AG from 1967 to 1968; chairman of Volkswagenwerk AG from 1968 to 1971.

He was born on September 18, 1912, in Lenderscheid near Ziegenhahn as one of nine children in a farming family. He graduated from the August Vilmar high school in Hamberg and joined the local police force in 1932, reaching lieutenant's rank in 1934. He was called up for military service in 1935. He was a Luftwaffe general staff officer from 1942 to 1945, coming out of World War II as a prisoner of war with the rank of major. In April 1946, he joined Brown, Boveri & Cie at Dortmund as a cost accountant, and took evening classes in industrial management. In 1947 he was transferred to Brown, Boveri & Cie German headquarters at Mannheim to work in cost-accounting and cost control, rising to assignments affecting the corporation's worldwide planning and budgeting. He was named head of the BBC business division in 1954, held a seat on the management board since 1957, and was elected chairman in 1963. Volkswagenwerk AG, desperately in need of modern management, recruited him in June 1967 as deputy chairman. He became head of the VW management board on May 1, 1968. He knew, as well as everyone in Wolfsburg, that the VW "Beetle" was at the end of its career. But there was no obvious replacement in the offing. He decided to broaden the product range while seeking fresh ideas. He signed the contract with Porsche for joint production of the 914 roadster/coupé, which began in 1969 and arranged the takeover of NSU Motorenwerke AG in 1969, merging it with VW's Audi subsidiary to form Audi NSU Auto Union AG. He appropriated the NSU K-70 and tooled up for its production at Salzgitter as a Volkswagen, only to drop it again since its production cost was higher than the Audi 100s, a roomier and more powerful model. The four-door VW 411, with its traditional air-cooled flat-four in the tail, developed before his arrival, was another failure in the market. Lotz had the good sense to reject some of the more outlandish ideas for a "Beetle" replacement such as the Porsche 266, with a water-cooled in-line four-cylinder engine under the back seat, but he had no solutions. His management also suffered from political maneuvering by Chancellor Willy Brandt and his Socialist Party of Lower Saxony, who packed his management board with left-wing informers and agitators. They forced Peter Frerk, a clerk in Hanover's City Hall, on him as personnel director, thereby robbing Lotz of any chance to appoint the middle-echelon executives of his choice. He resigned in protest on September 12, 1971.

From 1973 to 1976 he was attached to the Baden-Württemberg state government as an advisor of the environment. He held office as chairman of the World Wildlife Foundation for nine years, and was named chairman of the Heidelberg University Foundation in 1992.

Luraghi, Giuseppe (1905–1991)

Chairman and chief executive officer of Alfa Romeo S.p.A. from 1960 to 1974.

He was born on June 12, 1905, in Milan, studied economics and business administration at the Bocconi University, and graduated in 1926. He began his career in the textile industry and spent 20 years (1930 to 1950) with Pirelli, manufacturers of tires and cables, where he rose to the office of managing director. In 1950 he accepted an offer from IRI[1] to take charge of SIP, makers of hydro-electric power plants. Two years later he was named general manager of Finmeccanica, an IRI group of engineering industries which included Alfa Romeo S.p.A. From 1956 to 1960 he was chairman and managing director of Lanerossi S.p.A., a textiles group. He came to Alfa Romeo at a time when demand outstripped supply, and proposed to modernize the Portello works in Milan and to expand the 12-year-old Arese factory near Saronno. But the government's Interministerial Committee for Industrial Planning ruled that no plant construction projects would be approved unless the sites were located in southern Italy, where unemployment was rampant. The government did offer important subsidies for set-

ting up greenfield plants in the South, and he submitted a plan for a factory at Pomigliano d'Arco, inland from Naples, to create 14,000 jobs and make 1000 cars a day. The car was the Alfasud, ready for production early in 1972. Due to absenteeism, strikes, supplier problems, etc., the production goals were never met. Labor strife gained in intensity also at the Arese plant, upsetting the production schedule. He resigned from Alfa Romeo S.p.A. in January 1974 in protest over political interference ad appeasement vis-à-vis the labor unions.

1. Istituto per la Ricostruzione Industriale, a state-owned holding for failing enterprises, set up by Mussolini in 1933.

Luthe, Claus (1932–)

Director of the NSU styling department from 1967 to 1971; section leader in the Audi–NSU styling department from 1971 to 1976; styling development manager of BMW from 1976 to 1990.

He was born on December 8, 1932, at Wuppertal-Barmen and was trained as a panel beater and sheet-metal toolmaker. He graduated from high school in Würzburg in 1948 and continued his studies at the Kaiserslautern Technical College for Coachwork and Automobile Construction, and obtained a master's degree in 1954. He went to work in the body design department of Deutsche Fiat AG in Heilbronn in 1955 and joined NSU in Neckarsulm as a designer in 1956. He applied some professional retouching to Ewald Praxl's design for the Ro-80 body, and led the design team for the K-70. He had some influence on the 1974 Audi 50 and the 1975 Audi 80, and went to Munich in 1976. The 1977 BMW 7-Series was ready to go when he arrived, but he was the chief designer of the E-28 second-generation 5-Series which appeared in 1981. It was criticized for being too similar to the first-generation 5-Series, but he had been told not to break with a winning formula. He was in complete charge of styling the E 30 second-generation 3-Series, arriving late in 1983, the E 32 7-series for 1987, and the 850i coupé for 1989. He retired for private reasons in 1990.

Lutz, Robert Anthony (1932–)

General manager of Ford-Werke AG from 1974 to 1976; vice president of Ford's truck operations in Europe from 1976 to 1977; president of Ford of Europe from 1977 to 1979; chairman of Ford of Europe from 1979 to 1986; executive vice president of Chrysler Corporation from 1986 to 1998; vice chairman in charge of product development, General Motors, beginning in 2001.

He was born on February 12, 1932, in Zurich, Switzerland. His education was divided between Swiss and American schools, capped with a master's degree in business administration from the University of California at Berkeley. He served as a fighter pilot in the U.S. Marine Corps and began his career with General Motors in New York City in 1963. General Motors transferred him to Europe, where he held a series of management positions with Adam Opel AG and GM France. In the spring of 1972 he joined BMW as sales director and vice president. Ford lured him back two years later.

The integration of the British and Continental model ranges was fairly complete by 1975. He was able to concentrate on policy matters, production strategy, and mergers and takeovers. Under his direction, Ford of Europe was highly profitable. He held long talks with Fiat officials about combining Fiat's and Ford of Europe's passenger-car production in 1984–85, and with British government officials from 1983 to 1986 about Ford taking over Austin-Rover. Neither plan came to fruition.

His responsibilities with Chrysler included joint projects with Mitsubishi Motors, liaison with Maserati and Lamborghini, and worldwide truck operations.

He was miffed about having been left out of the merger talks with Daimler-Benz and resigned from the Chrysler board on July 1, 1998. Before the end of the year, he was named chairman and chief executive of Exide Corporation, the world's number one maker of storage batteries. On September 1, 2001, he joined General Motors as vice chairman in charge of product development.

Lutzmann, Friedrich (1860–1930)

Designer and builder of Pfeil cars in Dessau from 1895 to 1898; design engineer with Adam Opel from 1899 to 1901.

He was born in 1960 in Saxony and was apprenticed to a local workshop. He had no formal schooling in technical matters, but his mind was always occupied with inventions, and his work became so well known that he was appointed "master mechanic to the Court of Saxony."

In 1893 he bought a Benz Victoria and began planning to produce his own car. He had a test car, very similar to the Benz, on the road in 1894. Within a year he had developed it into a workable production model. He made about 20 cars in 1896, with a variety of body styles. His engine was Benz-like, with a horizontal single cylinder, automatic inlet valve, open crankcase, wick carburetor, and Ruhmkorff trembler-coil ignition. With 2370-cc displacement, it put out 4 hp at 350 rpm. In 1898 he sold his company, the Dessau Patent Motorwagenfabrik, including his patents, models, drawings and plans to Adam Opel for RM 116,000. The following year he was given a well-paid position on Opel's technical staff, but he misunderstood his duties. He spent most of his time dreaming up new inventions, not many directly applicable to automobiles, while the sons of Adam Opel expected him to take charge of vehicle production and develop new products.

Forced to resign from Opel, he became a door-to-door salesman — and fell into abject poverty. He spent his last years trying to invent a "miracle lock" that would bring him a fortune, but the working model was still unfinished when he died on April 23, 1930.

Lyons, William (1901–1985)

Founder of SS Cars, Ltd., in 1933, which he reorganized as Jaguar Cars, Ltd., in 1945; chairman and managing director of Jaguar Cars, Ltd., from 1945 to 1972; deputy chairman of British Leyland Motor Corporation from 1968 to 1972.

He was born on September 4, 1901, in Blackpool, the son of a professional musician, and was educated at Arnold House School in Blackpool. In partnership with William Walmesley, he established the Swallow Sidecar Company in Blackpool in 1923. He designed the sidecars, and soon began to draw car bodies. In 1927 the Swallow Sidecar Company was reorganized as the Swallow Coachbuilding Company. A year later, they moved to Coventry, leasing an idle shell-filling factory in Lockhurst Lane, Foleshill, which was later purchased. Swallow bodies attracted a great deal of attention for their styling, mounted on such chassis as the Austin Seven, Fiat 509, Swift Ten, and the Standard Nine. In 1931 Swallow built its first prototype chassis with a very low frame, powered by a 2-liter six-cylinder Standard engine. Rubery Owen made the frame, and R.W. Maudslay agreed to assemble the chassis in the Standard plant. The strikingly handsome body was produced by Swallow. The car was marketed as the SS I. The partnership with W. Walmesley broke up in November 1934, and Lyons bought his shares. The business changed its name again, becoming SS Cars, Ltd.

The Jaguar name first appeared in 1935 on a four-door sedan, the SS Jaguar 2½-Litre, which also introduced an overhead-valve version of the 20-hp Standard engine, with a cylinder head and manifolds developed by Harry Weslake. Also in 1935, the SS 90 roadster was added to the range (though it was not a Jaguar), being replaced by the SS 100 in 1936. The SS name disappeared when the company resumed car production after World War II, and the firm was renamed Jaguar Cars, Ltd. Lyons styled an interim sedan, the Mark V, the sleek XK 120 roadster, and the Mark VII sedan. He never lost his eye for beauty of line nor his hand in getting exactly the right look on the first attempt.

It was remarkable how he managed to respect Jaguar traditions, achieving an impression of continuity even in such modern designs as the 2.4-Litre in 1956, and the Mark X in 1962.

He sold Jaguar Cars, Ltd., to British Motor Corporation in 1966, remaining active as chairman and (unofficially) chief stylist, updating the Mark X into an XJ6 in 1968. He went into retirement on September 4, 1972, and died in his Warwickshire home on February 8, 1985.

Mackerle, Julius (1909–)

Director of the Tatra engineering office from 1948 to 1958.

He was born in 1909 in Moravia and held a degree in mechanical engineering from the Technical University of Brno. He began his career as an engine designer with Skoda in Pilsen and Prague. He made a brief guest appearance as a member of the Porsche drawing-office staff in Stuttgart in 1936–38, and then joined Tatra-Werke in Koprivnice, where he worked on spark-ignition and diesel engine projects. By 1944 he directed complete vehicle projects. He led the engineering team that created the T600 Tatraplan in 1946–47, a streamliner derived from the Type 87, with a rear-mounted air-cooled flat-four engine. He designed a sports car with a steel-tube space frame and the T600 powertrain in an inboard/rear location, but only a few test/demonstration cars were built. The Tatraplan became a profitable export article, and 3235 cars were produced at Koprivnice before final-assembly was transferred in mid-year 1951 to the Skoda works at Mlada Boleslav. Another 2099 units were built there until the end of 1952. From 1951 onwards the Koprivnice plant made dump trucks, military trucks, and trucks for the civilian market, all created under Mackerle's direction. In 1955 a small part of the factory was set aside to produce a new car he had designed, the 603. It was a big four-door fastback sedan, over 5 meters long, with a 2½-liter air-cooled V-8 mounted in the rear overhang. He created a single-seater derived from the 1950 roadster for testing the new engine at speeds up to 200 km/h. He also designed the tiny T605 streamlined roadster with an air-cooled 636-cc two-cylinder engine in 1956, and retired in 1958. He then took over the engine department at the Research Institute for Motor Vehicles in Prague, where he spent the rest of his career.

Macklin, Noel Campbell (1888–1946)

Managing director of Eric, Campbell & Co., Ltd., from 1919 to 1920; managing director of Silver Hawk Motors, Ltd., in 1921–22; managing director of Invicta Cars from 1924 to 1933; managing director of Railton Cars from 1933 to 1940.

He was born in 1888 in London as the son of a barrister and educated at Eton. He became proficient at driving his father's Panhard-Levassor at the age of 15 and raced Mercedes and Fiat cars at Brooklands from 1909 to 1914. He volunteered for military service in the Royal Horse Artillery and was sent into the front line in France, but suffered shell-shock and was invalided out with the rank of captain. He immediately signed up with the Royal Navy Voluntary Reserve as a lieutenant. Back in civilian life in 1919, he formed a partnership with Hugh Eric Orr-Ewing and made a prototype car in his garage at Glengariff, Cobham, Surrey. He showed the result to Handley Page, Ltd., and contracted for its production in their works at Cricklewood, London. They were nice, light cars with a 1½-liter side-valve Coventry Simplex engine, but he severed his connection with that venture in 1920. He modified an Eric-Campbell chassis into a sports/racing machine which he called the Silver Hawk, and built a few of them at Cobham. By 1923 he had laid more ambitious plans to make Invicta cars and won the financial backing of Oliver Lyle and Philip Lyle of the Tate & Lyle sugar empire, and Earl Fitzwilliam who had earlier sponsored the Sheffield-Simplex. He moved his home to the Fairmile at Cobham and made the first Invicta with a 2½-liter Coventry Simplex engine. He was introduced to Henry Meadows, a newcomer to the ranks of engine manufacturers, and they reached an agreement on a regular supply of engines. The first 2.6-liter six-cylinder Meadows-powered Invicta was put on the market in 1926, and a 3-liter model was added in 1927. The NLC touring car with a 4½-liter six was presented in 1929, followed by the low-chassis S 4½-liter roadster with two or four SU carburetors in 1930. A lower-priced Invicta 12/45 with Blackburne 1½-liter engine appeared in 1931. In 1933 he began producing the Railton car with Hudson engines and chassis at Cobham, which met with a certain commercial success. During World War II the Fairmile plant built patrol boats for the Royal Navy, to a total value of £35 million — but all the Admiralty ever paid him was £20,000.

MacPherson, Earle Steele (1891–1960)

Technical director, Ford Motor Company, Ltd., from 1947 to 1952.

He was born in 1891 at Highland Park, Illinois, and graduated from the University of Illinois in 1915. He began his career with the Chalmers Motor Company in Detroit, but spent most of World War I in Europe with the U.S. Army. From 1919 to 1922 he was an engineer with the Liberty Motor Car Company in Detroit, and then joined Hupmobile[1] as a design engineer, working on engines, transmissions, axles, steering and suspension. He was named assistant chief engineer of Hupmobile in 1931, but moved to General Motors' central engineering office in 1934 and became a design engineer with Chevrolet Motor Division in 1935. He was a member of the team that redesigned the six-cylinder engine for 1937, and then led several chassis engineering projects. In 1945 he was named chief engineer of Chevrolet's Light Car Project and led the design of the Cadet prototype with all-independent suspension. When the management of GM decided against putting the Cadet in production, he resigned. He joined the Ford Motor Company in September 1947, at Harold T. Youngren's invitation. He was delegated to Ford's subsidiary in England to prepare an entirely new generation of mid-market automobiles. He directed the design teams for the 1950 Ford Consul and Zephyr, which introduced unit-construction slab-sided bodies, short-stroke overhead-valve four- (Consul) and six- (Zephyr) cylinder engines, MacPherson's patented spring-leg front suspension, suspended pedals, and 13-inch wheels. He began engineering programs for new Anglia and Prefect models embodying the same principles, and returned to Dearborn in 1952 as vice president of engineering of the Ford Motor Company. He led the design of a modern six-cylinder engine for American-size cars, and laid down the basis concept for the Ford Falcon compact car. He retired in the spring of 1957 and died in 1960.

1. Hupp Motor Car Company, Detroit, Michigan.

Maioli, Mario (1931–)

Coordinator of vehicle architecture and styling for Fiat Auto and head of Fiat styling center's product planning office from 1980 to 1992.

He was born in 1931 at Valenza Po and educated as an architect. He began his career with a firm of architects and worked on a variety of civil-engineering projects until 1977, when he was invited to join Lancia as an idea man, to develop the purposes and shapes of new vehicle concepts. Transferred to Fiat Auto in 1980, he became the liaison officer for co-operation with outside styling studios. He laid out the Fiat Uno, defined its exact dimensions, and gave the styling contract to Ital Design (Giugiaro). The Lancia Thema and Fiat Croma were created the same way, both with Ital Design. He brought in I.DE.A. Institute in the concept stage of several new models, for the layout, space allocation and weight distribution as well as the styling. This led to the Fiat Tipo, Lancia Dedra and Fiat Tempra. He started a new project which grew into the Fiat Punto, and then retired from Fiat in 1992.

Marek, Tadek (1908–)

Engine development engineer with Aston Martin from 1954 to 1957; head of the engine design office from 1957 to 1963; research and development manager of Aston Martin Lagonda, Ltd., from 1963 to 1968.

He was born in 1908 in Poland into a wealthy and aristocratic family. He attended Polish schools until he was 20, and then went to Berlin for his engineering studies, graduating in 1933. He began his career with Polski-Fiat in Warsaw and was selected for a training scheme in Turin where he spent most of 1936–

37. He returned to Warsaw and worked on truck projects until the start of World War II. He escaped from Poland and by a very round-about route, found his way to France's unoccupied zone. For some time he made a living as a driving instructor, and later worked in an aircraft factory in Vichy. When the Germans occupied southern France in 1943, he escaped again, going across Spain to Morocco, where he never found steady work and was arrested on several occasions. He paid a Tangiers fisherman to take him across the Gibraltar, and was given passage to Britain by the Royal Navy. He joined the Polish Army and was assigned to an engineering group developing secret weapons.

His education and Polski-Fiat experience got him a job with Austin, where he was soon promoted to the engine design office. He worked mainly on the six-cylinder car and truck engines. Still, he was uncomfortable in such a big organization and began looking for another challenge.

He redesigned the Aston Martin 3-Liter DB 2/4 engine for 1957, and together with Edward Cutting, developed the DBR-3 racing engine. He was responsible for the DB-4 3.7-liter in-line six, later enlarged to 4.2, and the 5-liter V-8 for the DBS. These all-aluminum high-performance engines proved extremely rugged in service.

He left Aston Martin Lagonda Ltd. in July 1968. He refused an offer of a well-paid position with GM-Holden's in Australia and moved to Italy to work as an independent engine consultant.

Marendaz, Donald M.K. (1897–1988)

Director of Marseal Engineering Company, Ltd., from 1919 to 1922; director of Marseal Motors, Ltd., from 1922 to 1924; owner of D.M.K. Marendaz, Ltd., from 1926 to 1932; director of Marendaz Special Cars, Ltd., from 1932 to 1936.

He was born in 1897 in England, and broke off his formal schooling to serve an apprenticeship with Siddeley-Deasy. He volunteered for service with the Royal Flying Corps in 1915 and was discharged with the rank of captain at the end of the hostilities. In 1919 he linked up with the Danish-born J. Seelhaft, who had also been apprenticed to Siddeley-Deasy, to set up their own business. They moved into Atlantic Works, Harefield Road in Coventry, and began building motor scooters. The Marseel car appeared in 1920, powered by a 10.5-hp Coventry Simplex engine. When Seelhaft withdrew in 1921, he founded Marseal Motors., Ltd., in January 1922, financed by Horace Thompson, ship owner in Swansea, and Arthur Miles, engineer in Coventry. More than 1000 Marseal cars were produced before his company fell into receivership in 1924, to be liquidated the following year.

He moved to London and took over the garage of the General Cab Company in Brixton Road, where he built the first Marendaz Special Cars, most of them powered by Anzani engines. In the summer of 1932 he moved into a section of the Cordwallis Works at Maidenhead, just vacated by GWK,[1] and renewed his model range, now featuring Continental[2] and Coventry Climax engines. He built his last cars in 1936 and founded an aviator's school which over the next three years trained nearly 500 pilots, many of whom fought in the Battle of Britain in 1940–41. After the war, he went to live in South Africa and organized a new settlement, Marendaz Township, in Transvaal. He also set up a factory to build small industrial diesel engines. In the mid–1960s he returned to England and settled at Asterby Hall in Lincolnshire, where he died in 1988.

1. Grice, Wood & Keiller, makers of the GWK car.
2. Continental Motors, Detroit and Muskegon, Michigan.

Markland, Stanley (1903–)

Managing director of Standard-Triumph International from 1961 to 1963.

He was born on July 3, 1903, at Macclesfield, Lancashire, graduated from Chorley Grammar School, and joined Leyland Motors, Ltd., as an apprentice in 1920. Thanks to a scholarship grant from the company, he was able to study engineering and in 1924 became a junior assistant in the Leyland research department. He was promoted to research engineer in 1937, assistant chief engineer in 1942, and chief engineer in 1945. He was given a seat on the board of directors in 1946 and named joint general manager in 1950. When Leyland took over Albion Motors, Ltd., he was sent to Glasgow as managing director in 1957. Four years later, he went to Coventry as managing director of Standard-Triumph, which had also come under Leyland's control. Standard-Triumph was deep in debt, and he plunged into a ruthless economy drive, cutting the payroll and selling off unwanted facilities. He approved the Triumph 2000 program as a replacement for the Standard Vanguard, and rushed the Spitfire into production after seeing the prototype. Productivity improved, more cars were sold, and the break-even point was nearly reached when he suddenly announced his resignation, effective December 31, 1963. He felt he had been passed over for promotion to deputy chairman of Leyland, removed from the line of succession to the title of chairman. He settled on his farm at Chorley and took over his father's laundry business.

Marston, John (1836–1918)

Chairman of John Marston, Ltd., from 1871 to 1918; chairman of the Sunbeam Motor Car Company, Ltd., from 1905 to 1918.

He was born in 1836 at Ludlow and educated at Ludlow Grammar School and Christ's Hospital, London. In 1851 he was apprenticed to Edward Perry of the Jeddo Works, Penn road, Wolverhampton, to learn the art of japanware manufacturing. Eight years later he wanted to go into business for himself, purchased the shops of Daniel Smith Lester of Bilston, and became a manufacturer of tinplate and japanned goods. When Edward Perry died in 1871, Marston bought the Jeddo Works and moved the equipment from Bilston to Penn Road. In 1887 he founded the Sunbeamland Cycle Company and began producing Sunbeam bicycles, which made him a fortune. He thought motorcycles were inherently unsafe, but set up an automobile department on Upper Villiers Street in 1901, producing the Sunbeam-Mabley, and adding a 12-hp car with a Berliet engine in 1902. In 1905 he transformed the automobile department into the Sunbeam Motor Car Company, Ltd., holding the title of chairman until his death in March 1918.

Martin, Lionel Walter Birch (1878–1945)

Joint managing director of Bamford & Martin, Ltd., from 1913 to 1921; managing director of Bamford & Martin, Ltd., from 1921 to 1924; technical manager of Bamford & Martin, Ltd., from 1924 to 1925.

He was born in 1878 in London to wealthy parents and educated at Eton and Oxford. He raced bicycles for the Oxford University team, but he had to give up competitive cycling in 1905 due to lumbago. His interest turned to motoring and about 1908 he formed a partnership with Robert Bamford, a former engineer of Hess & Savory's Teddington Launch Works. They opened an auto repair shop in Callow Street, off Fulham Road, London, and became dealers for Singer, GWK and Calthorpe cars. He began racing a tuned Singer, and in 1913 rebuilt a 1908 Isotta Fraschini, repowering it with a four-cylinder Coventry Simplex engine. They sold the shop in 1914, as Martin was drafted into service in the Admiralty and Bamford into the Army Service Corps. They reopened in Abingdon Road, London, in 1918, and had a new prototype sports car running in 1919. But in 1920 the partnership broke up in a dispute. Martin secured the financial backing of Count Louis Vorow Zborowski (1895–1924) and presented the Aston Martin sports car in 1921, after moving to Henniker Mews, South Kensington. When the count was killed in a racing accident at Monza, he could not make ends meet, and in July 1924, sold the company to Lady Charnwood. He stayed on for a while but left in November 1925. In all, he had built 69 cars since 1921.

He settled in Kingston-on-Thames and took charge of the family business, Singleton Birch, Ltd., which owned and operated china clay mines in Cornwall. He died from injuries sustained in a traffic accident while cycling in the streets of Kingston in 1945.

Martin, Percy (1871–1958)

Managing director of the Daimler Motor Company (1904) Ltd. from 1906 to 1910; managing director of the Daimler Company, Ltd., from 1910 to 1929; chairman of the Daimler Company, Ltd., from 1934 to 1935.

He was born on June 19, 1871, in Columbus, Ohio, and held a degree in electrical engineering. Seeking experience in Europe, he went to Berlin in 1894 and found a position with the Union Elektrizitäts-Gesellschaft. In 1897 he was delegated to their Italian subsidiary and placed in charge of opening a branch in Milan. In 1900, he was on his way back to America when he stopped in London and met with Sir Edward Manville, electrical engineering consultant to government bodies and private industry, and chairman of the Daimler Motor Company, Ltd., in Coventry. Their conversation sparked an interest in automobiles in Martin's mind, and in October 1900, he went to Coventry as works manager of the Daimler factory. He directed the design of a new range of Daimler cars for 1902, including an 8-hp twin, and four-cylinder models of 12, 16 and 22 hp. In 1906 he was named chief executive, but continued to lay down the specifications of new products until 1909, when F.W. Lanchester came on the scene. Martin's signature also figured on the contract with Knight & Kilbourne for manufacturing rights to the sleeve-valve engine. During World War I he served in Britain's Ministry of Munitions as controller of petrol engines and held a seat on the government's Air Board, ending up as controller of its technical department. He set Daimler policy and ran the company's affairs until his resignation on September 20, 1929. He returned as chairman from January 1935 to October 1935 when he supervised the modernization of the plant with new machine tools and equipment. He settled on his Kenilworth farm in 1936 to breed pedigree cattle, and died in 1958.

Massacesi, Ettore (1921–1998)

Chairman of Alfa Romeo S.p.A. from 1981 to 1986.

He was born on March 6, 1921, in Pesaro and graduated from the Bocconi University in Milan with degrees in economics and commercial science. Massacesi began his career as an economics reporter on radio news broadcasts, pursued his studies in corporate management and industrial relations, published many articles and lectured at the university. He served as director of labor relations for the European Coal and Steel Union from 1955 to 1958 and became director of industrial relations at Finmeccanica, an IRI holding for a disparate group of engineering enterprises including Alfa Romeo S.p.A. From 1965 to 1968 he ran the IFAP institute for professional training, and joined Finsider, the iron and steel industry group within IRI, as manager of industrial relations. From 1974 to 1978 he headed industrial relations for all of IRI[1] and served as chairman of Intersind, the federation of state-owned industry employers, from 1976 to 1981. It is typical of the doctrinaire mindset of civil service bureaucrats, even at the highest level, that the heads of IRI could imagine *him* as the best possible leader for Alfa Romeo. He mistakenly thought the salvation for Alfa Romeo could be found in ARNA, a joint venture with Nissan Motor Company to utilize excess assembly capacity at Pomigliano d'Arco to build a car with the Alfasud powertrain and front suspension in the old model Nissan Cherry body shell. There was practically no demand for that kind of car in Europe. Production of the Arna began in 1983 and was halted in 1986, whereupon Massacesi retired. He died in Rome in mid–March 1998.

1. Istituto per la Ricostruzione Industriale, state-owned and perennially money losing holding set up by Mussolini in 1933.

Massimino, Alberto (1895–1975)

Designer of the Auto Avio 815 for Enzo Ferrari in 1940; director of machine tool and automobile engineering for Officine Alfieri Maserati from 1940 to 1953.

He was born on January 5, 1895, in Turin and studied engineering at a local technical college. He also went to a Swiss university and got a degree in mechanical engineering. Drafted for military service in the middle of World War I, he served for three years as a truck driver. When discharged, he went to work for E.M. Borgo, motorcycle maker in Turin and spent five years as a motorcycle designer. When he learned of a vacancy in Fiat's drawing office in 1924, he applied, was accepted and from 1925 to 1927 worked on racing car design under Carlo Cavalli and Tranquillo Zerbi. In 1930 he was named head of Fiat's aircraft-engine design office, which he led for seven years. Drawn to motor racing, in which Fiat took no part in that period, he resigned in 1937 to join Scuderia Ferrari, then running a team of Alfa Romeo cars. He designed for Enzo Ferrari a roadster with two four-cylinder Fiat engines mounted end to end, but then World War II broke out and he joined Maserati to design machine tools and battery-electric trucks. After the war, he designed new Maserati racing and sports cars, notably the 4 CLT and the A6G 1500 GT. He created the A6 GCS roadster and the A6 GCM Formula 2 racing car but resigned from Maserati in 1953. He worked for Ferrari as a design consultant in 1954–56, and set up his private design office in Modena. In 1959–60 he designed Formula Junior racing car chassis for Stanguelini and De Tomaso, and did a lot of work for them over the next few years. In 1962–63 he designed a 2½-liter V-8 engine for Count Giovanni Volpi di Misurata, owner of the Serenissima racing team, who wanted to produce sports cars.[1]

He died in Modena on November 27, 1975.

1. Automobili Turismo e Sport (ATS), Bologna.

Mathieu, Eugène (1869–1946)

Chief engineer and managing director of the Usines de Saventhem from 1903 to 1906; chief engineer of Automobiles Unic from 1919 to 1924; technical director of Automobiles Unic from 1924 to 1931.

He was born in Colmar in 1869 and was orphaned in infancy, during the Franco-Prussian war. After his graduation from high school, he studied architecture at the School of Decorative Arts[1] but was more interested in engineering and spent his meager savings on buying technical books. He was still a student when a friend asked him for help in fixing his mother's sewing machine. Mathieu made some changes and got it to run flawlessly.

He began his career with Hurtu, Hautin et Diligeon of Albert on the river Somme, makers of sewing machines and bicycles. The company began making cars in 1895 under a Léon Bollée license, and Mathieu was named engineer of the car department. Borrowing the general layout of the Benz Vélo, he designed a new Hurtu model for 1897. From 1901 to 1903 he took out a number of patents for improvements in automobile engineering. By 1902 he had saved enough to buy the former Delin factory at Louvain in Belgium, where he founded the Etablissements de Construction d'Automoteurs Système E. Mathieu. A year later he moved into more spacious premises at Saventhem, a suburb of Brussels and began building automobiles known variously as "Mathieu" or "U.S." (Usines de Saventhem).

On these cars, the propeller shaft to the rear axle ran inside a torque tube (a patented Mathieu design). Failing to prosper, his company was merged with Belgica in 1906, and Mathieu went to Paris where he held a series of technical posts until he was drafted into military service in 1914. He was delegated to the Unic factory at Puteaux to oversee war production, taking charge of their tooling, munitions, plant organization and methods.

After the war, the owners of Automobiles Unic asked him to stay, naming him their chief engineer. For the next 12 years he directed the design and development of all Unic cars and trucks. He resigned from Unic and was engaged by Maurice Houdaille to set up a laboratory for suspension systems, springs and dampers, in the Houdaille factory at Levallois. He retired in 1939 and died in the autumn of 1946.

   1. Ecole des Arts Décoratifs.

Mathis, Emile Ernest Charles (1880–1956)

Head of E.E.C. Mathis from 1898 to 1914; owner and managing director of Mathis & Cie from 1904 to 1907; owner and managing director of Mathis AG from 1914 to 1916; owner and managing director of Mathis S.A. from 1919 to 1940.

He was born on March 15, 1880, in Strasbourg as the son of a hotel owner/director. His boyhood passion for cars was nurtured by the fine automobiles of the hotel guests. He did not like going to school, and his father sent him to England at the age of 12 as a factory apprentice to learn work discipline and the language. Returning in 1898, the youngster went into business as an auto dealer, selling De Diétrich cars. He also became the importer of Fiat cars for Germany, Switzerland and Luxembourg, and held franchises for Richard-Brasier, Rochet-Schneider, Panhard-Levassor and Minerva. In 1902 he met Ettore Bugatti at the De Diétrich works in Niederbronn, and in 1904 they formed a partnership with Heinrich Esser, the former sales director of De Diétrich, to produce and sell Hermès touring cars. Since they did not have a factory, the cars Bugatti designed were built in the SACM[1] shops at Illkirch near Graffenstaden. Bugatti withdrew in 1906 and Hermès production was shut down.

Mathis had many irons in the fire. He started the Auto-Taxi Co., the Velograph Co. and in 1909, Aero GmbH. He had bought a tract of land at Neudorf on the south side of Strasbourg in 1905 and erected a factory that was completed in 1910. Most of the cars sold with the Mathis nameplate from 1907 to 1914 were Stoewer chassis bought complete from Stettin and fitted with Utermöhle or locally built coachwork. In 1911 the first pure Mathis cars appeared, designed by Ettore Bugatti. The 1913 Mathis Babylette 1100-cc four-cylinder cyclecar was based on the same Bugatti drawings that Peugeot used for the 855-cc four-cylinder Bébé.

In 1916 the Kaiser's government sent him on a mission to Switzerland and Italy, with a generous amount of cash, to buy second-hand trucks for military transport. When he failed to come through, he was declared a deserter. German agents searched for him in Switzerland, but he had moved on to Paris, where he volunteered for service in the French Army. In his absence, the German government requisitioned his factory and sold it to Lanz of Mannheim, who operated it as the Maschinenfabrik Neudorf AG. Winning repossession in 1919, Mathis AG became Mathis Société Anonyme, and car production resumed. In 1920–21 Mathis was making the 10 CV model at a daily rate of 130 units, i.e., faster than Citroën was producing his Type A car. Mathis had a proliferation of models, all powered by small four-cylinder engines up to 1928, when he made an across-the-board change to small sixes. A four-cylinder QM was added in 1930, and the 6 CV type PY in 1931. Mathis sold a license for the PY to W.C. Durant, who failed to produce any.

Mathis auto sales dropped precipitously from 1930 to 1934, and that year Mathis made an agreement to lease the Neudorf/Meinau factory to Ford for production of the Matford car with Ford V-8 engines and chassis, and steel bodies by Chausson. The final Mathis six-cylinder cars were made in 1934 and the last four-cylinder ones in 1936. Mathis poured the proceeds from the Ford contract into a new aviation division and set up an aircraft-engine factory at Gennevilliers. In August 1939 he gave orders to dismantle the Matford factory so the invading Germans would find it useless, and sailed for New York, where he set up munitions factories in Brooklyn and Long Island City. Matford S.A. was dissolved on June 10, 1941, and the plant was occupied by Junkers and retooled for aircraft-engine production. He returned to Paris in July 1946 and at the Paris Salon in October 1946 displayed the Mathis VEL 333 egg-shaped three-wheeler, based on a 1939 prototype by Jean Andreau. It was built at Gennevilliers and had a body by Antem. But he failed to get production started. Two years later he displayed a large car, the Mathis 666 (six cylinders, six passengers, six-speed transmission) which had been

designed in New York during the war and in many respects presaged the Citroën DS-19 which was still seven years away in the future. Production was planned for Strasbourg but was never implemented. He had introduced the Mathis motor scooter in 1945 and the Mathis-Moline farm tractor followed in 1950.

But none of these initiatives brought commercial success. He retired from business and died on August 3, 1956, after a fall from his seventh-floor hotel-room window in Geneva.

1. Société Alsacienne de Constructions Mécaniques, a De Diétrich affiliate whose main installations at Mulhouse produced steam locomotives.

Mattern, Ernest (1879–1952)

Technical director of Peugeot from 1928 to 1945.

He was born in eastern France in 1879 and held an engineering degree from the Ecole des Arts et Métiers. He began his career as a production engineer with Westinghouse at Le Havre. In addition to heavy electrical equipment, the plant also produced a few automobiles. He learned Westinghouse's cost-accounting techniques which were to prove valuable in later positions.

In 1906 he joined Peugeot and within three years had been named plant manager at Lille, where he cut the assembly time for one chassis from ten to three days. In 1912 Robert Peugeot put him in charge of modernizing the Audincourt plant and served as its director from 1914 to 1917. Towards the end of World War I the factory was turning out 5500 heavy artillery shells a day. He also coped with the return to civilian production in 1919 and developed new methods in car production.

He resigned from Peugeot in November 1923 and joined Citroën as a production specialist. He organized the production of steel pressings and body assembly and welding at Citroën's Saint-Ouen plant in 1924, and set up forge shops and foundries at the Clichy works.

In 1928 he returned to Peugeot with the title of technical director. In addition to holding responsibility for all the factories, he was in overall charge of the 201 program. The 201 was a medium-size family car, given top priority, as all earlier models were phased out.

It went into production at Sochaux in July 1929, and for three years the plant turned out 155 of them every working day. Several older plants were taken out of car production. The Lille plant had become a diesel-engine factory, C.L.M.[1] In the face of labor strife, socialist legislation, and a shaky national economy, the Peugeot plants ran with a minimum of downtime until the German occupation of 1940. Suddenly Mattern began working in reverse, minimizing all activity that could help the enemy. He did it so well that he was arrested for sabotage and deported to Germany in 1943. He survived but retired shortly after his return and died at Montbéliard in 1952.

1. Compagnie Lilloise des Moteurs.

Maudslay, Reginald Walter (1871–1934)

Chairman of the Standard Motor Car Company, Ltd., from 1903 to 1907; managing director from 1907 to 1933; chairman from 1916 to 1934.

He was born into a wealthy family in London on September 1, 1871, great grandson of Henry Maudslay, one of the world's pioneers of marine steam-engine construction. He was educated at St. David's School, Moffat, and Marlborough College. He became a pupil of Sir John Wolfe Barry, architect of the Tower Bridge in London, and later his assistant. In 1902 he decided to leave civil engineering and start building automobiles. With the financial backing of Sir John Wolfe Barry, he founded the Standard Motor Car Company, Ltd., on March 2, 1903, and moved into a small factory on Much Park Street in Coventry. He was careless about accounting, and profits eluded the company in its early years. Sales fell in 1905, but he kept production going, stockpiling the cars. Many were later sold at a loss, and he lost the chairmanship to Charles Friswell, who became sole distributor for Standard cars. In 1912, with financial support from Charles J. Band, Siegfried Bettmann, and others, he bought Friswell's shares and recovered the title of chairman in 1916.

Car production moved into a bigger factory at Bishopsgate Green in 1906, where the Model S was built, starting in 1913, on a big scale (75 cars a month). Most of the plant was converted to making munitions and trench mortars in World War I. In 1915 Maudslay bought land at Canley and erected a plant which built 1600 RE-8 biplanes before the Armistice. In 1920, the Canley works became the main factory for Standard cars, while the Bishopsgate Green plant built the bodywork only. By 1924, Standard was the most profitable car manufacturer in the Coventry area. Crisis loomed anew in 1926, under the effects of the general strike. Then an Australian order for some 5000 chassis was canceled. Standard's creditors were given new cars instead of money. Still, the company would have been declared bankrupt but for the intervention of William R. Morris, who had a personal friendship with Maudslay, telling Barclays Bank to be patient with Standard or he, Morris, would take his account elsewhere. It was not Maudslay, however, who assured Standard's recovery, but strict management by John Black, joint managing director since 1929, and new products created by Alfred Herbert Wilde, chief engineer since 1926. Maudslay took a back-seat approach to running the company from 1930 onwards. He fell ill in August 1934, had an operation in October, but succumbed on December 14, 1934.

Maybach, August Wilhelm (1846–1929)

Technical director of Daimler Motoren Gesellschaft from 1890 to 1907.

He was born on February 9, 1846, at Löwenstein near Heilbronn, the son of a cabinetmaker. The family moved to Stuttgart in 1847 and he was orphaned by the age of ten, as his mother died in 1853 and his father in 1856. He was given a home and schooling by a former clergyman who, in 1858, set up a religious institution for orphaned boys[1] with woodworking shops, machine shops, an agricultural department, and a paper mill. In 1861 he became an apprentice in the technical office for machinery, where he learned draftsmanship, foundry techniques, and mechanical skills. He graduated in 1865, under the eye of Gottlieb Daimler, and went on to college.[2] Daimler hired him for a machinery company[3] in Karlsruhe in 1869, and in 1872 he accompanied Daimler to the Otto & Langen engine works at Deutz, suburb of Cologne, where Maybach was named chief engine designer. In 1876 he spent six weeks in the United States, assisting the local Deutz agents with their display at the Philadelphia world's fair, and meeting with American industrialists. It was an eye-opening experience for a young engineer and reinforced his motivation. In 1880–81 he made extensive calculations of cylinder pressure and temperature, and developed four-stroke two-cylinder gas engines up to 60 hp. He left Deutz in September 1882, three months after Daimler's departure, and joined Daimler at Cannstatt, where their first engine was made in 1883. In 1888 he created the 20° V-twin engine with crankcase compression (and an inlet flap in the piston crown) and four-speed gearbox for a light wire-wheeled car[4] which they exhibited at the Paris world's fair in 1889 and demonstrated to Emile Levassin and Armand Peugeot. He invented the jet-nozzle carburetor in 1893 and designed the Phoenix vertical-twin engine in 1894. He developed the Daimler-Phoenix car in 1896 and ran Daimler's car-construction shops from 1897 onward. In 1900 he created the Mercedes car for Emil Jellinek, introduced the pressed-steel frame, honeycomb radiator and Bosch magneto ignition (all earlier Daimler engines had hot-tube ignition). The T-head four-cylinder Mercedes engine was designed by Otto Pfaender, but in 1901 Maybach redesigned it as the Mercedes-Simplex engine in sizes of 22, 40 and 45 hp. In 1902 the 40-hp Mercedes-Simplex engine propelled a speedboat to a new record of 35.5 km/h in the Nice International Regatta. He also designed airship engines for Count Zeppelin. In 1903 he was found to be suffering from a heart condition and went for treatment at Arco in Lake Garda. It lasted most of the year and was followed by months of convalescence at Churwalden in Switzerland. In his absence, most of the Cannstatt works had been destroyed by

fire, and a new plant was built at nearby Untertürkheim. Executive power had been assumed by Wilhelm Lorenz and Paul Daimler, who in 1906 accused him of neglecting the production facilities and running "an inventor's lab." His new overhead-camshaft six-cylinder Mercedes had teething problems, and his professional competence came under attack. He resigned on April 1, 1907. He spent that summer with his son Karl designing a powerful six-cylinder touring car which they planned to sell to Opel, but they never did see Fritz Opel about it. In the summer of 1908 they reoriented their plans, and on August 22, 1908, Wilhelm Maybach wrote to Count Zeppelin, with whom he was well acquainted, proposing to design and produce engines for his airships. With Zeppelin's financial backing, they formed a company[5] at Bissingen on March 23, 1909. The factory was moved to bigger premises in Friedrichshafen in 1912. He returned to Cannstatt in 1914, retaining his seat on the board of directors until 1918. License fees from his numerous patents and dividends from his 20 percent stake in the engine company assured him of a retirement without financial worries. He celebrated Christmas 1929 with his family, apparently in good health, but died on December 29, 1929, from a sudden and violent lung inflammation.

1. Verein zum Bruderhaus Reutlingen.
2. Reutlinger Fortbildungsschule, Reutlingen.
3. Maschinenbau-Gesellschaft Karlsruhe.
4. Stahlradwagen.
5. Luftfahrzeug-Motorenbau GmbH, 60 percent owned by Luftschiffbau Zeppelin.

Maybach, Karl Wilhelm (1879–1960)

Chief engineer of Luftfahrzeutg Motorenbau GmbH from 1909 to 1915; technical director from 1915 to 1918; technical director of Maybach Motorenbau GmbH from 1918 to 1945.

He was born on July 6, 1879, in Cologne as the son of Wilhelm Maybach. From the outset, he was a reluctant scholar, resisting authority and showing a sovereign disdain for all except technical subjects. He was happy only in the workshop. He was taken out of the Cannstatt high school at the age of 17, and used his freedom to purloin cars from the Daimler factory to go for joy rides. His father gave up on the idea of getting him to study engineering at the Stuttgart Polytechnic Institute, opting instead to let him go to work at Daimler Motoren Gesellschaft as a trainee. His joy rides stopped when he became an official test driver. Soon he had to drive every new car and sign it out for release to the sales department. He also revealed himself as a neat, accurate draftsman. In 1897 his father persuaded him to go to Maschinenfabrik Esslingen for a training program, learning to make patterns and templates, jigs and fixtures, blacksmithing and boiler-making, setting and operating lathes, drills and milling machines, toolmaking and fitting. After two and a half years, he was assigned to the electrotechnical department. He passed an examination at the Royal Construction Trades School[1] in 1901, getting a mechanical technician's diploma. Next he went to work in the drawing office of Ludwig Loewe & Co. in Berlin, a big contractor in the arms-and-munitions business. This gave him experience with mass production and an understanding of industrial finance. After six months with Loewe, a friend of his father's hired him as a test engineer for a research institute in Neu-Babelsberg. Due to his father's friendship with Count Henri de Lavalette, he won an entry to the Société des Ateliers de Construction de Lavalette at St. Ouen, component suppliers to Panhard since the days of Emile Levassor and a factor in the French auto-industry substructure. From 1906 to 1908, Karl was a member of an engineering group preparing an advanced high-performance touring car. He was to borrow elements from it for the project he entertained with his father of selling a modern car design to Opel.

In 1909 his attention turned to aircraft engines and he designed a water-cooled in-line six with an output of 150 hp at 1200 rpm, with a weight of no more than 450 kg. His factory produced a range of engines during World War I, but in 1919 the Treaty of Versailles outlawed German production of aircraft engines and powered aircraft. He had a big factory, an important payroll, and no market. It took him 18 months to design, test and develop an automotive engine, the W2, which he put into production in 1920. It was a water-cooled in-line six of 5740-cc, putting out 70 hp at 2200 rpm. Truck builders with no engine production of their own, such as Magirus, Faun, NAG, Vomag and Krauss-Maffei ordered numbers of them.

It was also chosen by Jacobus and Hendrik Spyker of Trompenburg in the Netherlands for their new touring car. Maybach delivered 150 units, but no payment came through. Maybach had a stock of at least 200 more, with passenger-car camshaft, cooling system, fuel system and electrical equipment. Until 1920, Karl had no intention of building complete car chassis, but now it looked like the way out of a troublesome situation. The first Maybach W3 chassis was hurriedly prepared for the 1921 Berlin auto show. The naked chassis weighed 1640 kg, putting it firmly in the luxury car class. It was also Germany's first car to feature four-wheel brakes. More peculiar, it had a one-speed transmission, for the designer had an obsessive aversion to shifting gears. There was a pedal-operated emergency low gear, and a separate pedal for reverse. But for backing up, the car was driven by the electric starter motor, not by the engine. In spite of unfortunate timing, Maybach was soon receiving orders for 20 chassis a month.

His desire for a no-shift transmission was to lead Maybach to develop the cleverest semi-automatic multi-speed gear systems yet seen anywhere in the world. The W5 car of 1926 had a new 120-hp engine and the same one-speed gearing, but the W5 SG of 1927 had an additional two-speed gearbox with vacuum-servo shifts. For the Maybach Zeppelin V-12 of 1929, he chose a conventional three-speed gearbox with an add-on overdrive. Suddenly he realized that he could eliminate the gearbox and stack any number of overdrive sets together, with dog-clutches and flick-a-switch gearshifts.

He introduced a five-speed "Doppel-schnellgang" with an emergency low on the Zeppelin V-12 in 1930, offering it also on the six-cylinder W6 in 1934–35. The SW 35 of

1935 was a smaller car with all-independent-suspension. Some were made with a four-speed ZF gearbox, then a six-speed "Doppelschnellgang" was adopted. A seven-speed version became available on the Zeppelin V-12 models in 1938. The company made more money on aircraft engines and railway-locomotive diesels than on its cars. The first military order was received in 1935, for engines for 500 half-track vehicles. From then on, the factory worked mainly for the military. Karl Maybach spent the war years in Friedrichshafen, and from 1943 onwards kept busy with the design of engines for postwar railway locomotives. The factory was 70 percent destroyed by Allied aerial bombardment in 1944, and he moved with his family to a small farming village in Allgäu. Making a comeback with new cars seemed impossible in 1945. He chose to offer his services to the French occupation authorities, and in 1946 he moved with a staff of 60 volunteers from the engine company, to an engineering office at Vernon on the lower Seine, preparing marine and industrial diesel-engine designs for French industry. He returned to Friedrichshafen in 1951 and retired to Garmisch-Partenkirchen in 1952. He died unexpectedly in Friedrichshafen on February 6, 1960.

1. Königliche Baugewerkschule, Stuttgart.

McCall White, David (1880–1950)

Chief engineer of All-British Car Co. in 1905; assistant chief engineer of Daimler in 1906; general manager of De Luca Daimler in 1907–08; general manager of Crossley Motors from 1909 to 1911.

He was born in Lanarkshire, served an apprenticeship with a Glasgow engineering firm, and received a formal technical education. He began his career with positions in steam-locomotive and marine engineering enterprises, but was eager to get into the motor industry. In 1902 he joined Arrol-Johnston as an assistant engineer, and in 1905 George Johnston named him chief engineer of his All-British venture. He left All-British before production got started and was signed up by Daimler in Coventry. Daimler sent him to Naples, Italy, where the De Luca enterprise had bought a license to produce Daimler cars. He ran their factory for two yeas and returned to Britain, where he was named general manager of the Crossley Brothers' car factory at Gorton, Manchester. He left Crossley to go to America, where he joined Cadillac as an engine designer in 1912. He was Cadillac's chief engineer from 1914 to 1917, and vice president of Cadillac from 1917 to 1919. He spent the next four years as engineering manager of Lafayette Motors in Indianapolis. He set up office as a consulting engineer in Detroit in 1924, but moved to Hartford, Connecticut, in 1928. For years, he counted United Aircraft among his clients, and one of his last assignments was to sort out some technical problems for Tucker in 1947–48. He died at his home in Hartford on January 29, 1950.

McCormack, Arthur J. (b. ca. 1880)

Technical manager of Wolseley from 1907 to 1911; joint general manager of Wolseley from 1911 to 1919; managing director of Wolseley from 1919 to 1923.

He was born about 1880 in the Midlands and had industrial experience from the Cycle Components Company at Bournville, Birmingham. The Vickers brothers hired him in 1907 to reorganize the production of Wolseley cars. Vickers had acquired a factory at Adderley Park, Birmingham, originally built by the Starley brothers and the Westwood Manufacturing Company, Ltd. He masterminded the transfer of Wolseley-Siddeley car production from Crayford, Kent, to Adderley Park in 1909. He then proceeded with an important expansion of car production. In 1914 he moved Stellite car production from Cheston Road in Aston, Birmingham, to a new plant in Drews Lane, Ward End, Birmingham, just erected by another Vickers subsidiary, the Electric & Ordnance Accessories Company. Title to the Ward End plant was finally transferred to Wolseley in September 1919. Wolseley made 12,000 cars in 1921 but next to no profit. Alfred Vickers blamed him, and he blamed high-level blundering by the Vickers executives. He resigned in protest in 1923. Three years later, Vickers sold Wolseley to W.R. Morris.

Mellde, Rolf Wilhelm (1922–)

Head of testing, Svenska Aeroplan Aktiebolaget from 1950 to 1962; technical director of Svenska Aeroplan AB from 1962 to 1965; technical director of SAAB Aktiebolaget from 1965 to 1971; chief engineer of development, AB Volvo from 1971 to 1980; chief engineer of development, Volvo Car Corporation from 1980 to 1987.

He was born on September 30, 1922, in Stockholm as the son of an instructor in a school for auto mechanics. He studied mechanical engineering, internal combustion engines and thermodynamics at Stockholm's Technical Institute, graduating in 1943. After military service in 1943–44, he began his career with Bergbolagen, a mining group based at Lindesberg, where he was mainly occupied with mine tramways and their prime movers. When Bergbolagen acquired Skandiaverken naval yards at Lysekil in 1945, he went there as a designer of two-stroke marine engines. A year later he answered a classified ad for a position with Svenska Aeroplan AB and joined Saab at Linköping on September 30, 1946, He was a member of a team that modified and developed the DKW two-cylinder two-stroke engine for the future Saab car. Engine production was contracted out to Hans Müller of Andernach, Germany. The first test cars ran in 1947 and he was named chief test engineer in 1950. He was a successful rally driver from 1948 to 1957 and held the title of Swedish rally master in 1953. He was the main force behind the original Saab Sonett roadster and built two Formula Junior racing cars in 1959. He developed the Saab 92 and 93, and had some design influence on the 96.

He warned the management that if Saab was to make more than 8000 cars a year, they would need a four-stroke engine. His advice was ignored. In 1964–65 Saab was stockpiling cars they could not sell. Against orders, Mellde had made contact with Ford executives about supplies of the Ford V-4 engine of the front-wheel-drive Taunus 12M. Ford was willing.

He converted a Saab 96 to Ford power, and got it approved. Most of the stockpiled cars were refitted with Ford engines, and the Saab 96 V-4 became a production model in 1966. His concept, the Saab 99, was still under development, and went into production late in 1967 with a Triumph-built four-cylinder engine. But the 96 V-4 was still being built in 1980. Despite his success with the V-4, Mellde found deaf ears for his ideas among Saab's management, and when he was approached by Svante Simonsson of Volvo about coming to Gothenburg, he accepted. He stopped all work on the planned successor of the 144 and hurriedly prepared the 240-series, which proved a highly profitable model for Volvo. He developed a New York taxicab prototype with front-wheel-drive and a low floor and led the team that developed the VESC safety car prototype. He began design work on the LCP in February 1979. It combined everything he saw as a future trend, including an all-aluminum structure, an Elsbett-designed direct-injection diesel engine, and front wheel drive. It was first put on public display in June 1983. But Volvo's production cars remained stubbornly orthodox. He retired to Lysekil in 1987.

Merosi, Giuseppe (1872–1956)

Head of the Bianchi technical office from 1906 to 1909; technical director of Alfa Romeo from 1914 to 1923; consulting engineer to Mathis from 1924 to 1927.

He was born on December 17, 1872, in Piacenza, and graduated in 1891 from the technical institute in his home town. After a year's military service, he was a partner in starting a bicycle factory, Ing. Bassi & Merosi, in Piacenza in 1893. He designed bicycles for five years and joined Orio & Marchand in 1898. Stefano Orio and the Marchand brothers had made a license agreement with Decauville for manufacturing rights to their voiturette, and hired Merosi to take charge of its production. He designed several more modern cars after Orio's death in 1899, with two- and four-cylinder engines, and motorcycles. He left Piacenza for Turin in 1904 to work on racing-car engineering for F.I.A.T. Only a year later, he was in Milan with a contract to design a touring car for the Lentz company. But only three prototypes were built.

Edoardo Bianchi engaged him in 1906, and he designed first a 16-hp model, then a 35-hp one, both with four-cylinder T-head engines. His 1907 Grand Prix car was powered by an 8-liter engine delivering 90 hp at 1600 rpm. In 1909 he was engaged to design two new cars for A.L.F.A.[1] of Milan, which had been building Darracq cars under a license agreement. They were typical for their day, with four-cylinder T-head engines of 15/20 and 25/30 hp, gearboxes mounted separately under the front seat, and shaft drive to the rear axles. The factory was then making 350 cars a year. When A.L.F.A. became Alfa Romeo, he was offered a permanent position and a director's title. Among his creations for Alfa Romeo, the RL-series of 1920–21 stands out for the adoption of overhead valves and its long production life.

After leaving Alfa Romeo to pursue a career as an automobile engineering consultant, he went to Strasbourg and designed a line of small side-valve six-cylinder engines for Mathis, returning to Milan in 1927.

He retired around 1940 and died in 1956.

1. Anonima Lombarda Fabbrica Automobili.

Mersheimer, Hans (1905–1982)

Chief engineer of Adam Opel AG from 1959 to 1970.

He was born on October 29, 1905, in a village near Mainz and graduated from the local high school.[1] He came to Opel in 1920 as a toolmaker's apprentice and showed so much promise that Opel sent him to a technical college in 1922. With a degree in mechanical engineering, he returned to Opel in 1926 as a draftsman. In 1930 he became a designer in the body-engineering department, and in 1934 he was named chief body engineer and head of the styling department. He was in charge of Opel's first unit-construction bodies, but did not personally do any styling. After World War II he played a major role in new-model development, modernizing the Olympia for 1950 and the Kapitän for 1951.

He was named director of body engineering in 1952 and created new unit-construction bodies for the 1954 Kapitän and the 1953 Rekord. Promoted to chief engineer in 1959, he devoted most of his attention to the new Kadett, a totally new car to be built in a brand-new factory at Bochum in the Ruhr. It was a tremendous success and annual production topped 300,000 units. He led the teams that renewed the Rekord for 1965, created the 1964 Kapitän-Admiral-Diplomat series, and the Manta and Ascona for 1970. He retired in 1970 and died on July 17, 1982.

1. Realschule.

Michaux, Gratien (1878–?)

Design engineer of SA des Automobiles Peugeot from 1896 to 1904, chief engineer of Les Fils des Peugeot Frères from 1904 to 1910.

He was born on December 15, 1878, at Pontarlier in the Jura mountains and graduated from the Ecole des Arts et Métiers at Chalons-sur-Marne with a degree in mechanical engineering. Armand Peugeot hired him to design a car engine, as he wanted to gain independence from the Daimler license and supplies from Panhard & Levassor, to make his own. Michaux designed a horizontal parallel-twin which was manufactured at Audincourt and first installed in Types 14, 15, 16, 17, 18, 19 and 20, beginning in 1897. A smaller 3-hp version was added for the Type 26 in 1899 and an enlarged 12-hp version for Types 28 and 29. He was partly responsible for moving the engine from the rear to the front of Peugeot cars, beginning in 1902.

In 1904 he signed up with Les Fils des Peugeot Frères and designed the single-cylinder Lion-Peugeot Type VA. About 1,000 VA cars were produced at Beaulieu from 1905 to 1908. He designed Type VC with a 1045 cc single-cylinder engine, produced from 1906 to 1909 and the Type VY roadster with an 1841 cc single-cylinder engine, built in 1908–09. In 1910 Robert Peugeot adopted a new V-twin engine designed by Lucien Verdet for the V2 chassis, and Michaux resigned.

Michelat, Arthur Léon (1883–1950)

Assistant chief engineer of Automobiles Impéria from 1906 to 1909; chief engineer of Automobiles Delage from 1910 to 1915 and 1919–20; engine designer for SA André Citroën from 1925 to 1928; chief engineer of Automobiles Delage from 1933 to 1935.

He was born in 1883 and studied engineering. He began his career as a member of the Panhard-Levassor engineering staff in 1902, worked briefly with Ariès in 1904 and with Clément-Bayard in 1905–06. Adrien G. Piedboeuf invited him to Nessonvaux near Liège, where he became assistant to Paul Henze. In 1908 he designed a small Impéria car with a one-piece four-cylinder block and L-head, 3-speed gearbox and shaft drive. He returned to Paris and joined Delage, where his first design was the AH with an L-head 2½-liter six. It was followed by the L-head four-cylinder AI in 1913 and BI in 1915, and the six-cylinder AK in 1913 and BK in 1915. He was drafted into military service in World War I but returned to Delage in 1919 and designed a 17 CV six which evolved into the 20 CV Type CO, and began to design a four-cylinder 11 CV Type DE. He quarreled with Louis Delage and left in 1920, and went to Marseille where he acted as a technical consultant to Léon Paulet and helped design a car with an overhead-camshaft six-cylinder engine. His first job for Citroën was the C4 engine, which could run indefinitely at 3000 rpm, compared with a maximum of 2300 rpm for the previous B-14 engine. He returned to Delage in 1933 and created the 2.7-liter D6-65 and 3.6-liter D8-85 in record time, reaching the dealer showrooms in 1934. He also developed a high-performance version of the 3.6-liter, which was named D8-105, and designed the 2.4-liter six-cylinder D6-60 in 1935. He left Delage before it was taken over by Delahaye. He spent World War II at Pamiers in the Pyrenees, inventing things and applying for patents. He returned to Paris in 1945 and ended his career with FAR,[1] makers of the Scammel "mechanical horse" three-wheeled tractor for semi-trailers under license.

1. Fritz, André, Raymond were the first names of the founders, Fritz Glaszmann, André Lagache, and Raymond Glaszmann, established in 1920 as a subsidiary of Chénard & Walcker.

Mickwausch, Günther (1908–1990)

Chief body designer of Auto Union AG from 1932 to 1945.

He was born in Dresden in 1908, attended local schools and revealed a natural talent for drawing. After graduating from high school in 1925, he felt he lacked qualifications to earn a living as an artist, and studied applied graphics under Professor Petzold in Dresden. He went on to the Academy of Arts[1] in 1931 and joined Horch-Werke AG in Zwickau a year later. His first design for Horch was a new radiator shell for the 830 models. He also designed the bodies for the DKW Schwebeklasse, Wanderer W-51, Audi Front UW-225, Horch 850-853 and 930, Audi 920 and DKW F-9. The F-9 broke with all Teutonic styling traditions, its shape being determined by the laws of aerodynamics, and was planned for production in four sizes; one for each make in the Auto Union stable. The Horch version, 930-S, was shown in 1939 but shelved when Germany went to war. In 1945 he chose to remain in East Germany and started his own design office in Dresden, working primarily in non-automotive industrial design and graphics. He died in Dresden on June 1, 1990.

1. Akademie für Kunstbewerbe.

Momberger, August (1905–1969)

Technical director of Goliath-Werke GmbH from 1950 to 1957.

He was born on June 26, 1905, in Wiesbaden as the son of a quarry operator and studied engineering at Darmstadt Technical University. From 1923 to 1925 he was a member of the NSU engineering staff and raced NSU cars with the factory team. He returned to the university and graduated in 1928. As a test engineer with Steyr in 1928–29 he assisted in the development of the 145-hp 4.9-liter "Klausen Sport" model. From 1929 to 1930 he was a member of the Simson engineering staff at Suhl in Thuringia, helping to prepare a new eight-cylinder model. He served as sports director of Wanderer in 1931, and the following year joined Deutsche Benzole Company to work on the development of new racing fuels. He returned to Wanderer in 1934 and took charge of the development of new sports models. He also became responsible for DKW's sports models and worked with the Auto Union racing team, occasionally driving in Grand Prix races. During World War II he worked on the development of a farm tractor with wood gas generator until William Werner assigned him to the production of a 700-hp V-12 tank engine (under Maybach license).

He fled to the West with his family in 1945 and joined the "INKA" group of automotive engineers based at Hude in Oldenburg, where he worked on car projects for C.F.W. Borgward. He designed the Goliath GP 700 front-wheel-drive economy car and its derivative, the GP 700 Sport Coupé. He created the 1956 Goliath 900V, the four-wheel-drive Jagdwagen, and the 1957 Goliath 1100. In 1957 he accepted an offer from Ford-Werke AG and went to Cologne as director of special projects. For the next two years he spent a lot of time in Dearborn, with a team of American and German engineers on the front-wheel-drive "Project Cardinal" which became the Ford Taunus M-12 in 1962. He left Ford in 1959 to become engineering manager of Henschel & Sohn, truck builders in Kassel, later adding the duties of works manager. He left Kassel for Bad Homburg in 1964, when Henschel was taken over by Rheinstahl-Hanomag, and became technical manager of PIV Antrieb Werner Reimers K-G, makers of industrial transmissions of the CVT (continuously variable) type for machine tools and other applications.

Momo, Cesare (1876–1966)

Chief engineer of SA Ansaldo Autobili from 1919 to 1932.

He was born in 1876 in Carrara to parents hailing from Vercelli. He was educated as a mining engineer, graduating from Turin University in 1901, and joined F.I.A.T. in 1902 as a member of Giovanni Enrico's staff. In 1904

he designed F.I.A.T.'s multi-plate oilbatch clutch, and a year later he patented a torque tube, cast as a unit with the rear axle casing, which became a feature of Fiat cars from 1909 to 1927. He left Fiat in 1906 to work for the Cantieri Aeronautici Ansaldo, makers of the SVA biplane in a factory on Corso Peschiera in Turin. He ran the factory during its busiest years, from 1912 to 1919. At the end of World War I the Ansaldo group management decided to retool the Turin plant for automobile production. Momo designed every Ansaldo car, beginning with the Type 4A, powered by an overhead-camshaft 1847-cc four-cylinder engine. Type 6A with a overhead-camshaft 1991-cc six appeared in 1922. He left Ansaldo when car production was given up and the factory sold to Viberti for building buses. He signed up with SPA,[1] truck builders in Turin and a Fiat subsidiary since 1927, where he spent the rest of his career. He died on February 9, 1966.

1. Società Ligure Piemontese di Automobili.

Montabone, Oscar (1913–1987)

Chief engineer of the Fiat technical design office from 1946 to 1956; director of the Simca technical center at Argenteuil from 1956 to 1962; co-director of Fiat motor vehicle projects and design from 1962 to 1973.

Born in Turin in 1913, he attended local schools and graduated from the Turin Polytechnic with degrees in mechanical and aeronautical engineering. He joined Fiat in 1937 as an aircraft-engine project engineer and remained in that department throughout World War II. In 1946 he was named chief engineer for Fiat cars, working directly under Giacosa's orders. Montabone led the design of the Fiat 1400 with its unit-construction body shell, its derivative, the 1900, the boxy and very lively 1100/103, the 8V sports model, and the 8001 experimental gas-turbine car of 1954. In 1956 he moved to Paris, on loan to Simca where he supervised the evolution of the Aronde and the Ford-based models. He created the Simca 1000 as a derivative of the Fiat 850 and the Simca 1300 and 1500, based on Fiat cars with the same identification.

Back in Turin in 1962, he inherited the 124

project which Montanari had started, and turned a corner in Fiat's car-design history. It was basically as successor to the 1100/103 and began life in 1966 as a 1200-cc four-door sedan which was to serve as the basis for the Autovaz Lada. The 124 TC sedan became available in 1972, with a 1608-cc twin-cam sports engine. A shortened 124 platform also gave birth to the 124 Sport Spider and Coupé, whose power units grew in size from the initial 1438 cc to 1998 cc.

He was also responsible for the 125 sedan, which combined a 1608-cc twin-cam engine and five-speed gearbox with the 1500 platform and a body shell sharing a maximum of 124 pressings, the big 130 sedan with its V-6 engine, and the Dino sports models. In 1970 he was named deputy general manager of Fiat Auto and director of Fiat's Research and Development Group. He led the planning of the new testing grounds at Nardo in Calabria, with its high-speed track. From 1969 to 1973 he also had a seat on the Ferrari board of directors. Retiring from Fiat in 1973, he became responsible for the automotive projects sponsored by Italy's CNR[1] which were mainly directed at fuel economy and environmental protection. Despite his body being ravaged by cancer, he maintained a dedicated work schedule right up to his death in April 1987.

1. Consiglia Nazionale delle Ricerche (National Research Council).

Montanari, Vittorio (1917–1996)

Head of experimental car design, Deutsche Fiat, from 1954 to 1956; leader of a Fiat car-planning group in Turin from 1956 to 1961; director of the Simca technical center from 1962 to 1965.

He was born in 1917 in Bologna, educated at local schools up to college level, and then studied at the Turin Polytechnic where he graduated in 1941 with a degree in aeronautical engineering. He joined Fiat in 1943 as an aircraft-engine designer. A brilliant mathematician, he became chief of the calculations office in Fiat's car engineering department in 1945. He was a member of the design team for the 1400 and in 1951 led the design of a 70° V8 engine for a Fiat sports car. He was trans-

ferred to Heilbronn in 1954 where he was involved with up to 20 research projects at any one time. Returning to Turin in 1956 he worked alternatively on car and truck design for several years. In 1961 he started project No. 124, the planned replacement for the aging 1100/103, laying down its basic specifications in considerable detail. Compared with the 1100 R, it was much wider (1625 mm vs. 1465 mm) and the director of testing, Carlo Salamano, protested, because he wanted Fiat cars to be narrow, suitable for slinking through gaps in big-city traffic. Montanari explained his wide-track choice to the managing director, Gaudenzio Bono, who overruled Salamano.

In 1962 Montanari replaced Montabone at the Simca technical center, where he initiated a front-wheel-drive concept which drew on his own and Ettore Cordiano's experiments at Heilbronn. It was developed and went into production in 1967 as the Simca 1100. He returned to Turin in 1965 and was put in charge of supervising a general renovation of Fiat's car and truck programs. That task was completed in four years, whereupon he was promoted to director of applied research, an office where he was free to exercise his imagination and experiment with advanced technology.

He retired from Fiat in 1982, and died on September 4, 1996.

Morgan, Henry Frederick Stanley (1881–1959)

Managing director of the Morgan Motor Company from 1912 to 1959.

He was born in 1881 at Stoke Lacy, Hereford, as the son of the local vicar and was educated at Stone House, Broadstairs, Marlborough, and the Crystal Palace Engineering College. He began his career by joining the Great Western Railway as a trainee draftsman at the Swindon locomotive works in 1900. After six years with steam engines, he left to open his own garage and workshop at Malvern Link in Worcestershire. Originally the Morgan car was just a three-wheeled runabout he had made for his own use in 1909, but he took the precaution of filing a patent application for its layout and design details in 1910. On the advice of several friends and the financial support of his father, he tooled up for production. The 1911 production model had a V-twin air-cooled JAP engine, sliding-pillar independent front suspension, two seats (staggered) and chain drive to the single rear wheel. In 1913–14 he was making 15 cars a week. In 1919 he opened a second plant on Pickersleigh Road, Malvern Link, which became his sole production site in 1923, building some 20 cars a week. All were three-wheelers, with MAG, JAP or Blackburne power. In 1933 he adopted Matchless engines for some models, and a year later, the four-cylinder Ford 8 engine. The Morgan 4/4, his first four-wheeled model, came on the market in 1936, with a Coventry Climax engine. No more three-wheelers were offered for sale after 1950.

Morris, William Richard (1877–1963)

Managing director of W.R.M. Motors from 1912 to 1919; chairman and chief executive of Morris Motors, Ltd., from 1919 to 1952.

He was born on October 10, 1877, at Comer Gardens, Worcestershire, as the son of a bookkeeper. The family moved to Oxford before he reached school age, and he was educated at the St. James Church of England School in Cowley. He left school at 14 to start an apprenticeship in a bicycle shop and became known as a good mechanic. In 1892 he opened his own bicycle repair shop, and soon began making complete bicycles. He became a motorcycle dealer in 1898, and by 1901 had designed his own motorcycle. He went into partnership with Joseph Cooper and produced motorcycles in Oxford from 1902 to 1905. In 1903 he established the Oxford Automobile & Cycle Agency and in 1906 opened the Morris Garage on Longwall, Street, Oxford. His next ambition was to build Morris cars, but he did not complete his design until 1912. He took his blueprints to London and went home with an order for 400 cars from Stewart & Ardern. He won the financial backing of Lord Macclesfield[1] who became president of W.R.M. Motors. The first cars were built in 1913 in the former premises of Hurst's Grammar School at Temple Cowley, with engines from White & Poppe, frames from Sankey, and bodies by Raworth. When the war ended in 1918, Morris was ready to expand with two low-priced models, Cowley and Oxford. Engines now came from Hotchkiss in Coventry, who produced a small Continental engine under license.

Morris now had capacity for 60 cars a week, but demand had shrunk to 68 cars a month. Soon he had a huge stockpile of unsold cars and over £84,000 in bank debts. He slashed his prices drastically and repeatedly, and the cars began to move. By April 1920, he was debt-free and had a credit balance of £80,000 in the bank. He speeded up production and turned out 23,000 cars in 1923. In 1922 he had purchased Lord Macclesfield's shares and began to seek control of his suppliers. In 1923 he bought the Hotchkiss engine factory, E.G. Wrigley (axles and steering gear), Osberton (radiators) and the coachbuilders Hollick & Pratt. In 1925 he set up a joint venture with the Budd Company to make sheet-metal stampings (The Pressed Steel Company, Ltd.,), bought the Léon Bollée enterprise at Le Mans, and refused a $45 million offer from General Motors for Morris Motors and subsidiaries. He still held a controlling interest in Morris Garages, Ltd., which grew into the MG Car Company, Ltd., and in 1927 he made a successful bid for ownership of Wolseley Motors, Ltd. That year he sailed to Australia, the first of many long vacations, and after 1933 he paid little attention to his business interests. In 1938 he sold his private holdings in MG, Wolseley and Riley to Morris Motors, Ltd., leading to the establishment of the Nuffield Organization as an umbrella for this vast industrial empire. He had been knighted as Baron Nuffield in 1934 and became Viscount Nuffield in 1938. He gave lavish donations to education and charities. He showed rising resistance to all forms of innovation in the cars, and drove a 1939 Wolseley Eight for the rest of his life. He was named chairman of the British Motor Corporation when it was formed in September 1952 by merging the Nuffield

Organization with the Austin Motor Company, Ltd., but tired and in ill health, he resigned on December 17, 1952, in favor of L.P. Lord. He died on August 22, 1963, at his home, Nuffield Place, Huntercombe, from cardiac asthma.

1. George Loveden William Henry Parker, Earl of Macclesfield.

Mors, Emile (b. 1859)

Joint managing director of the Société d'Electricité et d'Automobiles Mors from 1897 to 1907; chairman of Société Nouvelles des Automobiles Mors from 1908 to 1927.

He was born in 1859 in Paris as the son of Louis Mors, who in 1874 took over a wire-applications company founded by A. Mirand in 1851. He held a degree in electrical engineering from the Ecole Centrale des Arts et Manufactures in Paris, and in 1880 took over the parental enterprise in partnership with his older brother Louis, who was also a graduate of the Ecole Centrale des Arts et Manufactures. Their core business in 1880 was making electrical signal systems for the railroads. Their interest in automobiles first became evident when Louis Mors, Jr., built a steam car for his own use. He replaced it with a Panhard-Levassor in 1892, and three years later the brothers decided to build cars of their own. In 1896 they engaged Henri Brasier as chief engineer and Emile Cahen as production manager, and in 1907 reorganized the signals company into a new corporation with a capital of 2 million francs. The first Mors cars had a rear-mounted V-4 engine, and in 1898 a smaller model with a front-mounted flat-twin was added. Their factory at 48 rue du Théatre in Paris turned out some 200 cars in 1898. The twin and the V-4 were replaced in 1901 by a four-cylinder 1.7-liter 10 CV car with a coal-scuttle hood (like Renault's). They built six 60-hp racing cars in 1902, and in 1903 their 11.6-liter 70-hp car beat a field of nearly 200 entries by averaging 105.06 km/h from Paris to Bordeaux. In 1907 the car company was spun off from the electrical works, which Louis Mors took over, while Emile Mors remained on the car side. In 1908 he

brought in André Citroën as managing director, production volume multiplied, and Mors cars won an enviable reputation for quality and sturdiness. The racing program was canceled, its swan song having been the 1908 Grand Prix car with its overhead-valve 13-liter four-cylinder engine. In 1912 Mors discontinued engine production, and A. Citroën contracted for supplies of sleeve-valve engines from Minerva. From 1922 to 1927 Mors concentrated on two models, the 14/20 family car and the sporty 12/16. But the company, after three years of severe losses, was liquidated in 1927. In 1928 Emile Mors sold the car factory to SA André Citroën.

Moyet, Edmond (1895–1967)

Chief engineer of Amilcar from 1922 to 1936; head of the Citroën engine design office from 1936 to 1967.

He was born in 1895 and educated at the Navy Mechanics' School,[1] beginning his career with Borie & Cie as a draftsman of Le Zèbre cars, under the orders of Jules Salomon. In mid-year 1917 he accompanied Salomon to Citroën and assisted in the design of the Type A. He also worked on the small 5 CV model for Citroën but left the Quai de Javel in 1921. Financed by Joseph Lamy and Emile Akar, and assisted by André Morel, he opened a shop in rue du Chemin Vert near La Bastille in Paris, where he created the Amilcar. The chassis showed a close resemblance to Salomon's 5 CV Citroën, yet Amilcar made engines to Moyet's design from the outset. The first one was a modest side-valve 904-cc four-cylinder unit.

Amilcar moved into a bigger factory at St. Denis in 1922 and Morel formed a racing team which made its debut in the 1922 Bol d'Or, a 24-hour event at St. Germain-en-Laye. In 1924 Moyet set up a Sports Department and developed the CGS which had four-wheel brakes, and its lower-frame derivative, the CGSS with a 35-hp engine in a car that weighed only 610 kg complete. The 1923 E-type with its 1485-cc four-cylinder engine was the first Amilcar touring car, and the first to have a differential in the rear axle.

He designed a Grand Prix racing model for 1925, with a twin-cam supercharged 1100-cc 85-hp six-cylinder engine, and Amilcar built eight or nine of them. In 1926 he was ready with a touring-car derivative, also supercharged. The company went bankrupt in 1927 but was refinanced, and Moyet had two new models going into production in 1929, the four-cylinder 1188-cc L-type, and the CS 8 with its straight-eight single-overhead-camshaft two-liter engine. The 5 CV Amilcar C3 was produced from 1932 to 1934, beginning with a 621 cc engine and increasing to 877 cc.

He had very little to do with the Amilcar Pégase, conceived as a Salmson S4 rival in 1934, and left Amilcar. Arriving back at Citroën, he was assigned to the six-cylinder 2.8-liter engine for the 1939 15-Six. In 1944 he was placed in charge of developing the existing four- and six-cylinder engines. He designed new cylinder heads for the DS-19 and ID-19, and prepared enlarged versions of the four-cylinder engine.

He was found dead at the wheel of his car, still inside the underground garage of the engine design office in rue du Théâtre, Paris, one evening after leaving work.

1. Ecole des Mécaniciens de la Marine.

Müller, Josef (1900–1995)

Head of advanced engineering for Mercedes-Benz passenger cars from 1945 to 1950, director of Mercedes-Benz car engine design from 1950 to 1956 and chief engineer of passenger car design from 1956 to 1967.

He was born in 1900 at Stelzlhof in Lower Bavaria, attended college in Regensburg, and graduated from the Munich Technical University with a degree in mechanical engineering. He began his career as a junior engineer with Daimler Motoren Gesellschaft in 1922, became a design engineer in 1929 and helped design the side-valve four- and six-cylinder engines of the 170, 230 and 320. He also worked on chassis design for the last 770 and the planned V-8 and V-12 models. After 1945 he led a study group charged with defining a future model range and its evolution, and

started the W-120 project, which went into production as the 180 in 1953. He directed the design of the 30D SE and the 230 SL, the 250 S and the 300 SEL 6.3. He led the W-115 and W-114 programs, giving a full range of medium-sized cars, beginning with the 200 and 220 in 1967 and later including the 240-D, 230/6, 280 and 280 E.

In 1967 he retired to Prien on the Chiemsee, where he amused himself by designing bicycles. He died on April 26, 1995.

Mundy, Harry (1914–1988)

Senior designer of Morris Motors (Engine Branch) in 1939–40, chief designer of Coventry Climax engines from 1950 to 1955, engine design engineer with Jaguar Cars, Ltd. from 1967 to 1980.

He was born in 1914 in Coventry and educated at King Henry VIII School and Coventry Technical College. After serving an apprenticeship with Alvis, Ltd. he worked as a design engineer with ERA at Eastgate House, Bourne, Lincolnshire, from 1936 to 1939. During World War II he was an engineering officer in the RAF. He returned to ERA in 1945 and served as head of the BRM drawing office from 1947 to 1950. He was the principal designer of the Coventry Climax 1100-cc fire-pump engine, whose potential as motive power for sports and racing cars was soon discovered by John Cooper, Eric Broadley, Colin Chapman, Frank Nichols, and others.

He spent nine years as technical editor of *The Autocar*, raising the level of road-test reports and explaining new technological developments (such as the Wankel engine) with great lucidity to a general audience. Arriving at Jaguar Cars, Ltd., he was reunited with Wally Hassan and the two worked together on the design and development of the V-12 engine. Mundy also laid the groundwork for the AJ6 engine, built in 24-valve 3.6-liter form since 1977 and 12-valve 3.0-liter form since 1980. He retired in 1980 and died of a heart attack in June 1988.

Murray, Thomas Blackwood *see* Blackwood Murray, Thomas

Musgrove, Harold (1930–?)

Manufacturing director of Austin-Morris from 1978 to 1979, managing director of Austin-Morris from 1979 to 1982, chairman and chief executive of the Austin-Rover Group from 1982 to 1986.

He was born in 1930 and educated at King Edward Grammar School and obtained a national certificate of mechanical engineering from Birmingham Technical College. He joined the Austin Motor Company, Ltd. as an apprentice in 1945 and gained practical experience on the shop floor in the Longbridge, Bathgate, and Leyland plants. During his military service he was a navigator in the RAF. Returning to Austin, he worked his way up the ladder in production-engineering assignments.

He planned and organized production of the Austin Metro at the Longbridge works. He changed the Morris Marina to an Ital in 1980 and killed the Morris nameplate three years later. He authorized production of the LM 10 (Austin Maestro), which came on stream in March 1983, and its formal-sedan derivative, the LM 11 (Montego) introduced in 1984. He pushed the K-series Rover engine into production, and got the Rover 800 Sterling into production at the Cowley, Oxford, plant in 1986. Graham Day dismissed him on October 1, 1986.

Nallinger, Friedrich (1863–1937)

General manager of the Daimler factory at Marienfelde, Berlin, from 1909 to 1912; technical director of Benz & Cie from 1912 to 1926; member of the Daimler-Benz management board from 1926 to 1929; member of the Daimler-Benz supervisory board from 1929 to 1937.

He was born on May 23, 1863, in Stuttgart, and held an engineering degree from Stuttgart Technical University. In 1887 he joined Maschinenfabrik Esslingen as a designer of steam engines and steam boilers and in 1892 accepted a position with G. Kuhn of Stuttgart-Berg where his duties were mainly connected with steam machinery. In 1895 he joined the Royal Württemberg State Railroads[1] and became general manager of the Royal Wagon Works[2] at Cannstatt.

Forsaking the rails in favor of road vehicles, he joined Daimler[3] as a member of the management board in 1904, retaining his seat when going to Berlin in 1909 to take charge of Daimler's engine plant at Marienfelde. He joined Benz & Cie as technical director in 1912, revamped the machine shops in the Mannheim plant and installed new machine tools. Product design was in the hands of Hans Nibel and Max Wagner, under Nallinger's direction. Benz produced a model range of four-cylinder passenger cars from 22 to 60 hp. Nallinger also held responsibility for the truck plant at Gaggenau, with a combined payroll (Mannheim and Gaggenau) of 5380. Car production was suspended in August 1914, and the Mannheim plant produced truck engines, aircraft engines, marine diesel engines. A second plant at Waldhof also built aircraft engines.

Nallinger's responsibilities increased in 1915 when Benz & Cie took over Marta[4] and converted their motor-vehicle works at Arad, Hungary, to aircraft engine production. Benz & Cie also bought a stake in Aviatik Aircraft Works[5] whose Leipzig factory built 100 planes a month in 1917. Nallinger's activity also included an effort aimed at industry-wide standards. In 1919 he was named managing director of Benz-Sendling Motor Plows[6] established in June that year to make farm tractors. Nallinger gave strong support to Prosper L'Orange for his development of the prechamber diesel engine and directed Benz's resumption of car production, with the prewar Types W and GR, an 8/10 and a 10/30, respectively.

He was party to the first talks of cooperation between Benz and Daimler in 1919, at the initiative of one prominent banker, Karl Jahr of the Rheinische Creditbank who sat on the supervisory board of Benz & Cie, and Ernst Berge, a lawyer and businessman who was a director of Daimler Motoren Gesellschaft. Daimler's directors had secretly discussed a direct purchase of Benz & Cie since 1916.

After the merger was completed in 1926, Nallinger and Wilhelm Kissel were behind the ouster of Ferdinand Porsche. Nallinger's later contributions centered on policy matters and making full use of Benz developments, rather than those of Mercedes, in new products. After 1933 his influence diminished as company policy was reoriented towards giving priority to government contracts.

He died on February 17, 1937.

1. Königlichen Würtembergischen Staatseisenbahnen.
2. Königlichen Wagenwerkstätte Cannstatt.
3. Daimler Motoren Gesellschaft AG.
4. Magyar Automobil Reszveny Tarsasag Arad.
5. Aviatik Flugzeugwerken, Freiburg and Leipzig.
6. Benz-Sendling Motorpflüge GmbH, Berlin.

Nallinger, Friedrich "Fritz" (1898–1984)

Assistant technical director of Daimler-Benz AG from 1929 to 1945; technical director from 1948 to 1966.

He was born on August 6, 1898, the son of Friedrich Nallinger, an executive of the regional railroad company, at Esslingen on the Neckar. He had a thorough technical education, beginning with five years at the Realgymnasium in Stuttgart, graduating in 1912, followed by four years at the Realgymnasium in Mannheim.

He served in the Kaiser's air force in World War I, and from 1919 to 1922 attended the Karlsruhe Technical University, graduating with a diploma in mechanical engineering. He joined Benz & Cie as a design engineer in November 1922, but in August 1924 he transferred to Daimler Motoren Gesellschaft at Untertürkheim as experimental engineer. In 1928 he was placed in charge of the experimental department. His test reports analyses advanced the company's engine technology and chassis engineering in seven-league boots. In 1935 he was moved laterally to head all development work on marine engines, aircraft engines, railcar engines and engines for military equipment. He made the DB 601 inverted V-12 aircraft engine ready for service in record time and prepared the fuel-injection installation on the DB 601 in 1943–44.

In October 1945 he was arrested under the U.S. Military Government Law No. 8 for his contributions to the Nazi war effort but his trial was put aside under pressure from the French government who wanted him for its gas-turbine projects. From 1945 to 1948 he served as technical director of Turboméca of Bordes in the Pyrenées, developing small turbojets and industrial gas turbines. Returning to Stuttgart, he took charge of developing new cars and trucks, directing the engineering of the 1951 model 220 and 300 series. He had overall responsibility for the unit-construction 180 of 1953 and its derivatives, and led the teams that designed and developed the W-196 and 300 SLR racing cars in 1952–55. The racing team was disbanded in 1955 and the personnel placed in strategic positions for the creation of new production models. A direct result was the new W-111 and W-112 family, beginning with the 220, 220 S and 220 SE in 1959. The 230 SL was part of that family. He also directed the engineering of W-114 and W-115, the new "low" range of cars going into production in 1967.

He retired in 1966 and died in Stuttgart on June 4, 1984.

Napier, Montague Stanley (1870–1931)

Managing director of D. Napier & Son from 1895 to 1906; chairman of D. Napier & Son, Ltd., from 1906 to 1931.

He was born in 1870 in London as the son of James Murdoch Napier and grandson of David Napier, a Scots engineer who went to London in 1808 and set up a factory to make coin-weighing machines, bullet-making machines, cranes, printing presses and hydraulic machinery. When James Murdoch Napier died in 1895, he bought the assets of the industrial enterprise from the executors of his father's estate. He continued the production of printing presses and coin-weighing machines, and raced bicycles in his spare time. He was a member of the Bath Road Club as an amateur cyclist and met S.F. Edge, the pioneer motorist. When Edge wanted some modifications made on his Panhard in 1898, he asked Napier to do the work, and was impressed with Napier's quality. On Edge's suggestion, he decided to build automobiles in the old factory in Vine Street, Lambeth. He made his first engine in 1898, and began to design a car a year later. The first Napier car, built in 1900, had a two-cylinder engine, but before the end of the year, he had also completed a four-cylinder model. He had a contract with S.F. Edge for the distribution of his entire car production. In 1902 Edge won the Gordon Bennett Cup race driving a 30-hp four-cylinder Napier, and 250 Napier cars were built and sold in 1903. Towards the end of 1903, Napier also began producing an 18/30 model with a six-cylinder engine, and the 1906 Napier range was made of just two six-cylinder modes, the 5-liter Forty and the 7.8-liter Sixty. Car production was moved into better premises at Acton in 1904, and output came to 100 cars in 1907. The contract with S.F. Edge was rewritten in 1907 and he became director of S.F. Edge, Ltd., but resigned in 1909 during a conflict over the rights to sell Napier vans and trucks through other channels. It was settled in 1910, and in 1912 he bought control of S.F. Edge, Ltd., achieving undisputed control over sales. Annual production totaled approximately 700 cars, with a range of 15 and 20 hp four-cylinder models and 30/35 and 45 hp six-cylinder cars. The Acton plant delivered over 2000 trucks, vans and ambulances during World War I, and received its first orders for the Napier Lion W-12 aircraft engine in July 1916. He had a serious operation in 1915 and two years later his physicians advised him that the British climate had a bad effect on his health. He moved to Cannes on the French Riviera in 1917. He did not put an immediate stop to car production at Acton after the war, but he was determined that the company's future would be dedicated to aircraft engines. He directed the company's affairs by remote control from Cannes until his death on January 23, 1931.

Nash, Archibald Goodman Frazer *see* Frazer Nash, Archibald Goodman

Nazzaro, Felice (1881–1941)

Managing director of Nazzaro e C. Fabbrica di Automobili from 1911 to 1916; managing director of S.A. Automobili Nazzaro from 1919 to 1923.

He was born in 1881 in Turin as the son of a coal merchant and served an apprenticeship with Giovanni Ceirano from 1895 to 1899. When F.I.A.T. bought the Ceirano shops in 1899, he stayed on as a test driver and became one of F.I.A.T.'s top racing drivers. He was a friend of Vincenzo Lancia and looked admiringly at his former colleague's success as a car manufacturer. He decided to follow in Lancia's footsteps, resigned from Fiat in 1911, and set up his own factory in Turin in partnership with his brother-in-law, Pilado Masoero, and a local businessman, Maurizio Fabry. He engaged Arnold Zoller to design a car and production got under way in 1912. The first model, Tipo 2, was a high-performance touring car with a 4.4-liter L-head four-cylinder engine, four-speed gearbox and shaft drive. Nazzaro built chassis only, relying on coachbuilders such as Cav. Achicle Magliano to provide bodies. About 230 car chassis were made up to 1915, plus some 50 military trucks with 10-cylinder Anzani engines. The company was liquidated in 1916 and the plant sold to Stabilimenti Franco Tosi of Legnano. Nazzaro moved to Florence and set up a new car company. Tipo 5 was designed by Ottavio Fuscaldo and went into production in 1919 with a 3½-liter overhead-valve engine. Carrozzeria Lotti supplied most of the bodies. A little more than 200 chassis were built before the company went bankrupt. Felice Nazzaro returned to Turin and was a member of Fiat's racing team from 1924 through 1927. Giovanni Agnelli disbanded the racing team when Petro Bordino was killed, and Nazzaro was assigned to the testing department alongside Carlo Salamano.

Né, Jacques (1925–)

Project manager for the Citroën SM from 1965 to 1972.

He was born on December 17, 1925, at Le Havre and aimed for a career in naval engineering. He attended technical classes for petty officers until 1940 when all French naval yards were occupied by German forces. He spent the war years in the maritime repair and aircraft industry, and joined the French air force in 1945. He began his civilian career as a draftsman with Citroën in 1948 and went in for off-road motorcycle racing in his spare time. In 1949 he was sent to the patent office to study the prior art on disc brakes and began to propose new systems. He invented a disc brake with a floating caliper and two pistons on the same side, which became the basis for the brake system on the DS-19. He became a project engineer in the drawing office, under André Lefebvre's orders, in 1955 and was associated with the design of a streamliner with the 2 CV powertrain. He was named head of a project to create an ultra-low-profile two-to-three-passenger sports coupé on the DS-19 platform, powered by a 3-liter V-8. It was built, tested and abandoned. In 1965 he made a sports cabriolet with a twin-cam 2-liter four-cylinder engine on a shortened DS-19 platform. At that time Pierre Bercot asked him how much power one could safely put into the front wheels, and Né began building test cars with very powerful engines to find the limit. In 1968 Maserati engines became available, and he installed V-8 engines up to 380 hp in much-modified DS vehicles. He reported to Bercot that there was no limit, and was asked to develop a candidate production-model with a new Maserati V-6 engine. That became the SM, appearing in February 1970, with important innovations in steering, suspension and brakes. In 1967 he held talks with officials of Panhard and Chausson, who were seeking a way to keep the 24 CT body shell in production after Bercot had ordered a stop to Panhard car production. He adapted the 24 CT body shell to a DS chassis, with an experimental 16-valve 2-liter four-cylinder engine. It could have been an excellent lower-priced companion model for the SM, but Bercot turned it down. From 1972 to 1975 he was engaged on urban-car projects, proposing a lot of original two-passenger microcars on three and four wheels. All that was shelved when Peugeot took over Citroën in 1975 and he was transferred to the Automobiles Peugeot technical center at La Garenne. He became

involved in long-term research projects with no direct impact on near-term production models, and retired in 1985.

Neumeyer, Hans Friedrich "Fritz" (1875–1935)

Managing director of Zündapp Gesellschaft für den Bau von Spezialmaschinen GmbH from 1919 to 1928; chairman of Zündapp-Werke GmbH from 1928 to 1935.

He was born in 1875 in Nuremberg to well-to-do parents and was given a college education. He began his career in 1896 by opening a workshop to make flush toilets. In 1903 he established Bayerisches Hüttenwerk Fritz Neumeyer AG to produce metal pipes, stampings, and extruded parts. In 1917 he obtained the financial backing of Krupp to set up Zünder- und Apparatebau GmbH to make fuses and electrical equipment. He reorganized it under a new title in 1919 and added a line of starter motors, generators, and small refrigerators. The first Zündapp motorcycle appeared in 1921. He had also been a co-founder of BMW GmbH in Munich in 1918, and a minority shareholder. A new company, Fritz Neumeyer AG, took over the business of Bayerisches Hüttenwerk in 1919 and was in turn merged with the motorcycle and electrical equipment branches of Zündapp-Werke GmbH in 1928. He erected a very large factory at Nuremberg-Schweinau, combining the activities of four smaller plants, it had an annual capacity of 60,000 motorcycles. He also began to make plans for making automobiles. In 1931 he engaged the Porsche organization to design what he called a "Volksauto" and in 1932 Zündapp built three test cars with rear-mounted air-cooled radial 1.2-liter five-cylinder engines. In 1933 there was a big upswing in the German market for motorcycles and he quickly abandoned all plans for car production.

Newton, Noel Banner (1902–1966)

Chief engineer of John Newton Fabbrica Automobili in 1914–15; managing director of Newton & Bennett, Ltd., from 1916 to 1937.

He was born on February 22, 1902, at Sale,

Cheshire, as the son of the industrialist John Newton. He was educated at Oundle School and Manchester College of Technology. After serving an apprenticeship with the Metropolitan-Vickers works in Manchester, he was sent as a trainee to SCAT[1] in Turin, of which his father was a member of the management board. In 1913 John Newton purchased VALT[2] and its factory in Via Palmieri in Turin and the VALT 2.2-liter four-cylinder car was renamed NB.[3] Noel Newton set up a separate drawing office and experimental shop in Turin where he and R.O. Harper designed an 1100-cc light car which got no further than the prototype stage because they were called back to England at the outbreak of World War I. He returned to the Newton & Bennett, Ltd., headquarters in William Street, Salford, Manchester, and began importing Ceirano cars in 1924, sold in Britain with the Newton name. With the arrival of the 1½-liter S.150 in 1928, the Newton-Ceirano label was adopted. He began inventing hydraulic devices and in 1932 obtained a patent for a telescopic hydraulic damper, which Newton & Bennett, Ltd., put in mass production. He also designed the Newton "cushion-plate" clutch and the Newton centrifugal clutch which was patented in 1934 and became optional on several Riley models.

He resigned from the parental organization in 1937 to start Power Jacks, Ltd., and manufacture on-board hydraulic jacking systems for automobiles.

1. Società Ceirano Automobili Torino.
2. Vetture Automobili Leggere Torino.
3. Newton-Bennett, originally Newton, Bennett & Carlisle, Ltd.

Nibel, Hans (1880–1934)

Chief engineer of Benz & Cie from 1908 to 1926; technical director of Daimler-Benz AG from 1929 to 1934.

He was born on August 31, 1880, at Olleschau in Bohemia and studied engineering at Munich Technical College. He began his career with Benz & Cie as a design engineer in 1904, working under Fritz Erle and Georg Diehl. Four years later he was named chief engineer. He adopted shaft drive, beginning with the 10/18 and 18/28 PS models and created the Blitzen Benz 200-hp speed-record car in 1910. He designed a succession of powerful touring cars, including the 1912 29/60 and the 1913 33/75 and 39/100. During World War I small numbers of passenger-car chassis were built for military needs, but the main product line was aircraft engines designed by Dr. Arthur Berger. When car production resumed in 1919, it was with two small prewar models, the 6/18 and the 8/20. He led the design of the six-cylinder 2.9-liter 11/40 touring car and the 16/50 chassis, including the DS and DSS sports models. He purchased the rights to Edmund Rumpler's Tropfenwagen and designed Benz racing and sports models with a 2-liter six in an inboard/rear position and swing-axle rear suspension. He also made an inspection tour of the U.S. auto industry in 1922. After the merger with Daimler Motoren Gesellschaft in 1926, he kept a low profile, and quietly laid the groundwork for ten years or more of advanced chassis engineering. He filed a number of patent applications for independent suspension systems. Some front ends had transverse leaf springs, others coil springs. The rear ends were of the swing-axle type, with single or dual coil springs at each wheel. Several of these appeared on production models, beginning with the 170 in 1931. He adopted Schnellgang (overdrive) transmissions for some models, beginning in 1930, and encouraged the development of diesel engines for passenger cars. Assisted by Max Wagner and Albert Heess, he designed the W25A Grand Prix racing car for the 1934 season. He was given an honorary doctorate in engineering by the Karlsruhe Technical University in 1934. He died from a stroke on November 25, 1934.

Niefer, Werner (1928–1993)

Director of production, Daimler-Benz AG, from 1976 to 1986; chairman of Mercedes-Benz AG from 1988 to 1993.

He was born on August 26, 1928, at Plochingen on the Neckar and grew up at Notzingen near Kirchheim/ Teck. He came to Daimler-Benz as a toolmaker's apprentice at the age of 15 and was selected for higher education, with a scholarship grant from the company. He returned to the Stuttgart-Untertürkheim factory with a degree in mechanical engineering in 1952 and was assigned to the marine-diesel engine plant as a production engineer. He rose in rank to plant manager in 1958 and in 1962 was delegated to coordinate the diesel-engine program with Maybach Motorenbau, at Friedrichshafen. The Friedrichshafen works became part of MTU[1] in 1969, with Niefer as director, adding responsibility also for the Munich branch in 1971. Five years later he was placed in charge of all Mercedes-Benz vehicle production, with a seat on the management board. He planned and laid out new passenger-car factories in Bremen, coming on stream in 1983, and Rastatt, where assembly operations began in 1992.

In 1988 he was named chairman of a new subsidiary, Mercedes-Benz AG, formed to keep the motor vehicle operations separate from the parent organization's growing holdings in other industries (aircraft, electrical engineering, electronics, etc.).

Some of the passenger-car models showed quality problems during the rapid rise in production from 1983 to 1986. Niefer called them "design flaws" and blamed it all on the product-engineering head, Werner Breitschwerdt. Niefer also aligned himself with Edzard Reuter's relentless growth-by-acquisition, such as the Dornier purchase, which Brietschwerdt opposed, wanting to concentrate spending on the group's core activities. He retired on May 26, 1993, and died on September 12, 1993, after a lung cancer operation.

1. Motoren und Turbinen Union, a joint venture with MAN (Maschinenfabrik Augsburg-Nürnberg) which contributed its Allach gas-turbine factory near Munich.

Nordhoff, Heinz (1899–1968)

Managing director of Volkswagenwerk GmbH from 1948 to 1960; chairman and chief executive of Volkswagenwerk AG from 1960 to 1967.

He was born on January 6, 1899, at Hildesheim and attended high school in his home town. In 1917 he was drafted into the Kaiser's Army. He was wounded in action, discharged, and in 1921 continued his education at the Technical College in Berlin-Charlottenburg, graduating with a degree in mechanical engineering in 1925. He began his career as a draftsman in the aircraft engine design office of BMW in Munich, and signed up with Adam Opel AG as service manager in 1929. He was selected for training to higher management positions and made extensive visits to the United States, studying at the GM Institute and working in the car-division factories to learn American production techniques. In 1934 he was named technical assistant to Opel's sales director, and in 1937 he was elected to the Opel management board. Two years later he was delegated to the Berlin office which dealt with political and financial affairs, and in 1942 he was named plant director for the Opel truck factory at Brandenburg, where he served for the duration of World War II. In May 1945 he fled to the West, came to Rüsselsheim, and knocked on the door of Adam Opel AG. But the personnel office had a strict policy regarding workers who had held high office under the Third Reich period, and would not rehire him. He went to Hamburg and took temporary employment, running an auto repair shop for an old friend. In 1947 he was invited to Wolfsburg by the British Military Government who needed a man with the right qualifications to run the big auto plant, known under the temporary corporate identity of Volkswagenwerk GmbH while waiting for its exact ownership status to be determined. He took office on January 2, 1948, and doubled the output to 19,244 cars a year. He got the factory repaired and expanded capacity. A steady flow of refugees from East Germany provided an ample labor reserve. He introduced the Type 2 Transporter in 1950, and had a new plant built at Hanover in 1955 to take over its production. The one-millionth "Beetle" was built in 1955, and production of the Karmann-Ghia began at Osnabrück. VW Type 3, the 1500, was added to the program in 1961. On August 22, 1960, an agreement was signed with the federal and state governments, transforming Volkswagenwerk into a joint stock corporation with a capital stock of DM 600 million. The Federal Government in Bonn held 20 percent, the State of Lower Saxony 20 percent, and 60 percent was put on the market. Lawsuits from the original KdF-savers were still pending, but were settled over the next nine years. Nordhoff fully realized that he would sooner or later need a replacement for the "Beetle" and in 1963 acquired a set of drawings for a front-wheel-drive car from Hermann Klaue, formerly of the Argus Motorenwerke, Berlin, who also held a patent, issued in 1939, for a disc brake. But he was then making heavy investments in a new assembly plant in the port of Emden, and had no budget for developing Klaue's car or tooling up for its production. As late as 1966 the Wolfsburg plant was still turning out 3550 "Beetles" and 1300 Type 3 cars every day. But by 1969 the plant had so much excess capacity that it had to take in assembly of 700 Audi cars a day to keep from laying off workers. He had resigned on December 31, 1967, and died from heart failure in the Wolfsburg Hospital on April 12, 1968.

Oak, Albert Victor "Vic" (b. ca. 1890)

Chief engineer of Morris Motors, Ltd., from 1936 to 1941; director of the Nuffield Organization's research and development department from 1945 to 1952.

He was born about 1890 and educated as an engineer. He joined Wolseley Motors, Ltd., about 1925 and led the design of the 1931–35 Wolseley Hornet with its overhead-camshaft 1.3-liter six-cylinder engine. It became very popular and 37,000 were built, including 4500 Hornet Specials, a favorite with several coachbuilders. He directed the engineering of the 1934–38 Morris Eight. L.P. Lord gave him a Ford Eight and told him to make an engine of comparable cost and power/weight ratio, and the engine architecture and dimensions were very similar to Ford's. Morris built 250,000 Eights in its four-year career. Oak had its replacement, the E-series, ready for 1939, but production was suspended during World War II. The first postwar cars came off the Cowley assembly line in 1945, and over 120,000 were built until its replacement by the new Morris Minor in October 1948. He was associated with the start of the Mosquito project, which ripened into the Minor, but in 1945 was transferred to basic research and long-term car projects. He retired in 1952, before the merger of Morris Motors, Ltd., with Austin Motor Company, Ltd.

Oberhaus, Herbert (1929–)

Director of product development and engineering, Adam Opel AG from 1985 to 1991.

He was born on December 12, 1929, at Dienheim in the Rhineland-Lalatinate and joined Adam Opel AG in 1949 as an apprentice mechanic in the Rüsselsheim works. The company sponsored his engineering studies at Darmstadt Technical University and he returned to Opel after graduation. He was a member of the engineering team that prepared the chassis for the 1963 models, including the all-new Kadett. He developed the chassis for the Opel GT from the Kadett platform, and led the chassis engineering for the 1970 Ascona and 1971 Manta, with coil-spring suspension front and rear. He was instrumental in designing the de Dion–type rear suspension for the 1969 Admiral and Diplomat and designed the chassis for the 1972 Rekord. In 1976 he was named executive engineer for chassis and took charge of suspension, steering and brake systems for the 1982 Corsa and 1986 Omega. He was responsible for putting all-independent suspension on the Senator and Monza in 1978, and was named chief engineer of chassis and powertrain engineering in 1980. He directed the entire Vectra program, beginning in 1985 with production start-up in August 1988. He resigned on June 30, 1991, in protest against being passed over for promotion to technical director when Fritz Lohr retired, moved to America and set up a successful consulting practice.

Oberle, Georges (1933–)

Chief engineer of complete vehicle design, Automobiles Peugeot from 1977 to 1981; chief engineer of Peugeot top-of-the-line architecture from

1981 to 1986; chief engineer of Peugeot future car design from 1986 to 1989.

He was born in June 1933 in Strasbourg and graduated from the Ecole Nationale Supérieure des Arts et Industries in Strasbourg in 1955. As part of his studies he had been a trainee with Automobiles Bugatti from 1953 to 1955 and had some association with the design of the Type 251 Grand Prix racing car project. He joined Automobiles Peugeot at the Sochaux plant in September 1955, but was called up for military service in the French air force in 1956, returning to Sochaux in 1957. He was assigned to the vehicle layout office in 1959 and worked on the space allocation and disposition of the mechanical elements in the Peugeot 505, 305, 104 and 604. In 1977 he was given responsibility for liaison with Pininfarina over the layout of the 405 sedan. He worked on experimental car projects (Vera), preliminary studies for top-line Peugeot and Citroën projects, and aborted Talbot (ex-Chrysler) car concepts from 1981 to 1983. He began tackling the layout of the Peugeot 605, originally planned for rear-wheel-drive, in 1984, and rearranged for front-wheel-drive at a later stage. He retired in 1989.

Obländer, Kurt (1927–)

Director of Mercedes-Benz emission-control and engine testing from 1971 to 1985; director of Mercedes-Benz passenger-car testing from 1985 to 1988; director of all car-engine development from 1988 to 1991.

He was born in 1927 in Karlsruhe. In his youth he had ambitions to work in botanical and zoological research, and started out as a forestry pupil, but during World War II his class was canceled and he was trained as an auto mechanic. After the war, he studied mechanical engineering at the Karlsruhe Technical University. He joined Daimler-Benz AG in 1955 as a test engineer in the engine department at Untertürkheim and assisted in the development of the six-cylinder engines for the 300 SE, 230 SL, and the V-8 for the 600 and 350 SL. In 1964 he was granted a patent for a cylinder head design with four valves per cylinder. He perfected electronic ignition systems and electronic fuel injection for Mercedes-Benz car engines and matched them with new manual and automatic transmission systems. His career culminated with the development of the V-12 engine for the 1991 Mercedes-Benz 600. He retired in July 1991.

Oldfield, John (1937–)

Head of car product planning, Ford of Europe, Inc., from 1981 to 1986; product development program director, Ford of Europe, Inc., from 1986 to 1989; vice president of product development, Ford of Europe, Inc., from 1989 to 1993; vice president and director of car engineering and development, Ford of Europe, Inc., from 1993 to 1994; executive chairman of Aston Martin Lagonda, Ltd., from 1994 to 1997.

He was born in 1937 and held an engineering degree. He joined Ford of Britain in 1958 as a draftsman and was promoted to design engineer in 1964. His experience included some years with Ford do Brasil, developing the Corcel and local and export versions of the Escort. Upon his return to Europe, he worked in engineering management and was placed in charge of Ford of Europe, Inc.'s, product planning in 1981. He planned the Orion mainly as a replacement for the low-range Taunus models and the Sierra as a successor to the higher-grade Taunus/Cortina cars. The 1985 Scorpio was a step into a higher bracket, from which Ford had been absent since the Granada was taken out of production. He led the Mondeo program in 1990–92, a concept intended for worldwide production, with front-wheel drive, to replace the rear-wheel drive Sierra. He was then placed in charge of Aston Martin Lagonda, Ltd., where he did much to assure smooth cooperation with Jaguar. He retired in 1997.

Opel, Carl (1859–1927)

Managing director of Adam Opel KG from 1896 to 1927.

He was born in Rüsselsheim in 1869 as the oldest son of Adam Opel (1837– 1895) who had opened a sewing-machine factory in 1862 and started manufacturing bicycles in 1885. He served an apprenticeship in the Opel factories which turned out 2000 bicycles a year by 1890.

The Opel factory built 11 cars under Lutzmann license in 1898. Output increased year by year and reached 30 cars in 1901. Opel also made motorcycles from 1901 until 1907. In October 1901 Carl went to Paris for the Salon de l'Automobile, accompanied by his brothers Wilhelm and Friedrich, to look for a more modern vehicle than Lutzmann's. They signed a contract with Darracq and in 1902 began importing complete Darracq chassis which were fitted with Opel coachwork. Another brother, Heinrich (1873–1928), organized a dealer network for Opel cars in Germany. Opel began making engines and chassis of in-house design on a modest scale in 1902 while maintaining Opel-Darracq production up to 1906. Total output climbed from 178 cars in 1903 to 358 cars in 1905. Truck production was taken up in 1907.

On August 20, 1911, the factories were severely damaged by fire. The sewing-machine and bicycle shops were completely destroyed, and stockpiles of finished products, 2000 bicycles and 3000 sewing machines, were lost. A stock of 600 chassis was saved but covered in soot. The truck shop escaped with minor damage and Opel turned out 2251 motor vehicles in 1911. Sewing-machine production was discontinued, but the bicycle shops were rebuilt. In 1912 the company had a work force of 4500 and produced its ten-thousandth car. Towards the end of 1914 most of the factory was converted into giant repair shops for cars and trucks needed in the war. Carl was exempt from military service, since his leadership was essential for the proper running of the plant. Opel built 4400 three-ton trucks during the war, plus a number of military tractors and aircraft engines. In December 1918 Rüsselsheim was occupied by French troops seeking reparations, who removed a quantity of machine tools and material stores.

After 1921 Carl reduced his work place and stepped back from day-to-day executive duties but never retired. He died in 1927.

Opel, Friedrich "Fritz" (1875–1938)

Chief engineer of Opel from 1902 to 1929.

He was born on April 20, 1875, in Rüsselsheim as the fourth son of Adam Opel. Like his older brothers, he was a keen cyclist and served his apprenticeship with the Opel factories. He studied mechanical engineering at the Darmstadt Technical University (which awarded him an honorary doctorate in 1921).

He took charge of creating Opel's own car designs in 1902, though the two-cylinder 10/12 with shaft drive had a lot in common with the contemporary Darracq. Due to its success, he added a 12/14 derivative model later in the year. In 1903 he designed the first four-cylinder Opel, a 10/22, and the following year he designed a 120-hp racing car. He entered the Gordon Bennett Cup race in 1904, driving an 80-hp four-cylinder Opel, but retired with a broken crankshaft. Yet Opel cars won 30 other events that year, and in 1905 Fritz finished fifth overall on a 35-hp Opel in the Herkomer Trial. Opel made its last two-cylinder car, an 8/14, in 1908 and lined up a range of reasonably priced four-cylinder models: 10/16, 10/18 and 14/22 PS.

In 1909 he introduced the little 4/8 with an 1100-cc four-cylinder engine, which became popularly known as the Doktorwagen since it was a favorite of the medical profession in rural areas.

In 1910 he pioneered the "modular" car, with a set of related prefabricated bodies suitable for several different chassis with different engines. A year later he adopted the one-piece cast-iron cylinder block for the 6/16 and in 1912 he designed the 40/150 racing car with a 16-valve four-cylinder engine. That year Opel also won the team prize in the Austrian Alpine Trial, and the 5/12 replaced the 4/8 "Doktorwagen" which evolved into a 5/14 in 1914. The first six-cylinder Opel appeared in 1916 in the guise of a 4.7-liter 18/50 which soon led to two smaller companion models. In 1917 he ordered the construction of a test track which opened in 1919.

The postwar program began with the launch of the 8/25 in 1919, a medium-size touring car which remained in production until 1924. He added the 3.4-liter 14/38 in 1920 and the six-cylinder 21/50 in 1921. His brother Wilhelm kept asking for a simple economy car, and Fritz assigned it to an assistant named Hanau. What Hanau did first was to go out and buy a Citroën 5 CV Type C. The 1924 Opel 4/12 ended up with a too-obvious resemblance to the popular French model, and Citroën sued. Opel lost the litigation in a Berlin court in May 1926 but appealed and won the case in June 1927. Opel's lawyers pointed to a number of basic differences in the original design, and by 1927, the 4/12 had been modified with a longer wheelbase (and a higher price).

The 4/12 had been planned for a production rate of 25 cars a day, but the demand was so overwhelming that Rüsselsheim was soon building 120 a day. All models were periodically modernized, and on the 1925 Type 80 with a 10/45 engine, Opel also introduced four-wheel brakes. Hydraulic brakes were adopted for the Regent 24/110 in 1928, Opel's first eight-cylinder production model.

After selling the Opel car company to General Motors, the surviving Opel brothers were extremely rich. Fritz and Wilhelm invested wisely and took a majority stake in Continental Gummiwerke AG of Hanover. From 1932 to 1938 Fritz sat on the supervisory boards of both Continental and Adam Opel AG. When GM invited him and his assistant Gotthelf Paulus to Detroit for an extended visit in 1929, he may have expected to continue in his functions at Opel, but within months, GM had appointed Charles S. Crawford as Opel's new chief engineer.

Fritz died in 1938.

Opel, Wilhelm (1871–1948)

Production manger of Adam Opel KG from 1896 to 1901; director of production from 1901 to 1929.

He was born on May 15, 1871, in Rüsselsheim as the second son of Adam Opel. After local grade school and an apprenticeship in the Opel factories, he studied for an engineering degree at the Darmstadt Technical University, graduating in 1895. All the Opel brothers were keen cyclists and Wilhelm later distinguished himself as a rally driver. He was the overall winner of the 1909 Prince Henry Trials in a 2.6-liter Opel 10/20 PS car, an event where Opel cars filled five of the top ten places.

Late in 1909 he went to the United States, visiting industrial companies in Chicago, St. Louis and San Francisco, gaining a better understanding of mass production and rational manufacturing methods. On his return to Rüsselsheim, he tried to simplify the model range and speed up production. Opel had built 845 cars in 1909. The figure increased to 1615 in 1910 and 3202 in 1912.

He was called up for military service in 1914 but discharged a year later because the factory needed him for production vital to the war effort. Along with his older brother Carl, he was ennobled in 1917, becoming *von* Opel. After 1921 he emerged as the de facto chief executive officer of Adam Opel KG, having first organized Opel Brothers GmbH in 1919 (in partnership with his brothers Carl, Heinrich and Friedrich) as a holding for their industrial interests.

In the economic crisis that hit Germany in 1922–23, Opel met the payroll by printing its own money which was eagerly accepted as local currency in Rüsselsheim and beyond. Still, the failing market forced him to close the factories from July to December 1923. Up to 1925 the German market was protected by a virtual ban on imported cars, unless locally assembled. When the ban was lifted, he was worried about the risk of the market being flooded by low-priced American cars, and in August 1926, he wrote his first letter to General Motors, proposing various plans of cooperation. It landed on the desk of James D. Mooney, president of GM Export Company, and started a round of negotiations which led Opel Brothers GmbH to convert Adam Opel KG[1] into Adam Opel AG,[2] and selling 80 percent of the shares to GM for $25,967 in April 1929, with an option on the remaining 20 percent for $7395 valid until October 1931. It was taken up at that time.

In 1928 Opel was making 250 cars a day, with 8000 workers, and had a home-market share of 27.5 percent, by far Germany's biggest car company. Also in 1929 Opel took over the Elite Diamant Werke of Brand-Erbisdorf in Saxony and continued its motorcycle production up to 1933. After making 2½ million bicycles, Opel sold its bicycle operations to NSU in 1937.

Wilhelm von Opel held a seat on Opel's supervisory board under GM ownership until his retirement. He died in 1948.

1. KG = Kommanditgesellschaft = limited-liability partnership.
2. AG = Aktiengesellschaft = joint stock corporation.

Orsi, Omer (1918–1980)

Managing director of Maserati from 1948 to 1968.

He was born on May 12, 1918, the son of Adolfo Orsi (1888–1963), who had made his fortune in the scrap-iron business and ended up as a steel-works owner and one of Modena's biggest real-estate investors. The Orsi family also owned plants manufacturing farm implements and machinery and held concessions on streetcar lines and public transit in and around Modena.

Adolfo Orsi had also undertaken at an early date to help the Maserati brothers financially, and in 1937 he stepped discreetly into Maserati affairs with a ten-year commitment that would lead to full ownership of the enterprise. In October 1939, the Maserati factory was moved from Bologna to Orsi-owned buildings in Via Ciro Menotti, Modena. At that time, Maserati had two activities. One was making racing cars; the other was making spark plugs and batteries. The latter was maintained during World War II, and instead of racing cars, the company made machine tools and battery-driven three-wheel delivery vehicles. After the war, the racing team was revived, and Omer Orsi began a new policy of making Maserati touring cars, though on a very minor scale.

According to the terms of the contract, the Maserati brothers left in 1947 and returned to Bologna, and Orsi engaged A. Massimino to take charge of engineering.

In 1948 Omer Orsi introduced modern management at all levels. The machine tools became an autonomous department. Osvaldo Gorrini was hired away from Fiat's tractor division to run the day-to-day affairs of Maserati cars. Orsi originated a number of new programs, and in 1949 tried to get back into the commercial-vehicle market with the BC 20, a light cab-over-engine truck with alternative power systems, a two-stroke 547-cc spark-ignition engine or electric drive with a compound motor. It came to nothing, and he set new objectives and priorities with cars at the core. Yet from 1951 through 1957 Maserati built only 75 touring cars.

Maserati fell into receivership on April 1, 1950, and Adolfo Orsi kept it going only by selling the profitable machine tool branch and sacrificing big parts of his personal fortune. A big car, the 3500 GT, was ready for production in 1957, and the factory tooled up to build 20 cars a month. The cost put Maserati in receivership again in April 1958, but as sales picked up and the 5000 GT was added, Orsi was able to pay the creditors in full, and the receivership was lifted in May 1959. From 1957 to 1964, Maserati produced 1983 cars of 3500 GT and GTI specification, plus 245 3500 "spiders."

Maserati introduced its first four-door production model in 1963 and followed up with exciting new sports cars: the Sebring, Mistral, Mexico, Ghibli, and Indy.

In March 1967, Pierre Bercot approached Orsi about making Maserati engines for installation in future Citroën models, and an agreement was signed. In May 1968, Orsi sold a 60 percent stake in Maserati to Citroën. Under the management of Citroën's delegate, Guy Malleret, the work force at Via Ciro Menotti tripled, but production growth was negligible. In 1970 Citroën began producing the SM, with a Maserati V-6 engine. Maserati began building the Bora and the Merak. Orsi was shocked when Citroën announced the liquidation of Maserati on May 23, 1975, and withdrew from business. He died on June 6, 1980.

Osswald, Bernhard (1912–)

Technical manager and product planning director of BMW AG from 1965 to 1968; technical director of BMW AG from 1968 to 1975.

He was born in 1912 and held diplomas in mechanical engineering, aeronautical engineering, and a pilot's license. He undertook postgraduate studies in the laboratories of Professor W.A. Kamm at the Stuttgart Technical University, developed a rotary engine, and conducted experiments with liquid oxygen as a diesel-fuel additive. He began his career with the Jenbacher Werke in Salzburg, Austria, where he designed a small truck with short wheelbase and the engine between the seats. In 1952 Ford-Werke AG engaged him as head of chassis engineering and transmissions. He made gear-shifting easier on a number of Ford models and became known as a methodical engineer of great personal modesty. Ford placed him in charge of engine design and development in 1959. He led the design of the 60° V-6 for the 1964 Taunus 20M, and left Ford to accept an invitation from BMW AG. He led the teams that developed the 2002, 2000 tii and 2002 Turbo, and directed the design of the six-cylinder "Bavaria" sedans, 2500 and 2800, launched in 1968. He held overall responsibility for the 3.0 S and all its derivatives including the 3.3 L of 1974. He retired at the end of 1975.

Palmer, Gerald Marley (1911–1999)

Chief designer of Jowett Cars, Ltd., from 1942 to 1949; chief engineer of British Motor Corporation from 1952 to 1955; assistant chief engineer of Vauxhall Motors, Ltd., from 1955 to 1966; executive engineer of Vauxhall Motors, Ltd., from 1966 to 1972.

He was born on January 30, 1911, in Rhodesia as the son of a district engineer with the Beira, Mashonaland, and Rhodesia Railway. After a general education in local schools, he sailed to England in 1927 and served an apprenticeship with Scammell Lorries, Ltd., at Watford. He also took evening classes in engineering and received a degree from the University of London in 1931. Serving as a member of Scammell's engineering staff, he privately undertook the design of a sports car for A.G.A. Fisher, who set up the Deroy Car Company at Penge, Surrey. Only one Deroy roadster was built. When it was demonstrated to the leaders of the MG Car Company, Ltd., Palmer was invited to join Morris Motors, Ltd., as a draftsman. He was quickly promoted and was the chief designer of the 1938 MG Y-series sedan. In September 1939, he was called upon to convert a Morris plant to a production line for Tiger Moth airplanes. When he arrived at Bradford, Yorkshire, in 1942, he was briefed to create a modern postwar car. The Jowett Javelin fastback sedan, with its 1½-liter flat-four engine in the front overhang and rear-axle drive, was ready for road-testing before the end of 1944. Production began in 1945. He returned to Morris Motors in 1949 and designed the MG Magnette, Wolseley 4/44, Riley Pathfinder and Wolseley 6/90. He proposed an updated Morris Minor which instead was chosen as the basis for the 1957 Riley 1.5 and Wolseley 1500. With his mind strictly concentrated on the products, he neglected the administrative and executive duties that went with the chief engineer's title, and L.P. Lord sacked him in 1955. He was welcomed by Vauxhall Motors, Ltd., at Luton, Bedfordshire, and was associated with the design of the 1957 Victor. He worked on the Viva, the Viva GT, the VX 4/90, Ventora, and the Firenza. He retired in 1972 and died on June 23, 1999.

Panhard, Adrien Hippolyte François (1870–1955)

Chairman of Société Anonyme des Anciens Etablissements Panhard & Levassor from 1916 to 1946.

He was born on October 5, 1870, at Hyères on the French Riviera as the son of René Panhard. He attended school in Paris and took a keen interest in the various activities of his father's factory. In July 1891 he accompanied Emile Levassor on a drive from Ivry to the Normandy coast, and in March 1893, he completed a test-and-promotion tour from Paris to Nice and back. He was elected to the board[1] of Société Anonyme Panhard & Levassor in 1897 and proved a tireless worker. With his fellow director René de Knyff he provided an ample budget and technical support for the racing team, which they both regarded as the most effective means of advertising. At René de Knyff's investigation, he also secured a license for the Knight sleeve-valve engine in 1909. That same year, the factory made its first aircraft engines. He was mobilized as a captain in the army reserves in 1914, becoming head of a motor transport group stationed near Paris. He was discharged in 1916 to take over the duties of chairman of the company's board.

After 1920 he set the major guidelines for company policy and had the final word in all important decisions, but did not interfere in day-to-day affairs. He retired in 1946 and died in 1955.

Panhard, Jean Joseph Léon (1913–)

Technical director of the Société Anonyme des Anciens Etablissements Panhard & Levassor from 1937 to 1948; secretary and general manager in 1948–49; joint managing director from 1949 to 1955; managing director in 1955–56; managing director and vice president from 1957 to 1965.

He was born on June 12, 1913, in Paris as the son of Paul Panhard. After a general education, he was accepted at the Ecole Polytechnique and graduated in 1937. He was the family's first engineer, which they celebrated by naming him technical director. From 1940 to 1944 he was at Tarbes with the engineering staff, and on his return to Paris took charge of the Dyna economy-car program. The government's "Pons Plan" had grouped the company with two truck builders, and allowed Panhard to produce a limited number of cars and light-duty trucks, while Somua was to make trucks in the 5-to-10 ton range and Willème, trucks over 10 tons. The plan proved unworkable and collapsed in 1949. He switched from sleeve-valves to poppet valves in the diesel engines, and maintained truck production up to 1964. He signed the first contract for cooperation

with Citroën in 1955 and undertook assembly of the Citroën 2 CV van at Avenue d'Ivry. The Citroën takeover was completed in 1965, and he became president and managing director of Société des Constructions Mécaniques Panhard & Levassor, formed to take over the production of military vehicles. He retired in 1983 but remained active in business associations and managed the State's acquisition of the Schlumpf collection, which he turned into France's National Automobile Museum in Mulhouse.

Panhard, Joseph Paul René (1881–1969)

Managing director of Société des Anciens Etablissements Panhard & Levassor from 1916 to 1951.

He was born on August 1, 1881, in Versailles as the son of a State Council lawyer, Léon Panhard, and nephew of René Panhard. He received a business education and joined the family firm in Avenue d'Ivry in 1906, working in different departments until he was named sales director in 1913. At the start of World War I he was mobilized as a driver for the General Staff in Paris, but released in 1916 to return to the factory. Among its wartime products were machine tools, shells, trucks, artillery tractors, and V-12 aircraft engines. Civilian production resumed in 1919 with 10 CV and 16 CV four-cylinder cars. During the next decade, the passenger-car model range included four-, six- and eight-cylinder units, all with sleeve-valve engines. Panhard trucks and buses also had sleeve-valve engines, and when diesel-engine production was taken up in 1930, they were also of the sleeve-valve type. The engineering office was removed to Tarbes in the Pyrenées in 1940, where the agenda was filled with postwar projects, while the factories at Ivry were given orders by the Occupation forces. German orders were so badly filled that in 1945, Paul Panhard was awarded the Gold Medal of the French *Résistance* for his wartime record. He relinquished his title in 1951 but kept coming to his office every day until his death in March 1969.

Panhard, Louis François René (1841–1908)

Chairman of Société Panhard & Levassor (later SA Panhard & Levassor) from 1886 to 1908.

He was born on May 27, 1841, in Paris, into a family of Savoyard origins. He was the son of Adrien Panhard, a coachbuilder with the trademark "Prieur," who also kept a fleet of carriages for rent. He graduated from the Ecole Centrale des Arts et Manufactures in Paris in 1864. Three years later he joined J.L. Périn, manufacturer of bandsaws in rue du Faubourg Saint-Antoine in Paris. At his initiative they also began making metal-cutting tools and machines. The business was dormant from 1870 to 1873, due to war and civil uprising. He moved the factory to Avenue d'Ivry in 1873, greatly assisted by his new associate, Emile Levassor, and set up two separate departments, one for woodworking machinery, the second for metalworking tools and mechanical engineering. He secured a license for Deutz gas engines and added a third department to the factory. In 1887 Panhard & Levassor received an order for a Daimler engine destined for motorboat installation, which led to a Daimler license and engine production for Peugeot and other customers. They began making their own cars in 1890 and built their first truck in 1897. He maintained the other departments, though engines and automobiles became dominant in 1902. The first heavy trucks were made in 1906. He died on July 16, 1908, at La Bourboule while undergoing health treatment.

Parayre, Jean-Paul (1937–)

Chairman of Peugeot SA from 1977 to 1984.

He was born in 1937 in Paris as the son of a naval engineer, Louis Parayre. He graduated from the Ecole Polytechnique with a degree in bridge and highways engineering and worked on motorway projects from 1963 to 1967. He remained in the Ministry of Equipment from 1967 to 1970, when he was given a position in the Ministry of Industrial and Scientific Development. He had his first taste of the automobile business when appointed to the administrative council of the Régie Nationale des Usines Renault in 1971. In 1974 François Gautier engaged him at director's level in the Peugeot management, and he was a member of the team that negotiated Peugeot's takeover of Citroën later that year. In 1976 he became director of Automobiles Peugeot, and a year later, chairman of the parent company. In 1978 he paid $200 million plus a 14.4 percent stake in the Peugeot SA capital for Chrysler's French (ex–Simca), Spanish (ex–Barreiros) and British (ex–Rootes) branches. The Chrysler models from the Poissy plant were renamed Talbot in 1979. He saw holes in the model lineup of both Citroën and Peugeot, pushed the Citroën BX into production late in 1982 and the Peugeot 205 towards the end of 1983. He also wanted a new engine plant that would produce power units for all three makes, and inaugurated the Trémery works in Lorraine in June 1982. In August 1984 he came into conflict with "new broom" Jacques Calvet, who blamed him for the high cost of integrating the ex–Chrysler operations with Peugeot's, and he resigned in September 1984. He joined Dumez SA, a leading construction firm where his uncle, André Chaufour, was chairman of the supervisory board, and became president of Dumez SA in 1988. He resigned in July 1992, and in January 1994 was appointed president of Bolloré Technologies. He was in charge of Bolloré's Albatros holdings from 1994 to 1996, when he became president of Saga, a joint venture between Bolloré Technologies and CMB, a Belgo–South African shipping group.

Park, Alex (1926–)

Chief executive officer of British Leyland from 1975 to 1977.

He was born on November 16, 1926, in Yorkshire and held a degree in mechanical engineering from Constantine College. He went in for postgraduate studies in accounting, finance and prognosis, and began his career with Monsanto Chemicals as a planning engineer in 1957. He left Monsanto in 1967 to become finance director of Cummins Engines, which had two manufacturing plants in Britain. Within a year, however, he deserted Cummins to become finance director of Rank-Xerox, where he was promoted to managing director in 1972. He joined British Leyland in January 1974, replacing John Barber as finance director, and in April 1975 became chief executive. Jaguar introduced the XJS V-12 in September 1975, and during 1976 Triumph began production of the TR7 at Speke Road and Rover started building the SD1 (3500) at Solihull. The Austin Allegro had been on the market since 1973 and the Austin Princess arrived in 1976. But the group's accounting system was so clumsily organized that Park could never find out which cars were making a profit, and vice versa. British Leyland soon lost its leadership position in the British market. As late as 1975 BL was selling three cars to every two Fords. Two years later, the situation was reversed. He resigned in December 1977. In February 1978 he joined Lonrho[1] as head of a subsidiary that imported VW cars and commercial vehicles to the United Kingdom.

1. London & Rhodesia Mining and Trading Co., Ltd.

Parkes, Michael Johnson "Mike" (1931–1977)

Design and development engineer with the Rootes Group from 1956 to 1961; test and development engineer with Ferrari from 1962 to 1968; consultant to Ferrari in 1969–70; technical director of Lancia from 1973 to 1977.

He was born on December 24, 1931, in Richmond, Surrey, as the son of J.J. Parkes who later became chairman of Alvis, Ltd. He graduated from Haileybury School in 1949 and began a five-year apprenticeship with the Rootes Group. He was a co-designer of a minicar, known literally as the Slug, with a rear-mounted air-cooled flat-twin. The concept ripened into the Apex prototype of 1958, still with an air-cooled flat-twin. It was redesigned with an all-aluminum water-cooled overhead-camshaft four-cylinder engine, and went into production as the Hillman Imp in 1963. He linked up with the Ferrari racing

team while testing a Sunbeam at Le Mans in 1961 and moved to Maranello in 1962. He was a superb driver and helped develop the 250/LM, the 330/P3, and 330/P4. In 1970 he joined Scuderia Filipinetti as head of the group's workshops in Modena, and after the death of Georges Filipinetti, Parkes began collaborating with Alejandro De Tomaso, mainly to develop a racing version of the Pantera. Pier Ugo Gobbato engaged him in 1973 to take charge of developing Lancia's high-performance and racing cars. He was responsible for the Stratos HF and the Stratos Turbo, which won the world rally championship in three consecutive years (1974-75-76). His career came to a brutal stop when he was killed in a freak traffic accident at Moncalieri on August 29, 1977.

Parnell, Frank Gordon (b. 1892)

Chief engineer of the Triumph Cycle Company, Ltd., from 1923 to 1929; chief engineer of the Rootes Group's light car section from 1930 to 1932.

He was born on June 28, 1892, in London and educated at the City of London School and Northampton Institute. He began his career as a draftsman with Clement Talbot Ltd. in 1912, worked for Straker-Squire in 1913–14 and Arrol-Johnston in 1914–15. During World War I he was attached to the Royal Aircraft Factory, where he conducted experiments with supercharging aircraft engines. In 1919 he signed up with Ricardo & Co., Ltd., consulting engineers, and was sent on a mission to the United States in 1920–21. He became familiar with Lockheed hydraulic brakes during his stay in America. On his return to Britain, Ricardo had been engaged to design a small four-cylinder engine for Triumph, and in 1923 Siegfried Bettmann engaged him to design new cars for Triumph. He created the 12.8-hp model in 1924, and put Lockheed hydraulic brakes on the front wheels. Within two years all Triumph models were equipped with four-wheel hydraulic brakes. In 1926 he brought out the 15/50, and the following year the Super Seven was launched with an 832-cc four-cylinder engine, not a Ricardo design but drawn up by A.A. Sykes in Triumph's own drawing office. He left Triumph in 1929 and joined the Rootes Group, where he assisted Norman Wishart in the design of the original Hillman Minx in 1931. But he resigned in 1932 to become a sales engineer for Automotive Products, which held the British rights to the Lockheed patents, and spent the rest of his career with that group. In 1947 he was a specialist on hydraulic power systems for heavy trucks, and in 1955 he delivered a paper on disc brakes to the Institution of Mechanical Engineers.

Parry-Jones, Richard (1951–)

Vice president of car development, Ford of Europe, from 1994 to 1997; vice president of product development, Ford of Europe, from 1997 to 2000; Ford Motor Company vice president in charge of global product development since 2000.

He was born in 1951 in Wales and educated at Salford University. He joined Ford in 1969 and worked in product planning, engineering and manufacturing positions in Europe and the United States. He led the development of the 1981 Escort and Sierra, and became manager of Ford of Europe's small-car programs in 1982. From 1985 to 1988 he served as executive engineer of Ford's technological research, and then transferred to Dearborn as director of vehicle concept engineering. He returned to Europe in 1990 as director of manufacturing operations at Ford-Werke AG's production center at Cologne, and served as chief engineer, Ford of Europe, from 1991 to 1994. He drew up the master plan for model renewals and outlined conceptual changes for generations of Fiesta, Escort, and Mondeo cars. He went to Dearborn in 2000 to take charge of worldwide product development, but returned to Britain in November 2001 to take on the additional duties of product engineering and development for Ford's Premier Automotive Group (Volvo, Land Rover, Jaguar, Aston Martin).

Parry Thomas, John Godfrey (1885–1927)

Chief engineer of Leyland Motors, Ltd., from 1917 to 1923.

He was born in 1885 at Wrexham as the son of a vicar and attended the City & Guilds Engineering College in London. He began his career with the British branch of Siemens as a trainee in electrical engineering, and then joined Clayton & Shuttleworth, makers of steam engines and steam traction engines. His association with Leyland dated from 1909, when Leyland agreed to build and test an electric transmission he had invented. He set up his own engineering office in Kensington High Street, London, and received design contracts for such projects as a 200-hp railcar for New Zealand and a military road-train for Armstrong-Whitworth. In 1915 he designed the X-8 "Maltese Cross" engine, in which Leyland professed an interest but never acted upon. He joined Leyland in 1917 to design aircraft engines and in 1919 submitted designs for a luxury car that Leyland agreed to put in production. The Leyland Eight was an extravagant design with triple-eccentric drive to a single overhead camshaft, hemispherical combustion chambers, tubular steel connecting rods, and laminated leaf springs to close the valves. The initial version had 6920 cc displacement, later increased to 7266 cc. Perhaps 20 of them were actually completed. He resigned from Leyland because the top management had decided to take over the Croydon factory that built the Trojan, an economy car of odd specifications, for which he expressed strong disapproval.

He set up a private design office in "The Hermitage" at Brooklands race track, and from 1924 to 1927 he worked closely with Thomson & Taylor, where among other tasks, he assisted Reid Railton in the development of the Brooklands edition of the Riley Nine. He also created a speed-record car he named "Babs" by rebuilding Count Louis Vorov Zborowski's Higham Special with its 27-liter Liberty V-12 aircraft engine, Benz gearbox, and chain drive. He was killed during a speed-record attempt on Pendine Sands on March 3, 1927, when a driving chain broke and the loose end smashed his skull.

Perlo, Giuseppe (1945–)

Chief platform manager for the Lancia Thema and Fiat Croma from 1981 to 1985; vice president of products for Fiat Auto S.p.A. from 1990 to 2001; vice president of corporate development, Fiat Auto S.p.A. beginning in 2001.

He was born in 1945 in Turin and joined Fiat at the age of 16 as a trainee. Fiat sent him to study economics and business administration at the Massachusetts Institute of Technology, and gave him various assignments in sales and marketing when he returned to Turin. He became marketing manager for Fiat Auto, and in 1981 moved into the product area with responsibility for the group's biggest cars. They were laid out by Mario Maioli, engineered by Giovanni Canavese, styled by Giorgetto Giugiaro, and developed by Bruno Cena. They were assembled at Mirafiori and had long production lives, the Thema being phased out in 1994 and the Croma in 1997. His promotion in 1990 gave him responsibility for product planning, definition, price and volume targets. The Fiat 176 (Punto), planned as a replacement for the 156 (Uno), was then well into its final development stage, and went into production at the Melfi plant in 1993. But he led the 182 projects (Bravo/Brava and Marea) from the inception. They were compact and mid-size family cars, vital to Fiat's place in Italy and many export markets. None of them came close to meeting their sales targets and for years the Cassino plant was running below capacity. He was removed from products in May 2001.

Peronnin, Jean (1924–)

Director of industrial operations, Chrysler-France from 1970 to 1976; executive director of Chrysler Industrial Operations Europe from 1976 to 1978; managing director of Chrysler-France in 1978–79.

He was born on May 13, 1924, at Commentry (Allier) and held an engineering degree from the Ecole des Arts et Métiers of Cluny. He joined Simca in May 1958 as head of the stamping plant and body assembly at Poissy, making body shells for the Simca Aronde, Ariane and Chambord. He was promoted to director of methods, installations and plant construction in 1965, and became director of the entire Poissy works in 1968. He raised production from 1400 cars a day in 1968 to 2100 by 1970. In 1970 he took charge of all Simca production, supervised the implementation of the decentralization plan and set up parts factories in the provinces. He directed the entire production sequence for the Simca 1307/1308 and the Horizon. In 1976 he was given responsibility also for production at Chrysler UK (ex–Rootes) and Chrysler-España (ex–Barreiros), including purchasing, standardization, manufacturing engineering, production control and quality. On February 24, 1978, he became managing director of Chrysler-France and a vice president and administrator of Chrysler International SA in Geneva. Just months later, Chrysler Corporation sold its European subsidiaries to Peugeot SA, and Jean-Paul Parayre tried to accommodate him in the PSA organization, but it did not work out. It was a younger ex–Simca engineer, Christian Cardoux, who was given the task of coordinating the ten major plants of Peugeot, Citroën and Chrysler-France.

Perrett, John Bertram (1912–)

Chief designer of AFN Limited from 1945 to 1952; technical manager of Trojan from 1952 to 1965.

He was born on February 20, 1912, in London, and educated at the Kingston Technical College and the Regent Street Polytechnic. After serving an apprenticeship with G. Wailes & Company, he worked as a draftsman with Thomson Taylor of Brooklands, Byfleet, Surrey, constructors of speed-record cars and closely involved with the racing activities of Clement Talbot Ltd. He joined the MG Car Company, Ltd., as a design engineer in 1933 and moved on to Luton in 1936 as a member of Vauxhall's engineering staff. In 1939 he was named chief engineer of Robot Gears Limited, and during World War II served as a design engineer with Vickers-Armstrong at the Fighting Vehicle Research Establishment. He joined AFN Limited at Isleworth, Middlesex, in 1945 and was responsible for a new generation of Frazer Nash cars, with a lot of guidance from Fritz Fiedler, who was also acting as a consultant to the Bristol Aeroplane Company, Ltd. He went to Trojan, Ltd., in Croydon in 1952, just after the company had introduced a new van powered by Perkins diesel engines. In 1960 he was placed in charge of producing the Heinkel Kabinenroller, marketed as the Trojan 200. He retired in 1965.

Perrot, Henri (1883–1960)

Head of the drawing office, Société des Automobiles Brasier, from 1906 to 1909; chief engineer of Argyll Motors, Ltd., from 1909 to 1914.

He was born on August 21, 1883, in Paris and held an engineering diploma from the Ecole des Arts et Métiers of Chalons-sur-Marne. He joined Brasier as a machine setter and was quickly promoted to tool designer, draftsman, and automobile designer. He was partly responsible for the Richard-Brasier racing cars that won the Gordon Bennett Cup in 1904 and 1905. In 1909 he moved to Glasgow and designed a new range of Argyll cars. He filed a patent application for a mechanical front wheel brake system on March 18, 1910, and put his system on the 1911 Argyll 12 hp car. He also sold licenses to Delage and Peugeot, who put Perrot front-wheel brakes on their 1914 racing cars. He returned to Paris in 1914 and was drafted into the French Army as a corporal. But he never went to the front, since he was taken out of the ranks and detached to Lorraine-Diétrich to work in aircraft-engine production. At the end of the war, he opened an office as an independent engineering consultant in Paris, and Perrot brakes were adopted by Hispano-Suiza in 1919, Talbot in 1920, and Renault in 1921. Delage began fitting Perrot brakes on its 1919 production cars. Perrot also patented a servo-brake mechanism in 1919, and the following year established Perrot Freins in partnership with Gérard Piganeau, and by 1925 Perrot brakes were fitted by more than 50 automakers. In 1923 he was a partner with Vincent Bendix in starting Bendix Brake Corporation, but sold his part to Bendix a year later. He collected royalties from Bendix until the end of 1948. Louis Coatalen engaged him as a consulting engineer for Lockheed and he began experimenting with disc brakes. About 1953

he retired to Monte Carlo, where he died on September 22, 1960.

Perry, Percival Lea Dewhurst (1878–1957)

Managing director of the Ford Motor Company (England), Ltd., from 1911 to 1919; chairman and managing director of the Ford Motor Company, Ltd., from 1928 to 1948.

He was born in 1878 in Bristol and grew up in Birmingham. He was educated at King Edward's School, Birmingham, and only the family's poverty kept him from going to college to study law. Instead he went to London, answered a "help wanted" advertisement, and was hired by H.J. Lawson in 1896. But a year later he left the motor business for his uncle's printing shop at Kingston-upon-Hull, where he spent four years. Returning to London, he met a man who held the Ford franchise for Europe. They founded the American Motor Car Company in 1904, changed it to Central Motor Car Company in 1905, and in 1907 to Perry, Thornton & Schreiber. The partners quarreled and Perry resigned. In 1908 he visited Henry Ford in Dearborn, returning with a Ford-backed plan to make vehicles in Britain. An assembly plant at Trafford Park, Manchester, came on stream in 1911. He was knighted in 1918 for his war efforts, but had ruined his health and resigned in 1919. He had an income from part ownership of a company dealing in war-surplus automobiles and through it, the Slough Trading Company, operators of an industrial park, and moved his residence to Guernsey, off the Brittany coast. He maintained friendly relations with Ford's top production man, Charles E. Sorensen, continually lobbying for Ford to set up a manufacturing plant at Southampton. Ford picked another site, Dagenham on the Thames estuary, where construction began in May 1929. Perry's health had recovered enough to let him resume full executive duties. The first AA truck was completed in October 1931, followed by Model A cars early in 1932 and the small 8-hp Y-Model later that year.

He became Lord Perry of Stock Harvard in

1938, saw Ford through World War II, and retired in 1948. He died on June 17, 1957.

Perry, Richard William "Dick" (1930–)

Plant director and general manager of the Longbridge Works, British Motor Corporation, from 1964 to 1969; manufacturing manager for Leyland Cars from 1969 to 1974; managing director of Rover-Triumph in 1974–75 and managing director of Austin-Morris in 1976–77; managing director of Mulliner Park Ward from 1977 to 1982; managing director of Rolls-Royce Motors, Ltd., from 1982 to 1984; chief executive officer of Rolls-Royce Motors, Ltd., from 1984 to 1987; chairman of Rolls-Royce Motors, Ltd., from 1987 to 1994.

He was born in 1930 at King's Heath, Birmingham, and educated at Oxford. He joined Austin Motor Co., Ltd., as an apprentice in 1948, saw two years' military service as a pilot in the Fleet Air Arm in 1949–51, and resumed the apprenticeship. He became a production manager and held assignments in engine and transmission manufacturing and final assembly of Austin cars. He directed the tooling-up for the Austin 1800, which went into production in September 1964. He organized production of the rear-wheel-drive 3-Litre in 1967. The Austin Maxi, assembled at Longbridge since October 1968, introduced the new E-series engine made at Cofton Hackett, and the Maxi body shells came from Pressed Steel Fisher Division of British Leyland at Cowley. He spent only £13 million at Longbridge for assembly of the 1973 Austin Allegro, whose body shells came from the Austin-Morris Body Plant at Swindon, and remodeled the engine manufacturing shops at Longbridge for the O-series engine which replaced the old (1954) BMC B-series engine in 1978. He left British Leyland in 1977 to take command of Rolls-Royce's coachbuilding subsidiary in London and was brought to the Crewe headquarters in 1982. The Silver Spirit had replaced the Silver Shadow in October 1980. He pushed the development of the Corniche II convertible with its Mulliner Park body, which came on the market early in 1986. He retired in 1994.

Persson, Jan Christer Bernhard (1943–)

Vice president of Volvo Car Company and its manager of product engineering from 1986 to 1993.

He was born on September 2, 1943, and graduated from Chalmers Technical Institute in 1968 with a master's degree in mechanical engineering. In 1968–69 he was a technical attaché representing the Swedish Academy of Engineering Science at the Swedish Embassy in Washington, D.C. He returned to Gothenburg in 1970 and joined Volvo ABH as a project coordinator in the passenger-car division, and worked as a car planning engineer from 1972 to 1975. From 1975 to 1978 he held the office of materials manager, and then served three years as manager of component purchasing and sales. He was named manager of gasoline engine production at the Skövde plant in 1981 and two years later was promoted to general manager of Volvo Components Engine Division. In 1986 he began preparing an entirely new line of modern cars with front wheel drive and bodies revealing some regard for aerodynamics. Production of the Volvo 850 did not begin until the spring of 1991, however. He also bore responsibility for the heavy and boxy 960, launched in the summer of 1990, the last of Volvo's rear-wheel-drive cars. In 1993 the Volvo AB management brought in Martin Rybeck to take charge of passenger-car engineering, and Persson was transferred to another office as chief engineer of power unit engineering and production. He resigned from Volvo at the end of 1995 to take up a teaching position at the IVF Institute of Mechanical Engineering in Gothenburg.

Peter, Wolfgang (1941–)

Chief engineer of passenger-car development, Daimler-Benz AG, from 1985 to 1992.

He was born on April 29, 1941, at Otterbach in Hessen. Armed with a degree in mechanical engineering from the Darmstadt Technical University, he joined Daimler-Benz AG as a test engineer on November 1, 1968. He was promoted to head of a department in the engine test section in 1973 and was named head of passenger-car testing in 1984. The new S-Class was on the drawing board when he became chief engineer, the compact W-201 had been in production for 18 months, and the medium-size W-124 for less than a year. He led the development of the 300 SL and 500 SL, which shared most of the chassis elements of the W-124, but introduced new engines. The new S-Class, W-140, was his big opportunity. But its development took 18 months longer than planned and exceeded the budget by 50 percent. The car was launched in March 1991 and was greeted with mixed opinion. The previous S-Class (W-126) had enjoyed an 800,000-unit production run over an 11-year span. Sales of the W-140 started off at half the volume. It cost him his job. He left Daimler-Benz AG in March 1992 and joined the management board of Mannesmann AG in Düsseldorf. For the next four years he was in charge of coordinating Mannesmann's automotive divisions: Boge dampers and level-control equipment, Fichtel & Sachs clutch and transmission, Kienzle chrono-tachographs and VDO instruments.

Petiet, Charles (1879–1958)

President and general manager of Société des Automobiles Ariès from 1903 to 1936.

He was born on January 20, 1879, in Paris and educated at Ecole Fénelon and Ecole Duvignan de Lanneau. He studied mechanical engineering at the Ecole Centrale des Arts et Manufactures in Paris, graduating in 1901. After military service in the 30th Artillery Regiment, he joined Panhard & Levassor, starting on the shop floor, but was soon to be invited into the drawing office. He founded the Ariès company in April 1903 with a capital of 500,000 francs and a factory at Villeneuve-la-Garenne. The first Ariès car had a two-cylinder Aster engine and shaft drive, with its own version of de Dion–type rear suspension. Four-cylinder models, also powered by Aster, arrived in 1904, offering customers a choice of final drive (shaft or chains). Truck production also began in 1904. Baron (an inherited title) Petiet directed all activities, including policy-making, engineering and production, purchasing, sales and accounting. He gathered a respectable engineering staff and Ariès cars had a reputation for quality. From 1905 to 1910 he followed the Mercedes style of layout and proportions. Aster supplied the 10/14, 15/18 and 40/50 side-valve sixes for several models built between 1908 and 1912, as well as a 20/30 engine with Knight-patented sleeve valves in 1912. Seeing the success of the "Bébé" Peugeot ad Le Zèbre, he launched Model S with a 9 CV single-cylinder 1040-cc Aster engine in 1914.

During World War I the Ariès factory built its R-series truck, rated for four-ton payloads, with Aster and Ballot engines, in important numbers, and from 1916 onwards, Hispano-Suiza V-8 aircraft engines. After the war, Baron Petiet moved Ariès production to a bigger plant at Courbevoie, where Ariès manufactured its own engines for the first time. He introduced a new four-cylinder 5/8 CV model with an overhead-camshaft engine and prepared a new 3-liter model, which was produced in touring and sports versions. The latter scored a number of racing successes in 1925–26. The truck and bus models were also continually updated. In 1930 Ariès simplified its car program to a single 1500-cc 9 CV model, which grew to 1600 cc in 1933 and 1800 in 1934.

Baron Petiet had always maintained good relations with the leaders of the motor industry, and in 1938–40 had been a close advisor on industrial and political matters to Louis Renault. In 1931 he was elected president of the trade-and-industry federation[1] and organized chambers of commerce for all sections of France's automotive industries. In 1934 he became president of the CSCA[2] and president of the Salon de l'Automobile, Ariès brought out a new car, the 10/50, but very few were built, the last one in 1938.

After leaving Ariès, Baron Petiet continued in several trade-and-industry offices, and remained president of the Salon de l'Automobile until his death on October 2, 1958.

1. Fédération Nationale de l'Automobile, de Cycle, de l'Aéronautique et des Transports.
2. Chambre Syndicale des Constructeurs d'Automobiles.

Petit, Emile (1883–1974)

Chief engineer of Moteurs Salmson from 1921 to 1928; technical director of Automobiles Ariès from 1929 to 1934.

He was born on August 24, 1883, at Saint-Mandé on the east side of Paris and studied at a technical school in Brie-Comte-Robert. He held a degree in mechanical engineering from the Conservatoire des Arts et Métiers in Paris. He began his career with Pluton at Levallois in 1901, a company which produced about 100 cars that year. In 1903 he designed single- and two-cylinder engines for Société des Moteurs Monarque. In 1905 he joined Clément-Bayard as a tool designer, and rose in rank to be named chief engine designer before leaving in 1910. Ernest Ballot engaged him as production manager and head of testing. He designed cars for Fernand Charron in 1912–13 and in 1914 was named technical director of J. & A. Niclausse, whose steam-boiler works had added an automobile department in 1906. When war broke out he volunteered for military service and was assigned to a firearms engineering office. In 1919 he signed up with Atéliers Rhône-et-Seine as technical director. The company had a strong line in axles and other vehicle components, and Petit designed an air-cooled 1087-cc four-cylinder engine that went into production and was installed in some Salmson cars. Invited to join Salmson in 1921, he designed some remarkable engines, starting with the 1922 10/15 twin-cam four. He also updated the Salmson chassis from its G.N. origins. In 1927 he put two small four-cylinder units together to make an 1100-cc straight-eight developing 140 hp!

He resigned from Salmson in 1928 and went to work for Ariès on January 1, 1929. His first assignment was a new F-head engine for the CB-series car. After that, he found himself in charge of truck engineering, for which he had no passion. In his spare time, he designed the SEFAC[1] racing car (which never raced) and left Ariès to become a consulting engineer. From 1935 to 1939 he was a technical advisor to Debard Foundries. As a captain in the Army reserve, he reported for duty in 1939 and was

stationed at the materiel test establishment at Vincennes. In 1949–50 he designed a primitive minicar "Voiture du Bled" aimed at colonial markets, planned for production in France by Panhard and in Morocco by Georges Irat. But it remained in the prototype stage.

He died in Paris on June 24, 1974.

1. Société d'Etudes et de Fabrication d'Automobiles de Course.

Petri, Helmut (1940–)

Chief engineer of passenger-car development, Daimler-Benz AG, from July 1, 1995.

He was born in 1940 at Neunkirchen in Siegerland and apprenticed to a local auto mechanic after leaving grade school. He studied production technology at the Siegen-Gommersbach Technical College from 1962 to 1965 and began his career by joining Waldrich as a test manager. In 1969 he joined Hanomag-Henschel as a member of the central planning staff and was named manager of production preparation in 1974. The truck operations of Hanomag-Henschel had been taken over by Daimler-Benz AG and in 1982 he was transferred to the parent company as technical manager of the Berlin-Marienfelde works. He arrived at Sindelfingen in May 1990 as technical manager of the company's main passenger-car assembly plant. His tasks were to improve productivity and quality control simultaneously. In 1995 he took over the new S-Class project, with a tight timetable, for it was urgent to replace the W-140 launched in 1991. He directed the development of lighter luxury cars with new technical advances, with a range of V-6, V-8 and V-12 engines, with production start-up in October 1998.

Peugeot, Armand Godefroy Pierre (1849–1915)

Managing director of SA des Automobiles Peugeot from 1896 to 1910.

He was born on June 18, 1849, at Beaulieu (Doubs), the son of Emile Peugeot (1815–1874), one of the partners in Société Peugeot Frères, steel works and makers of hand tools and miscellaneous hardware. His father sent him to school in Paris, and then to a steel-maker in Leeds. He returned to France in 1871 and went to work in the family enterprise. He initiated bicycle production in 1885 and opened a new plant at Beaulieu, where he introduced the "safety" bicycle in 1887. He became interested in automobiles and met Amédée Bollée, Jr., and Léon Serpollet. In 1889 he built three three-wheeled steam cars designed by Léon Serpollet in the Valentigney plant, but when Gottlieb Daimler and René Levassor demonstrated a four-wheeled car powered by a V-twin Daimler internal combustion engine to him, he abandoned all thoughts of steam power, secured a Daimler engine license and made arrangements for a supply of engines from Panhard & Levassor.

The first two Peugeot cars were built at Valentigney in 1890, five were built in 1891 and 29 in 1892. He was eager to expand the automobile department, but the Peugeot family did not share his enthusiasm and it came to a break in February 1896.

Financed by Robert Fallot,[1] Adolphe Kreiss and Léon Sahler, Armand set up his own company and built an automobile factory at Audincourt in 1897. In 1898 he bought a factory in rue Gutenberg, Fives, Lille, and retooled it for car production. His plants completed 156 cars in 1898 and 500 in 1900. In 1902 he moved his headquarters to Levallois, on the doorstep of Paris. He made a sizeable fortune, but incurred heavy losses by speculation in real-estate development in the Doubs and at Morgat in Brittany. By 1908 he was ready for a reconciliation with the family and prepared to sell his holdings in the automobile business.

The merger of SA des Automobiles Peugeot and Les Fils des Peugeot Frères was completed in 1910, and he played no executive role in the newly established SA des Automobiles et Cycles Peugeot. He died on February 4, 1915, at his home in Neuilly-sur-Seine.

1. The Fallot and Peugeot families had been linked by intermarriage for generations. Louis Fallot had been instrumental in securing Swiss bank loans for Société Peugeot Frères in 1848. Robert Fallot owned textile mills at Tourcoing near Lille. Adolphe Kreiss was a brewery owner and Léon Sahler owned textile mills at Montbéliard.

Peugeot, Jean Pierre (1896–1966)

Chairman of SA des Automobiles Peugeot from 1944 to 1966.

He was born in 1896 at Montbéliard, the son of Robert Peugeot. He attended high school in Paris and graduated from the Ecole Centrale des Arts et Manufactures in 1921. He joined SA des Automobiles et Cycles Peugeot at the Paris headquarters in 1921, with a succession of assignments in planning, accounting, production and sales. He was given a seat on the board of directors in 1929 and became deputy chairman in 1941. When he took command in 1944, Sochaux was in ruins and he made its restoration his Number One priority. In the longer term, he also had to define the scope of Peugeot's activities, its methods of management, product philosophy and principles of operation.

Car production made a restart with the 202, which was replaced by the 203 at the end of 1948. He decided against suggestions of making a small car, since that is where Renault's strength lay. By 1950 the Sochaux plant was turning out 250 cars a day and in the year 1954, it produced 105,011 cars. The company then enjoyed a 7.5-percent profit margin on the car side. When a second model was added, it was the 403 with its Pinin Farina–styled body, launched in April 1955 and priced above the 203. When the 203 was taken out of production in March 1960, a total output of 685,828 had been built (sedan, coupé, four-passenger cabriolet, two-passenger cabriolet, station wagon, van, and pickup truck). The 403 was replaced by the 404 in May 1960, after a total output of 1,619,787 units. For the next ten years, Peugeot's pre-tax earnings averaged nearly 10 percent.

In 1961 he bought land on Napoleon Isle outside Mulhouse and built a machine shop for gears. Forges and a foundry were added in 1963, a stamping plant and body-welding and paint shops in 1968. The first car assembled at Mulhouse was the 304, beginning in 1971.

In 1961 he also approved plans for a small car, not basic transportation like Renault's R4L, but a quality four-door sedan of tight dimensions. It became the 204, Peugeot's first front-wheel-drive model, with production startup in 1965. He retired in 1964, due to health problems, and died on October 18, 1966.

Peugeot, Robert (1873–1944)

Chairman of Les Fils des Peugeot Frères from 1907 to 1910, chairman of SA des Automobiles et Cycles Peugeot from 1910 to 1926, chairman of SA des Automobiles Peugeot from 1926 to 1944.

He was born on July 21, 1873, at Hérimoncourt, the son of Eugène Peugeot (1844–1907), chairman of Société Peugeot Frères. He studied engineering and graduated from the Ecole Centrale des Arts et Manufactures in 1895. He went to work in the family enterprise, which made 8,000 bicycles in 1892. By 1900, annual production of bicycles and pedal tricycles reached 20,000 units.

In 1896 Eugène and Armand had signed an agreement proscribing Les Fils des Peugeot Frères from making automobiles. His company built the first Peugeot motorcycle in 1899 and also a few motor quadricycles with a two-passenger bench between the front wheels and the driver on a high, central perch, with a steering tiller behind. In 1905 Eugène wanted to cancel the agreement and ended up paying Armand one million francs for the right to make automobiles.

Robert took charge of Lion-Peugeot car production at Beaulieu and Valentigney. They were light cars with single-cylinder engines and the production volume never reached 1000 units a year. When his father died in 1907 he took over the chairmanship of the company and masterminded the merger with SA des Automobiles Peugeot in 1910. He provided dynamic leadership and launched a small flood of initiatives. He ordered the construction of a big plant, including foundries and forges, at Sochaux; he installed three racing engineers/drivers (Paul Zuccarelli, Jules Goux and Georges Boillot) in the experimental shop at Levallois and bought a racing-car design from Ernest Henry. And on November 16, 1911, he signed a contract with Ettore Bugatti for manufacturing rights to a small car that went into production at Beaulieu as Type BP1, better known as the Bébé Peugeot.

During World War I, the Beaulieu, Montbéliard, Valentigney, Audincourt and Sochaux plants furnished the French Army with 10,000 motorcycles, 3,000 cars, 6,000 trucks, and 60,000 bicycles. Peugeot opened new plants at Clichy and Issy-les-Moulineaux which turned out 5,000 Hispano-Suiza V-8 aircraft engines.

The return to civilian production was made quickly, with the Type 159 "Quadrilette" going into production in 1919 and Type 156, a high-priced touring car with a 5.9-liter six-cylinder sleeve-valve engine, becoming the first car assembled at Sochaux.

In 1922 Peugeot stumbled into a financial crisis and Robert hired Lucien Rosengart as managing director. The next few years were hectic, but the company pulled through. In 1926 Robert undertook a full corporate restructuring, spinning off the hand tools, coffee grinders and pepper mills into a new company, Peugeot & Cie, while the motorcycles and bicycles were combined to form Cycles Peugeot. The automobiles were put under a new company, SA des Automobiles Peugeot. The plants at Levallois, Clichy and Issy-les-Moulineaux were sold and the Lille plant reorganized as Compagnie Lilloise des Moteurs. In 1928 Rosengart's contract expired and in 1929, Banque Oustric, trustee of the Peugeot finances, collapsed.

It looked like the end for SA des Automobiles Peugeot, but it was saved by the family, with cousins and in-laws providing fresh capital. Type 201 went into production in July 1929 and soon proved highly profitable. Robert directed the company's affairs throughout France's political and economic upheavals, but was helpless against the German occupation forces' looting of machine tools and equipment from Sochaux during World War II. He died in 1944 at Bannot (Doubs).

Picard, Fernand (1906–1994)

Chief engineer of the Renault passenger car drawing office from 1941 to 1945, director of Renault car engineering and research from 1945 to 1966.

He was born on February 22, 1906, at Chennevières-sur-Marne and graduated from the Ecole des Arts et Métiers in Lille in 1923. After military service, he joined Delage in 1927 as a tool designer. By 1935 he was head of the Delage methods department, but lost his position when the new owners (Delahaye) closed the Delage factory at Courbevoie.

Auguste Riolfo brought him to Renault in 1936 as a test engineer. In 1939 he was assigned to Renault-Caudron (aircraft engines) and then to the Arsenal at Tulle to supervise the production of Hispano-Suiza cannon. Renault managed to get him back to Paris in 1940 and named him chief engineer for power units.

During World War II he spent most of his time on a small economy car project. Louis Renault had seen the KdF at the Berlin auto show in 1939 and told him to put the engine in the tail end. The first prototype was ready for road-testing on December 1, 1942, but the German occupants of Billancourt had outlawed all car activity and all development work had to be concealed. Still, the final product, identified as the Renault 4 CV, was displayed at the Paris Salon in October 1946 and production was underway at Billancourt. Picard also led the engineering team for the Frégate, a medium size sedan with an old 2-liter engine, all-independent suspension, and Renault's own optional Transfluide automatic transmission. He extrapolated the Dauphine from the 4 CV with its derivatives, the R8 and R 10 sedans and the Floride and Caravelle sports models. He left full-time employment with the Régie Nationale des Usines Renault in 1966, but remained a technical advisor until his final retirement in 1969. He died in January 1994.

Piccone, Alessandro (1943–)

Chief engineer of engine testing and design for Alfa Romeo S.p.A. from 1987 to 1994, director of engine design and development, Fiat Auto S.p.A. from 1994 to 2000.

He was born on April 6, 1943, in San Remo, the son of a lawyer with interests in olive oil. He graduated from the Turin Polytechnic in 1967 with a degree in mechanical engineering and spent 18 months in the Italian Army. He joined Alfa Romeo in the engine testing de-partment in 1968. His first major challenge at Alfa Romeo was to develop the SPICA[1] fuel injection system to meet U.S. emission-control standards for the Alfetta, the GTV and the 2000 Spider. By 1985 he was in charge of all engine testing for Alfa Romeo.

His duties expanded in 1987, with responsibility for engine design, redesign, and engine/vehicle coordination. He redesigned the V-6 from timing chains to belt-driven camshafts and led the design team for the 24-valve V-6 for the Alfa 164. He directed the design of the in-line five-cylinder 2.4-liter Lancia engine in both spark ignition and turbodiesel versions and developed the Twin Spark concept for several engine sizes. He planned a new generation of iron-block engines for mass production at Pratola Serra and phased out older Fiat and Alfa Romeo engines. He resigned for family reasons in April 2000 and now has homes in Milan and Lake Maggiore.

1. Società Pompe d'Iniezione Cassani, La Spezia.

Piëch, Ferdinand (1937–)

Technical director of Dr. Ing. e.h. F. Porsche K-G in 1971; technical director of Audi NSU Auto Nation AG from 1975 to 1983; vice chairman of Audio NSU Auto Nation AG from 1983 to 1988 and its chairman from 1988 to 1993; chairman of Volkswagen AG from 1993 to 2002.

He was born on April 17, 1937, in Vienna as the son of a lawyer, Anton Piëch, and his wife Louise, née Porsche. He graduated in 1963 with a degree in mechanical engineering from the ETH[1] of Zurich and joined Porsche in the engine design office. He led the development of the air-cooled flat-six for the 911 and was named test manager in 1966. Two years later he became chief engineer of product development. From the outset he had led the team that created the 917 racing cars, culminating with an 1100-hp twin-turbo 12-cylinder version for the CanAm series. He left Porsche when his uncle, Ferry Porsche, and his mother decided that members of the family should not hold leading positions in the company. He joined Audi in 1972 as head of a department for special projects and was named test manager in 1973. He gets credit for master-minding Audi's development of five-cylinder engines, full-time four-wheel-drive, turbo-in-tercooling, aerodynamic body shapes, and full-zinc body coating. Taking charge at Wolfsburg in January 1993, he launched a wholesale purge of directors and executives, replacing them with new heads of his own choice. In his analysis, VW was "wobbling under a bloated cost structure" and had become "a duck that was too fat to fly." He reneged on his predecessor's commitments on major investments in Skoda but passed record-high budgets for Audi and Seat. He eliminated six layers of management, leaving just three for the entire VW group, speeded up the company's dismal time-to-market record, and raised productivity. He hired José Ignacio Lopez de Arriortua away from his purchasing vice presidency with General Motors, only to run into an avalanche of litigation when Lopez brought crates and crates of confidential GM pricing and planning documents to Wolfs-burg. VW settled GM's case out of court, and the criminal case (under German law) petered out. But Lopez's career was over. Piëch's reputation was tarnished. He changed tactics for expansion, opening an engine plant at Györ, Hungary, in 1994, and tightening VW's grip on Skoda (70 percent) in 1995. In June 1993 he tried to buy Rolls-Royce Motor Cars, but ended up paying $715 million for the Bentley name and a factory at Crewe, Cheshire, while BMW secured rights to the Rolls-Royce name for $60 million. A month later, VW paid $175 million for Cosworth's engine division. He arranged VW's purchase of Automobili Lamborghini, which became an Audi subsidiary, and bought the rights to the Bugatti name, setting up Bugatti Automobiles SAS at Dorl-isheim in Alsace as a subsidiary of VW–France. Seeking to expand into heavy trucks, he made VW buy a 34 percent stake in Scania in 2000.

He pushed the VW brand relentlessly up-market, showing a four-wheel-drive car with a V-10 turbodiesel in 1999, a Passat W8 4-Motion in 2001, and a VW V-12 Phaeton in 2002. In November 2000 he overhauled his platform strategy, opting instead for new ways of sharing all sorts of components, replacing the VW group's four platforms with 11 modules, enabling each model to offer a wider range of engines, drive lines, suspension systems and body elements. He reached the age limit in April 2002 and left his chief executive office, but continued as chairman of the VW supervisory board.

1. Eidgenössische Technische Hochschule (Federal Technical University).

Pigozzi, Henri Théodore (1898–1964)

Founder of Simca (Société Industrielle de Mécanique et de Carrosserie Automobile) in 1934 and its president and chief executive officer from 1934 to 1964.

He was born on June 26, 1898, in Turin into a family which gave him dual nationality (Italian and French). He began his career by selling British and American motorcycles in Italy, but in 1920 he found a position with an Italian import company that represented the Sarre (Saar) coal mines. He gained intimate knowledge of the French mining industry. In 1924 he started his own company to recuperate scrap iron and steel in France and selling it in Italy. Fiat became a main customer and he made friends with the Agnelli family.

In 1926 he purchased the business of Ernest Loste, who had been the Fiat importer in France since 1907. But France had high customs duties on foreign cars, and trade with Mussolini's Italy came under a quota system. Pigozzi leased a factory at Suresnes and began assembling Fiat cars in 1929. At first, all components came from Italy. Fiat sent engineers to help start local production, relying on reputable suppliers. By 1932 Maison Chaize was making Fiat engines and gearboxes in Paris. The rear axles came from the arms factories of Saint-Etienne. Manessius of Levallois built the bodies. In 1934 Pigozzi found out that the Donnet automobile factory at Nanterre was for sale for 8,050,000 francs. He bought it and founded Simca.

Now Pigozzi had the industrial means of making the complete car. Production of the

Simca 5 (a copy of the Fiat 500) began in April 1936, followed by the Simca 8 (a copy of the Fiat 508C) in September 1937.

In 1945 he narrowly avoided being put on trial for collaboration with the Germans during the war, and a year later he successfully undermined a French government plan to use the Simca factories for production of the AF-Grégoire two-cylinder economy car. In 1949 he bought majority control of Société Nouvelle des Usines Unic, truck makers of Puteaux. He also took over a farm-tractor company at Bourbon-Lancy (Soméca).

With the launch of the Aronde sedan in 1951, Simca cars no longer corresponded exactly to models in the Fiat range, though they were designed with Fiat's help. Introduced in 1950, the Simca 8 Sport had an elegant Facel body. The biggest coup Pigozzi ever made was the purchase of the production facilities of Ford S.A.F. in 1954. The Poissy plant had turned out over 20,000 V-8 powered Vedette cars a year. Vedette bodies were made by Chausson. In 1958 Simca bought the assets of Automobiles Talbot at Suresnes. At that time, Fiat had begun to sell Simca shares, and Chrysler bought a 25 percent stake in Simca in 1958, increasing its ownership to 64 percent in 1963. Pigozzi resigned in May 1964 and died of a heart attack in Paris on November 18, 1964.

Pilain, Emile (1880–1958)

Technical director of the Etablissements Rolland et Pilain from 1905 to 1920; joint managing director of SA des Etablissements Rolland & Emile Pilain from 1920 to 1927; managing director of SA des Etablissements Emile Pilain from 1927 through 1931.

He was born on July 28, 1880, at Macon, the son of a mechanic. In 1898 he joined his uncle, François Pilain, in setting up Vermorel's automobile works at Villefranche-sur-Saône and stayed on with Vermorel until 1904. In 1904–05 he drove Pilain cars in several races, and moved to Tours, where he met François Rolland, a prosperous wine merchant and keen automobilist. Rolland financed their enterprise and Pilain contributed his knowledge,

skills, and his time. The 20 CV Type A and 40 CV Type B Rolland-Pilain models appeared in 1907, and Rolland-Pilain's first big seller, the 12/16 Type C, went into production in 1908. The company was also active in voiturette racing, with considerable success. Type D of 1909 was their entry-level model, an 8/10 two-liter, which evolved into Type RP in 1913. During World War I a big section of the plant was leased to Pierre Clerget for making air-cooled radial aircraft engines. In another section, Rolland-Pilain manufactured its own water-cooled aircraft engines. After the war, François Rolland retired, and his son Lucien Rolland (1888–1957) became joint managing director. Emile Pilain designed a 2.3-liter overhead-valve engine for the 1922 B-25, and went to an overhead camshaft for the 1923 C-23, which also featured four-wheel-brakes, four-speed gearbox and torque-tube drive. They failed to see the coming cash-flow crisis, lost control of the company, and left Tours. In partnership with Lucien Rolland and André Pezon, Emile Pilain set up a new company at Levallois-Perret in June 1927 to build a high-grade 919-cc 5 CV car, but it failed in the market due to its high price. He prepared a 7 CV for 1932, but the company was bankrupt. It was dissolved on December 15, 1931. He remained in the Paris area for the rest of his life, keeping busy with a wide range of inventions. He was still applying for patents in 1949.

He died on December 18, 1958, in Paris.

Pilain, François (1859–1924)

Chairman of the Société des Voitures Automobiles F. Pilain & Cie from 1893 to 1897; founder of the Société des Voitures Automobiles et Moteurs (F. Pilain & Cie) in 1897; managing director of Société des Automobiles François Pilain in 1914; chairman of the Société Lyonnaise d'Industrie Mécanique from 1919 to 1924.

He was born on September 7, 1859, at Saint-Bérin (Saône-et-Loire), and at the age of 13 began an apprenticeship with Pinette's machine shop in Chalon-sur-Saône. He stayed on with Pinette, working mainly as a draftsman, until 1887. The following year he went to Paris and became a test-driver and drafts-man for Léon Serpollet, steam-vehicle pioneer. He returned to Lyon in 1890 and signed up with La Buire as a design engineer on automobiles in 1892. A year later, he left to open his own shop and began to design his own automobile. He built and sold a small number of cars from 1896 to 1898, when he was engaged by Edouard Vermorel to set up an automobile branch at Villefranche-sur-Saône. He made a good start for Vermorel in the auto industry, but by 1901 he had saved enough money to start another company, making the SAP or Pilain car in Lyon. His cars were aimed at the high-priced market, boasting of quality, power and speed. He completed 350 chassis in 1907, but resigned in 1909 after a quarrel with his board of directors.

In 1912 he opened a drawing office and prototype construction shop in Lyon, which two years later evolved into Société des Automobiles François Pilain. In 1917 his factory was requisitioned for national defense work and placed under Hotchkiss administration to produce machine guns. The plant was sold to Saurer, the Swiss truck builders, after the war. In 1919 Pilain set up yet another company with works at Caluire to build cars sold as SLIM or Pilain, with interesting specifications and high prices. He led the enterprise until his death in 1924.

Pischetsrieder, Bernd (1948–)

Head of technical planning for BMW from 1987 to 1990; BMW management board member for production from 1991 to 1993; chairman of the BMW management board from 1993 to 1999; chairman of the VW Group management board since 2002.

He was born on February 15, 1948, in Munich and attended local schools. He studied Greek and Latin history, literature and language in high school, then decided to aim for a career in industry, and graduated from the Technical University of Munich in 1973 with a degree in mechanical engineering. He joined BMW as a production planning engineer that same year and advanced into assignments of production management, purchasing and lo-gistics. From 1982 to 1985 he was with BMW South Africa as director of production and development. He returned to Munich in 1985 and was named head of quality assurance. In 1987 he was placed in charge of technical planning, and was appointed deputy member of the management board in June 1990. In July 1991, he became a full member. Later that year he was sent to the United States to select a factory site for production of the Z1 roadster, recently developed by BMW Technik GmbH in Munich, a subsidiary for experimental cars. He secured a site at Spartanburg, South Carolina, where BMW invested $625 million to assemble 70,000 cars a year. He was elected chairman in May 1993 and led the purchase of Rover from British Aerospace in 1994. Throughout BMW's operations, he cultivated teamwork and streamlined the hierarchical organization. BMW was consistently profitable, but Rover proved to be a serious drain of resources. He saw Rover's future as a full-line car maker, competing in the mass market, and was opposed by Wolfgang Reitzle, who wanted Rover to make fewer cars, each model aimed at specialist markets. It ended with the dismissal of both Pischetsrieder and Reitzle in February 1999. In 2000 he became a member of Volkswagen's management board, and in April 2002 succeeded F. Piëch as chairman of the VW management board.

Piziali, Alfred P. (1910–)

Vice president of car engineering, Ford of Europe, Inc., from June 1968 to March 1972.

He was born on May 6, 1910, at Negaunee, Michigan, and joined Ford Motor Company in 1927 as a tool-and-die maker in the River Rouge plant. His superiors realized his potential and sent him through the Lawrence Institute of Technology and Wayne State University. After graduation, he returned to Ford as a body engineer. He held a number of appointments, from chief of truck body engineering and executive engineer of body design to chief engineer of the metal stamping office. In January 1967, he was named chief engineer of the body engineering office in Dearborn, but transferred to Europe within 18 months.

He organized production of the Ford Capri in two plants, Cologne and Dagenham, and worked hard to coordinate the British and German product lines. He had nothing to do with the original Escort, but held overall responsibility for its 1975 successor. He masterminded the Consul/Taunus and Granada programs, with extensive sharing of chassis sub-assemblies and major body elements among a group of cars produced in both Britain and Germany. He was not given time to rationalize the engine programs, being called back to Dearborn in March 1972 as executive director of engineering, Ford Motor Company.

Place, René (1938–)

Director of body manufacturing at Peugeot's Mulhouse plant from 1980 to 1984, and at the Sochaux plant from 1984 to 1988; director of Peugeot's body-manufacturing methods from 1988 to 1990; director of the Sochaux industrial complex since 1990.

He was born on March 23, 1938, at St. Georges-sur-Allier and educated at the Ecole Nationale Supérieure des Arts et Métiers. He began his career with Peugeot in December 1962 as an engineer in the stamping plant at Sochaux. In 1969 he was attached to the body-structure planning office at Sochaux, and a year later was transferred to the corporate headquarters in Paris for a position in personnel and labor relations. He returned to Sochaux in 1975 as deputy director of the body plant, transferred to Mulhouse in 1980, returning to Sochaux four years later.

When he took full charge of the Sochaux complex, it had a work force of 25,000 and a daily output of 1650 cars (Types 205, 405 and 605). Nine years later, the model lineup had changed to Types 306, 406, with the 607 ready to replace the 605). Productivity was increased, quality was improved, but due to reduced demand, the plant was working well below capacity.

Plastow, David Arnold Stuart (1932–)

Managing director of the Motor Car Division

of Rolls-Royce, Ltd., from 1971 to 1972; managing director of Rolls-Royce Motors, Ltd., from 1972 to 1987.

He was born on May 9, 1932, in Lincolnshire as the son of a car dealer, and educated at Culford School, Bury St. Edmunds. After serving a commercial apprenticeship with Vauxhall Motors, Ltd., at Luton, he joined Rolls-Royce in 1958 as head of the Scotland sales branch. He held a number of sales and marketing positions before being promoted to marketing director in 1967. He took an active interest in product planning, but the first model stemming from his ideas was the Camargue, which he rammed through and put in production against considerable internal opposition. In 1980 he masterminded the sales of Rolls-Royce Motors, Ltd., to Vickers PLC for £38 million, and in 1987 he resigned from Rolls-Royce in order to take over as chairman of Vickers PLC, defense contractors for tanks, torpedo boats, and other military equipment. In May 1992, he left Vickers and was named chairman of Inchcape, worldwide motor vehicle distributors, in 1993. He retired on December 31, 1995.

Platt, Maurice (1898–1993)

Power unit engineer of Vauxhall Motors from 1940 to 1946; passenger-vehicle engineer from 1946 to 1949; executive engineer from 1949 to 1953; chief engineer and a member of Vauxhall's board of directors from 1953 to 1963.

He was born in 1898 in Yorkshire and graduated from Sheffield University in 1919 with a bachelor's degree in mechanical engineering and later received his master's degree from the same institution. He began his career with Albion, truck builders at Scotstoun near Glasgow, but moved to London in 1924 to work as technical editor of *The Motor*, a weekly car magazine. During the next 13 years he made great strides in road-testing techniques and technical journalism for a wide public. He joined Vauxhall Motors in 1937 as a sales and service contact engineer. When passenger-car production was halted in 1940, he was put to work on the adaptation of production-model Vauxhall and Bedford engines to military vehicles.

After the war he was a member of the team that created the 1948 Wyvern and Velox, and held major responsibility for their 1951–52 replacements. He was in overall charge of the projects that led to the 1954 Cresta and 1957 Velox. He argued in vain against building the 1957 Vauxhall Victor which was pushed into production at the insistence of GM president Harlow Curtice. In 1963 he was named director of engineering for Vauxhall but announced his retirement a few months later. He died on August 19, 1993.

Pollich, Carl (1897–1972)

Co-designer of the Hanomag 2/10 PS minicar in 1924; chief engineer of Hanomag from 1927 to 1951.

He was born on April 7, 1897, in Berlin and educated at a technical college. With a fellow student, Fidelis Böhler, he developed an unusual small-car concept, with a single-cylinder vertical engine mounted in an inboard-rear position, and a full-width slab-sided body produced in a number of styles, from roadster to taxicab, coupé to tourer. They sold it to Hanomag of Hanover, who had built steam locomotives and trucks, farm and road tractors, and was looking for a new type of product. It was Germany's first car selling for less than RM 2000 but was quickly outmoded. Still, more than 10,500 Hanomag 2/10 PS cars were built from 1924 to 1928, Pollich and Böhler even created a mini-truck using the same chassis. In 1930 they were preparing a potential replacement model with a rear-mounted water-cooled flat-four engine, but the Hanomag management wanted a purely conventional car, on the lines of the ex–Dixi BMW. Böhler left Hanomag, and Polich designed the 3/16 with a 750-cc four-cylinder engine mounted in front. Bigger cars, the 1933 Kurier and 1934 Garant, soon followed, with the best-selling Rekord made from 1934 to 1940, alongside the six-cylinder Sturm. In 1935 he began to design a four-passenger streamliner that went into production in 1939 as the Hanomag 1.3. After World War II he designed new vans and light transporters that

went into production in 1950. The next Hanomag car was the Partner, displayed in 1951, with a two-cylinder two-stroke engine and front-wheel-drive. But it never got beyond the prototype stage, as Hanomag gave priority to its new line of light trucks. He retired about 1960 and died in 1972.

Pomeroy, Laurence Henry (1883–1941)

Works manager of Vauxhall from 1909 to 1912; chief engineer of Vauxhall from 1912 to 1918; chief engineer of Daimler from 1926 to 1929; managing director of Daimler from 1929 to 1936.

He was born on March 13, 1883, in London, and at the age of 16 was apprenticed to the Northwestern Locomotive Works at Bow. He began his career with Humphreys & Company, civil engineers, in 1904, but left within a year to go with Thornycroft at Basingstoke where he only spent a few months. He joined Vauxhall, who had just moved from London to Luton, in 1905, as a design engineer. His first complete car design was a special 3-Litre model made for the 2000-Miles Trial of 1908. In 1910 he designed the Prince Henry model, a splendid rally car, and in 1913 the immortal 30/98. In 1914 he even designed a racing car for the Tourist Trophy.

In 1919 he sailed to America and approached Aluminum Manufacturers, Inc., of Cleveland, Ohio (later Aluminum Company of America, then Alcoa), with a scheme to promote their light alloys by making an all-aluminum car. It was accepted. Parts were made, including fame, engine block, head and crankcase, clutch housing, gearbox casing, rear axle, and body, in light alloy. Several prototypes were assembled in the Pierce-Arrow factory at Buffalo, New York, in 1922–23. He returned to England in 1925 and joined Daimler in Coventry. He was responsible for Daimler's adoption of the Fluid Flywheel and preselector planetary transmission in 1930–31, and designed new poppet-valve engines to replace the old sleeve-valve units.

He resigned from Daimler in 1936 and joined De Havilland Aircraft Company, Ltd., as general manager of the engine division. In

1940 he became a director of H.M. Hobson (Aircraft & Motor) Components, Ltd., in Coventry, but died in 1941.

Poole, Stephen Carey (b. 1888)

Chief engineer of Clyno from 1918 to 1920; chief designer of Singer cars from 1920 to 1928; chief designer of Jowett from 1928 to 1934; director of research and development for Jowett from 1934 to 1938.

He was born on December 15, 1888, in Coventry, and educated at Bablake School, Coventry, and Birmingham University. In 1907 he joined Wolseley as a draftsman, leaving in 1910 for a position in the Standard Motor Company's drawing office, where he spent the next two years. Clyno engaged him in 1913 to design a small car with a four-cylinder engine that would also serve as the power unit for a motorcycle/sidecar combination.

He designed the six-cylinder 15-hp Singer introduced in 1922 and the 1927 Singer Junior with its 848-cc single-overhead-camshaft-engine. Leaving Coventry for Bradford in the West Riding of Yorkshire in 1928, his first action was to put four-wheel brakes on Jowett cars. He redesigned the flat-twin engine with detachable cylinder heads, and developed a delivery van on the passenger-car chassis. He introduced a four-speed gearbox in 1934, centrifugal clutch and free-wheeling in 1935. He created a prototype car with a vertical in-line four-cylinder engine, but the Jowett brothers refused their approval since they felt any new engine should be related to their flat-twin. Consequently, Poole designed a flat-four for the new family sedan introduced in 1936. He took the opportunity to resign in 1938 when a new management team was installed at Jowett.

Popp, Franz Josef (1886–1954)

Technical director of the Bavarian Motor Works[1] in Munich from 1917 to 1918; managing director of the Bavarian Motor Works[2] from 1918 to 1942.

He was born on January 14, 1886, in Vienna and held degrees in mechanical and electrical engineering from the Technical University of Brünn (now Brno in the Czech Republic). He was a reserve officer in the Austro-Hungarian Navy and served at the Pola Naval Base in the Adriatic for one year at the start of World War I. In 1915 he was released from military duty to work on electrical projects with Austro-Daimler in Wiener-Neustadt, but in November 1916 was assigned to a small factory in Munich that held a license to produce Austro-Daimler aircraft engines. Founded in 1913 as the Rapp Motor Works by a former Mercedes engineer, Karl Rapp, it had merged in March 1916 with the Gustav Otto Flying-Machine Factory[3] to form the Bavarian Aeroplane Works.[4] Gustav Otto was the son of N.A. Otto, co-founder of the Deutz Gas Engine Factory.[5] Popp closed the small Rapp factory and transferred engine production to the Gustav Otto plant opposite the Oberwiesenfelde, and incorporated the Bavarian Motor Works (trademark BMW) on July 20, 1917, making it a joint stock company on August 13, 1918. The stock was held by a consortium of Munich banks and two private industrialists. Popp held no shares but served as the salaried chief executive.

At the end of 1918, Popp found himself facing a payroll of 3400 and no orders to fill. He converted part of the plant into auto repair shops and in another part, began manufacturing farm tractors. A four-cylinder overhead-camshaft engine designed for aircraft was revamped for marine and truck applications, deliveries beginning in 1920. A flat-twin motorcycle engine went into production in 1921, the biggest customer being Martin Stolle. Popp also secured the Austro-Daimler car franchise for Germany. BMW began building complete motorcycles in 1922 and added a line of six-cylinder and V-12 aircraft engines in 1923.

The success of the motorcycle branch gave Popp the idea of producing a small economy car. He was tempted by the SHW[6] front-wheel-drive prototype designed by Professor W. Kamm, but officials of the Deutsche Bank, a major shareholder, pointed out that BMW could save cost and time by taking over an existing production car.

In June 1926, Popp was invited to take a seat on the supervisory board of Daimler-Benz AG, and in reciprocation Wilhelm Kissel from Stuttgart joined the BMW supervisory board. This exchange brought Popp in direct contact with Jakob Schapiro, who held ten-year manufacturing rights to the Austin Seven. One of his companies, the Dixi Works of Eisenbach, began producing it towards the end of 1927. On November 18, 1928, BMW bought the Dixi Works and the Austin license, and in 1930 the Dixi badge was replaced by the BMW trademark.

By 1932 the BMW product was practically free of Austin ingredients, and Popp visited Sir Herbert Austin in Birmingham, who graciously wavered royalties for the remaining 4½ years of the contract.

After 1936 the German aircraft industry wanted more engines than BMW could produce, and in 1939 BMW purchased the Brandenburg Motor Works from the Siemens electrical company, with a big factory in Berlin making Bramo air-cooled radial engines.

Popp resigned from BMW in 1942 and lived quietly in retirement until 1954.

1. Bayerische Motoren Werke GmbH.
2. Bayerische Motoren Werke AG.
3. Gustav Otto Flugzeugmaschinen Werke.
4. Bayerische Flugzeug-Werke AG.
5. Gasmotorenfabrik Deutz, Cologne.
6. Süddeutsche Hütten-Werke = South German Steel Works.

Poppe, Peter August (1870–1933)

Technical director of White & Poppe, Ltd., from 1899 to 1922; works manager and chief engineer of the Rover Company, Ltd., from 1923 to 1929.

He was born on August 21, 1870, on a farm in northern Norway and attended Horten (naval engineering) technical school. He joined a Norwegian company engaged in the production of navigational instruments as a trainee. In 1897 the government of Norway sent him on a mission to Waffenfabrik Steyr in Austria to supervise the fulfillment of an order for small firearms. In Austria he met A.J. White, one of the directors of Swift of Coventry, who was trying to sell the Steyr manage-

ment a license to produce Swift bicycles. They agreed to go into business together as makers of engines, carburetors, and other automotive components, and White & Poppe, Ltd., was registered on September 30, 1899, with a small factory in Drake Street, Coventry. A year later they moved into bigger premises in Lockhurst Lane, Coventry, and added a line of firearms designed by Poppe.

Engine production began in 1905, with a family of modular single-, twin-, three- and four-cylinder units. A six was added in 1906. The company supplied engines to Swift, Singer, Clyde, Horbick, Fairfax, Academy, West, Rothwell, Horley, Globe, and Calthorpe. In 1911 he was asked to design an engine and gearbox for W.R. Morris. It was destined for the Morris Oxford and produced in considerable numbers.

The partnership began to break up in 1919 when A.J. White wanted to make engines exclusively for Dennis (who eventually took over their factory). Poppe left in 1922, became an engine consultant to Maudslay, and joined Rover a year later. He designed the Rover 14/45 with its single-overhead-camshaft engine and splayed valves, which was replaced by the bigger 16/50 in 1926–27. He put an end to the production of the Rover 8 with its air-cooled flat-twin in 1925, replacing it with the four-cylinder 9/20, which grew into a 10/25 in 1928. He created the six-cylinder 15.7-hp model in 1928 and prepared the 20-hp Meteor for 1930. He was dismissed in 1929 after a quarrel with some of the Rover directors.

Porsche, Ferdinand Anton Ernst "Ferry" (1909–1998)

Office manager of Porsche KG at Stuttgart-Zuffenhausen from 1939 to 1944, and at Gmünd, Katschberg, in Austria from 1944 to 1950; managing director of Porsche KG from 1950 to 1972; chairman of the supervisory board of Porsche AG from 1972 to 1993.

He was born on September 19, 1909, in Wiener Neustadt, Austria, as the son of Ferdinand Porsche and his wife Aloisia, née Kaes. He was not a particularly good scholar, but showed a remarkable talent for mathematics.

He finished grade school in Vienna and enrolled in a Stuttgart high school in 1923. In 1928 he was apprenticed to Bosch but left in 1931 to join his father's new business as a draftsman. He handled external relations and became involved with the test program for the KdF-Wagen. In 1936 he was named KdF test manager and was made an honorary SS officer. In 1938 he was given a seat on the Porsche KG management board and prepared designs for KdF-based competition cars, turning his mind to military projects when no other contracts were coming in.

His office was moved to Austria in 1944, and in 1945–46 he was held under arrest, first by U.S. authorities, later by the French. The Porsche organization had been preparing a four-wheel-drive racing car for Auto Union AG before the war. It became the basis for the Porsche 360, which he sold to Piero Dusio of Cisitalia and provided the cash needed to bail his father out of French prison.

He revived his prewar sports-car plans and created the Porsche 356 roadster and coupe, using a VW–based engine and other VW components. Heinz Nordhoff of VW agreed to pay Porsche a royalty on every "Beetle" and to keep Porsche busy with enough small-car projects to block Porsche from working for other clients. Those contracts were valid until Rudolf Leiding, in despair over Porsche's failure to offer an acceptable successor model to the "Beetle," canceled them in 1971.

Porsche responded by developing a worldwide clientele for its engineering and design projects. "Ferry" held personal responsibility for the Porsche 911 and 914 models. He was not enthusiastic about the 928 but allowed it to go into production. The 924 was a joint program with Audi but he paid $825,000 to prevent Leiding from selling it with VW and Audi badges. Before his retirement he focused the company's attention on continued evolution of the 911. He retired to his landed estate at Zell am See in Austria in 1993 and died there on March 27, 1998.

Porsche, Ferdinand (1875–1951)

Chief engineer of Jacob Lohner GmbH from

1898 to 1906; technical director of Oesterreichische Daimler-Werke AG from 1906 to 1916; general manager of Oesterreichische Daimler-Werke AG from 1916 to 1922; technical director of Daimler Motoren Gesellschaft from 1923 to 1926; technical director of Daimler-Benz AG from 1926 to 1928; head of Porsche Konstruktionene GmbH from 1931; joint managing director of Volkswagenwerk GmbH beginning in 1938.

He was born on September 4, 1875, at Maffersdorf near Reichenberg in Bohemia as the son of a tinsmith, Anton Porsche. At the age of 15 he was apprenticed to Willy Ginskey, another tinsmith, and simultaneously attended trade school at Reichenberg where he learned the rudiments of electrical engineering. He was only 18 when he installed a generator set in his parents' home and wired it for electric lighting.

On April 18, 1894, he left Maffersdorf to take employment in Bela Egger's United Electrical Company in Vienna. He also attended evening classes at the Technical University as an unregistered "crasher." Four years later he was made head of Bela Egger's experimental shop. In 1898 he joined Jacob Lohner in Vienna as a design engineer, working on both battery-electric and gasoline-electric vehicles. He patented his system of mounting an electric generator of the engine and electric motors in the wheel hubs. The Lohner-Porsche "Mixte" car displayed at the Paris Universal Exposition in 1900 had front-wheel drive.

One of his first tasks at Austro-Daimler was to draw up an airship engine. He also designed the high-performance touring cars with advanced engines and became an outstanding rally driver. From 1911 to 1918 he was mainly occupied with military projects, notably creating a line of heavy vehicles with gasoline engines and electric drive, for production by the Skoda works at Jungbunzlau. In 1917 he was the recipient of an honorary doctorate in engineering from the Vienna Technical University. He directed the design of the first postwar cars and created the 15-PS 1100-cc "Sascha" light car with a double-overhead-camshaft four-cylinder engine.

He left Austro-Daimler during an argument with the management board over financial matters and moved to Stuttgart in April 1923 as technical director of the company that built Mercedes cars. He designed the supercharged straight-eight Targa Florio racing car and the six-cylinder 15/70/100 touring car with an aluminum-block supercharged engine. He was also responsible for the 200 Stuttgart with its smooth L-head six, and the high-performance Types K, S, SS, and SSK.

After the Daimler-Benz merger, his position was shaky, for powerful ex–Benz directors Wilhelm Kissel and Friedrich Nallinger wanted to promote former Benz engineers, and many of Porsche's decisions were severely criticized. He left before the end of 1928, returned to Vienna, where Steyr-Werke AG announced his appointment as technical director on January 1, 1929. He designed a luxury car, the Steyr Austria, which he accompanied to Paris for the 1929 Salon. During his stay in Paris, he read in a newspaper that Steyr had terminated his employment.

He assembled a group of reputable engineers and set up a firm of automotive consulting engineers in Stuttgart. They designed two six-cylinder cars for Wanderer, the 7/35 and 8/40, in 1933. Porsche also proposed an eight-cylinder Wanderer, but Auto Union rejected it, and the prototype served as Porsche's personal car for years. He sold suspension-system designs to Mathis and the 16-cylinder P-Wagen racing car to Auto Union.

Vague ideas about a mass-produced economy car occupied his mind, evolving into an image with a rear-mounted engine and low-cost specifications. The Porsche Type 12 was a complete design for such a car, which he supplied to Fritz Neumayer of Zündapp in 1932. But Zündapp never put it into production. Porsche Type 32 was a revised design made for NSU in 1933, its fate coming to a dead end when the heads of NSU remembered that their contract with Fiat barred them from passenger-car production for a number of years.

In a meeting with Reichs Chancellor Adolf Hitler in May 1934, Porsche talked at length about the economy-car project, which resulted in a design contract signed on June 22, 1934, for a car with an air-cooled engine and four seats, capable of cruising at 100 km/h. From then on, the government paid all costs and salaries associated with the people's car.

It was to be sold as the KdF-Wagen,[1] its production to take place in a government-owned plant. In 1937 Hitler gave Porsche the meaningless title of "Reichs-Auto Konstrukteur" and a year later he joined the Volkswagenwerk AG management team. The Porsche organization developed the Kübelwagen (Type 62) for the military, and later the Schwimmwage (Type 128), the amphibious version, based on the KdF-Wagen.

In 1940 Porsche was named head of the Panzer Commission, with responsibility for the design of armored vehicles and military transport, an office he held until 1943, when he had a falling out with Albert Speer, minister of armaments production. In 1944 Speer arranged to have Porsche's office removed to Austria, ostensibly to keep him in a low-risk area from allied bombardment. He was lured by the French occupation forces to a meeting at Baden-Baden on December 15, 1945, to discuss Volkswagen production in France. Instead, he was arrested on charges of having sent Peugeot workers against their will to Wolfsburg and other German plants. He was never put on trial, but released on bail posted by his son Ferry via a circuitous delivery route. On August 1, 1947, he was free to return to Austria.

But he was only a spectator to rather than a participant in the construction of the first car to bear the Porsche name, Type 356, first revealed in the summer of 1949. His health was broken and he died on January 30, 1951.

1. KdF = Kraft durch Freude (Strength through Joy), a leisure-time organization of the Deutsche Arbeitsfront (German Labor Front), not a labor union but a government office which financed workers' pension plans and a number of other social-spending programs.

Praxl, Ewald (1911–)

Director of testing and development, NSU Motorenwerke AG, from 1962 to 1971.

He was born on July 5, 1911, at Postelberg

on the Eger River in Bohemia and attended local colleges until 1930 when he went to Prague to study engineering at the Technical Institute. He graduated in 1937 and joined NSU in 1939 as a design engineer in the off-highway vehicle drawing office. He was the principal designer of the Kettenrad, a lightweight half-track steered by a single front wheel, powered by a four-cylinder 1½-liter Opel engine, with a six-speed gearbox. He was promoted to chief draftsman in 1943 and deputy chief engineer in 1945. He worked on the development of the NSU Quick motorcycle and helped design the NSU Fox. In 1950 he was named head of testing and racing, leading the NSU to work champion titles in 1953 and 1954. He proposed the Prinz minicar concept and began design work in 1956, with production start-up in 1958 after a most thorough test program. He developed the Prinz, the Sport-Prinz, the Prinz Spider and the Prinz 4. He started a front-wheel drive project which evolved into an up-market four-door sedan, powered by a twin-rotor Wankel engine with a semi-automatic 3-speed transmission, which went into production in 1967 as the NSU Ro-80. The engine had seal-tip wear and reliability problems, blocking its way to commercial success, for the reputation lived on long after the problems had been eliminated.

Prinz, Gerhard (1929–1983)

President of Audi NSU Auto Union AG from 1971 to 1974; president of the management board of Daimler-Benz AG from 1980 to 1983.

He was born in 1929 in Solingen and educated as a lawyer. After holding positions as a barrister and notary, he became interested in economics and business. He went to work in the steel industry in 1958 and became assistant to the chief executive of Klöckner AG in Duisburg in 1960. He left Klöckner in 1963 to accept a better offer from Brown, Boveri & Cie at Mannheim, working under the orders of Kurt Lotz. He joined Audi as a director in 1967 and became a member of the Volkswagen management board in 1968. He left the VW group because he felt passed over for promotion to chief executive when Rudolf Leiding resigned in 1974. He was immediately invited to Daimler-Benz AG as purchasing director, and succeeded Joachim Zahn into the president's chair in 1980. It was Prinz's decision to expand the model range downwards by adding a compact car, the W-201 or 190-series, which made its debut in 1983. He died in October 1983 at his Stuttgart home of sudden heart failure.

Puch, Johann (1862–1914)

Owner and managing director of Puchwerken AG of Graz, Austria, from 1909 to 1912.

He was born on June 27, 1862, at St. Lorenzen near Pottau in Untersteyermark and trained as a mechanic. He set up his own shop in 1891, with one mechanic and one apprentice, operating as the Styria Bicycle Works. It grew into a factory which he sold to Nikolaus Dürkopp in 1897. He went into partnership with his former co-worker, Anton Werner, starting a rival bicycle factory at Graz in Werner's name,[1] which was then reorganized in 1899 in Puch's name.[2] A line of single-cylinder motorcycles was added in 1901, and they also built a prototype car that year, but shelved to build up their motorcycle production. In 1905 Johann Puch bought a license for the Heldé automobile from Lévèque & Dobenrieder of Boulogne-sur-Seine and a year later began building the Puch voiturette with a 9-hp 904-cc water-cooled V-twin.

In 1909 Johann Puch repurchased the original Styria works and merged it into his current enterprise, to form the Puchwerken AG. The first four-cylinder Puch cars came on the market in 1909, designed by Karl Slevogt, though the company's main business was to remain motorcycles. Puch retired in 1912 for health reasons and died in the summer of 1914 from a heart attack.

1. Grazer Fahrradwerke Anton Werner & Co.
2. Johann Puch, Erste Steiermärkische Fahrrad-Fabrik AG.

Puiseux, Robert (1892–1991)

President of SA André Citroën from 1950 to 1958.

He was born in 1892 in Paris as the son and grandson of astronomers, but he was more interested in mechanical engineering and was accepted at the Ecole Polytechnique. He had to break off his studies when called up for military service in World War I, and fought first in the Artillery, later in the Air Force. He married Anne Michelin in 1921 and was invited to join the tire manufacturers at Clermont-Ferrand. He held a succession of administrative positions before being placed in charge of technical research. As a trusted family member, he was named joint chief executive alongside Pierre Boulanger when André Michelin died in 1940. He provided an impetus for the development of the Michelin X tire for production, supported by an exceptional budget.

He was called upon to replace Pierre Boulanger as president of Citroën upon the latter's death in 1950, without relinquishing his duties with Michelin. Since he could not be in two places at once, he promoted Pierre Bercot and Antoine Brueder to run Citroën from day to day while he alternated between Clermont-Ferrand and Paris. He pushed the VGD project, which became the DS-19 in 1955, through to reality, against all obstacles.

Retiring in 1958, he held the title of honorary president of Citroën up to 1974, and died in 1991.

Puleo, Giuseppe (1926–1999)

Chief engineer of the Autobianchi A-112; project manager for the Fiat X 1/9 sports car; director of Fiat engine design from 1980 to 1983.

He was born on August 7, 1926, at Cinisi near Palermo, Sicily, into a well-to-do family who sent him to Turin for his engineering studies. He graduated from the Turin Polytechnic in 1950 and immediately joined Fiat. He worked on production cars and advanced projects, including all-wheel-drive military vehicles, for ten years. In 1960, Giacosa transferred him to Deutsche Fiat at Heilbronn to plan and design experimental cars. Returning to Turin in 1964, he became a member of Cordiano's front-wheel-drive engineering team. The Autobianchi Primula was ready for production, but he was placed in charge of the A-

112 small-car project. He had the idea of a sports car using the Fiat 128 powertrain installed between the rear wheels. It went into production in 1971 with a Bertone body. He drew up new Campagnola 4 × 4 variants, and a prototype three-wheeled pickup truck named Pully. He succeeded Aurelio Lampredi as director of engine design in 1980 but retired in 1983.

He died in Turin on November 7, 1999.

Puliga, Orazio Satta *see* Satta Puliga, Orazio

Pullinger, Thomas Charles Willis "T.C." (1867–1945)

Works manager for the Humber factory at Beeston, Nottingham, from 1904 to 1909; general manager of the New Arrol-Johnston works at Paisley near Glasgow from 1909 to 1913; managing director of Arrol-Johnston from 1913 to 1926.

He was born in 1867 in London as the son of a Royal Navy Fleet Paymaster and educated at Dartford Grammar School. After an apprenticeship with J. & E. Hall, engineers of Dartford, Kent, he began his career as a draftsman in the Royal Gun Factory, Woolwich Arsenal, but left in 1889 to set up a bicycle shop in Brockley Road, London. Within a year, he had started production of Parade bicycles in a factory at New Cross, London. M.D. Rucker of Humber, considering whether to set up a branch factory in France, asked him to go to Paris and do some market research. Pullinger met Alexandre Darracq, who was then producing Gladiator bicycles, and became Darracq's personal assistant. Humber never set up a French branch, and in 1894 he joined Herbert O. Duncan and Louis Suberbie as works manager of their Croissy plant, and tooled it up to build motorcycles under license from Hildebrand & Wolfmüller in Munich. He designed a small car with a two-cylinder engine and two-speed transmission, and sold the design to Auguste Teste and Jules Moret of Lyon. Teste, Moret & Cie were makers of chassis frames for several car companies including Rochet-Schneider, Decauville, and Cottereau.

He moved to Lyon and organized production of his car, which was marketed as La Mouche. Returning to England in 1901, he went to work for Sunbeam in Wolverhampton as director and works manager. He put an end to production of the Sunbeam-Mabley and began importing Berliet chassis which were rebadged as Sunbeams. He had a new four-cylinder model ready in 1903 and was planning a six. He left in 1904 when he got a better offer from Humber. At Beeston, he supervised the production of powerful and expensive cars, while the Coventry-built Humbers were mainly light, low-powered and moderately priced. When the two Humber companies were merged, he went to Scotland. He did not design the Arrol-Johnston cars. The four- and six-cylinder models of 1909–13 were the work of Theo Biggs, and the 1919 "Victory" model came from W.G.A. Brown's drawing board. He ran the company through the war years and the difficult period that followed. The "Victory" turned into a blot on the make's reputation and was replaced by a new 3-Litre in 1922. A new 3.3-liter "Empire" followed in 1924 and a new 2.4-liter "Dominion" in 1926.

That year Pullinger retired to his residence on Jersey, biggest of the British Channel Islands. He died in 1945.

Quandt, Harald (1921–1967)

Chairman of BMW from 1960 to 1967.

He was born in 1921 as the younger son of Günther Quandt, head of an industrial holding. He was only seven years old when his parents divorced, and he was raised in Berlin by his mother. When World War II broke out, he was working as a trainee in Poland, in a factory part-owned by his father. He volunteered for military service with the paratroopers and saw action on Crete and Sicily before being taken prisoner at Monte Cassino. He returned to Germany in 1947 and worked as a mason, mechanic, welder and foundry worker. In 1949 he was able to go to Hanover and study engineering. He moved to Stuttgart before graduating but continued his studies and obtained an engineering degree from Stuttgart

Technical University. When his father died in 1954, he inherited one-half of the Quandt industrial holding. In May 1958, Quandt purchased BMW bonds worth DM 15 million. A year later, BMW declared heavy losses, and the biggest creditor, Deutsche Bank, proposed to merge it with Daimler-Benz AG. The Quandt group, with a maximum of discretion, began buying BMW shares on the financial markets and by 1960 held a majority of 52 percent. BMW was "saved." Quandt installed new leadership at BMW and financed the planning and preparation of new products. BMW returned to profits in 1963. Harald Quandt was killed when his executive jet crashed on a flight to the French Riviera on September 22, 1967.

Quandt, Herbert (1910–1982)

Member of the Daimler-Benz supervisory board from 1956 to 1974; member of the BMW supervisory board from 1960 to 1982.

He was born in 1910 as the first son of Günther Quandt (1881–1954), an industrial investor who made his start by providing cash-for-shares in the years following World War I. After an education which included commercial and law studies, he went to work in the administration of his father's business. He became familiar with the inner workings of companies in a variety of industries, including Wintershall potash and shale oil, Mauser firearms and ammunition, Berlin-Karlsruher Industrie-Werke AG machinery and arms, Accumulator-Fabrik AG batteries and Petrix Chemical Works petrochemicals. Shortly before his never-elucidated death in Cairo late in 1954, Günther Quandt had bought a 3.85 percent stake in Daimler-Benz AG, and in 1956 Herbert was named Second Deputy Chairman (the First was Hermann Abs of Deutsche Bank). The Quandt group's ownership in Daimler-Benz AG peaked in 1960 at 15.04 percent. It was still 14 percent in 1974 when it was sold to the Kuwait Investment Office. Quandt had begun investing in BMW in 1958, and by the end of 1960 Herbert held 26 percent of the capital in BMW and his half-brother Harald an equal amount. Herbert continued to buy until he held a 40 percent stake in BMW. He took over Harald's BMW stock upon his death in 1967, and personally held 65 percent of Varta batteries and 40 percent of IWKA[1] machinery and industrial robots. In June 1970, he turned down an offer of DM 750 million for his BMW shares from Kurt Lotz of Volkswagenwerk AG. He took a keen interest in BMW's affairs and had the final voice in all executive appointments.

He never retired and died on June 3, 1982.

1. Industrie-Werke Karlsruhe-Augsburg.

Quaroni, Francesco "Franco" (1908–?)

Managing director of Alfa Romeo S.p.A. from 1951 to 1958, managing director of Société Automobiles Simca from 1959 to 1963.

He was born on March 28, 1908, at Stradella near Voghera and graduated from the Turin Polytechnic in 1932. He joined Pirelli as a sales engineer and rose in the ranks to a tire sales executive title. In 1951 he was called upon to succeed Antonio Alessio as chief executive officer of Alfa Romeo S.p.A. in Milan. He saw the need for spreading the product line into lower price classes and authorized the program that led to the production of the Giulietta with its 1290 cc twin-cam all-aluminum four-cylinder engine in 1954. He also recognized that Alfa Romeo needed alliances to raise its turnover and signed an agreement with Régie Nationale des Usines Renault for production of the Dauphine under license, and also the Saviem[1] Goëland and Goëlette light commercial vehicles. He bought a stake in Brazil's Fabrica Nacional de Motores and started production there of diesel engines, trucks and a 2½-liter four-cylinder passenger car. He understood that the old Portello works in Milan, partly due to the downtown location, faced rising inefficiency in the years ahead. He therefore planned a new industrial complex at Arese, outside Milan, and began construction. By 1958 he had quadrupled the company's turnover, but there was always friction with the IRI and Finmeccanica[2] executives. He resigned in 1958 along with Rudolf Hruska. Fiat's director Gaudenzio Bono invited them both into the Fiat organization and sent them to Simca at Poissy.

Quaroni presided over a thorough model-renewal process, but left when the Chrysler Corporation began buying shares in Simca in 1963. He returned to Italy and worked for two years with Pozzi ceramics, returning to IRI in 1966 as director of its Finsider steelmaking subsidiary, Sant'-Eustacchio in Brescia. In 1970 he was back in the private sector as chairman of San Giorgio Elettrodomestici[3] at La Spezia, an office he held for eight years. From 1978 to 1980 he was chairman of Masoneilan, a Studebaker-Worthington subsidiary and concluded his career with 18 months as chairman of ITT's[4] IAO subsidiary in Turin.

1. Société Anonyme de Véhicules Industriels et Equipements Mécaniques, a Renault subsidiary created in 1957 by the merger of Automobiles Industriels Latil and Société d'Outillage Mécanique et d'Usinage d'Artillerie with Renault's truck and bus division.
2. Istituto per la Ricostruzione Industriale, state-owned holding, grouped its engine and automotive interests into the Finmeccanica group.
3. Electrical home appliances.
4. International Telegraph & Telephone.

Quement, Patrick Le *see* Le Quement, Patrick

Rabe, Karl (1895–1968)

Chief engineer of Austro-Daimler from 1923 to 1930; head of the Porsche drawing office from 1930 to 1965.

He was born on October 29, 1895, at Pottendorf in Austria and had a minimum of formal schooling when he went to work for Austro-Daimler in 1913 as a trainee draftsman. He took evening classes in mechanical engineering and was promoted to design engineer. In 1923 he succeeded Dr. F. Porsche as chief engineer. He designed the ADR 11/70 and 12/100 with overhead-camshaft 3-liter six-cylinder engines, central tube "backbone" frame and rear swing axles. The ADR 6 with a 3.6-liter engine was added in 1929, followed by the ADR 8, with a straight-eight 4.6-liter engine, in 1930. He left Wiener-Neustadt in 1930 to join the newly established Porsche

Konstruktions-GmbH in Stuttgart, becoming head of the drawing office. He worked on the early "people's car" proposals for Zündapp and NSU, and led the studies for the KdF-Wagen, which evolved into the production-model Volkswagen by 1939. He remained at his drawing board throughout World War II, mainly occupied with military vehicle projects, and in 1945 accompanied the Porsche family to Gmünd, Katschberg, in Austria. He assisted Ferry Porsche in the design of the four-wheel-drive Cisitalia racing car (Porsche 360) and helped create the VW–based Porsche 356 sports car in 1948. After the return of the Porsche headquarters to Stuttgart-Zuffenhausen in 1950, he took charge of all projects for outside clients and had little to do with the subsequent evolution of the Porsche product line. He retired in 1965 and died on October 28, 1968.

Radermacher, Karlheinz (1931–)

BMW management board member for research and development from 1976 to 1983.

He was born on August 10, 1931, in Dortmund and graduated from the Karlsruhe Technical University with a degree in mechanical engineering. He stayed on at Karlsruhe for postgraduate studies, leading a group conducting research in machine-tool and automobile engineering. In 1962 he signed up with Siemens & Halske in Berlin as head of a test laboratory, and two years later he joined Rheinstahl-Hanomag as chief engineer of all testing programs, including the engine and vehicle departments of Rheinstahl-Henschel AG in Kassel. He arrived at director's level in the Rheinstahl group in 1967 but resigned a year later to join SKF (ball bearings) as technical director for its German branch factories. He also took on responsibility for SKF marketing and sales in Germany in 1970. The Quandt family engaged him late in 1972 as deputy director of BMW's research laboratories. He was the mastermind behind the 745i, 535i and 525i in 1980–81, and planned the 325i, 324d, and 325iX. He left BMW on August 1, 1983, to join ZF[1] as deputy chairman and director of the ZF technical center. In May 1986, he moved to Pierburg[2] as chairman

of the management board, and retired in July 1987. He then founded Techno-Consult GmbH in Munich and opened branch offices in Berlin and the United States.

1. Zahnradfabrik Friedrichshafen.
2. Formerly Deutsche Vergaser GmbH.

Railton, Reid Anthony (1895–1977)

Assistant design engineer, Leyland Motors Ltd. from 1915 to 1923, engineering consultant to Arab Motors Ltd. in 1926-27, chief engineer of Thomson & Taylor from 1929 to 1939.

He was born on June 24, 1895, at Alderley Edge, the son of a Manchester stockbroker. He attended Rugby School and graduated from Manchester University with a degree in science. He began his career in 1915 with Leyland Motors Ltd. under the direction of J.G. Parry Thomas and worked on military vehicle design during World War I. After the war, he was Parry Thomas's assistant in creating the Leyland 8 luxury car of 1921–24.

He resigned from Leyland in 1923 to accompany Parry Thomas in setting up his racing-car shop at Brooklands, near Byfleet, Surrey. He designed the Arab sports/touring car with a single-overhead-camshaft engine (copying the eccentric-arm camshaft drive from the Leyland), produced in a factory at Letchworth from 1926 to 1928. He completed Parry Thomas's preparation of the Brooklands-type Riley Nine for Thomson & Taylor of Surbiton, who took over the Brooklands shop when Parry Thomas was killed. In 1931 he designed a Blue Bird speed-record car for Malcolm Campbell showing a remarkable understanding of aerodynamics.

In 1933 he was associated with the chassis design of the ERA racing car and lent his name to Noel Macklin's Hudson-chassis touring and sports cars. In 1938 he designed a streamlined speed-record car for MG and the remarkable Napier-Railton four-wheel-drive, twin-engine, speed-record car for John Cobb. He was a member of Cobb's team when the car set a new World Land Speed Record in 1939 on the Bonneville Salt Flats in Utah — and stayed in the U.S. He worked as an engineer on speed-boat engines with the Hall-Scott Motor Company from 1939 to 1945 and then became a consulting engineer. He was a consultant to the Hudson Motor Car Company from 1948 to 1956. He settled in Berkeley, California, where he died in August 1977.

Randle, James Neville "Jim" (1938–)

Chief vehicle research engineer of Jaguar Cars, Ltd., from 1972 to 1978; director of Jaguar vehicle engineering from 1978 to 1980; director of Jaguar product engineering from 1980 to 1990; director of engineering, Jaguar vehicles and concepts, from 1990 to 1992.

He was born on April 24, 1938, and educated at Waverly College, Birmingham, and the University of Aston. From 1954 to 1959 he was an engineering trainee with Rover, became a technical assistant with Rover in 1959 and later qualified as a project engineer. He arrived at Jaguar's Browns Lane factory on September 5, 1965, and was assigned to the XJ 40 program as a project engineer. That program led to the XJ 6 of 1968.

He directed the teams that created the XJ 12 of 1972 and the XJS of 1975 and was responsible for their subsequent development. In the 1980s he was working on advanced and hybrid-electric powertrains for Jaguar. He left Browns Lane in 1992 to become director of the automotive engineering center at Birmingham University, turning into a voluble advocate of environmental protection. In 1996 he designed the Hermes taxicab, financed in part by London Taxis International but mainly by himself, to promote a gas-turbine hybrid-electric power system.

Rasmussen, Jörgen Skafte (1878–1964)

Owner and managing director of Zschopauer Motorenwerke J.S. Rasmussen from 1923 to 1932; Auto Union AG management board member for technical affairs from 1932 to 1934.

He was born on July 30, 1878, at Nakskov, Denmark, and was apprenticed to a smith in Copenhagen. He studied engineering at Mittweida Technical College in Saxony and graduated in 1900. He began his career with the

Rheinischen Maschinenfabrik in Düsseldorf, but returned to Saxony in 1903 and set up shop as Rasmussen & Ernst GmbH at Chemnitz to make heaters for steam boilers and steam condensers. In 1907 he bought an idle textile factory in the Zschopau valley and founded the Zschopauer Maschinenfabrik. His business prospered up to 1914, when it was converted to making war materiel. He began experimenting with steam-powered trucks[1] in 1916, and also built some electric vehicles. When the war ended, he was primed for expansion. He acquired 12 plants from companies that were in trouble, and soon had his own screw works, stamping plant, and an extrusion plant, later adding forges and aluminum and iron foundries. Production of a two-stroke toy engine began in 1919, followed two years later by a cyclomotor (auxiliary power unit for bicycles), and a motor scooter. In 1921 he began financing the minicar experiments of Slaby & Beringer in Berlin, and set up a company in Saxony to make Framo[2] three-wheel delivery vehicles. Testing of the SB minicar began in 1923, and within a year, Rasmussen bought their company and started production in Berlin-Spandau.

The first proper DKW motorcycle, Type ZM with a 175-cc two-stroke engine, went into production at Zschopau in 1924. Within three years, he was the biggest motorcycle manufacturer in Europe, growing bigger still in 1928 by taking over the Schüttoff works, whose motorcycles had four-stroke engines. In 1926 he made a study tour of the American auto industry and discovered the Rickenbacker car. A year later, he purchased the jigs, tools, dies, designs and patterns of the Rickenbacker Motor Company and shipped it all to Zwickau, where it was installed in the Audi factory he acquired in 1928. By that time, the Spandau factory was building DKW cars with two-cylinder two-stroke engines, and added a two-stroke V-4 model in 1930, while Rasmussen was drawing up a front-wheel-drive car.

In 1928, however, Rasmussen alerted the State Bank of Saxony to the looming crisis in German industry, and the bank took a 25 percent stake in his company. In 1929 the bank acquired the remaining 75 percent. Rasmussen then purchased the ailing Horch company as his personal property. Production of the front-wheel-drive DKW began in 1931 in the Audi plant at Zwickau, while Audi car production was moved to the Scharfenstein works at Chemnitz. The DKW, Audi and Horch companies were absorbed into Auto Union AG, established on June 29, 1932, financed by the State Bank of Saxony. It was an industrial co-operative society, formed with a view to privatization if and when the shares rose to fair value. Later in the year, Auto Union AG purchased the Wanderer company and its works at Siegmar-Schönau. Rasmussen was ill at ease in his Auto Union AG position, however, and withdrew at the end of 1934. He devoted himself to his other properties, including Metallwerk Zöblitz, Flugzeugfabrik Erla, and Framo-Werke GmbH, directing them until 1945, assisted by his three sons. In 1948 he retired to Copenhagen, still a Danish citizen, and died on August 12, 1964.

1. Campf Kraft Wagen, shortened to the initials DKW. They were given other meanings for later products, Des Knaben Wunsch for the toy engine, and Das kleine Wunder for the cyclomotor.
2. Framo-Werke GmbH, originally at Frankenberg, later moved to Hainichen in Saxony.

Ravenel, Raymond (1926–)

Managing director of Automobiles Citroën from 1968 to 1970, chairman of the Automobiles Citroën management board from 1970 to 1974.

He was born on May 5, 1926, at Noyen (Sarthe) and educated at the Ecoles des Arts et Métiers of Angers and Lille. He joined Citroën in 1949 in production timing studies and collaborated on tooling matters. He became involved with professional training schemes and was later attached to the personnel department, where one of his responsibilities was to place engineers where their talents could be put to the most advantageous use. In 1966 he was attached to the management board.

This was a difficult time, when Citroën had no mid-range cars. He assisted in the planning of the GS sedan with its air-cooled 1016 cc flat-four, power steering and hydro-pneu-matic suspension, introduced in 1970. It was a time when the Wankel engine loomed large in Citroën's lens and a merger with NSU seemed possible. He tried to develop synergistic effects from the alliance with Berliet and presided over negotiations with Michelin and Fiat officials during most of 1968, with a view to establishing a balanced Fiat-Citroën partnership. The Citroën CX and the Lancia Gamma came out of a joint program in which conflicting engineering philosophies made any meaningful collaboration impossible. His signature also appeared on documents concerning Citroën's purchase of Officine Alfieri Maserati, S.p.A. and their industrial and commercial integration. He resigned when Citroën was taken over by Peugeot at the end of 1974.

Ravigneaux, Pol (1873–1953)

Technical director of de Dion–Bouton from 1906 to 1914.

He was born in 1873 in the Ile-de-France into a well-to-do family. After graduation from the Ecole Polytechnique, where he revealed a fertile imagination in addition to proving himself an excellent mathematician, he went to Canada and worked as a civil engineer with the Dominion Bridge Company. News of a burgeoning automobile industry in France brought him back to Paris in 1900, where he was promptly signed up as a draftsman by de Dion–Bouton. For the next several yeas he delivered learned papers on gearing and gear systems to French engineering societies. He also established a formula for calculating the diameter of exhaust pipes as a function of required power output. In 1905 he invented a planetary free-wheel mechanism, and a year later he became technical director of de Dion–Bouton.

He did not invent the V-8 engine (Léon Levavasseur and Clément Ader had preceded him) but he was the first to assure economical production of several sizes of V-8 engines by designing them to share the cylinder blocks, valve gear, camshafts and a number of minor parts, with concurrent four-cylinder units. De Dion–Bouton offered cars with V-8 engines of his design from 1910 through 1922. In 1911 he devised a four-speed (and reverse) planetary transmission with no more than eight toothed wheels. Various forms of the Ravigneaux gear system were used in multiple applications for years, and in 1935 he perfected an improved control system for planetary transmissions. Ravigneaux-type gear combinations held a predominant position in the area of semi- and automatic transmissions until it was displaced by the Simpson system in the 1950s.

He resigned from de Dion–Bouton in 1914 to work for Société Mayen, manufacturers of industrial gear systems.

Rayner, Herbert (b. 1898)

Chief draftsman of Morris Motors, Ltd., from 1949 to 1952; chief designer of Morris, Riley and Wolseley cars, British Motor Corporation, from 1952 to 1963.

He was born on April 8, 1898, in Nottingham and educated at King Edward VI grammar school in Grantham. He studied engineering at the Lincoln Technical School after serving an apprenticeship with Ruston & Hornsby in Grantham, became a Royal Flying Corps cadet in 1918 and trained as a pilot. He returned to Ruston & Hornsby as a designer of heavy transport equipment, agricultural and construction equipment, from 1919 to 1929. He joined Morris Motors, Ltd., at Cowley in 1929 as a section leader in the passenger-car drawing office, working on both chassis and body engineering projects. Much of his work went into the 1934 Ten-Six, 1935 Twelve-Four and Sixteen, and the 1939 Morris Eight. He was on Active Service during World War II, in charge of an RAF repairs unit, and went back to the Morris drawing office in 1945. He was involved with the 1948 Morris Six and the new Minor, and was promoted to chief draftsman. Late in 1952 he was named chief designer, with responsibility also for Wolseley and Riley. He had a major design role in the 1955 Morris Oxford and the 1959 Riley 1.5 and Wolseley 1500. He was sidelined when the Austin Design Office took over the engineering of Morris, Riley and Wolseley cars, and retired in 1963.

Rebling, Arthur (1873–1953)

Technical manager of Automobilfabrik Fritz Scheibler in Aachen from 1901 to 1902; works manager of August Horch & Cie in Zwickau in 1902–03; chief engineer of Maurer Union from 1903 to 1907; managing director of Automobilfabrik Turicum AG of Uster near Zurich from 1907 to 1914.

He was born on February 26, 1873, at Gotha in Saxony and studied engineering at the Karlsruhe Technical University. Upon graduation, he went to America to gain industrial experience and worked for the Western Wheel Works in Chicago, a main supplier to the automobile industry.

Returning to Germany in 1897, he was engaged to start car production at Fahrzeugfabrik Eisenach under a license from Decauville. Their car was named Wartburg, the first model being a straight Decauville copy with its air-cooled parallel-twin rear-mounted engine, two-speed gearbox, and sliding-pillar independent front suspension. It also borrowed from Decauville the absence of any rear-axle suspension. For 1899, Rebling put the rear axle on semi-elliptic springs and adopted water-cooling for the engine.

He left Eisenach to join Fritz Scheibler and designed a light car with a front-mounted flat-twin engine, friction drive, and final drive by chain to the rear axle. His Maurer Union was a single-cylinder mini-car with friction drive, a formula which he adopted also for Turicum.

He left the auto industry at the outbreak of World War I and never designed another car.

Rédélé, Jean (1922–)

Originator of the Alpine (Renault-Alpine) car in 1952; managing director of Automobiles Alpine from 1952 to 1973; beginning in 1973, chairman of Automobiles Alpine.

He was born on May 17, 1922, in Dieppe, the son of a garage owner. He grew up in and under cars, but chose a business education, graduating from the Ecole des Hautes Etudes Commerciales in 1945. His final exam paper was a treatise on Renault's industrial activities. It was read by Renault's management and an interview ensued which led to a dealer contract in 1946. Within five years he ran five dealerships. In 1950 he started driving Renault 4 CV cars in competition and introduced modifications to give them more speed. In 1952 he opened a shop in rue Forest, Paris 18, to produce five-speed gearboxes for the little Renault, and showed the first Alpine car, a Renault-based sports model. In 1955 he introduced the Alpine Coach, using predominantly Renault components, with a Michelotti-designed body, reproduced in fiberglass-reinforced plastic by Chappe Frères & Gessalin of St. Maur.

When Willys-Overland signed an agreement with Renault to assemble the Dauphine in Brazil, Renault found that the local market was more interested in the Alpine, which led to production of the Alpine A-106 (sold as the Willys Interlagos) at Sao Paulo from 1958 until Willys-Overland of Brazil was taken over by Ford in 1969.

He moved Automobiles Alpine to bigger premises in Dieppe in 1969, continued production of the Alpine A-108 and A-110 "berlinette." In 1963 Rédélé built a team of Renault-Gordini–powered racing cars for Le Mans, and in 1967 the French Ministry of Industry awarded Automobiles Alpine a grant of 700,000 francs. Renault also began investing in the company with a view to expanding production. By 1976, Renault held majority control.

The Alpine A-310 was launched in March 1971 with a four-cylinder Renault engine, and became available with a Renault V-6 in 1976. The A-110 was taken out of production in 1977. The A-310 was replaced by the Renault-Alpine V-6 GT in 1984, augmented by the arrival of the V-6 Turbo in 1986.

Reichstein, Carl (1847–1931)

General manager of Reichstein Brothers, Brennaborwerke, from 1871 to 1910; chairman of Reichstein Brothers, Brennaborwerke, from 1910 to 1931.

He was born in 1847 in Brandenburg, the third son of an artisan-craftsman. Together with his brothers Adolf (1839–1910) and Hermann (1841–1913) he set up a mechanical workshop in 1871. Their first product line was baby carriages. By 1874 their shop was too small and they erected a new factory alongside the main railway station in Brandenburg. In 1883 Carl took the initiative to open a bicycle department in 1883 and led it to great commercial success. In 1900 the factory had 800 workers, and Carl discovered automobiles. He set up an experimental shop to build test cars in 1903. His son, Carl Reichstein, Jr. (1885–1955), completed the first one, with an 8-hp Fafnir parallel-twin, in 1906. In 1908 he began production of a different machine, the three-wheeled Brennaborette with a 452-cc single-cylinder engine and chain drive to the rear axle. It was produced up to 1911. Two four-wheeled cars were developed in 1908–10, a two-cylinder 904-cc Fafnir-powered model and a bigger vehicle with Brennabor's own 1.3-liter four-cylinder engine. He became chairman in 1910, succeeding his brother Adolf, and when Hermann died three years later, he became the sole owner of the Brennabor Works. He started a major expansion of the automobile department in 1910 and put his sons Carl Jr. and Walter (1886–1977) in charge of it. In 1914, production was four to six cars a day. That year they opened their own body works including a stamping plant and machine shops. In 1920 they purchased the derelict Brandenburg gas works and rebuilt it as a foundry, with facilities for hardening and other processes. His youngest son Eduard returned from eight years in the American auto industry and designed the 1921 Brennabor 6/20. The factories then possessed 2000 machines and occupied 3500 workers. Brennaborwerke linked up with NAG and Hansa-Lloyd to combine their sales networks as the GDA[1] which probably helped maintain the cash flow of its members during the years of hyperinflation in Germany. Carl Sr. gave less attention to company affairs and devoted himself to civic projects. He set up the Carl Reichstein Foundation in Brandenburg for the care of crippled children. In 1925 his sons installed a moving assembly line for the 6/24 and 8/32, and GDA was pushing its members towards joint purchasing and more rational product planning, with fewer and not mutually competing types of cars. But there were too many problems and the GDA was dissolved in 1928. Brennaborwerke had over 8000 workers in 1929 and was turning out 100 cars a day, plus 1000 bicycles and 2000 baby carriages.

1. Gemeinschaft deutscher Automobilfabriken.

Reitzle, Wolfgang (1949–)

Head of BMW's planning office from 1984 to 1986; leader of BMW's product development group from 1986 to 1987; BMW director of marketing and product in 1988; BMW management board member for research and development from 1988 to 1999; chairman of Ford's Premium Automotive Group (Jaguar, Land-Rover, Lincoln, Mercury and Volvo) from 1999 to 2002.

He was born on March 7, 1949, in Bavaria, and graduated with a degree in mechanical engineering from Munich Technical University, going on to postgraduate studies in economics and labor science. He joined BMW in January 1976 as a specialist in production technology, and was named head of the methods testing department in April 1976. By April 1977, he was head of process engineering, and in November 1981 took charge of the foundry division, engine production, and parts production. In 1983 he was the closest assistant to the technical director, Karlheinz Radermacher, and was groomed to succeed him.

He supervised the development of the 1987 7-series cars and the 1988 5-series cars. He replaced Bernd Pischetsrieder as chairman of Rover in September 1995 and opposed BMW's plans to make Rover a mainstream high-volume auto maker. He saw Rover as a distinctive niche-market brand and resigned in protest following a boardroom quarrel over Rover strategy. Jacques Nasser brought him to Ford, but Reitzle lost his main support when Nasser was ousted late in 2001, and left Ford in April 2002 to become chief executive of Linde AG (forklift trucks and gas-industry equipment) in Germany.

Remington, Alfred Arthur (1877–1922)

Chief draftsman of Wolseley from 1902 to 1904; chief designer of Wolseley from 1904 to 1908; chief engineer of Wolseley from 1908 to 1920.

He was born in 1877 at Sutton Coldfield in the Midlands and served an apprenticeship with Kynoch, Ltd., of Birmingham. He studied engineering and joined the Wolseley Sheep Shearing Machine Company as a draftsman in 1900. A year later he accompanied Herbert Austin to the Wolseley Tool & Motor Car Company at Adderley Park, Birmingham. He stayed on when Austin left, and came under the orders of John D. Siddeley who ordered a phasing-out of all models with single-two and four-cylinder horizontal engines in favor of new Wolseley-Siddeley cars with parallel-twin and four-in-line vertical engines, produced in a Vickers factory at Crayford, Kent. After Siddeley's resignation in 1909, all the parts and equipment were transferred to Adderley Park, and Remington was given the task of creating an entire range of new Wolseley cars.

Five new four-cylinder models from 12/16 to 40 hp made their debut in 1910, alongside three six-cylinder cars from 24/30 to 60 hp. He introduced worm-drive rear axles on the 12/16, 16/20 and 24/30. Most of the engines had dual ignition. He designed Wolseley trucks and armored vehicles for the War Department, and prepared for a return to civilian production by preparing the 10-hp light car and the 15-hp touring car, both powered by overhead-camshaft four-cylinder engines. Both went into production in 1921, more than a year after he had resigned from Wolseley.

He joined Karrier Motors, truck manufacturers at Huddersfield, Yorkshire, as chief engineer in 1920, but resigned in 1921 to set himself up as a consulting engineer in Birmingham.

He died unexpectedly in the summer of 1922 at Wylde Green, Birmingham.

Renault, Louis (1877–1944)

Partner in Renault Frères from 1899 to 1908; owner and chairman of Automobiles Renault,

L. Renault Constructeur, from 1908 to 1922; chairman and majority owner of the Société Anonyme des Usines Renault from 1922 to 1940.

He was born on November 12, 1877, in Paris as the son of a button and drapery maker. He had a talent for practical inventions but cared little for book learning, and was regarded as a poor pupil at the Lycée Condorcet. Yet he was only 11 when he installed a battery and an electric lamp in his room and only 15 when he invented a device to prevent train collisions. He prepared for the Ecole Centrale des Arts et Manufactures, but broke off his studies to join Delaunay-Belleville as a draftsman. Called up for military service, he joined the 106th infantry regiment at Chalons-sur-Marne, where he was noted for inventing a system of moving targets. Back in civilian life, he bought a de Dion–Bouton tricycle and began making sketches for an automobile. He built his first car in his own workshop in 1898–99, with a single-cylinder de Dion–Bouton engine, three-speed gearbox with direct drive on top gear, and shaft drive to the rear axle. He told his older brothers Fernand (1864–1909) and Marcel (1874–1903) about his plan to start car production, and they put up the capital (FRF 60,000) and formed a company, Renault Frères, on March 21, 1899. They set up shops alongside the family property at Billancourt, and produced 350 cars in 1900.

He was granted French patent (no. 285,753) on February 9, 1899, for the direct-drive gearbox, and in 1903 he sued Berliet, Brasier, Clément-Bayard, Gobron-Brillié and Mors for patent infringement. Renault filed 74 patent applications from 1899 to 1904. His own engine factory became operational in 1902, introducing parallel-twin and four-cylinder engines. The first four-cylinder Renault car was the 20/30 hp 1902 Type K.

After Marcel's accidental death in the 1903 Paris–Madrid road race, Louis purchased his shares and signed a new contract with Fernand. When Fernand's health began failing, Louis cajoled him to sell his shares, too, and Louis became sole owner in 1908. In the following years, great strides were made in productivity. With 2000 workers, Renault made 2500 cars in 1907, and in 1910, 3200 workers completed 6800 units. Renault's first six, the 54 CV 9½-liter Type AR, appeared in 1908. That year the plant also began quantity production of the 8 CV 1205-cc two-cylinder Type AX. To see mass production with his own eyes, Louis Renault visited the Ford plants in Detroit in April 1911. In March 1917 Renault installed two short, side-by-side moving assembly lines at Billancourt, but they were not used for cars. They were used for building tanks. During World War I, the Renault factories also produced important numbers of aircraft engines, guns and shells.

In 1918 he purchased a tract of land on the south side of Le Mans, expecting further defense contracts. But the war ended, and the land lay idle for years. He knew he would need it sooner or later, but it was not until 1936 that he installed foundries, machine shops, and farm-tractor production at Le Mans. The company was reorganized as a joint stock corporation on March 17, 1922. That year Renault had a range of 12 models, including the KZ which was kept in production until 1933. But he realized that too many models restricted plant capacity, and cut the lineup to six in 1923. His 15,000-man work force had built 9500 cars in 1922. He hired another 5000, and turned out 23,600 cars in 1923. The 6 CV Type NN went into production in 1924 and produced up to 1930. In April 1928 Louis Renault took his engineers Edmond Serre and Emile Tordet along for a factory tour of the U.S. auto industry. The model range spread again, with the difference from the earlier decade that by 1932, many models were created by making new combinations of existing engines, chassis and bodies. The Monaquatre was 7 CV, its chassis serving as a basis for the 10 CV Primaquatre and Vivaquatre. The 8 CV Monastella was a small six, based on the Primaquatre. The 15 CV Primastella and Vivastella were sisters. The Nervastella had a small (24 CV) straight-eight, its chassis serving also for the big eight-cylinder 40 CV Reinastella.

By 1938 Renault was a diversified manufacturer, producing machine tools (mainly for its own use), marine engines, tractors, railcars, aircraft engines and airplanes. The result was a lot of extra costs with little or no profits. Louis Renault decided to return to the core activities: cars and trucks. But the outbreak of World War II changed everything. In May 1940 Louis Renault flew to the United States on a French government mission to advise on tank production, returning to France on July 3 and to Paris on July 23. He then found his factories working under the orders of three German commissars, Karl Schippert, Wilhelm von Urach, and Alfred Vischer, all from Daimler-Benz AG. He resigned on December 4, 1940, holding on only to the vehicle concept, design and production departments for himself. If the Occupation was troublesome, the Liberation multiplied his problems. The Renault plants were requisitioned on September 27, 1944, by order of General Charles de Gaulle, and Louis Renault was arrested on October 7, 1944, on false charges of collaboration with the enemy. After ten days in the medical ward at Ville Evrard, he was placed in the Clinique Saint-Jean de Dieu, where he died during the night of October 24, 1944, under circumstances that remain unclear.

Reuter, Edzard (1928–)

Daimler-Benz AG finance director from 1980 to 1987; vice chairman of the management board in 1987; chairman of the management board from 1987 to 1995.

He was born on February 16, 1928, in Berlin the son of Ernst Reuter, a former communist but still left-wing politician who became mayor of Magdeburg in 1933. The family fled from Nazi rule to Turkey, where Edzard grew up. They returned to Berlin in 1946, and his father was elected mayor of West Berlin. Edzard studied at Göttingen and in West Berlin, graduating in law and physics. He spent two years, 1954–56, as an assistant in the Law Faculty of the Berlin Free University, and began his career with Ufa (motion pictures). He also worked for Bertelsmann (travel agents, publishers) for a few years prior to joining Daimler-Benz AG in 1964. He worked in secretarial offices, in the financial department, and in

corporate planning. He was head of planning by 1971 and became deputy board member for planning and organization in 1973. In the spring of 1976 he became a full member of the management board and thought he would become the next financial chairman. But the supervisory board, worried about his political leanings, chose Werner Breitschwerdt to succeed Joachim Zahn in 1979. For the next eight years he did everything he could do to discredit and undermine Breitschwerdt, supported by the labor-union representatives on the board and the Deutsche Bank, the biggest single shareholder in Daimler-Benz AG.

The company was still making glorious profits when he became chairman. But he did not want to invest in its core activities. He announced a new diversification strategy in 1984, and the following year, bought MAN's half of MTU[1] and majority control of Dornier (aircraft). He secured majority control of AEG[2] in 1986 and in 1989 acquired 51 percent of MBB[3] from the government. By March 1991, Daimler-Benz AG held 80 percent of MBB. By 1994 he had invested over DM 8 billion in diversification, without having any clear idea of how the parent organization could benefit from this sort of expansion. In fact, it did not. Daimler-Benz AG declared losses of DM 65.8 million in 1991–92 and DM 1.8 billion in 1992–93. Irate shareholders called him "the biggest destroyer of capital in German history" and he was forced out of office in May 1995, though he was able to hold a seat on the supervisory board until March 1996.

1. Motoren und Turbinen Union.
2. Allgemeine Elektrizitäts-Gesellschaft.
3. Messerschmitt-Bölkow-Blohm (aircraft).

Revelli de Beaumont, Mario (1907–1985)

International design consultant.

He was born on June 25, 1907, in Rome, into a noble Piedmont family as the son of the inventor of the Piat automatic gun, used by the Italian Army in the conquest of Ethiopia and by the British forces in World War II. He attended a technical school but was only interested in inventions. He was only 16 when he designed a racing motorcycle which he built, with the help of his brother Gino, and to his own and general amazement, rode to victory in the 500-cc class of the World Cup race at Monza in 1925. He made a lot of car-body designs which he showed to Italy's coachbuilders, leading to several short-term contracts, and a steady relationship with Stabilimenti Farina in 1927. He also designed cars for Ghia, Sala, Castagna, Montescani, Casaro, and Carrozzeria Moderna. The combination of family connections and artistic success made him a welcome member of Italy's high society, and he became friends with Giovanni Agnelli, who bought some of his designs privately and pushed them on the Fiat body-engineering office. That led to the gradual modernization of the Fiat car shape, beginning with the 1931 524C and the 1933 518C Ardita. In 1932 he made some wildly radical sketches which Fiat's body engineer Rodolfo Schaeffer turned into the 1935 "1500" production car. He held a patent on the flat-vent window, dated 1927, which Fisher Body acquired to introduce "No-Draft Ventilation" in the 1933-model GM bodies, and a patent on sill- and roof-locks for car doors, to eliminate the B-post. It was first exploited by Bertone for an Alfa Romeo body in 1934 and then on the Fiat 1500 and Lancia Aprilia. From 1932 to 1939 he had a very close relationship with Viotti, but continued to make designs for Stabilimenti Farina and Pinin Farina, Ghia and Bertone. In 1931 he designed a minibus, anticipating the Matra Espace concept of 50 years later, and in 1933 he proposed a sports car with a transverse engine mounted in an inboard/rear position, anticipating the Lamborghini Miura by 33 years. In 1938 he designed a one-box taxicab, offering alternative engine locations, and in 1940 a forward-control battery-electric pickup truck. During World War II he also designed military and colonial trucks and trailers, which led to his indictment for collaboration with the enemy in 1945 and a prison sentence. Released in 1948, he designed a neat little coupe for Stabilimenti Farina, which Jean Daninos bought

and produced as the Simca Sport. He also renewed ties with Viotti, Pinin Farina, CANSA and Fiat. He spent the years 1952–54 in Michigan as a styling consultant to General Motors, helping to find definitions for their future compact cars (Corvair, Tempest, Special and F-85). He lived in Paris from 1955 to 1963, creating a studio for advanced design for Simca, and then set up his own studio at Grugliasco, outside Turin. He acted as an advisor to the Art Center College of Design in Pasadena, California, and served as chairman of the examination board with the School of Applied Art and Design in Turin. He started preparing for the Revelli Foundation, to preserve his works and educate new talent in automobile architecture and styling, but died on May 29, 1985.

Ribeyrolles, Paul (b. 1874)

Chief engineer of A. Darracq & Cie from 1899 to 1905.

He was born in December 1874 in Paris and graduated from the Ecole Centrale des Arts et Manufactures at Chalons-sur-Marne in 1896. He began his career in the bicycle factory (Usines Perfecta at Suresnes) of Société Aucoc & Darracq, and was their chief draftsman by the time the first Perfecta motorcycles were designed. Darracq made a number of cars under Bollée license from 1898 to 1901, but Ribeyrolles worked on original Darracq automobile projects from 1897 onwards. When the Bollée license expired in 1901, Ribeyrolles felt free to draw on his own imagination to a much greater extent. He adopted shaft drive for the 12-hp voiturette of 1901 and put mechanically-operated inlet valves on the 24-hp four-cylinder racing car for the 1902 Paris–Vienna road race. For the 1903 20-hp production model, he abandoned the reinforced wood frame in favor of a pressed-steel frame. In 1905 he designed a speed-record car with a 22,518-cc V-8 engine with overhead valves, which delivered 200 hp at 1200 rpm. He resigned from Darracq in 1905 when Alexandre Darracq bought the manufacturing rights to a new Léon Bollée car design.

Ricart, Wilfredo Pelayo y Medina (1897–1974)

Founder of Ricart-Espana in 1927; designer of the 1929–31 Nacional Pescara; technical consultant to Alfa Romeo from 1936 to 1940; head of Alfa Romeo special projects from 1940 to 1945; technical director of ENASA[1] and chief designer of the Pegaso Z-102 and Z-103.

He was born on May 15, 1897, in Barcelona and graduated in 1920 from the Barcelona Engineering College with a degree in industrial engineering. That same year he joined Vallet & Fiol as a director, and in 1921 established the Sociedad de Construccion dos Motores Rex. A year later he created the Ricart & Perez sports car with a dual-ignition 16-valve four-cylinder engine that ran up to 5500 rpm.

In 1926 he set up a company, Industria Nacional Metalurgica (later Motores y Automoviles Ricart) in Barcelona and built about 950 Ricart-Espana six-cylinder cars over a 4.5-year period. He designed the eight-cylinder 2.8-liter Nacional Pescara 15/70 produced from 1929 to 1931. A second model, the 16/85/125 with its supercharged in-line 10-cylinder 4.2-liter engine, did not get beyond the prototype stage.

From 1932 to 1934 he acted as technical consultant to several companies and designed, among others, a supercharged two-stroke flat-twin diesel engine, the Nacional SuperDiesel. In 1935 he established the Valencia Public Transport Company and organized its bus lines.

He left Spain in 1936 and was signed up as a technical consultant by Alfa Romeo, where he designed a V-6 diesel engine in 1937 and worked on racing-car projects, earning the everlasting enmity of Enzo Ferrari. During World War II he designed a 3000-hp 28-cylinder aircraft engine and in 1943 led the team that designed the radically advanced Gazzella passenger-car prototype, aimed at postwar production.

At the end of March 1945, he returned to Barcelona and in 1946 established the Centro de Estudios Tecnicos de Automocion, with a team of consulting engineers under his direction. When the Franco government's INI[2] decided that Spain must have its own truck-and-bus industry, Ricart was consulted and later named technical director of ENASA. Generalissimo Franco also wanted Spain to have a car of its own, imagining an all-new design and an all-new factory. Ricart talked him out of that and steered him into making a deal with Fiat, offering a ready-made model range. That led to a joint venture between INI and Fiat under the name Seat.[3] In 1950 Ricart also revived the Gazzella project as a training scheme for a new generation of engineers, who developed it into the Pegaso Z-102 sports and grand-touring car. Giving vent to his spleen, he crowed: "Ferrari's horse only rears; mine flies." The Z-102 and its successor, the Z-103, were made in small numbers up to 1954.

Ricart was dismissed from ENASA in 1957, moved to Paris and served as president of SA Lockheed for two years, 1959–61. He stayed on as technical consultant to Lockheed, Ducellier, Bendix and Air Equipement from 1961 to 1965, and then retired to Barcelona, where he died on August 19, 1974.

1. Empresa Nacional. de Autocamiones, Sociedad Anonima.
2. Instituto Nacional de Industria.
3. Sociedad Espanola de Autmoviles de Turismo, S.A.

Richard, Carl Hermann Ludwig (1880–1948)

Works manager of Deutsche Automobil-Industrie Friedrich Hering from 1900 to 1908; vice chairman of Deutsche Automobil-Industrie Hering & Richard AG from 1908 to 1920; chairman of Carl Richard & Co. GmbH from 1926 to 1934.

He was born on November 12, 1880, in Berlin and educated locally and in Paris, where he met the industrialist Friedrich Hering as a mere boy of ten. Returning to Germany he visited the Hering family at Ronneburg in Saxony and became familiar with its production of bicycle parts. He developed a friendship with Friedrich Hering's son Max[1] and steered him towards production of automobile components. Soon they were making frames, axles and wheels for the fledgling auto industry. They decided to build complete cars, based on the de Dion–Bouton Vis-a-Vis, in 1900, naming it the Rex Simplex.[2] Production began in 1901, with engines supplied from de Dion–Bouton or Max Cudell, who held the German manufacturing rights. In 1902 Rex Simplex adopted the de Dion–Bouton Populaire design as a basis for its production cars. The cars sold well in Germany and were also in considerable demand in Russia. The automobile department became a joint stock corporation in 1908, with Max Hering as majority shareholder. Carl Richard became vice chairman, while Alfred Hering was named managing director. They opened an engine plant and added a four-cylinder 2.1-liter 9/16 model in 1909. In 1914 Richard was called up for military service and sent to Montmedy, Belgium, as head of a motor pool. The Ronneburg works built trucks and trailers during World War I, and when he returned the resumption of passenger-car production, with prewar models, was slow and unprofitable. In 1920 Hering & Richard AG was merged with Elite Diamantwerke AG of Brand-Erbisdorf to form Elitewagen AG, with headquarters in Berlin. The Rex Simplex make was discontinued and Richard resigned. Max and Alfred Hering also resigned, to run the wheel company they had set up as a sideline in 1914. For five years Carl Richard held a succession of appointments in industry, building a cash reserve. He was ready to make his move when Elitewagen AG was sold to Adam Opel AG in 1926, and repurchased the Ronneburg plant. Trading as Carl Richard & Co. GmbH, he began producing motor trucks as well as bodywork for passenger-car and truck chassis. The last trucks were built in 1928, and the auto-body plant was closed in 1934.

1. Max Karl Traugott Hering (1874–1949).
2. "King of Simplicity."

Richard, Georges (1863–1922)

Managing director of Société de Construction de Cycles et d'Automobiles La Marque Georges Richard from 1895 to 1899; chairman of Société des Anciens Etablissements Georges Richard from 1899 to 1904; managing director of Georges Richard & Cie from 1904 to 1906; managing director of SA des Automobiles Unic from 1906 to 1922.

He was born in 1863 in Paris. After a general education he went into partnership with his two brothers, Felix Maxime and Jules, starting a small enterprise to make electrical instruments for meteorological measurements. In 1893 he opened a shop on rue d'Angoulême in Paris to make bicycles, operating as Cycles Georges Richard. He began experimental work on a car in 1896. Completed in 1897, it looked like the Benz Vélo, but was powered by a flat-twin engine. In 1898 he bought a license for the Vivinus car and set up a new factory in rue Galilée at Ivry-Port, but his bicycle business did not generate an adequate cash flow to support the automobile branch. His company was reorganized in 1899 and refinanced by Henri de Rothschild, drawing on fresh funds from the Union Financière de Genève. He began building the Poney quadricycle in 1900, with a front-mounted air-cooled single-cylinder engine, two-speed belt drive, and handlebar steering. In November 1901, he went into partnership with Charles-Henry Brasier, who designed new shaft-driven models with two-cylinder engines. His factory had a modern machine shop with electrically powered machine tools and gave work to 300–400 men. Most chassis components were supplied from outside: frames from Arbel, axles from Lemoine, and springs and dampers from Truffault. In 1903–04 the cars from Ivry-Port featured a four-leafed clover trademark and were sold as Richard-Brasier. The partnership with Brasier broke up in 1904 and an ugly law suit was fought in 1905. Once again, Henri de Rothschild came forward to provide financial backing for majority control of Georges Richard & Cie, founded on October 28, 1904, with a small factory in rue Saint-Maur, Paris, and introducing the Unic trademark. The company built a solid business in taxicabs and light commercial vehicles, and became SA des Automobiles Unic in 1906. In 1905 he had been able to purchase the former Bardon factory on Quai National, Puteaux, where Unic production was installed. A line of four-cylinder cars was added in 1908 and production doubled within four years. During World War I the Unic factory made ambulances, trucks, and ammunition. Car production resumed in 1920 with a one-model program, the 13/24 low-priced family car. He died on June 14, 1922, from injuries sustained in a traffic accident on the road to Les Andelys on June 11.

Richardson, Percy (ca. 1878–?)

Managing director of Brotherhood-Crocker Motor Company, Ltd., from 1904 to 1907; technical director of Sheffield Simplex Motor Works, Ltd., from 1907 to 1914.

He was born about 1878 and trained as an engineer. He entered the motor industry with Daimler in Coventry, becoming manager of the London sales-and-service branch about 1900.

In 1904 Peter Brotherhood, who had started an engineering firm in Belvedere Road, Lambeth, in 1867, engaged him to design a range of top-quality cars. Jonathan Crocker and Stanley Brotherhood were directors of the company, and set up a separate motor car branch in Norwood Road, West Norwood. The first Brotherhood car was powered by a 12/16 hp four cylinder T-head engine, soon joined by 20-hp and 40-hp models. All were low-built, elegant touring cars, notable for their two-pedal control, clutch and brake being combined (braking began when the clutch was fully released). Earl Fitzwilliam bought a controlling interest in the company after trying in vain to persuade Henry Royce to build a car to his (Fitzwilliam's) specifications, and discussed his ideas with Richardson, who was then briefed to design "the best car in the world." The Earl erected a purpose-designed factory at Tinsley, Sheffield and the London plants were closed. The parent Brotherhood company moved to Peterborough, while the car branch was later sold to AC (Auto Carriers, Ltd.). Richardson moved to Sheffield and took charge of operations. The cars were renamed Sheffield-Simplex and the first model was identical with the 20-hp Brotherhood. In 1907 he designed a six cylinder 20/30 model, followed by a lower-priced four-cylinder 14/20 hp. In 1909 he introduced the 45-hp luxury car with an L-head 7-liter six, two-pedal control, and a two-speed rear axle. Except for a forward/reverse gear, there was no gearbox. That was remedied in 1911, by combining a three-speed gearbox with the final drive gears in the rear axle. The 45 hp car was replaced in 1913 with a more modern 30-hp 4.7-liter L-head six, with a three speed gearbox mounted centrally in the chassis, and a worm drive rear axle. Richardson left the company at the beginning of World War I, when the factory was converted to munitions work.

Riecken, Christian (1880–1950)

Chief engineer of NAG[1] from 1919 to 1927.

He was born in 1880 in Germany and trained as an engineer. He went to Belgium and worked on Ernst Lehmann's staff in the drawing office of L'Auto Métallurgique at Marchienne-au-Pont from 1904 to 1911. He joined Minerva in Antwerp as an engine designer and in 1912 created a racing voiturette powered by a 2.6-liter sleeve-valve engine. He drove one of the three team cars in the 1912 Belgian Grand Prix. He returned to Germany in 1914. In 1919 he was engaged by NAG and designed a new car, the C4 medium-size touring car, powered by a 33-hp side-valve 2.5-liter four-cylinder engine. He drove one of them to victory in the Grunewald race of 1921. In 1922 he created the Cob with a lower, lightweight frame and 40-hp engine. NAG entered a team of Cob cars with sports torpedo bodies in the Monza 24-hour race, which Riecken won. Given four-wheel-brakes and a 50-hp engine, it was renamed C4m in 1925. He also designed the D4 and D6 family cars, built from 1923 to 1926. He made up his mind to leave NAG after its acquisition of the Protos factory in Berlin (and its range of cars) in 1926, and resigned in 1927. NAG director Oskar Knoop then replaced the D6 with two new sixes, Types 201 and 204, their engines designed by Gabriel Lienhard.

1. Nationale Automobil-Gesellschaft, Berlin.

Rigoulot, Louis (ca. 1848–?)

Chief engineer of the motor car department of Les Fils des Peugeot Frères from 1889 to 1896, chief engineer of SA des Automobiles Peugeot from 1896 to 1910, engine design engineer with SA des Anciens Etablissements Panhard & Levassor from 1911 to about 1924.

He was born about 1848 and studied engineering at the Ecole des Arts et Métiers of Angers, later graduating from the Ecole Centrale des Arts et Manufactures. He joined Société Peugeot Frères in 1872 and was named head of the bicycle department of Les Fils des Peugeot Frères in 1886. In 1889 Armand Peugeot put him in charge of constructing the three-wheeled steam cars designed by Léon Serpollet. The steam engines came from La Buire[1] in Lyon; he tried to lighten the chassis as much as possible, but Armand Peugeot was disappointed in their performance.

He was then given a set of drawings for the Daimler wire-wheeled car, a primitive quadricycle with no springs and fork-guidance of the front wheels. What he kept was the light tubular steel frame. He designed a tubular front axle with stub-axle steering and a transverse leaf spring. The V-twin Daimler engine was mounted below the seat with a three-speed gearbox and chain drive to the rear wheels, which were carried on parallel leaf springs. This car set the pattern for a decade. He designed a four-speed gearbox in 1891 and adopted a horizontal parallel-twin engine in 1897. His first design with a front-mounted engine was the single-cylinder Type 36 of 1901, and he made the change to shaft drive on the Type 39 of 1902, which was also his first four-cylinder design. He also created the two-cylinder 1.8-liter Type 77 of 1905 and the four-cylinder 2.2-liter Type 81 of 1906.

He left the company when Armand Peugeot left and Robert Peugeot took over and moved to Paris where he became an engine designer for Panhard, working mainly on sleeve-valve engines. During World War I he worked on a variety of aircraft engines with both sleeve and poppet valves. In 1919 he came under the direction of Eugène Gorju, newly appointed technical director, and retired about 1924.

1. Chantiers de L'Horme et de La Buire, Lyon.

Riley, Percy (1882–1941)

Chief engineer of the Riley Engine Company from 1903 to 1931; managing director of Riley (Coventry) Ltd. Engines Branch from 1931 to 1938.

He was born on November 8, 1882, in Coventry, as the third son of William Riley. He was only ten years old when he made his first sketches of cars. He attended King Henry VIII grammar school from 1892 to 1896, during which time he designed a complete motor car but his father would not let him have the money to build it. He built it anyway, after leaving school, with a front-mounted 2½-hp engine, belt drive, and a steering wheel. It was completed and successfully tested in 1898. Still without parental assistance, he set up the Riley Engine Company in Cook Street, Coventry, in partnership with his brothers Victor and Allan. They produced considerable numbers of air-cooled single-cylinder engines for the Riley Cycle Company, Ltd., and the Singer Company. He designed the Riley tri-car in 1903, and started production. It can be described as a motorcycle with two front wheels, and a passenger's seat between them. The first model had handlebar steering and belt drive to the rear wheel. The 1905 model had a steering wheel and a bucket seat for the driver. In 1906 he applied for a patent for a detachable wire wheel and moved the Riley Engine Co. into bigger premises, the Castle Works in Aldbourne Street, Coventry. He designed the Riley 17/30 with a 3-liter in-line four-cylinder side-valve engine, ready for production as a 1913 model. He would make the engines and Castle Works, and the Riley Cycle Co., Ltd., was to assemble the car at the King Street works. But William Riley decided to supplement wire-wheel production in St. Nicholas Street, Coventry, by converting the motor department in King Street to making wire wheels. Percy, Victor and Allan Riley responded by organizing the Riley Motor Manufacturing Company to take over car production, and moved into a factory in Aldbourne Street, next door to Riley Engine Company. The 17/30 was built in the new plant, alongside the V-twin 12/18. In 1914 the Riley family placed all their factories at the disposal of the Ministry of Munitions, and all orders were filled strictly at cost, so that at the end of 1918, their financial reserves were at a critically low point. In 1919 the Riley Motor Manufacturing Company was taken over by Riley (Coventry) Ltd. and renamed Midland Motor Body Company. Percy spent most of his time experimenting with new engine designs. This work resulted in a 1087-cc in-line four with one camshaft on each side of the block and overhead valves splayed at 45° over hemispherical combustion chambers. It went into production in 1926 as the Riley Nine. In 1931 Riley Engine Company was taken over by Riley (Coventry) Ltd. He had designed a six-cylinder 1633-cc version of the Nine, which went into production in 1928 and enjoyed a 10-year run. He resigned when Riley (Coventry) Ltd. fell into receivership in February 1938, and died on August 23, 1941.

Riley, Stanley (1885–1952)

Chief engineer of the Riley Cycle Company, Ltd., from 1904 to 1908; managing director of Riley Cycle Co., Ltd., from 1908 to 1912; director of Riley's wire-wheel production from 1912 to 1919; chief designer of the Midland Motor Body Company from 1919 to 1931; director of Riley (Coventry) Ltd. from 1931 to 1938.

He was born in 1885 in Coventry as the fourth son of William Riley and educated at King Henry VIII grammar school. He joined the Riley Cycle Company, Ltd., in 1904 and designed its first four-wheeled car, a 9-hp model produced from 1907 to 1909. He also designed the 12/18 with a two-liter V-twin, that was produced from 1907 to 1914. During the years when he was in charge of the wire-wheel business, Riley had about 150 regular customers and was one of the biggest wire-wheel suppliers in the world. He designed a lightweight ladder-type chassis for the new engine (Nine) that Percy Riley was preparing and created a range of new body types. The

Riley Nine prototype was ready in 1924 but their brothers Victor and Allan wanted no part of it. The Nine was thoroughly tested during 1925, with expeditions to Switzerland and Austria, Scotland and Ireland. Stanley and Percy Riley put it in production in 1926, the power unit being made by the Riley Engine Company and the vehicle itself, with final assembly, by the Midland Motor Body Company, both on Aldbourne Road, Coventry.

The Monaco close-coupled sedan was greeted with enthusiasm, and was soon supplemented by the San Remo and Biarritz models. The Riley Nine made waves in the market unmatched by any of the contemporary Rileys, such as the 11/40, the 12 and the 17. In 1931 Riley (Coventry) Ltd. took over the production of the Nine by swallowing up the companies that built it. Stanley Riley modernized the Monaco styling almost every year, designed the Gamecock, Lynx and Trinity convertibles, and created the Kestrel fastback four-door sedan in 1933. The 1935 Merlin was the first Riley featuring a pressed-steel welded-up body shell.

He resigned from Riley (Coventry) Ltd. in 1938 and lived in retirement until 1952.

Riley, William (1851–1944)

Chairman of the Riley Cycle Company, Ltd., from 1896 to 1912; chairman of Riley (Coventry) Ltd. from 1912 to 1923.

He was born in 1851 in Coventry as the son of a ribbon weaver and was only 19 when he took charge of the family business, William Riley & Son. In 1890 he took over a firm of bicycle manufacturers, Bonnick & Co., of King Street, Coventry, and six years later merged the weaving and bicycle factories to form Riley Cycle Company, Ltd. At the outset, he was opposed to his sons' experiments with automobiles but was in favor of starting motorcycle production in 1901. Over the years, his attitude to cars changed, and in 1908 he became a shareholder in Riley Engine Co. He financed setting up a new plant in St. Nicholas Street to make the detachable wire wheel patented by his son Percy. In 1911 he built his last bicycles and motorcycles, and in March 1912, the Riley Cycle Company, Ltd., was reorganized as Riley (Coventry) Ltd. During World War I he devoted all his energy to orders from the Ministry of Munitions, and acquired a new factory at Foleshill, Coventry. He retired in 1923 but maintained his seat on the Riley (Coventry) Ltd. board of directors until 1938. He died at his home in Kenilworth on January 17, 1944.

Riley, William Victor (1876–1958)

Works manager of the Riley Cycle Company, Ltd., from 1898 to 1912; managing director of Riley (Coventry) Ltd. from 1912 to 1923; chairman and managing director of Riley (Coventry) Ltd. from 1923 to 1938.

He was born on January 5, 1876, in Coventry, as the eldest son of William Riley, and educated at King Henry VIII grammar school up to the age of 14, when he went to work in the parental bicycle shops. He was a co-founder (with his father) of the Riley Cycle Company, Ltd., in 1896 and took the initiative in starting motorcycle production, with a line of motor tricycles and quadricycles, powered by de Dion–Bouton single-cylinder engines. In 1903 he was a co-founder (with his brothers Percy and Allan) of the Riley Engine Company. He introduced the first four-wheeled Riley in 1907, with a 9-hp 60° V-twin and chain drive. He founded the Nero Engine Company in 1913 to produce the four-cylinder engine for the new Riley Ten, and sold it to Riley (Coventry) Ltd. in 1919. He was also a co-founder of the Riley Motor Manufacturing Company in 1913, set up to build the Riley 10-hp V-twin and 17/30, which became the Midland Motor Body Company in 1919. He moved Riley (Coventry) Ltd. to Durbar Road, Foleshill, Coventry, and in 1921 introduced the 11-hp Riley with a four-cylinder side-valve engine designed by Harry Rush. He stage-managed the merger of Riley Engine Company and Midland Motor Body Company into Riley (Coventry) Ltd. in 1931, which thereby became the makers of the Riley Nine. Late in 1936 he organized Autovia Cars, Ltd., to make an up market V-8 powered sports car with bodies by Arthur Mulliner, but it folded about 18 months later. Riley (Coventry) Ltd. fell into

receivership in 1938 and was sold to Morris Motors, Ltd. Victor Riley was invited to stay on, and held the title of managing director until 1947, but never held any real authority in the Nuffield Organization. He resigned when given orders to close Riley's main plant and transfer the final assembly of Riley cars to the MG works at Abingdon-on-Thames, and in 1948 settled at Stow-on-the Wold. He died in an Oxford nursing home on February 9, 1958.

Riolfo, Auguste (1894–1985)

Director of testing, SA des Usines Renault, from 1933 to 1964.

He was born in 1894 and fought as an infantryman in the front lines in World War I, and was decorated with the Croix de Guerre. He went back to school after being discharged and graduated from the Ecole des Arts et Métiers with a degree in mechanical engineering. He joined Delage and became head of the drawing office at the time Delage was preparing the DR 70 and the DMS. He left Delage in May 1930, to join Renault as a test engineer in the experimental department. In 1933 he was named director of testing for cars, trucks and buses, agricultural tractors and rail-car engines. Formally he reported to Serre, but he was in daily contact with Louis Renault. He was also responsible for quality control and made spot checks by picking cars at random off the assembly line for extended tests. He enjoyed driving cars, testing new models on the road and at Montlhery. Louis Renault would not authorize the start of production until Riolfo had given his written approval. It has been said that he let cars get by with poor brakes. Perhaps some were under-dimensioned, but it was a common defect of most mechanical brake systems that they needed frequent adjustment. He was able to get hydraulic brakes for the 4 CV of 1946 and all subsequent models. The 4 CV, Dauphine and later models with outboard/rear-mounted engines had quirky handling, but they passed the test criteria he had laid down. Their record in international rallying also proved that in professional hands, they were both fast and safe. He put Renault's first front-wheel-drive models through full test programs, beginning with the Estafette van and later the R4 economy car. He retired to Beauvallon on the French Riviera in 1964 and died in a Toulon hospital in July 1985.

Rivolta, Renzo (1908–1966)

Chairman of Iso Automotoveicoli S.p.A. from 1948 to 1966.

He was born on September 5, 1908, at Desio near Milan into a family of wood and lumber traders. After leaving school he worked in the family enterprise until he had saved sufficient funds to go into business for himself. In 1940 he started a small factory at Bolzaneto near Genoa to make water heaters, radiators and refrigerators. He moved into more spacious premises at Bresso near Milan in 1942, launching the Isothermos trademark. In 1948 he became the first in Italy to make more than 50 refrigerators a day, and launched the Isomoto motor scooter with a small two-stroke engine. He sensed that a market existed for a scooter with weather protection, and designed a bubble-shaped vehicle on four wheels, weighting 350 kg, powered by a 198-cc two-stroke engine mounted behind the seats. The Isetta had no side doors — the front panel opened to give access, with the steering wheel on a swivel column. Isetta production at Bresso began in 1953 and ended in 1956. He sold the manufacturing rights, however, to BMW, Vélam, Borgward-Iso Española, and Iso-Romi in Brazil. By 1962, more than 160,000 units had been produced. He had done some speed-boat racing as a young man, and now his ambition was to build a grand touring car. He was inspired by the Gordon GT, which he went to England to inspect in 1961. He made arrangements with GM–Italia for a supply of V-8 Chevrolet engines, engaged Pierluigi Raggi to design a chassis, and contracted with Bertone for body styling and construction. The first Iso Rivolta was shown in June 1962, and the Bresso works built 799 units over an eight-year span. Scooter production was discontinued in 1963, and the Iso Grifo appeared in 1963. Over an 11-year span, Iso produced 412 Grifo coupés with Bertone

bodies. A four-door model, the S4, was introduced in September 1967 with a Ghia body. He died on August 19, 1966, in the Pavia Polyclinic, following a heart attack.

Robinson, Geoffrey (1938–)

Managing director of Jaguar Cars, Ltd., from 1973 to 1975.

He was born on May 25, 1938, read Russian and German at Cambridge, and graduated from Yale University with a degree in economics. When he returned to London in 1963, he became a research assistant at Transport House and a protégé of Labour party leader Harold Wilson. Lord Stokes of Leyland found him a senior position in the Industrial Reorganization Corporation, a government unit set up to "rationalize" British industry. The IRC was dissolved in 1971 by the new Tory government, and Lord Stokes invited him to British Leyland as financial controller. A year later, Lord Stokes dispatched him to Milan as managing director of Leyland Innocenti but called him back late in 1973 to take over the helm at Jaguar from F.R.W. England. He boosted Jaguar production but lost product quality. He was forced to resign when it came to light that he had ordered a $40 million paint shop from Interlack of Italy without asking approval from the head office. He left in May 1975 and immediately founded TransTec, financed by Joska Bourgeois, the Jaguar agent in Brussels, and Dr. Fawzi El-Menshawy. He became chairman of Hollis Industries PLC in 1988, a holding for engineering companies formerly owned by press tycoon Robert Maxwell. He was appointed postmaster-general in Tony Blair's new Labour government in 1997 but was forced to resign over allegations of illegal financial dealings. Yet he was still a member of Parliament (Labour, Coventry North) in 2002.

Robotham, William Arthur "Roy" (1899–1980)

Chief experimental engineer of Rolls-Royce, Ltd., from 1936 to 1940; chief engineer of the Rolls-Royce Motor Car Division from 1945 to 1950.

He was born on November 26, 1899, in Derby and educated at Repton School. He was commissioned into the Royal Artillery in 1918, stationed at Lydd Camp in Kent, and discharged in 1919. He returned to Derby and joined Rolls-Royce, Ltd., as a premium apprentice. In 1923 he became a technical assistant in the experimental department, and was named head of the chassis experimental department in 1930. Top management had a plan to introduce an 18-hp model, priced below the 20/25, and Robotham designed a shorter, lightweight chassis for it. Known as the Peregrine, it was shelved due to the low output of its 2½-liter engine. In 1932 he was a member of a Rolls-Royce team making a study tour of U.S. industry. Upon his return, he brought out the Peregrine chassis and mounted a Rolls-Royce 20/25 engine in it. A year later it was cleared for production as the Bentley 3½-Litre, the first Bentley made at Derby. He visited the United States again in 1934 and 1937, and accompanied A.G. Elliott on his mission to the United States in 1938. He reported to E.W. Hives that the Buick Roadmaster, priced at £925 compared with £2,575 for the Rolls-Royce Phantom III, was faster in both acceleration and top speed. In 1940 he was delegated to lead a military-vehicle engineering team in temporary quarters at Belper, outside Derby.

In 1941 he was attached to the Ministry of Supply as the government's chief engineer for tank design, and worked in close liaison with Leyland engineers on the development of the Centaur tank. He visited the United States again in 1942 as a member of a tank-production mission. He left government service in 1943, returned to Belper, and devised a family of modular military-vehicle engines in four-, six-, and eight-cylinder versions. In 1946 he went to Detroit to negotiate for manufacturing rights to the Hydramatic transmission with C.E. Wilson, president of General Motors. In 1948 he designed an experimental four-cylinder Bentley, which did not meet the board's approval. He also realized that Britain's industry was not making a lightweight, high-speed diesel engine of 200 to 300 hp as would

be needed for tomorrow's trucks. He designed a 10-liter six which passed its first test in September 1948. In 1949 he was given a seat on Rolls-Royce's main board of directors, and a year later he was named managing director of the Oil Engine Division, which began production at Derby in February 1952. In 1957 he directed the transfer of diesel-engine production into the former Sentinel truck factory at Shrewsbury, and led the division until his retirement in January 1963. He moved to his farm at Great Engeham, Kent, which he had bought in 1945, and died in 1980.

Rochet, Edouard (1867–1945)

Joint managing director of the Société Lyonnaise de Vélocipèdes et d'Automobiles Rochet Schneider from 1895 to 1898, joint managing director of the Société Lyonnaise de Construction d'Automobiles from 1898 to 1904, managing director of Rochet & Schneider, Ltd. from 1905 to 1908, managing director of SA des Etablissements Lyonnais Rochet Schneider from 1908 to 1910 and 1914-15, chairman of SA des Etablissements Lyonnais Rochet Schneider from 1915 to 1933.

He was born on February 7, 1867, in the Saint-Paul neighborhood of Lyon, the son of Jean-François Rochet, machine-setter with the PLM[1] at Oullins. He graduated with an engineering degree from La Martinière Ecóle de Science et Arts Industriels in Lyon in 1884 and went to work for the PLM at Oullins. He set up his own shop on Place Saint-Pothain in Lyon in 1888 and went into partnership with Theo Schneider to make bicycles and tricycles in 1889, operating as the Société des Constructions Vélocipèdiques du Rhône. They moved into a factory in rue Paul Bert in Lyon in 1892 and built their first test cars in 1895, with rear-mounted horizontal single-cylinder engines and chain drive. They made 40 to 45 cars in 1896-97. A new company, financed by Demetrios Zafiropulo, banker in Marseilles, was organized in 1898 and in 1899 they bought a large factory site in the Chemin Feuillat, Lyon-Monplaisir, where operations began in 1901. The first model with a vertical four-cylinder appeared in 1900 and shaft drive was adopted in 1902. He sold manufacturing rights to Rochet Schneider cars to Alden Sampson of Pittsfield, Massachusetts, in 1902, Max de Martini of Neuchatel, Switzerland in 1903, Nagant of Liège in 1903 (who ceded it to FN[2] in 1905), and Florentia[3] in 1904. In January 1905, the company's assets were taken over by an international consortium headquartered above Moorgate Station, London, and Paul Horvath was named chairman of Rochet & Schneider Ltd. Rochet retained his office and remained at the factory. The London-based company collapsed in 1908 and ownership returned to French hands. The principal shareholders were Demetrius and Georges Zafiropulo, and Jules Fisch, director of Bouhey machine tools. At the start of World War I the factory was requisitioned by Raymond Poincaré's government and retooled to produce Renault V-8 aircraft engines. Rochet-Schneider also built over 1,300 military 1½-ton trucks and 120 18 CV "colonial" type cars. Rochet took an active role in the company's affairs with the return to production for civilian markets, but began stepping back in 1930. He resigned the chairmanship on April 28, 1933, and severed his formal ties with the company in 1936. He died on April 12, 1945, in the Clinique Jeanne d'Arc in Lyon.

1. Compagnie des Chemins de Fer de Paris-Lyon-Méditerranée.
2. Fabrique Nationale d'Armes de Guerre, Herstal-Iez-Liège.
3. Fabbrica Toscana di Automobili, Florence, Italy.

Roesch, Georges Henri (1891–1969)

Chief engineer of Clement Talbot, Ltd. from 1916 to 1923 and 1925 to 1935.

He was born on April 15, 1891, in Geneva, the son of a German blacksmith and wagon repairman. He attended local schools and served an apprenticeship in his father's auto repair shop, returned to school and graduated from the College of Geneva in 1909.

Roesch went to stay with his mother's family in Paris and found employment with Automobiles Grégoire at Poissy. He was a draftsman with Delaunay-Belleville at Saint-Denis in 1910 and joined Renault in 1911.

He met Henry W. Watts, chief engineer of the Coventry Chain Company, who suggested he would find better opportunities in England. He gave notice at Renault and arrived in Coventry in February 1914. Watts introduced him to A.E. Berriman, chief engineer of Daimler, who engaged him as a design engineer. Within months, Daimler stopped making cars in order to produce military tractors, tank engines, artillery shells, aircraft engines, and trucks.

Roesch began looking for a way to get back to designing cars, and in 1916 he answered a newspaper advertisement from Clement Talbot Ltd. in London. He was named chief engineer, but the factory was not making cars at that time. He was expected to design aircraft engines. While Napier, Sunbeam and others were busy with multi-cylinder W- and X-formation engine projects, Roesch had a burst of originality and proposed an 18-cylinder "barrel" engine (two rows of nine axially arranged cylinders around a central shaft) and "swashplate" drive. It was built and tested, but the War Office did not order any.

He had a new light car with a 1750 cc sidevalve four-cylinder engine ready and four test prototypes were built in 1919. But then the company was sold to A. Darracq & Co. Ltd. and the Darracq board refused to put it into production. Instead, he was instructed to develop the French-designed 8/18 which he did to such good purpose that it won considerable sales success as the Talbot 10/23. In 1923 Louis Coatalen, group chief engineer of Sunbeam-Talbot-Darracq Motors Ltd. sent him to the design office at Suresnes. He returned to Ladbroke Grove in 1925 and patented a new type of overhead valve train, eliminating the rocker-arm shaft by mounting the rocker arms on individual inverted pivot studs, with pushrods as thin as knitting needles. He also directed a lot of experimental work on higher compression ratios. His valve gear became part of the 1665 cc high-compression (6.3:1 in 1927, 7.0:1 in the 1930 Sports model) six of the 14/45, which set the stage for all later Talbots from London. It was renamed the Talbot 65 in 1932 and a year later became available with a preselector gearbox and centrifugal clutch plus free-wheeling. It was produced until the end of 1935. The 2.3-liter Type 75 was introduced in 1930 and Type 90 (with the same engine) in 1931. The 3-liter Type 105 appeared in September 1931 and a 105 Speed model followed in 1933. In January 1935 the company was sold to Rootes Securities Ltd. and the 75, 90 and 105 were discontinued at the end of 1937. Roesch's final masterpiece, the 3.4-liter Talbot 110, was built from September 1934 to June 1938. He resigned from the Rootes Group in 1938, and held the title of chief engineer, David Brown Tractors Ltd. from 1939 to 1942. He joined Power Jets Ltd. to work for Frank Whittle on the development of turbojet engines and in 1944 went to the National Gas Turbine Establishment as chief mechanical engineer. From 1950 to 1959 he was an official advisor to the Ministry of Supply on small industrial gas turbines.

He died on November 7, 1969.

Roger, Emile (1850–1897)

Owner and director of Emile Roger, Ingénieur-Constructeur, from 1883 to 1897.

He was born in 1850 in Paris and opened a mechanical workshop in rue des Dames in the Batignolles district of Paris. In 1883 he secured the agency for Benz two-stroke gas engines, and manufacturing rights. He subcontracted for their production to the Société des Forges d'Aubrives in the Ardennes, but in 1888 switched the contract to the Société Anonyme des Anciens Etablissements Panhard & Levassor. In 1888 he brought a Benz three-wheeled car to Paris, securing sole distribution rights not only for France but for all markets outside Germany. In addition to selling complete Benz cars, he began building his own chassis, powered by Benz engines. He made his first four-wheeled model in 1889, at a time when Benz was only making three-wheelers. When Carl Benz accused him of breach of contract, Roger avoided litigation by painting an exaggerated picture of how France's high import duties made foreign cars too expensive, forcing him to build his own chassis. Benz could also take comfort from the fact that Roger

produced far fewer cars than he ordered from Mannheim. In 1895 he made no more than 20 cars, but sold 49 complete Benz cars in France. In 1895 he also extended a Benz license to Léon L'Hollier of Digbeth, Birmingham, which led to the establishment of the Anglo-French Motor Carriage Company, and local production. He also began selling Benz cars in the United States, imported by the R. H. Macy department store in New York City. In 1897 he made arrangements with A. Diligeon & Cie of Albert in the Somme for local production of the Benz Vélo, which was sold under the Hurtu trademark. But Roger died suddenly towards the end of 1897 from a syncope following a seizure of paralysis.

Röhr, Hans Gustav (1895–1937)

Technical manager and joint managing director of Priamus Automobilwerke AG from 1918 to 1921; managing director and technical director of Röhr Auto AG from 1926 to 1930; chief engineer of Adlerwerke AG from 1931 to 1934; technical director of Adlerwerke AG in 1934–35; technical director of Daimler-Benz AG from 1935 to 1937.

He was born on February 10, 1895, at Uerdingen as the son of a small lead-products manufacturer, graduated from college in 1914 and joined the Rheinischen Aerowerke as a trainee. He was drafted into the Royal Prussian Flying Troops in 1915 and was shot down over France in 1918. He spent the rest of the war in a field hospital. He linked up with Fritz Kallenbach, business manager of Priamus Automobilwerke AG, and designed a radical economy car with an air-cooled four-cylinder engine and independent front suspension, and a weight of only 346 kg. But the company went bankrupt in 1921, and he moved to Berlin, designing new cars, and building prototypes in space rented from Bolle & Fiedler, engine manufacturers. His 1924 prototype had independent front suspension, a very low frame, and weighed only 845 kg including its six-cylinder engine. The following year he obtained the backing of Dr. Hugo Greffenius of MIAG[1] in Frankfurt to take over the failing Falcon Automobilwerke in Ober-Ramstadt. It was reorganized as Röhr Auto AG on October 10, 1926, and production of the Röhr 8 began in 1927. It had all-independent suspension and a sheet-steel platform frame and a 50-hp 2250-cc overhead-valve straight-eight engine. It was priced just above the undistinguished Mercedes-Benz 260 Stuttgart. His enterprise folded in 1930, but he managed to sell the assets to a Swiss consortium. Arriving in Frankfurt in May 1931, he brought a full engineering team including Joseph Dauben, Willy Syring, Otto Winkelmann, Laurenz Niessen and Oberingenieur Engel from Ober-Ramstadt. They designed the front-wheel-drive Adler Trumpf and Trumpf Junior, which were also produced under license by Rosengart in Paris and Imperia at Nessonvaux. He began testing the Trilok automatic transmission (with hydraulic torque converter) in the Adler Diplomat in 1933. He resigned after quarreling with Ernst Hagemeier, the managing director, over his share of the license fees. He explained the situation to Emil Georg von Stauss (1877–1942), head of Deutsche Bank and chairman of the supervisory board of Daimler-Benz AG. On September 17, 1935, he took office at Untertürkheim, with a private staff of ex–Röhr, ex–Adler engineers. He wanted to make front-wheel-drive cars, and they designed the W-144, W-145 and W-146 with horizontally opposed four-, six- and eight-cylinder engines, but they never reached production. He also directed the design teams for the 1938 Mercedes-Benz 230 and 770, and began a V-8 project. But his life was cut short on his way to the German Grand Prix on the Nürburgring in the summer of 1937 when he caught a cold which turned to lung inflammation. He died on August 10, 1937, in the Marienhof Hospital at Koblenz.

1. Mühlenbau und Industrie Aktien-Gesellschaft.

Rolls, Charles Stewart (1877–1910)

Technical managing director of Rolls-Royce, Ltd., from 1906 to 1910.

He was born on August 27, 1877, in Monmouth as the third son of Lord Llangattock. He was educated privately and at Eton College. He graduated from Trinity College,

Cambridge, with a degree in engineering. He raced bicycles in the contests between Cambridge and Oxford, apparently never rode motorcycles, but bought a Peugeot 3.5-hp car in 1896. He started motor-racing in 1899 with a de Dion–Bouton tricycle. In 1900 he drove a Panhard in the 1000-Miles Trial and won a gold medal. His finished 18th in a Mors in the 1901 Paris-Berlin road race, came eighth in a Wolseley in the 1905 Gordon-Bennett Trophy race, and won the 1906 Tourist Trophy in a Rolls-Royce.

In 1902 his father had backed him in going into the motor trade in London, importing Mors and Panhard cars, later adding the Gardner-Serpollet steamer and the Minerva. When first test-driving a two-cylinder Royce, he was not impressed with it. His business partner, Claude Johnson, thought better of the Royce, and his friend, Henry Edmunds, pioneer motorist and founder of the Royal Automobile Club, arranged for Rolls to visit Manchester and meet Henry Royce in May 1904. By Christmas that year, Rolls had contracted to sell every car Royce could produce, the products then taking the trademark Rolls-Royce.

In 1906 C.S. Rolls & Co. cancelled all its other franchises and merged with Royce, Ltd., to form Rolls-Royce, Ltd. Since 1904 Rolls had developed a market-oriented product-planning technique and given Royce many useful ideas. After the creation of the Silver Ghost in 1906 and the opening of the Derby works in 1908, his interest in cars was relegated to second place in favor of flying. He revealed tremendous talent as an amateur aviator, was a founder-member of the Aeronautical Club, and in the spring of 1910 was assigned to the Royal Arsenal on Hounslow Heath as a test pilot. He made the first two-way flight across the Channel on June 2, 1910, and was killed on July 11, 1910, when his plane crashed during a contest at Bournemouth.

Romeo, Nicola (1876–1938)

Chief executive officer of S.A. Italiana Ing. Nicola Romeo e C. from 1918 to 1925; chairman of the company from 1925 to 1929.

He was born on April 18, 1876, at Sant'Antimo near Naples, as the son of an elementary school teacher. He worked his way to a degree in civil engineering at the Naples Technical Institute, graduating in 1899. He opened a civil-engineering office in 1900, and pursued his scientific studies at the Liège Polytechnic, and classes at German and French universities. In 1902 he set up an agency in Naples to sell Ingersoll-Rand, Hadfield and Blackwell mining equipment and other products. To remedy the lack of a service organization, he set up shop in Milan in 1909, to repair equipment imported from America. In 1911 he also began assembly of machine parts shipped from the U.S. in his factory in the Portello district of Milan. In 1915 he invited orders for war supplies and won contracts for artillery shells and air compressors. He undertook a vast expansion at Portello, and in 1916 changed his private firm[1] into a joint stock corporation.[2] The government placed factories at Saronno, Rome and Naples at his disposal, and gave him contracts for the production of Titan military tractors and Isotta Fraschini aircraft engines.

His connection with automobiles was established on December 2, 1915, when a Tribunal of Commerce appointed him receiver of A.L.F.A.,[3] established on January 1, 1910, to take over the business of S.A. Italiana Darracq, put into liquidation in 1909. Alexandre Darracq sold his shares to the Banca Italiana di Sconto, which became majority owner of the A.L.F.A. car company. When civilian production resumed in 1919, the A.L.F.A. cars were renamed Alfa-Romeo. The bank, now known as Banca Nazionale di Sconto, declared bankruptcy in 1921, and the funds required to maintain car production were put up by a banking consortium and a government agency. Romeo had a passion for motor racing, and gave high priority to the Alfa Romeo racing team — expensive, but most effective in building up the name. In 1925 he was given the title of chairman, which reduced his authority while Mussolini tightened his grip on Italian industries with war-materiel potential. In 1929 Alfa Romeo cars won every race they entered, but the company faced financial ruin

once again. Romeo was removed, and the shares were placed with the Institute of Liquidation, an office of government receivers. The company was re-organized as Alfa Romeo S.A. in 1930 and formally nationalized in 1933 as part of IRI,[4] a government holding for failed enterprises in steel-making, shipbuilding, chemicals, mechanical and general engineering. Prospero Gianferrari, a prominent lawyer from Trento, was named managing director of Alfa Romeo S.A. and Nicola Romeo retired to his villa overlooking Lake Como, where he died in 1938.

1. Società in accommandita semplice Ing. Nicola Romeo e C.
2. Società Anonima Ing. Nicola Romeo e C.
3. Anonima Lombarda Fabbrica di Automobili.
4. Istituto per la Ricostruzione Industriale.

Romiti, Cesare (1923–)

Managing director of Fiat S.p.A. from 1976 to 1996; chairman of Fiat S.p.A. from 1996 to 1998.

He was born on June 24, 1923, in Rome, and graduated in 1945 with degrees in economics and commerce. He began his career with the Bombrini, Parodi, Delfino group of bankers in 1947 and spent 21 years with them. In 1968 he joined SNIA-Viscosa as financial manager, going to Alitalia as managing director in 1970. He served for over a year as head of Italstat, the financial arm of IRI[1] before moving to Turin in November 1974 as chief of finance, planning and control of Fiat S.p.A. Fiat was on the brink of financial collapse, and it was Enrico Cuccia, chairman of Mediobanca[2] who suggested to the Agnelli brothers that Romiti might be able to save Fiat.

Romiti directed a complete reorganization of Fiat's activities, with separate subsidiary groups for cars (Fiat Auto S.p.A.), trucks and buses (Iveco)[3], construction machines (Fiat Allis, later Fiat Geotech), farming equipment (Fiat New Holland), motor components (Gilardini and Magneti Marelli, which merged in 1993), machine tools and automation (Comau), steel and aluminum foundries (Teksid).

He was ready to raise Fiat's stake in Seat[4] but in anger over the duplicity of Spain's socialist government, he pulled Fiat out of Seat altogether in 1980. He coped brilliantly with labor strife at Fiat's Turin-based plants, ending a 35-day strike in October 1980, by staging a protest march against the labor unions, when 40,000 office personnel, plant managers, production engineers, shop foremen and factory workers walked peacefully through the streets of Turin. In 1977 he had raised $400 million by selling a 15 percent stake in Fiat S.p.A. to Libyan bankers. He was able to buy it back for $3.15 billion in 1986. In 1987 he staved off Ford's bid for Alfa Romeo by bringing it into the Fiat fold, and in 1990 he broke off a series of aimless talks about collaboration with Chrysler. He secured 90 percent control of Ferrari in 1988 and 100 percent control of Maserati in 1993. After retiring from Fiat in 1998, he became chairman of a publishing company, RCS Editori S.p.A.

1. Istituto per la Ricostruzione Industriale.
2. Mediobanca was Fiat's main credit line inside Italy (Lazard Frères negotiated most of Fiat's international financing).
3. Industrial Vehicles Corporation was established in 1974 by a merger of Fiat's truck-and-bus division, OM, Unic and Magirus-Deutz.
4. Sociedad Española de Automoviles de Turismo.

Ronayne, Michael (1905–)

Chief engineer, Ford of Britain, from 1952 to 1963.

He was born on February 9, 1905, in Cork, Ireland, and educated at the Christian Brothers School, North Monastery Technical School, and Sharman Crawford Technical School. He joined Henry Ford & Son, Ltd., at the farm-tractor plant in Cork as an apprentice pattern maker, and later graduated to the drawing office. In 1932 he was transferred to Dagenham, where he worked on experimental cars and trucks up to 1939. During World War II he worked on the design of Ford military vehicles, worked on new vehicle projects from 1944 to 1948, and then became general manager of Ford's tractor division. Four years later he returned to Dagenham and took charge of modernizing the Consul and Zephyr for 1956 and the Popular,

Anglia and Prefect for 1959. He led the Classic and Capri programs, and put the Cortina on the drawing board.

Rootes, William Edward (1894–1964)

Managing director of Rootes, Ltd., from 1919 to 1932; chairman of Rootes Securities, Ltd., from 1932 to 1950; chairman and chief executive of Rootes Motors, Ltd., from 1950 to 1964.

He was born on August 17, 1894, at Goudhurst, Kent, as the son of a cycle-shop owner who became an automobile dealer in 1902. He attended Cranbrook School but left at 16 to become an independent chicken farmer. He served an apprenticeship with Singer & Co. in Coventry from 1913 to 1916, reported to the Royal Naval Voluntary Reserve and served as a sub-lieutenant in the Royal Naval Air Service during 1916–17. He was reassigned to the civilian side with a contact for aircraft-engine reconditioning and organized Rootes Maidstone in 1917. In 1919 he took over his father's franchises, also obtaining a £10,000 loan from the same source, and established Rootes, Ltd., in Maidstone. He opened a London branch in 1923 and set up an export department in 1924, with a shipping depot at Chiswick. In 1925 Rootes, Ltd., took over the coachbuilders Thrupp & Maberly of London, and a year later moved its headquarters to Devonshire House, an office block on Piccadilly. Rootes, Ltd., became the sole exporter of Hillman and Humber cars and Commer commercial vehicles. With the financial backing of the Prudential Assurance Company, he bought control of Hillman Motor Car Company, Ltd., in 1927 and Humber, Ltd., with its subsidiary Commer Cars, Ltd., in 1928.

In 1932 he established Rootes Securities, Ltd., as a holding for his industrial properties, and bought a truck builder, Karrier Motors, Ltd., in 1933. Two years later he took over the assets of Clement Talbot Ltd. of London and the Sunbeam Motor Car Company of Wolverhampton. He was always active in business associations and sat on the Board of Trade Advisory Council from 1931 to 1940. He became head of the government's Shadow Industry Plan to multiply the nation's capacity for making aircraft and aircraft engines in 1936 and chairman of the government's Supply Council in 1939. In 1940 he was also named chairman of the Joint Aero Engine Committee. Rootes operated two Shadow Factories, one on Speke Road, Liverpool, for bomber plane assembly, and one at Ryton-on-Dunsmore outside Coventry, for building aircraft engines. During World War II, Rootes made nearly 14 percent of all bomber planes built in Britain, 30 percent of the scout cars and 60 percent of the armored cars, plus thousands of vehicles based on prewar civilian cars and trucks. In addition, Rootes made 50,000 aircraft engines and repaired another 28,000. In 1946–48, the former Shadow Factory at Ryton-on-Dunsmore became the sole assembly point for Hillman and Humber. Knighted in 1942, Sir William criss-crossed the world to promote British exports and never tired of telling his technical managers to create new models for world markets, which brought forth the Sunbeam-Talbot Alpine in 1953 and the Hillman Super-Minx in 1961.

He became Baron Rootes of Ramsbury in 1959, and devoted more time to his livestock farms in Hampshire and Perthshire. The fortunes of Rootes Motors, Ltd., went steeply downhill in 1961–63, and in 1964 he sold a big stake in its to Chrysler Corporation. He never retired, but fell ill and died from cancer of the liver on December 12, 1964.

Rose, Raymond Hugh (b. 1886)

Chief engineer of Guy Motors from 1914 to 1918; chief engineer of Riley from 1931 to 1935; chief engineer of Lea-Francis from 1935 to 1939; technical director of Lea-Francis from 1939 to 1949.

He was born in 1886 in Southampton as the son of an engraver and educated at Taunton School in Somerset. He served an apprenticeship with Humber in Coventry and was hired as a draftsman when Humber began building cars. He accompanied Coatalen to Hillman in 1907 and went with him also to Sunbeam in 1909. He worked on the engines and drive lines for the four-cylinder 12/16 and

16/20 hp models introduced in 1912 and the six-cylinder 25/30 of 1912–14. When Sydney S. Guy left Sunbeam, he asked Rose to come with him. He designed the first-generation Guy cars and the Guy 4-liter side-valve V-8 engine. After brief stints with Crossley and Belsize, he returned to Sunbeam in 1920 to design trucks. He also laid the foundations for the Sunbeam trolleybus business which was eventually sold to Guy Motors. In 1924 he was designing new cars for Calthorpe, and later that year signed up as a technical advisor to Harper, Sons & Bean, Ltd. From 1927 to 1931 he was a member of the engineering staff of Bean Cars, Ltd., but resigned in protest when an advanced overhead-valve engine of his design was not approved for production. He went to Riley and designed a new engine for the 1933 Riley 12, and did the preliminary design work on the four-cylinder 2½-Litre and the 2.2-liter V-8. He joined Lea-Francis early in 1935 and designed 12- and 14-hp four-cylinder engines with Riley-style valve gear (splayed overhead valves with dual side camshafts). He prepared a modernized postwar model, which was given independent front suspension in 1949, and retired in 1950. He remained an engine consultant to Lea-Francis until 1960, and designed a line of small outboard-marine and industrial engines.

Rosengart, Lucien (1881–1976)

Special director in charge of finance at SA André Citroën from 1920 to 1923, managing director of SA des Automobiles et Cycles Peugeot from 1923 to 1928, chairman of Société des Automobiles Rosengart from 1928 to 1936, chairman of SIOP[1] from 1936 to 1955.

He was born in Paris on January 11, 1881, the son of the owner of a small precision-engineering shop. He was an inattentive pupil at the Lycée Charlemagne but showed outstanding aptitude for machine-shop tasks. His formal schooling ended when he was 15. Soon afterwards he began to invent methods for simplifying and speeding up manufacturing processes. He was the *de facto* works manager in the paternal business from 1896 to 1901, leaving only when called up for military service. He was assigned to the 17th Battalion of the "Chasseurs à Pied" and volunteered for a staff position in the Arab Office at Biskra, Algeria. He returned to Paris in 1905 and set up his own business a year later, making buttons, washers, nuts and bolts. Later he was one of the first to make electric lighting sets for bicycles, and he invented a flashlight with integral manual generator, which found a ready market. In 1914 he invented an artillery projectile that would explode a split-second prior to impact, and during World War I Citroën produced over 100,000 of them. After a few months of military duty he was discharged and given contracts for making munitions, for which his own facilities were too small. He took over two other factories and employed 4500 people.

André Citroën admired him for his grasp of economics and lively financial imagination and invited him to join him when preparing to go into the automobile business with a new product. Rosengart set up a credit organization for Citroën dealers and founded a taxicab company running an all–Citroën fleet. He left in 1923 when Robert Peugeot was in financial straits and asked him to help. He brought in an immediate cash injection of five million francs and set up the SADIF credit company to finance installment plan sales. He contracted with Société des Usines Bellanger to make vans on Peugeot passenger car chassis, opening an untapped market. He arranged a partnership between Peugeot and the Compagnie de Taxis-Transport to put a fleet of Peugeot taxicabs on the streets of Paris. In 1926 he set up Peugeot Maritime on the Quai de Passy, which made more than 4,000 motorboats with engines from Peugeot's Lille factory in its two-year career. He also handled Peugeot's advertising campaigns and directed the motor sports activity (which was cut off in 1926). He was a member of a French auto industry delegation to the US in 1926 and went back in 1927 to try to break into the American market with the 5 CV Peugeot 172M. Durant Motors Inc. considered a plan to build it at Elizabeth, New Jersey, but withdrew from the talks. When his five-year contract with

Peugeot expired, he took over the Bellanger plant in Levallois, purchased a license to produce the Austin Seven, and began production of the 5 CV Rosengart in 1928.

In 1931 he met Hans Gustav Röhr and bought a license to build the Adler Trumpf in France. Adler-Werke supplied the engines, front-wheel drive train and front suspension and production began at Levallois in 1932. His company was on the brink of bankruptcy in 1936, which led him to transfer all its assets to a new corporation, SIOP. He let the Adler license lapse and introduced the Supertraction, with the Citroën 11 CV powertrain, and a high-style body. He fled to America when the Germans invaded France in 1940, returning after the Liberation in 1944. His factory was occupied by the nationalized Farman company. He showed prototypes of the Supertraction fitted with a Mercury V-8, which he named "Supertrahuit" but found no means of getting it into production. He tried to modernize the Austin-based 5 CV into a 4 CV, but got no further than a few prototypes. He sold his interest in SIOP and retired to Villefranche-sur-Mer, where he died on July 27, 1976.

1. Société Industrielle de l'Ouest Parisien.

Rouge, Francis (1921–1976)

Joint managing director of Automobiles Peugeot from 1966 to 1973; president of Automobiles Peugeot from 1973 to 1976.

He was born on September 30, 1921, in Paris, and graduated from the Ecole Polytechnique in 1942. He fled from France and joined General de Gaulle's Free French Forces in Africa, seeing action in Italy and the liberation of France. He was stationed in Indochina from 1945 to 1948 and discharged with the rank of Captain and two medals. He spent four years in the administration of the Compagnie Générale des Colonies, and joined Peugeot in 1953 as an engineer attached to the board of directors. In 1957 he was named director of the Sochaux factories, where he supervised the preparations for building the 404 and the front-wheel-drive 204. He was transferred to the Peugeot headquarters in Paris in 1963, helping to plan new factories and overseeing the evolution of the model range. He played a central part in the takeover of Citroën, but fell ill and died on July 4, 1976.

Rouvier, Arturo Elizalde y *see* Elizalde y Rouvier, Arturo

Rowledge, Arthur J. (1876–1957)

Assistant chief designer of Wolseley from 1905 to 1908; chief designer of Wolseley from 1908 to 1913; chief draftsman of D. Napier & Son from 1913 to 1918; chief designer of Napier cars from 1918 to 1921.

He was born on July 30, 1876, at Peterborough and attended local schools. Even in his boyhood he revealed a strong interest in science and the arts. After serving an apprenticeship with Barford & Perkins in his home town, he went to London in 1898, where he gained valuable work experience in the fields of printing machinery and heavy engineering.

In 1901 he was engaged by M. S. Napier to work on car projects and helped design the first six-cylinder engines. In 1905 he moved to Birmingham to design new cars for Wolseley, returning to Napier in 1913. He designed the Napier Lion W-12 aircraft engine, a watercooled 450-hp 32-liter unit with a patented reduction gear for the airscrew. After World War I he was placed in charge of the new-car program, and designed the T 75 model, aimed right at the Rolls-Royce market. But it failed to attract the expected amount of orders, which led to his dismissal in 1921. Henry Royce knew him by reputation and hired him as his chief assistant. He became head of aircraft-engine design at the Derby plant, and developed the Kestrel and Buzzard. He was also instrumental in starting the Merlin V-12 project, the engine that was to power the Supermarine Spitfire in World War II. But he fell victim to a serious illness in 1933, causing the company to relieve him of full-time duties. Instead, he was appointed chief consultant to Rolls-Royce, and had considerable input in all subsequent Rolls-Royce piston engines for aircraft until his retirement in 1945. He died on December 12, 1957.

Royce, Frederick Henry (1863–1933)

Maker of Royce and Rolls-Royce cars from 1904 to 1906; chief engineer and works manager of Rolls-Royce Ltd. from 1906 to 1911; senior technical partner in Rolls-Royce Ltd. from 1911 to 1933.

He was born on March 27, 1863, at Alwalton in Huntingdonshire as the son of James Royce, operator of a rented flour mill who left his debts and most of his family behind when moving to London in 1867 for a post with the London Flour Company. Henry went to school in London until his father died in 1872 and he had to work. He began by selling newspapers for W.H. Smith & Sons, later serving as a telegram messenger for Her Majesty's Post Office in Mayfair. In 1876 a friendly aunt provided funds for him to be taken on as an apprentice with the Great Northern Railway at the Doncaster shops, but served only three out of four years when his aunt's money ran out. He was only 17 years old when he landed a post as toolmaker with Greenwood & Batley in Leeds.

Next he worked briefly for Edison & Swan in London, joined the Electric Light & Power Company, but soon transferred to the Lancashire Maxim & Western Electricity Co. in Liverpool as chief electrician. With financial backing from Ernest A. Claremont, he set up Royce & Co. in a tiny shop in Cooke Street, Manchester, to make electrical appliances.

By 1890 the Royce name was recognized as a maker of high-quality electric cranes. The business was put on the stock market in 1894 as Royce, Ltd., and prospered. Reduced demand for cranes, partly due to competition from lower-price equipment made on the Continent, coincided with Royce's awakening interest in automobiles at the turn of the century.

In 1902 Royce bought a second-hand Decauville "voiturelle" which gave rise to grave complaints. He began solving the problems in a systematic manner, and decided to build a better car himself. The Decauville was taken apart and Ernie Wooler was given the task of sketching every part. Discontented with the single-cylinder Decauville engine, Royce decided to make a vertical parallel-twin, which was designed by A.J. Adams, his chief draftsman, under Royce's supervision. The first 10-hp Royce engine was tested on September 16, 1903. It was mounted in a chassis that owed a lot to Decauville practice, the first test drive being made on April 1, 1904. Car production began on a modest scale at Cooke Street. Royce put R.D. Spinney on the job of designing a three-cylinder derivative of the basic twin, and a four-cylinder version soon followed.

Royce's health problems began in 1902 when he collapsed as a result of malnutrition and overwork, clouding the rest of his career.

The Royce car became Rolls-Royce near the end of 1904 when the London agents, C.S. Rolls & Co., secured selling rights to Royce's total output. Royce designed a slow-running V-8 for a town car sold as the "Legalimit" in 1905 but his heart was set on a six-cylinder car of generous dimensions, a 40/50 hp model which was given the unofficial name of Silver Ghost. It went into production in 1906.

The Silver Ghost became famous, and the small shop at Cooke Street could not cope with the demand. In 1907 Rolls-Royce secured a site at Derby and erected a new factory, strictly for automobiles, though it was soon be expanded to produce aircraft engines.

Royce fell seriously ill in 1911, and his doctors discouraged him from returning to the Derby climate. He set up drawing offices at country houses in West Wittering, Sussex, and Le Canadel on the French Riviera, where small, select design teams joined him to turn his ideas into new products. This arrangement was effective until his death on April 22, 1933.

Rudd, Anthony Cyril "Tony" (b. 1923)

Powertrain manager for Lotus Cars, Ltd., from 1969 to 1970; engineering director of Lotus Cars, Ltd., in charge of research, development production engineering and quality from 1970 to 1974; Lotus Group engineering director for cars, racing cars and power boats from 1974 to 1978;

corporate research director of Lotus Cars, Ltd., from 1978 to 1991.

He was born on March 8, 1923, at Stony Stratford, Buckinghamshire, and educated at Ratcliffe School, Wolverton. He obtained his engineering degree from Derby Technical College and served an apprenticeship with Rolls-Royce at Derby. From 1942 to 1944 he worked in aircraft-engine defects investigation for Rolls-Royce, and in 1945 he was assigned to the Hucknall plant to develop a quality-control program. He left Rolls-Royce in 1951 to work on engine development for BRM[1] and a year later went on the Rubery, Owen & Co. Ltd. payroll as BRM chief designer and team manager. In 1969 Sir Alfred Owen, dissatisfied with BRM's poor racing record, requested his resignation. Almost immediately, Colin Chapman brought him to Hethel in Norfolk to develop engines and transmissions for Lotus Cars, Ltd. He was responsible for the Lotus engine that powered the Jensen-Healey and directed the engineering of the 1974 Lotus Elite and the 1975 Eclat. He was one of the men who built the Lotus Engineering Consultancy into a high-profile position in advanced automotive technology. He retired in April 1991.

1. British Racing Motors, Bourne, Lincolnshire.

Rumpler, Edmund (1872–1940)

Managing director of Rumpler Motoren-GmbH from 1919 to 1928; managing director of Vornantrieb and Vertriebs GmbH from 1929 to 1933.

He was born on January 4, 1872, in Vienna as the son of a small shopkeeper. He was educated locally and graduated with a degree in mechanical engineering from the Technical Institute of Vienna at the end of 1895. In 1897 he was working as a trainee engineer with Nesselsdorfer WagenbauFabriks-Gesellschaft, helping in the design of a two-cylinder automobile engine. In 1898 he became office manager of AMG,[1] a conglomerate of auto parts manufacturing, trading, and patent-brokering. For a brief spell in 1899 he held the title of technical director for a company named Centaur, and then returned to AMG where he became head of the drawing office. He resigned when AMG was taken over by Daimler Motoren Gesellschaft in 1902 and went to Heinrich Kleyer in Frankfurt where he designed two Adler cars, a 14hp twin and a 28-hp four-cylinder model. He patented a swing-axle suspension for driving wheels which he tried to put on Adler cars in 1903 but failed. He went to Amsterdam in 1904 and did some work for Spyker in 1905–06. He returned to Berlin, obtained a patent for a roller-cam tappet, and set up his own engineering-consultant's office. In 1907 he also founded a company[2] to make and market electric welding equipment. He became interested in aviation and founded an aircraft company[3] in 1908. Having learned that the Etrich & Wels "Taube" airplane was not patented in Germany, he copied the design and made a prototype in 1910.

He also designed and built a 60-hp V-8 "Aeolus" aircraft engine. During World War I he produced important numbers of the "Taube" airplane and ran a factory that made refrigerators. He toyed with unconventional car designs and in 1919 applied for a patent covering a teardrop-shaped body with an inboard/rear engine. He then formed Rumpler Motoren GmbH to produce it in his former airplane works at Berlin-Johannistal. He called it the "Tropfenwagen" and sold a license to Benz & Cie. He built about 20 Tropfenwagen from 1921 to 1924, most of them powered by a W-6 engine of his design, produced by Siemens & Halske, and the final ones with in-line four-cylinder power units. He won a doctor's degree in engineering in 1921 for his thesis on a 1000-hp 28-cylinder radial aircraft engine but found no takers for a license. In 1925 he turned his ideas on chassis layout back to front, and a year later displayed a low-slung front-wheel-drive touring car, the 6A 104, with a 50-hp four-cylinder engine and "Lautal" light-alloy frame. He liquidated the motor company in 1928 and founded an engineering office in Berlin-Charlottenburg, offering front-wheel-drive truck, bus, passenger car and special vehicle applications to the industry. He sold a bus-design license to Henschel & Sohn of Kassel, who made a prototype in

1931. But he was forced to close his office in 1933 due to lack of business. From 1935 onwards he kept trying to get himself and his family out of Germany, but without success. He died on September 7, 1940, during a visit to Wismar.

1. Allgemeiner Motorwagen GmbH, Berlin-Marienfelde.
2. Autogena Schweiss-Industrielle GmbH, Berlin.
3. E. Rumpler Luftfahrzeugbau GmbH, Berlin-Johannistal.

Ruppe, Oskar Berthold (1854–1932)

Owner and managing director of A. Ruppe & Sohn from 1887 to 1908; chairman of the supervisory board of A. Ruppe & Sohn AG from 1908 to 1910, chairman of the supervisory board of Apollo-Werke AG from 1910 to 1927.

He was born on June 11, 1854, in Dornburg on the Saale River, as the son of Artur Ruppe, maker of farming tools and equipment. He began experimenting with steam cars while still a teenager and drove his home-made steam car through the streets of Weimar in 1869, only to have the local authorities ban his vehicle from public roads. The family moved to Apolda in Thuringia when his father bought an iron foundry and machine shops there. He predicted the demise of steam power for road transport, and began taking an interest in small motor vehicles. About 1895 he bought several motor tricycles for testing and encouraged his son Hugo to experiment with engines. The company began making Apoldania motorcycles in 1902 and Piccolo cars in 1904. He was a founder-member of the Mid-German Automobile Club and in 1908 was given the title of Kommerzienrat (business councillor) by Grand Duke Ernst of Saxony-Weimar. In 1920 he organized the takeover of Markranstädter Automobil-Fabrik which he merged into the Apollo-Werke AG. The postwar models were moderately priced touring cars of good concept and good quality, but the company failed to make profits. Apollo car production ended in 1927.

Ruppe, Paul Hugo (1879–1949)

Chief engineer for engines, motorcycles and automobiles, A. Ruppe & Sohn from 1900 to 1907; technical director of Markranstädter Automobil-Fabrik from 1907 to 1917; chief engine designer of Zschopauer Maschininfabrok J.S. Rasmussen from 1919 to 1923.

He was born on August 15, 1879, at Apolda in Thuringia, as the son of Oskar Berthold Ruppe. He graduated with a degree in mechanical engineering from the Technikum in Ilmenau in 1892 and went to work in the paternal factories. He designed a single-cylinder engine and a motorcycle, which went into production in 1902 as the Apoldania. In 1903–04 he designed a light car with a 5 PS 704-cc V-twin engine mounted in front, two-speed gearbox, and shaft drive to the rear axle. It came on the market in 1904 with a Piccolo badge, and over 1,000 units were sold within 20 months. He left the family firm in 1907 to set up his own car company at Markranstädt near Leipzig and introduced the MAF 5/12 and 6/14 with air-cooled four-cylinder engines in 1908. He got into financial straits in 1911 and secured fresh capital from Friedrich W. Mithoff, who became joint managing director of the reorganized firm, Markranstädter Automobilfabrik vorm. Hugo Ruppe AG. Production of MAF cars continued until the outbreak of World War I. Mithoff and Ruppe resigned in 1917. A year later he designed a 25-cc two-stroke toy engine which he demonstrated to J.S. Rasmussen, who put it in production as the DKW (des Knaben Wunsch).

Ruppe designed air-cooled two-stroke DKW industrial engines from 1919 onwards, a 118-cc auxiliary engine for bicycles in 1920, and engines for the 1921 DKW Golem and 1922 DKW Lomos motor scooters. He left Zschopau in 1923 and moved to Berlin, where he designed two-stroke motorcycle engines for Motoren und Apparate-Fabrik Wilhelm Baier.

Sacco, Bruno (1933–)

Chief engineer for passenger-car bodies, Daimler-Benz AG, from 1974 to 1975; director of Mercedes-Benz passenger-car styling from 1975 to 1999.

He was born on November 12, 1933, at Udine in Italy's Friuli (northeast) region. After a general education, he studied engineering at the Turin Polytechnic, joined Carrozzeria Ghia and learned the rudiments of coachbuilding. He joined Daimler-Benz AG in 1958 and worked on the evolution of the basic 220-series bodies and had a hand in the 220 SE coupé and convertible. He became an assistant to Karl Wilfert and helped design the 230 SL and the 600 bodies. He also worked with Bela Barenyi on structural safety, new engineering concepts, and body structure pre-development. All Mercedes-Benz cars designed between 1970 and 1999 bear Sacco's stamp. He encouraged his crew to evolve new grille forms for special series and directed the continual evolution of the traditional fake-honeycomb grille. He approved the triangular tail-light clusters (which were soon abandoned) and Joseph Gallitzendörfer's Siamesed dual-headlamp combination. When the company's organizational chart was redrawn in 1987, he finally reached director's level in terms of authority and salary. He retired in the spring of 1999.

Sailer, Max (1882–1964)

General manager of the Daimler-Benz AG factories at Stuttgart-Untertürkheim from 1929 to 1934; head of the Mercedes-Benz drawing office, testing and development departments from 1934 to 1937; technical director of Daimler-Benz AG from 1937 to 1942.

He was born on December 20, 1882, in Esslingen and joined Daimler Motoren Gesellschaft in Cannstatt as a trainee engineer in 1902. He left in 1905 to broaden his experience by working for other companies, but returned to Daimler in 1910. He worked as a test driver, racing driver, and development engineer from 1910 to 1925. He was never on good terms with Ferdinand Porsche, who banished him to Berlin-Marienfelde as works manager in 1926, at a time when the future engine plant was nothing more than repair shops and a parts depot. He returned to Untertürkheim as soon as Porsche was gone, and took charge of all manufacturing, engine production, and testing. He was intrigued with all the "people's car" projects under study in Germany and neighboring countries in those years, and created the Mercedes-Benz 130 H, using a 1.3-liter current production engine, mounted in the tail end of a monotube "backbone" chassis with all-independent suspension. More than 4000 units wee produced from 1934 to 1936, plus some 300 150 H's and 1500 170 H's from 1935 to 1939. He also designed the 150 S roadster, using the 130 H chassis with the engine in an inboard/rear installation, but no more than 20 were built. He was also responsible for some of the experimental diesel-powered cars of 1933–36. As early as 1932 he began testing semi-automatic transmissions with the Vickers-Sinclair hydraulic coupling and all-electric shifting or Knorr vacuum-shift in Mercedes-Benz chassis. In 1937 he succeeded Hans Gustav Röhr as technical director and full member of the board (he had been a deputy member since 1935). He resigned in 1942.

Saintigny, Henri (1944–)

Head of the Automobiles Peugeot drawing office from 1988 to 1992; joint technical director of Peugeot SA in 1992–93; technical director of Peugeot SA beginning in 1993.

He was born in 1944 and held a diploma from the Ecole Centrale des Arts et Manufactures in Paris. He began his career in 1968 with Automobiles Peugeot at the Sochaux complex and held a succession of positions in product engineering. The main concern of the drawing office at Sochaux was to get each new car from the prototype stage to production, and he worked on the development of the 505, 205, and 605. From 1988 onwards, his task was mainly administrative and not creative, but he bore heavy responsibility for making the right decisions. He was also closely associated with Michel Durin in combining the functions of the design offices at La Garenne and Sochaux, and integrating them with Citroën's. Durin selected Saintigny to succeed him as technical director, and he moved to La Garenne in 1992. He inherited Durin's team of technical leaders, but also made some

changes, such as placing Frédéric Dieu in charge of body synthesis project engineering for Peugeot and Citroën in 1993 and naming Claude Durand as executive engineer for the design and product management group. His team was responsible for the planning and development of the Peugeot 206, Peugeot 607, Citroën C5 and C3, and the Peugeot 307. In 2002 he named Bruno de Guibert director of products, Automobiles Peugeot, and Vincent Besson director of products, Automobiles Citroën.

Sainturat, Maurice (1881–1967)

Chief designer of Hotchkiss from 1913 to 1914; technical director of Hotchkiss from 1919 to 1922; design engineer with Delage from 1922 to 1925; design engineer with Donnet from 1925 to 1928; design engineer with Chenard-Walcker from 1929 to 1932.

A graduate of the Ecole des Arts et Métiers at Angers, he went to work for Hotchkiss in 1904, moonlighting as a journalist with some of the earliest French motoring magazines such as *La Vie Automobile, Omnia, La Technique Automobile et Aérienne.* In 1905 he presented a treatise on body roll, shimmy and steering stability. Leaving Hotchkiss in 1909, he was called back in 1913 but left a year later to design aircraft engines for Delaunay-Belleville.

Hotchkiss invited him back in 1919. He modernized and simplified the model lineup and designed the magnificent six-cylinder AK prototype. When the Hotchkiss directors refused to put the AK in production, he walked out and joined Delage, where he created the GL model in record time. He accepted an offer from Donnet in 1925 and designed the 14 CV Donnet. Then he was faced with the task of preparing two all-new models in less than nine months, planned for a combined production of 5000 units a year. The results were a four-cylinder 7 CV and six-cylinder 10 CV using essentially the same chassis. But Job One was in reality just a prototype, untested, without the benefit of development. It ran into quality problems and nearly killed the Donnet reputation.

He joined Chenard-Walcker towards the end of 1929 and had a role in combining the two model ranges of Delahaye and Chenard-Walcker into a coherent lineup. When their cooperation ended in 1932, he joined Citroën and designed the 7 CV and 11 CV overhead-valve engines for the "Traction." In 1935 he was a member of a technical delegation from Citroën that toured U.S. industry. For the next 20 years he was associated with a multitude of Citroën projects, and he ended his career by participating in the creation of the DS-19. He spent his retirement years in Neuilly on the western doorstep of Paris and died in 1967.

Salmson, Emile (1859–1917)

Chairman of Société des Moteurs Salmson from 1913 to 1917.

He was born on September 12, 1859, in Paris. He prepared for the Ecole Polytechnique but broke off his studies. He became an engineer with Piltzer, importers of gas equipment. In 1890 he founded his own company, Société E. Salmson et Cie, in Paris, to manufacture gas generators based on his own patents, and Bollinckx gas engines under license. Over the next ten years, he expanded into making centrifugal pumps, hydraulic wheels, Girard turbines, and Gould-patented hydraulic buffers. He designed and built a helicopter prior to 1910, but it may never have been put to a test flight. In 1909 he took over the manufacturing rights to a seven-cylinder air-cooled radial aircraft engine developed by Georges Canton and Georges Unne. In 1911 he designed his own 11-cylinder version. In 1914 the government requisitioned the Bichler works in Lyon for the production of Salmson aircraft engines, and Salmson's Billancourt factory was tooled up for making propellers. Salmson produced approx. 15,500 engines during World War I. After the war, the company became part of the automobile industry, securing a license to the GN[1] cycle-car. Salmson car production began in 1921.

Emile Salmson, however, had died on September 21, 1917.

1. GN for H.R. Godfrey and A. Frazer Nash, of Kingston-on-Thames.

Salomon, Jules (1870–1964)

Chief engineer of chassis design for Georges Richard (Unic) from 1903 to 1909; chief engineer of Le Zèbre from 1909 to 1917; Citroën design engineer from 1917 to 1926; chief engineer of Automobiles Rosengart from 1927 to 1940.

He was born in 1870 at Cahors and attended college at Brive-la-Gaillarde. He failed his entry examination to the Ecole des Arts et Metiers, and went instead to the Ecole de Commerce in Bordeaux. He began his career with Rouart's gas engine factory, leaving for military service in 1888. He continued his career with Thirion in Paris and then with Thomson-Houston, leading makers of electric streetcars. He attended evening classes at Lycée Condorcet and was hired as an engineer by the "Economic" railway company[1]. Next he went to the CGT[2], prominent makers of electric motors and generators. In 1898 he became manager of Niel's gas-engine factory, best known for their three-cylinder two-stroke unit, patented in 1895.

In 1903 he joined Delaunay-Belleville but left the same day to link up with Georges Richard. He designed engines and chassis for Unic cars, notably the four-cylinder 10/12 CV model with one-piece block and a two-bearing crankshaft. He had ideas for a smaller and lighter car than Georges Richard would make and in 1909 formed a partnership with Jacques Bizet to produce it. Bizet was a Unic dealer and a friend of Baron Henri de Rothschild, who financed them in the Le Zèbre venture.

The low-priced single-cylinder Le Zèbre became quite popular. He modernized the design with a four-cylinder engine without losing its basic simplicity, and offered it to Andre Citroën, who engaged him in February 1917. Salomon and Edmond Moyet designed the Citroën 10 CV Types A and B2, and then the best-selling 5 CV Type C. Salomon then directed the team that designed the B 12 and B 14. Citroën dismissed him in 1926, and he sought refuge with Peugeot, where he met Lucien Rosengart. They left Peugeot together when Rosengart secured manufacturing rights to the Austin Seven and bought the former Bellanger factory. He converted the Austin drawings to the metric system, reworked the material specifications to match the capabilities of the French supplier industries, and was Lucien Rosengart's top technical advisor. In 1931 Rosengart acquired manufacturing rights to the Adler Trumpf with front-wheel drive, a sound design which Salomon developed over the years into the Supertraction, which had modern styling by M. Robin. Privately, with other backers, he designed the 1932 Cabri cycle-car with 387-cc single-cylinder engine and a dry weight of only 330 kg. He retired in 1940 and died in 1964.

1. Société des Chemins de Fer Economiques.

Sampietro, Achille C. "Sam" (1905–1980)

Co-founder of the Healey Motor Company in Warwick in 1945; design engineer with Healey from 1945 to 1948.

He was born on October 26, 1905, into a family of Italian hotel-keepers at Rapallo, the Ligurian coastal resort, and graduated from the Liège Polytechnic[1] in Belgium in 1930, with an honors degree in electrical engineering. Later, he obtained degrees in mathematics and physics from the Royal Technical Institute of Sondrio, Italy.

He began his career in England, joining the British agents for Alfa Romeo. Next he went to Thomson & Taylor at Brooklands, where he became assistant to the chief engineer, Reid A. Railton. In 1937 he joined the Rootes Group as deputy chief engineer for Sunbeam-Talbot, and spent the war years as a research engineer with Humber, Ltd., in Coventry, where he met Donald Healey.

In association with Benjamin G. Bowden, the body engineer, they planned a postwar sports car. Sampietro designed the chassis and Riley agreed to supply the engines. He left Warwick in 1948 to become technical director of D.R. Robertson, Ltd., an electronics firm. In 1960 he moved to America and was named chief engineer of Willys-Overland Motors in Toledo, Ohio (later Kaiser Jeep Corporation) but joined Ford Motor Company in 1964 as manager of the chassis components department and head of the products research office. Leaving Ford in 1970 he became an international consulting engineer whose list of clients included TRW,[2] Chrysler France, and British Leyland.

1. Institut Polytechnique de la Ville de Liège.
2. Formerly Thompson-Ramo-Wooldridge, Cleveland, Ohio.

Sangster, Charles (1872–1935)

Managing director of the Ariel Cycle Company, Ltd., from 1902 to 1920; chairman of Swift of Coventry, Ltd., from 1920 to 1931.

He was born in 1872 in Lanarkshire, attended local schools, and began his career in the shops of New Hope Cycle Company in Glasgow. At the age of 20 he moved to Coventry where he worked briefly for Rudge-Whitworth before joining James Starley of the Coventry Machinists' Company, makers of sewing machines and Swift bicycles. The company was controlled by W. & G. Du Cros who also owned Components, Ltd., and in 1895 Sangster was transferred to Components, Ltd., in Birmingham which later took over the company that made Ariel bicycles. He introduced the Ariel motorcycle, which soon became the company's main line of business. Ariel prospered and produced single-cylinder and V-twin military motorcycles in World War I.

Harvey Du Cros, Jr., managing director of Swift of Coventry, put Sangster in charge of Swift in 1920. Swift had just begun production of a postwar car, the 12-hp tourer, which replaced the 15-hp prewar model in 1921. The Swift Ten from 1915 was periodically modernized and remained in production until 1931. The 12 evolved into a 14/40 in 1927. But Swift had been badly hurt financially during the general strike of 1926, and the aging machinery in the Quinton Street plant suffered frequent breakdowns, halting production while makeshift repairs were made.

Sangster could be blamed for not giving Swift his undivided attention. He also held interests in the Rover Cycle Company, Midland Tube & Forging Company, and the Endless Rim Company. Swift of Coventry Ltd.

was declared bankrupt in April 1931, and Sangster sold the factory to Alfred Herbert (machine tools) and the name, patents, designs and other assets to R.H. Collier & Company, who had purchased the Clyno Motor Company.

Sangster remained an industrial investor but withdrew from active participation in business, and died at his home at Moseley, Birmingham, on March 18, 1935.

Sangster, John Young "Jack" (1896–1977)

Chairman of the Triumph Engineering Company, Ltd., from 1936 to 1951; chairman of the Daimler Company from 1956 to 1960.

He was born in 1896 at King's Norton as the son of Charles Sangster and educated at Hurst College, Hurstpierpoint. He journeyed on the Continent, working in French and German factories to gain engineering experience. Returning to Britain in 1914, he served with the 14th Royal Warwickshire Regiment in World War I. After his discharge, he privately designed a low-cost light car with an air-cooled flat-twin engine, three-speed gearbox, open propeller shaft and worm-drive rear axle, a single rear-wheel brake and disc wheels. He leased a factory at Tyseley, Birmingham, for its production, which would be financed by his father. While he was tooling up, Rover's managing director, Harry Smith, became interested in it. The car was developed into the Rover 8, and the Rover company also took over the Tyseley plant, where Sangster served as assistant works manager from 1919 to 1922.

He joined Ariel Motors, Ltd., in Birmingham as assistant managing director, designed a new car with a water-cooled flat-twin, and supervised the production of 700 Ariel cars from 1922 to 1925. He was named joint managing director of Ariel in 1924 and sole managing director in 1930. He led Ariel until he sold it to BSA[1] in 1944.

With his private fortune, he purchased the motorcycle branch of Triumph in Coventry in 1935, and led the Triumph Engineering Company, Ltd., until he sold it to BSA in 1951. That transaction won him a seat on the BSA board of directors, with oversight of the affairs of the Daimler Company, a BSA affiliate since 1910. Sangster is given the main credit for the ouster in 1956 of Sir Bernard Docker, the wayward chairman of both BSA and Daimler. He installed a new management team at Daimler and in 1960 sold the Daimler Company to Sir William Lyons.

He retired from business soon afterwards and died in 1977.

1. Birmingham Small Arms Company, Ltd., Small Heath, Birmingham.

Sarazin, Edouard Auguste (1839–1887)

Patent lawyer and industrial broker, founder of the Compagnie Française des Moteurs à Gaz et des Constructions Mécaniques in Paris in 1879.

He was born on August 20, 1839, in Liège as the son of a Belgian artillery captain. Orphaned at a tender age, he went to work as an office boy in the Cockerill works at Seraing, then making steam engines, locomotives, and machine tools. He took evening classes in engineering and law, receiving a degree in civil engineering from the Ecole Polytechnique of Liège. He spent three years supervising the installation of a Cockerill plant at Saint Petersburg from 1866 to 1869. He resigned from Cockerill and went to Paris, where he registered as a law student at the Sorbonne. He also opened an office and spread the word that he could handle litigation over patent rights and other technical matters. Eugen Langen, chairman of Gasmotorenfabrik Deutz in Cologne, went to see him, which led to the establishment of a French subsidiary and the start of France's gas-engine industry. In 1886 he filed several applications for French patents on behalf of Gottlieb Daimler and reached a Daimler-license agreement with Emile Levassor, director of Pèrin, Panhard & Cie in Paris, for production of Daimler engines.

Sarazin took ill in November 1887 and died on December 24, 1887. His widow Louise (née Cayrol) visited Gottlieb Daimler in Cannstatt, where he signed over to her all his French and Belgian patents "for an amount to be determined, payable within three years." In November 1889, she sold them to Panhard & Levassor against a 20 percent license fee. And on May 17, 1890, she married Emile Levassor. After his death, she filed suit against Panhard & Levassor because they did not pay her any royalties on the Phénix and other engines of their own design. The case was settled out of court, and in 1914 Henri Sarazin (son of Edouard and Louise) was elected to the Panhard & Levassor board of directors. Louise Sarazin died in 1917 at the age of 70.

Sarre, Claude-Alain (1928–)

Chief executive officer of Automobiles Citroën from 1968 to 1970.

He was born in 1928 and was educated privately (degree in history) and at the Paris Institute of Political Studies,[1] graduating in 1953. He worked in the administration of Air France in 1954–55 and joined Citroën in 1956 in the sales department. By 1960 he was head of Citroën's commercial methods department. He was named joint sales director for the French market in 1967 and general sales manager in 1968. In August 1968, he succeeded Pierre Bercot (who became chairman) as president and chief executive. Sarre cut Citroën's operating losses in half, but still ended the year 1969 with a staggering deficit of $11,500,000. Bercot stifled many of his initiatives to get the cobwebs out of Citroën's organization. He saw his position undermined by the proposed merger with Fiat and resigned in April 1970, just as deliveries of the SM were beginning and the CX was taking form.

In 1972 he became president of the Roubaix textile mills,[2] completed its restructuring inside 18 months, and made big profits in 1973–74. He was then loaned out to do the same job for Boussac textiles, 1974–76, and served as president of Nobel-Bozel from 1977 to 1982. In 1983 he accepted the position of director of economic services for the French employers' association.[3]

1. Institut d'Etudes Politiques à Paris.
2. La Lainière de Roubaix.
3. Comité National du Patronat Français.

Satta Puliga, Orazio (1910–1974)

Manager of the Alfa Romeo automobile drawing office from 1946 to 1959; technical director of Alfa Romeo from 1959 to 1969; deputy managing director of Alfa Romeo from 1969 to 1974.

He was born on October 6, 1910, in Turin, served an apprenticeship with Alfa Romeo from 1925 to 1928, and went back to school, graduating from the Turin Polytechnic in 1933. After military service, he obtained a diploma in aeronautical engineering in 1935. He stayed on for post-graduate studies at the Turin Polytechnic and became an assistant to the chair of the aeronautical laboratory. He joined Alfa Romeo as a design engineer on May 2, 1938, and worked on a number of passenger-car projects and the evolution of the Tipo 158 Grand Prix racing car. After World War II he developed the 159 and 159A racing cars and the 6C 2500 production model. He led the design teams for the 1900 and the Giulietta, the Disco Volante competition cars, the 2000 and 2600, the Giulia and 1750, and the Montreal V-8. He fell ill and died on March 29, 1974, in the Milan Polyclinic.

Savey, Dominique (1932–)

Director of products and planning, Automobiles Peugeot, from 1976 to 1991, joint managing director of Automobiles Peugeot from 1992 to 1997.

He was born in Paris in 1932 and held a diploma from the Institut d'Economies de Paris. He began his career in 1957 with the Banque Européenne et Financière and two years later joined Peugeot in the general secretariat, which reported to the board of directors. In 1969 he set up a new office in Peugeot's management structure, Economic Forecasting and Planning, which he led until 1975. As director of products and planning, he was mainly concerned with market trends and pointing out the market segments and price levels where Peugeot should aim its cars and not the actual vehicle specifications.

From 1979 to 1981 he was also a director of Automobiles Talbot (ex–Chrysler France). His briefings enabled Peugeot to upgrade the 304 to a 305 in 1977 and the 504 into a 505 in 1979. His analysis of the bottom segment led Peugeot to upgrade the 104 in 1982 and the slot left open by ending 204 production in 1976 was filled at the end of 1982 by the arrival of the 205. His recommendations determined the price brackets of the new 406 in 1987 and the 605 in 1989.

From 1992 Savey had shared responsibility with Jean-Yves Helmer for Automobiles Peugeot operations and he continued to ensure the coherence of the product line until his retirement in 1997.

Savonuzzi, Giovanni (1911–1987)

Chief engineer of Cisitalia Automobile S.p.A. from 1945 to 1949, technical director of Automotores Argentinos SA from 1949 to 1951, technical director of Carrozzeria Ghia S.p.A. from 1954 to 1957, director of engineering for Chrysler Corporation's research department from 1962 to 1968.

He was born on January 28, 1911, in Ferrara, the son of a medical doctor. He attended local schools and went to college in San Marino. The family wanted him to study medicine, but he filled book after book with sketches of cars and airplanes. The family agreed to let him study mathematics at the University of Ferrara, thinking he would fail — and then fall back on medicine. He did not fail and was accepted at the Turin Polytechnic, where he graduated in 1939. He joined Fiat as a test engineer in the Aviation Section, where he worked throughout World War II.

In 1945 he joined Piero Dusio and organized production of the single-seater Cisitalia D-46 racing car (designed by Dante Giacosa). He designed the Cisitalia 202 chassis and a streamlined prototype. He directed the production of Cisitalia sports and racing cars until the company failed and then accompanied Piero Dusio to Argentina, where he designed the Autoar economy car. But the project ran aground due to the lack of a supplier industry and he returned to Italy in 1951. He became engineering director of LPM S.p.A., a plastics company. He designed a small racing car for SVA[1] and in 1952, when Piero Dusio

returned, designed a new Cisitalia powered by a 2.8-liter four-cylinder BPM[2] engine.

He joined Ghia at Luigi Segre's call, made some design proposals and organized the production of special cars for Chrysler. He joined Chrysler in Detroit in 1957 as a research engineer and designed the A-831 gas-turbine power unit. He was promoted in 1962 and directed a multitude of research projects. He returned to Turin in 1968 and Gaudenzio Bono invited him to join Fiat. He took over Fiat's gas-turbine project for heavy trucks and buses. He retired in 1976 but remained a consultant to Fiat Aviazione. In 1980 he began working on a helicopter engine project for Messerschmitt-Bölkow-Blohm (a turbocharged two-cylinder two-stroke design).

He died on February 18, 1987, in the Moncalieri Hospital.

1. Società Valdostana Automotori, financed by an American midget-car racing group.
2. Botta & Puricelli, Milano, manufacturers of speed-boat engines.

Savoye, Raymond (1943–)

Program manager for the Renault Laguna from 1989 to 1994.

He was born in 1943 and graduated from an engineering college in Lyon. He joined Renault as a tool-and-die engineer in 1966 and held a succession of appointments in manufacturing and production organization. From 1985 to 1989 he was plant director at the Sandouville works, which assembled the R 25 and R 21. The X-56 project was planned by Jacques Cheinisse and was vital to Renault's survival in the market segment of the Ford Mondeo and Peugeot 405. It was to be Renault's first experience of "simultaneous engineering."

Production methods, sites, quality, cost and timing targets were put on paper in 1989. The logistics were drawn up in 1990 and the details of sourcing supplies, fabrication and subassembly methods added in 1991. The main outside suppliers were brought into the program in 1989 and the first running Laguna prototype was ready for the road in April 1991. It replaced the R 21 at Sandouville in December 1993.

Savoye's next assignment was to handle the liaison with Automobiles Matra for the Espace II program. He was put in charge of Renault's industrial cooperation schemes, such as Karmann's production of the Mégane cabriolet and joint production of the Trafic and Master commercial vehicles with Iveco[1] and General Motors. In January 1997 he was delegated to Renault Industrial Vehicles as director of quality.

1. Industrial Vehicles Corporation, formed in 1975 as a subsidiary of Fiat S.p.A. to combine the truck and bus operations of Fiat S.p.A. Unic and Magirus-Deutz.

Schaefer, Herbert (1932–)

Chief designer of the Volkswagen styling studio from 1972 to 1993.

He was born in 1932 and studied at the Kaiserslautern Technical College, graduating in 1952 as master craftsman of body arts. He began his career with Auto Union GmbH at Ingolstadt, helping to modernize the DKW bodies, and was promoted to design manager for a studio. He worked briefly in the body design department of Daimler-Benz AG and moved to Wolfsburg in 1961 as a project designer with Volkswagenwerk. For several years he worked mainly on preliminary designs. Then came a period when he was associated with Ulrich Seiffert and the design of experimental safety cars. The styling contract for the VW Passat had been given to Ital Design before he was named chief designer. Nor did he work on the first-generation Scirocco and Golf, also contracted out to Ital Design. He restyled the Scirocco for 1981 and the Golf for 1983, and created the Jetta sedan for 1984. He led the design teams for the second-generation Passat and its companion-model, the Santana, and the Polo C, all three launched in September 1981. He is also credited with the design of the third-generation Passat of 1988, and gave the guiding light for the Corrado which replaced the Scirocco in 1988. He prepared the third-generation Golf, launched in 1991, and the Vento which replaced the Jetta in 1992. He retired in 1993.

Schaeffer, Rodolfo (1893–1964)

Director of body engineering for Fiat from 1929 to 1945.

He was born on November 11, 1893, in Turin and attended local schools. He had hardly begun his technical studies when he was called up for military service in the Artillery in 1916. Before long he was transferred to an aeronautical engineering group, and by the end of World War I, he was a major on the technical staff of the Air Force, GARI.[1] Upon discharge, he resumed his studies at the Turin Polytechnic and graduated in November 1920. He joined Fiat on January 1, 1921, as a test engineer, and within six months was assigned to the body engineering section. On June 1, 1925, he was promoted to director of the Special Body Section, and on March 1, 1929, he was named director of body engineering, which included responsibility for body design. He modernized the appearance of Fiat cars rather timidly at first, in keeping with American styling trends. That changed dramatically in 1932 when Giovanni Agnelli brought in Mario Revelli de Beaumont as design consultant. Revelli provided the original drawings for the car that became the Fiat 1500 in 1935, but the body engineering and detail design were Schaeffer's responsibility. He applied the same general theme to an entire generation of Fiat cars, including the 500 in 1936, 2800 and 508 C in 1937. He remained at Lingotto during World War II, and in 1945 accusations were made against him for having been a little too ready to execute German orders.

He was suspended from his functions and never worked on car design again. In 1948 Fiat made him director of its CANSA[2] subsidiary in Novara, and in 1952 he returned to Turin as technical manager of the Railway Equipment Office. He retired on December 31, 1960, and died in Turin on February 22, 1964.

1. Genio Aeronautico Ruolo Ingegneri.
2. Carrozzeria Automobili Novara, S.A.

Schapiro, Jakob (b. 1885)

Business speculator and industrial investor who in 1927 owned one-third of Germany's motor vehicle industry.

He was born on November 6, 1885, in Odessa, and began an apprenticeship with a toolmaker. The family fled Odessa in the nationalist revolt in Georgia in 1905 and made their way to Vienna. He studied engineering and obtained a diploma. He went to Berlin in the early days of World War I and started a driving school which was incorporated in 1918 as Automobilhaus Jakob Schapiro, Handel- und Reparaturwerkstätte, an auto repair shop and used-car dealership. In 1919 he bought Carrosserie Schebera GmbH of Berlin-Tempelhof where coachbuilding was almost at a standstill. He saw the start of Germany's galloping inflation, and made it work for him. He bought cars, paying with bonds which he renewed until the amount due had become insignificant in real value. He bought 200 Benz 8/20 PS chassis, paid in bonds, and put Schebera bodies on them in the spring of 1921. At the same time he bought Heilbronner Fahrzeugfabrik GmbH and reorganized it as Süddeutsche Carrosseriewerke Schebera AG. The Berlin factory was turned into a sales company, Schebera Automobil-Werke AG, acting as distributors for NAG, NSU, Protos and Mercedes cars. The Heilbronn plant was retooled to make bodies for Benz, Protos and NSU. In November 1922, he announced to the press that he held 40 million of the original 96 million shares in Benz & Cie, and in January 1923, Schebera took over the general distribution of Benz motor vehicles. In July 1923, Schapiro joined the supervisory board of Benz & Cie. By 1924 he held 60 percent ownership of Benz & Cie. Between 1921 and 1925 he bought 30 percent of Hansa Automobil-AG of Varel in Oldenburg, and in 1922 acquired majority control of Cyklon Maschinenfabrik GmbH of Berlin and Mylau, Cyklon and Cyklonette sales taken over by the Schebera network. In 1925 he secured control of Gothaer Waggonfabrik AG including its subsidiary, Dixi Automobilwerke AG of Eisenach, and a year later he prevailed on Sir Herbert Austin to grant him the German manufacturing rights to the Austin Seven. He planned an output of 300 cars a week, to be sold with a Dixi label, and guaranteed Austin a volume of 2000 units in the first year. He also gave birth to a new Cyklon model with a six-cylinder Dixi engine. He then merged Cyklon Kraftfahrzeugwerke AG with Gothaer Waggonfabrik AG, and in November 1926 forced the merger of Neckarsulmer Fahrzeugwerke AG with Schebera Automobil-Werke AG to create the NSU Vereinigte Fahrzeugwerke AG based at the Heilbronn plant. His Benz & Cie holding gave him a seat on the supervisory board of Daimler-Benz AG, an unwelcome presence in the eyes of many other board members and notably the representatives of Deutsche Bank. Schapiro could no longer operate as before, since German finance minister Hjalmar Schacht had tamed the inflation, and Schapiro had to face real debts. He sold his stock in Hansa Automobilwerke AG to Dutch investors[1] and disposed of Dixi Automobilwerke AG to BMW, which took over the Eisenach plant and the production of the Austin Seven. He sold NSU Vereinigte Fahrzeugwerke AG to Fiat but was obliged to repurchase the ex–Schebera real estate in Berlin. Finally, he cashed in his Daimler-Benz AG stock in 1929. He still held many other interests, including majority control of Rudolf Chillingworth AG, makers of car and truck frames in Nuremberg, the Kandelhardt AG taxicab fleet in Berlin, Panzer AG garage rentals and motor traders in Berlin, Metrum Apparatebau AG and the Georg Grauert AG iron foundries and machine works. When Hitler came to power in 1933, he began to prepare his exit from Nazi Germany, and sold for whatever price he could get. Finally, Gebrüder Schapiro Automobilhaus AG in Berlin was liquidated on December 1, 1933. But Jakob Schapiro was already in Paris.

1. Industrielle Disconto Maatschappij, Amsterdam.

Scheele, Nicholas V. "Nick" (1944–)

Chairman of Jaguar Cars, Ltd., from 1992 to 1999; chairman of Ford of Europe, GmbH from 1999 to 2001.

He was born on January 3, 1944, at Brentwood, Essex, and studied modern languages at

Durham University. He joined Ford of Britain as a trainee in 1966, and held a succession of assignments in purchasing and supply. He was managing director of Ford of Mexico from 1989 to 1992. Ford appointed him deputy chairman of Jaguars Cars, Ltd., in January 1992 and chairman two months later. The company was then losing one million dollars a day. He cut the payroll in half and put in a new final-assembly line at Browns Lane. By the end of 1995, Jaguar was making money again. He served briefly as Ford of Europe's vice president of sales and service before being named chairman. He moved Ford's European headquarters from Brentwood, Essex, to Cologne in August 1999 and arranged the purchase of Volvo's car division for $4 billion. In March 2000, he negotiated the purchase of Land Rover from BMW for $2.85 billion, and later that year engaged W. Reitzle from BMW to run a new subsidiary, Premier Automotive Group. He ordered an end to Escort assembly at Halewood and Fiesta assembly at Dagenham, had Halewood retooled to build the Jaguar X-Type, and started revamping Dagenham for diesel-engine production. In July 2001, he was transferred to Dearborn as vice president in charge of Ford's North American Operations.

Scheibler, Fritz (b. 1845)

Founder and chairman of Motorenfabrik Fritz Scheibler from 1898 to 1907.

He was born on July 9, 1845, in Brand near Aachen (Aix-la-Chapelle) and educated as an engineer. Without seeking any license agreement or importing parts, he started engine production in Aachen in 1898 and won a gold medal and a special award of honor for his flat-twin engine at the International Automobile Show in Frankfurt am Main in 1900. He then decided to expand into production of cars and trucks, and in 1903 Motorenfabrik Fritz Scheibler was reorganized as Scheibler Automobil-Industrie GmbH. His cars had a front-mounted flat-twin engine, friction-type stepless drive, and chain drive to the rear wheels. In 1903 he gave buyers the option of a three-speed gearbox, and added four-cylinder models in 1904.

Scheibler cars became increasingly powerful and expensive, and were in low demand. The last models were made in 1907. Scheibler truck production, led by his son Kurt Scheibler (1874–1958), continued until the company was taken over by Mannesmann-Mulag in 1913.

Scherenberg, Hans (1910–)

Chief engineer of the passenger-car design office, Daimler-Benz AG, from 1952 to 1965; technical director of Daimler-Benz AG from 1965 to 1977.

He was born on October 28, 1910, in Dresden but grew up in Heidelberg and studied mechanical engineering at the Technical Universities of Stuttgart and Karlsruhe. He began his career with Daimler-Benz in the experimental department in 1935, worked on passenger-car diesel engine projects and fuel-injection systems for aircraft engines. He spent the war years in the technical laboratories.

In 1946 he went to Wilhelm Gutbrod at Plichingen, makers of garden, farm and forestry equipment who had built the Moto Standard economy car before the war, as technical director. Gutbrod returned to the auto industry by making a tiny cab-over-engine truck named Atlas 800, and in 1949 introduced the Superior 600 mini-car.

He returned to Daimler-Benz in 1952 and got acquainted with the 220 and 300-series, the 300 SL, and the 180. In 1955 he was named deputy member of the supervisory board and given responsibility for industrial engines, aircraft engines, and central development works. In 1965 he succeeded Fritz Nallinger as director of vehicle development. He directed the programs that led to the 1972 S-Class and the 1976 W-123 family. He retired in 1978.

Schmücker, Toni (1921–1996)

President and chief executive officer, Volkswagenwerk AG, from 1975 to 1981.

He was born in 1921 at Frechen near Cologne as the son of a Ford-Werke employee. At the age of 16 he joined Ford as a sales trainee, spent six years in the purchasing department and in 1961 was given the title of purchasing director, with a seat on the management board. In 1967 he was named sales director of Ford-Werke AG. He left Ford in 1968 in protest of Henry Ford, II's, manner of passing over German executives in selecting leaders for the newly formed Ford of Europe, Ltd. He became president and chief executive of Rheinstahl in 1968 and sold off its unprofitable divisions, disposing of Hanomag-Henschel to Daimler-Benz AG and Hanomag Engineering (tractors and earth-moving vehicles) to Massey-Ferguson. Then he successfully maneuvered the sale of Rheinstahl to the Thyssen steel group.

On February 10, 1975, he was named chairman of Volkswagenwerk AG, who had 25,000 redundant workers and an empty cash box. The stockpile of unsold cars counted 600,000. He took many initiatives, arranging VW's purchase of Chrysler do Brasil in 1978 and spending $350 million to start VW assembly at Westmoreland, Pennsylvania. He bought Triumph-Adler office machines in 1979 after failing to get control of Nixdorf Computer. In 1980 he set up a factory to produce the VW Golf at Sterling Heights, Michigan (a plant that was sold to Chrysler in 1983).

But he proved ineffective in cutting the production costs of VW cars, shortening their time-to-market, and improving productivity in the German plants. He resigned from VW after suffering a heart attack in 1981 and died in November 1996.

Schneider, Théodore (1862–1950)

Co-founder of Rochet-Schneider in Lyon in 1896; director of Rochet-Schneider from 1896 to 1910; maker of Th. Schneider cars from 1911 to 1931.

Sometimes referred to as "Théophile," he was born on January 18, 1862, into a silk-industry family at Bourg-Argental and was a keen cyclist during his school days. Needing a new bicycle in 1888, he went to see Jean-François Rochet in Lyon, who had a reputation for making the best. He proposed setting up a parallel sales organization for Rochet two-wheelers, but was turned down. But he knew there was another Rochet not connected

with the family business, by the name of Edouard, and sought him out. They got along well, applied for patents on bicycle and tricycle design, and in 1891 established the Constructions Vélocipédiques du Rhône to start production.

Their first prototype automobile was completed in 1895 and the following year they incorporated the Société Lyonnaise de Vélocipèdes et d'Automobiles (abridged to omit the velocipedes in 1901). Schneider lost influence during a period of international ownership from 1905 to 1908, when the corporate power was vested in Rochet & Schneider, Ltd., of Moorgate, London. The headquarters returned on November 1, 1908, with the founding of S.A. des Etablissements Lyonnais Rochet-Schneider in Lyon.

He resigned in April 1910, moved to Besançon in the Jura, and formed a partnership with an engineer, Joseph Ravel. From a modest start in 1911 they brought out a line of robust cars with lively performance, and set up a racing team in 1912.

Soon they were building 500 cars a year, notably the four-cylinder 10/12 CV with its coal-scuttle hood and radiator behind the engine, which was produced up to 1920. In 1913 Th. Schneider introduced a 15 CV six (with two blocks of three cylinders) and a 26 CV racing car. That year he also moved final assembly to the R.E.P.[1] airplane works at Boulogne-Billancourt, while engine production continued at Besançon. During World War I, the engine plant made shells, and the assembly plant was sold back to the aircraft industry in 1916. Joseph Ravel had left the partnership in 1914. Th. Schneider revived car production at Besançon but fell into receivership in 1921. Still he managed to bring out a remarkable 10 CV model in 1923, its 2-liter four-cylinder engine having splayed overhead valves, and the front-wheel brakes having power assist. By 1925 it was joined by a four-cylinder 15 CV and a six-cylinder 20 CV. The company came under control of Paris-Nord Automobile and its Lille-based chief executive, Robert Poirier, in 1924. A lower-priced 7 CV came on the market in 1927, but the sales curve went into a decline. The last cars were built in 1931. Th. Schneider sold the factory to SADIM[2] and retired. He died in 1950.

1. Robert Esnault-Pelterie.
2. Société Anonyme d'Instruments de Motoculture.

Schrempp, Jürgen (1944–)

Chairman of the Daimler-Benz AG management board from 1995 to 1998; chairman of the DaimlerChrysler management board since 1998.

He was born on September 15, 1944, at Freiburg in Breisgau, served an apprenticeship with Daimler-Benz AG and put himself through college in Offenburg by playing jazz trumpet in local bands. After joining Daimler-Benz AG in 1967, he passed his first test as chief-executive material by bringing the Euclid subsidiary (earth-moving vehicles) out of its path to bankruptcy and back into profits. He ran Mercedes-Benz of South Africa from 1985 to 1987, and was named director of the truck division on his return to Stuttgart. In 1989 Edzard Reuter placed him in charge of solving the problems of Deutsche Aerospace, where he pruned the payroll and cut losses. In June 1996 he sold Dornier Aviation to Fairchild, spit up AEG into three parts which were sold separately, and disposed of Fokker Flugzeugwerke. In November 1998 he merged Daimler-Benz AG with Chrysler Corporation, which enabled him to bring down Deutsche Bank's stake in DaimlerChrysler to 12 percent and the Kuwait Investment Office's to 7 percent. In 1999 he bought a controlling stake in Mitsubishi Motor Corporation, and purchased Detroit Diesel Corporation from Roger Penske. He then sold Deutsche Aerospace to the EADS[1] joint venture and announced spending plans of $47.6 billion to bring out 34 new vehicles in a 36-month time span. In 2001 he bought a 10 percent stake in Hyundai Motors Corporation and in 2002 prepared to buy a minority stake in the Fuso truck-and-bus division of Mitsubishi.

1. European Aeronautic Defense and Space Company.

Schürmann, Gustav (1872–1962)

Technical director of the Automobile Department, Polyphon Musikwerke AG, from 1908 to 1915; technical director of Dux Automobilwerke AG from 1915 to 1927.

He was born in 1872 and held an engineering diploma. He occupied a technical position with the Gaggenau Iron Works[1] from 1896 to 1899, and then joined Helios Fahrradwerken in Austria and designed a small car with a two-cylinder engine mounted under the seat. He then had a brief association with Heinrich Ehrhardt at Eisenach[2] before returning to Gaggenau in 1903 as works manager of SAF[3] and designed a touring car with a four-cylinder overhead-camshaft 4.7-liter engine, produced from 1906 to 1911. In 1908 he moved to Leipzig-Wahren at the invitation of the management of Polyphone Musikwerke AG, a diversified company producing musical instruments, phonographs, machine tools and automobiles. The automobile department, which had been building the "curved dash" Oldsmobile under license since 1905 and the curious little Polymobil with its F-head parallel-twin mounted horizontally under the front seat since 1907, had high costs and little income. Both were replaced in 1909 by the Dux E 12, with a four-cylinder L-head 6/12 hp engine having a one-piece block, 3-speed gearbox and shaft drive. He broadened the Dux range in 1911 with the similar but bigger 6/18 and the 2-liter 8/21 with four-speed gearbox. The Leipzig-Wahren factory produced trucks during World War I and resumed car production in 1920 with the Type S, for which he had designed a 4.7-liter 17/50 PS engine. The company was then a member of DAK,[4] a sales-and-purchasing cooperative society for Dux, Presto, Magirus and Vomag, which lasted until 1935. In 1924 he replaced the Type S with the six-cylinder 4½-liter Type R, priced just below the Maybach W3 luxury car. It was the wrong move, and in April 1926, Dux Automobilwerke AG came under control of Presto-Werke AG of Chemnitz, which then belonged to its body supplier, Ernst Dietzsch of Glauchau in Saxony. In November 1927 Presto-Werke AG was in turn taken over by NAG.[5] Schürmann held on to his position at Leipzig-Wahren, and in 1931 was named works director of the plant, which he retooled for production of light-duty Büssing–NAG trucks. He became a deputy member of the Büssing–NAG management board in 1935 and retired in 1937.

1. Eisenwerke Gaggenau, owned by Theodor Begmann, makers of the Orient Express automobile designed by Joseph Vollmer.
2. Fahrzeugfabrik Eisenach, makers of the Wartburg car under Decauville license.
3. Süddeutsche Automobilfabrik GmbH, founded by Georg Wiss after taking over the automobile division of Bermann's Industriewerke in 1904.
4. Deutsche Automobil-Konzern GmbH, a joint venture of Dux Automobilwerke AG, C.D. Magirus of Ulm, Presto-Werke AG of Chemnitz, Vomag (Vogtländische Maschinenfabrik) of Plauen. Magirus made fire trucks; Vomag built heavy transport vehicles.
5. Nationale Automobil-Gesellschaft AG, Berlin-Oberschöneweide, united with Automobilwerke H. Büssing AG of Brunswick in 1928, and they were merged into Büssing–NAG Vereinigte Nutzkraftwagen AG in 1931.

Schutz, Peter W. (1930–)

Chairman of Dr. Ing. h.c. F. Porsche K-G from 1981 to 1987.

He was born on April 20, 1930, in Berlin as the son of a pediatrician. His parents escaped Hitler's "final solution" by fleeing to America in 1938, and they spent two years in Cuba before being admitted to the United States. He grew up on Chicago's South Side and graduated from the Illinois Institute of Technology with a degree in mechanical engineering in 1952. That same year he joined Caterpillar in Peoria, Illinois, as a test engineer, but was called up for military service. Two years later he returned to Caterpillar as a development engineer, and later became a project engineer. In 1966 he went to Cummins Engine Company of Columbus, Indiana, as director of technical planning, and three years later was promoted to vice president in charge of sales and service. He returned to Germany in 1978 as director of power train research and development and a member of the Klöckner-Humboldt-Deutz AG management board. Porsche head-hunted him in 1980, seeking a replacement for Ernst Fuhrmann, and he arrived at Zuffenhausen on January 1, 1981, without ever having driven a Porsche car. He did not make waves. He took some long-needed actions, such as ordering a $35 million paint shop, but neglected other needs, such as the labor-intensive under-mechanized assembly of the 911 and 928 body shells. The Porsche family accused him of making the company over-dependent on the U.S. market, which accounted for 60 percent of Porsche car sales, and pushed him out in December 1987.

Schwarz, Georg (1862–1929)

Works manager of Fahrzeugfabrik Eisenach AG from 1898 to 1903; technical director of Neckarsulmer Fahrzeugfabrik AG from 1912 to 1927.

He was born on December 20, 1862, at Bolheim and held an engineering diploma. When Fahrzeugfabrik Eisenach AG bought exclusive rights to make the Decauville car in Germany, he was engaged to organize production. The cars were called Wartburg and exported under the name Cosmobil. After leaving the company, he wrote technical papers for trade publications and became a prominent rally driver. Fritz Gehr brought him to Neckarsulm in 1912 with a brief to develop new cars and take charge of the factories. He modernized the popular NSU 5/12 as a 5/15 for 1914, and in 1920 developed it into a 5/20. He designed an expensive touring car, the 14/40 in 1921, with a 3.6-liter engine and four-wheel-brakes, followed in 1923 by the NSU M-Type luxury car with a 70-hp 4.7-liter six-cylinder engine, aimed at a market that had just been wiped out by hyperinflation. Also in 1923 he designed a sports car based on the 5/20, a lower frame and Roots-blown 1.3-liter side-valve engine, which did well in racing. During 1925–25 NSU was producing about 1000 cars a year and some 25,000 motorcycles with a 4000-man work force. In 1925 Schwarz designed a racing car with a 60-hp Roots-blown 1-liter side-valve six, which became a regular class winner in major races. It led him to prepare two six-cylinder production models, the 6/30 and 7/34, which went into production at the new Heilbronn plant in 1927. He retired that year, due to ill health, and died in 1929.

Schweitzer, Louis (1942–)

Managing director of the Régie Nationale des Usines Renault from 1990 to 1992; chairman of the Régie Nationale des Usines Renault from 1992 to 1996; chairman of Renault SA beginning in 1996.

He was born on July 8, 1942, in Geneva as the son of Pierre-Paul Schweitzer, who became a director of the French Treasury and managing director of the International Monetary Fund. He was educated at the Lycée de Saint-Cloud, obtained a law degree from the Institut d'Etudes Politiques in Paris. He graduated from the Ecole Nationale d'Administration in 1970 and went into government service as a finance inspector. He served as a budget official from 1972 to 1979, and was cabinet manager for Laurent Fabius, minister of the budget and industry, from 1981 to 1986. He joined the Régie Nationale des Usines Renault in 1986 at director's level, and became assistant to Georges Besse. He became Renault's finance director in 1988, joint managing director in 1989, and managing director in 1990. He masterminded the merger pact with AB Volvo, which was soon aborted. Later, he put Renault's bus division into a joint venture with Iveco and sold the Renault truck division to AB Volvo. In 1999 he bought a controlling stake in Nissan Motor Company, Ltd., and took over the Dacia enterprise which had been building Renault cars at Pitesti, Romania. In 2000 he bought a controlling interest in the bankrupt Samsung Motor, Inc., of Korea.

Schwenke, Robert (1873–1944)

Technical director of Vulkan Automobil GmbH from 1899 to 1905; chief engineer of the vehicle-engineering office of Siemens-Schuckertwerke GmbH from 1907 to 1927.

He was born on November 6, 1873, at Rybnik near Ratibor and held degrees in electrical and mechanical engineering. He was chief engineer of the Watt Akkumulator Werke in Berlin from 1896 to 1898, and in 1898 designed a front-wheel-drive battery-electric car for N. Israel in Berlin. He was co-founder and part owner of the Vulkan Automobil GmbH, and in 1902 filed patent applications for a front-wheel-drive vehicle with transverse engine installation between the front wheels, open drive shafts and double universal joints. His 1906 prototype featured an air-cooled 60° V-4 engine, leaving ample space for sharp steering angles. In 1907 he joined Siemens-Schuckertwerke GmbH in Berlin-Siemensstadt as chief engineer of electric traction vehicles and designed battery-electric delivery trucks, railway-station baggage carriers, municipal vehicles, and other special-purpose applications. Even after Siemens-Schuckertwerke GmbH had bought Protos Automobile GmbH of Berlin-Reinickendorf, Schwenke found no car at management board level for production of his front-wheel-drive gasoline engine vehicle. In 1909 he designed an all-wheel-drive gasoline-engine vehicle with a combination of shaft and chain drive that Siemens-Schuckert produced for the Dutch colonial office. About 1924 he came in contact with an inventor, Iwan Freiherr von Stietencron, who held patents on new layouts of front-wheel-drive power trains, gearshifts, and other chassis details. When Schwenke resigned from Siemens-Schuckert, he went to Freiherr von Stietencron's office at Welsede near Hamelin and designed a prototype three-passenger car powered by a 4/30 flat-twin installed with its crankshaft running transversely, front wheel drive and independent front suspension. But in the German business climate of 1928–29, they found no industrial enterprise willing to put it in production.

Scolari, Paolo (1939–)

Technical director of Fiat Auto S.p.A. from 1980 to 1991.

He was born in Turin in 1939, grew up in the Piedmont countryside and returned to Turin for his technical studies. He began his career as an engineer with Fiat Tractors in Modena in 1964 and later designed earth-moving vehicles for Fiat's public works equipment branch which became Fiat-allis after the merger with Allis-Chalmers. He was still working on farm tractors when he met Vittorio Ghidella, an ambitious production engineer with Fiat's RIV ball-bearing subsidiary. When Ghidella took over as managing director of Fiat Auto S.p.A. in 1979, he invited Scolari to join him in Turin.

Scolari was instructed to discard all existing plans for future products and start all over from scratch. "His" first car was the Fiat Uno, introduced in 1983, which topped Europe's bestseller list for years. By 1987 he had replaced the old model range of eight basically different car lines with a new model line-up of four platforms and four basic body shells. Not known as a creative engineer, his greatest value was having the ability to conduct a meeting with men of different minds in such a manner as to reach an agreement. He was too closely linked with Ghidella to be allowed to keep his title when Ghidella walked out, but too valuable to risk losing him to a rival enterprise. In October 1991, he was promoted to Fiat S.p.A. vice president for environmental and industrial policy.

Seck, Willy (1868–1955)

Chief engineer of Motorenfabrik Oberursel from 1890 to 1899, technical director of Fahrzeugfabrik Eisenach from 1901 to 1904, engineering consultant with SAF[1] in 1904–05, engineering director of the Berlin Motorwagenfabrik from 1906 to 1909.

He was born on May 27, 1868, in Frankfurt am Main and educated as a mechanical engineer. He designed the first Gnom flat-twin engine for Motorenfabrik Oberursel in 1890 and in 1896 designed a Gnom car with flat-twin engine and his own patented friction drive, which went into production a year later. It was renamed Scheibler when Fritz Scheibler took over the company in 1899 and produced with that nameplate until 1903. He went to Eisenach and designed a four-cylinder Wartburg racing car with a five-speed gearbox for the 1902 Paris–Vienna road race. He designed the Dixi S.12 for 1904, with an F-head 16/20 four-cylinder engine and tubular steel frame. He adopted pressed-steel-girder ladder-type frames for the T.17 of 1905 and left Fahrzeugfabrik Eisenach with a stack of complete car designs

including the T.7 single-cylinder model, the T.14 parallel-twin, and the 6.8-liter U.35 which was built up to 1914. He created the 600 cc Liliput which was produced by SAF at Gaggenau from 1904 to 1907, with his latest friction-drive and coil-spring front suspension.

He went to Berlin and joined Curt Bendix in 1906 as engineering director of the Berlin Motorwagenfabrik and designed the Oryx Type X with Karl Golliasch as his assistant. Type X was a 10-PS light car with a four-cylinder 1½-liter engine and two-speed pedal-shift planetary transmission. He resigned when the Berlin Motorwagenfabrik was taken over by Dürkopp-Werke AG of Bielefeld in 1909. His career in the auto industry was ended, but he lived on until 1955.

Sée, Marcel (1891–1977)

Technical director of Société Nouvelle de l'Automobile from 1922 to 1924; technical director of Société Anonyme Française d'Automobiles from 1924 to 1927; managing director of SA Française d'Automobiles from 1927 to 1940.

He was born in 1891 and graduated from the Ecole Polytechnique. He began his career in stock-broking and banking in Paris, switching to the industrial sector in 1922 by joining the company that made the Amilcar cyclecar in a small factory in rue du Chemin Vert near La Bastille in Paris. It was exceptional for cyclecar builders to produce their own engines, but Amilcar did. Sée left product engineering in the hands of Edmond Moyet but organized production along more efficient lines. In 1924 the company took over the Forges et Ateliers de la Fournaise at St. Denis, where Sée masterminded the methods and tooling in larger premises. The cyclecars grew into voiturettes, usually with sports car bodies with engines from 1004 to 1380 cc. A racing rivalry with Salmson was sustained for years, ending unresolved. The plant turned out 4800 cars in 1926, but was losing money. The founders and chief executives, Emile Akar and Joseph Lamy, sold their interests in the company, which was refinanced by Albert Neubauer, the biggest Peugeot dealer in Paris, and other investors.

Marcel Sée replaced Emile Akar as managing director. A series of 50 touring cars with six-cylinder engines had been built in 1926, and in 1929 the Amilcar CS8 was launched as a compact luxury car with a 2-liter straight-eight overhead-camshaft engine. The four-cylinder sports cars were still keeping the company afloat, but in 1931 Sée decided to discontinue the CS8 and to launch a 5 CV economy car, the M3 with a four-cylinder 621-cc engine. It came up against Rosengart, Mathis, La Licorne, all suffering from the fact that Peugeot and Renault were offering bigger cars at the same prices. Marcel Sée ordered the closing of the St. Denis plant and moved Amilcar into smaller facilities at Boulogne-sur Seine in 1934. Amilcar sales and distribution were handed over to a finance company, SOFIA[1] which had been set up by Henry M. Ainsworth, managing director of Hotchkiss, to revitalize that company's retail business. SOFIA even took an equity position in the SAF d'Automobiles. Following Salmson's example, Sée decided to move Amilcar into the elite of the compact touring car market and launched the Pégase in October 1934, with a four-cylinder overhead-valve 2-liter engine designed by Pierre Grillot. Production costs for this engine proved excessive, and it was replaced by a Delahaye engine. In 1937, however, Neubauer and the other main shareholders sold out to Automobiles Hotchkiss. Marcel Sée stayed on in his capacity until 1940, when he retired to Grenoble. He was a pioneer of the "maquis" underground movement under the code name "Sultan" but fled to England in 1942 to avoid capture. He was parachuted back, led some operations, and fled to Spain but was caught and imprisoned. Returning to Paris in 1944, he acted as a management consultant for several years in non-automotive enterprises.

He died on May 18, 1977, in Paris.

1. Société Financière pour l'Automobile, Boulogne-Billancourt.

Segre, Luigi (1919–1963)

Sales director of Carrozzeria Ghia from 1950 to 1953; majority owner and president of Carrozzeria Ghia from 1953 to 1963; co-founder of OSI[1] in 1960.

He was born on November 8, 1919, into a family with strong connections to Turin's financial world. He was given a general and commercial education, and at his own initiative, studied for a degree in mechanical engineering to satisfy his ambitions about a place in the auto industry. During World War II he led a clandestine existence, moving about in secrecy, while serving as a liaison officer between the Italian partisans and the U.S. Army. He quickly readjusted to civilian life in 1945, and joined his friend Giorgio Ambrosini, owner of SIATA[2] as sales manager for custom-built sports cars and a line of performance equipment. In 1949 he left Ambrosini to join Vittorino Viotti to run the business side for the coachbuilder. But Segre demanded a high salary, which he justified by boundless energy. Viotti did not want to do without him, but did not need him on a full-time basis. That led Viotti to approach Mario Felice Boano with a proposal for sharing the salary and services of Luigi Segre. Soon Segre was selling Ghia bodies all over Italy, and said "goodbye" to Viotti. He masterminded Ghia's merger with Monviso and cultivated its contacts with Chrysler, which led to a lot of design work and prototype construction by Ghia. Segre also put some of his friends on the Ghia payroll, which Boano could not tolerate. The conflict was settled by their lawyers and ended with Segre buying Boano's shares in Ghia, but losing his head of styling. Segre was no designer himself. He hired Giovanni Savonuzzi as technical director, and also employed other stylists (Michelotti, Frua, Sartorelli, Coggiola, Tjaarda). Realizing that Ghia's long-term survival depended on having manufacturing capacity to back up the design studio, he formed OSI in 1960 as a joint venture with Arrigo Olivetti. OSI became a big supplier of sheet-metal stampings and produced complete bodies for some Fiat and Ford platforms.

In the winter of 1963 he went to a clinic for a gallstone operation, but died on the operating table on February 28, 1963.

1. Officina Stampaggi Industriale.

2. Società Italiana Applicazione Trasformazione Automobilistiche.

Seiffert, Ulrich (1941–)

Director of research, Volkswagen AG, from 1979 to 1987; technical director of Volkswagen AG from 1988 to 1993.

He was born in 1941 at Waldenberg in Silesia and graduated in 1966 from the University of Brunswick where he had studied power-unit and vehicle engineering. He joined Volkswagen in 1966 as a safety test engineer and was named head of the VW safety test department in 1969. He served as coordinator of safety programs and handled the liaison between VW and Audi. He won a doctor's degree in mechanical engineering from the University of Berlin in 1975 for his thesis on automobile safety problems and was promoted to head of the vehicle safety department. He developed experimental safety cars such as the RSVW and ESVW, and was named director of research in 1979. He organized the research effort into three separate departments, and in 1981 began lecturing on vehicle safety, structural optimization, and occupant protection at the Braunschweig Technical University. In 1987–88 he was general manager of power units, and was named technical director in July 1988. He tackled coordination of components for Seat and Skoda with VW and Audi. The group was building more than 30 distinct models on 18 different platforms. He devised a platform-sharing plan with no more than four basic platforms for 40 to 50 models.

In March 1993, F. Piëch demoted him to VW board member in charge of the VW brand and pushed him out at the end of 1995.

Seiffert continued his lectures at the University, and in 1996 joined Witech Engineering GmbH.

von Selve, Walther (1876–1950)

Chairman of the management board of Nord-deutsche Automobilwerke GmbH from 1907 to 1918, chairman of the management board of Selve Automobilwerke GmbH from 1918 to 1932.

He was born on July 25, 1876, at Altena in Westfalia, the only son of Gustav Selve, founder in 1861 of a metalworking shop which grew into an industrial empire. He was privately educated, attending schools on all five continents, and worked as a trainee with FN.[1] He took a keen interest in automobiles and became a prominent rally driver.

In 1907 he founded a car company in Hamelin on the Weser and began production of the 8 hp Colibri designed by Adolf G. von Löwe. The 6/15 hp four-cylinder Sperber, designed by Hans Hartmann, was added in 1911. Upon the death of his father in 1909, he became managing director of the family holding, Selve AG of Altena. He initiated aircraft-engine production in the Basse & Selve plant at Hünengraben in 1909; they were fitted with aluminum pistons produced in the Basse & Selve die-casting foundry next door. The Hamelin plant also took up aircraft-engine repairwork in 1917. On August 3, 1918, he sold the Norddeutsche Automobilwerke GmbH to Basse & Selve and reorganized it as the Selve Automobilwerke GmbH, installing Ernst Lehmann as technical director. New Selve automobiles went into production in Hamelin in 1920 with four-cylinder engines from the Altena works. In addition to directing 12 factories with a combined work force of 5,000 men, von Selve remained an active rally driver at the age of 50.

The Altena and Hünengraben plants also supplied cylinder-block and head castings to Gebrüder Körting, and complete engines to Beckmann and Mannesmann-Mulag. But the automobile subsidiary ran up big losses in 1928–29. Car production was almost halted in 1929 and the capital stock written down (one share for 10). The Selve Automobilwerke GmbH was officially liquidated on June 25, 1932.He ran the Selve AG affairs until 1945 and died in 1950.

1. Fabrique Nationale d'Armes de Guerre, Herstal-lez-Liège.

Sénéchal, Robert (1892–1985)

Managing director of Sénéchal & Cie from 1920 to 1923.

He was born on May 5, 1892, at Rosières-en-Santerre in Picardy, the son of a wealthy cereal-farmer and merchant. He was educated at a religious boarding school in Paris and prepared to study engineering at the Ecole Centrale des Arts et Manufactures when he caught typhoid fever. He recovered in time for military service in World War I and joined a Dragoons Cavalry regiment but applied for transfer to the Air Force and held a military pilot's license.

In 1919 he went into the war-surplus liquidation business and became quite rich. In 1920 he took over Cordier & Lebeau, makers of the Eclair cyclecar at Courbevoie, and renamed it Sénéchal. In 1921 he transferred production to an idle plant at Gennevilliers which belonged to a subsidiary of Chénard & Walcker. He began racing both Sénéchal and Chénard-Walcker cars and in 1922 he entered 30 races, of which he won 21.

He sold his company in 1923 to Chénard & Walcker, who renamed it Société Industrielle et Commerciale, maintaining the Sénéchal make. It is estimated that 5000 Sénéchal cars were produced from 1921 to 1927. He became a Chénard-Walcker dealer, later adding the Bugatti and Delage franchises. Sénéchal became a member of the Delage racing team in 1926– 27. In 1939 he reported for duty in the Air Force but saw little action. He survived the war and was active in motorcycling and motor racing associations for the rest of his life.

Sensaud de Lavaud, Dimitri (b. 1884)

Director of Société d'Expansion Technique D. Sensaud de Lavaud from 1920 to 1940.

He was born in 1884 in Valladoid, Spain, to a French father and Russian mother. His parents moved to Brazil in 1900 and set up a factory to make ceramic tubes. Soon the materials, processes and equipment held no secrets from him, and he began inventing new methods for making tubes. He built his own experimental airplane in 1909 — and made test flights himself. He invented a clever machine to make seamless cast-iron tubes, and made a vast fortune from selling manufacturing rights to his patent all over the world. He went to the United States and set up a steel-tube manufacturing plant, but after a while he found the

business tedious. In 1920 he moved to Paris to amuse himself and re-invent the automobile. In 1926 he built some experimental cars with a CVT (continuously variable transmission) of his own invention, which changed torque multiplication ratios by means of eccentric arms and a swashplate. He also patented several types of independent suspension systems, some using coil springs and others having rubber spring elements, differential gear systems, including limited-slip versions, and other types of gearing. He invented a hydraulic torque converter and persuaded André Citroën to buy the manufacturing rights for it. André Citroën wanted it on the front-wheel-drive 1934 models, but Pierre Prévost, Citroën's director of testing, scornfully rejected it. Sensaud de Lavaud had paid little attention to power units until then, but now he began to study unconventional engine types. He drew up geometrical curves for a rotary-piston engine, and his work in this area resulted in the granting of a French patent in 1938.

Serpollet, Léon Emmanuel (1858–1907)

Technical director of Société des Automobiles Serpollet from 1900 to 1907.

He was born on October 4, 1858, at Culoz (Ain) as the son of a blacksmith, from whom he learned carpentry and forging. He became intrigued with steam power and experimented with steam engines as a teenager. He moved to Paris in 1881 and went to work in a shop in Montmartre as a pattern maker and metalworker. After a few years, he set up his own engineering shop and in 1887 converted a pedal tricycle to steam propulsion. Next, he designed a complete three-wheeled steam car with a single, small, fork-steered wheel in front. He ordered three sets of boilers, expanders, and ancillary equipment from the Chantiers de L'Horme et de La Buire in Lyon, and persuaded Armand Peugeot to build three cars for him in 1889. A year later, Serpollet found financial backing from Ernest Archdeacon, and contracted with Charles Jeantaud, carriage builder in Paris and pioneer of electric vehicles, to build another three steam-powered three-wheelers, also with the main engine components from Chantiers de L'Horme et de La Buire. In 1891 he began to design a four-wheeled car and devised a new steam engine to run on liquid fuel (kerosene) instead of coke. The first Serpollet four-wheelers were built in 1894 by Etablissements Decauville at Corbeil, old-established makers of narrow-gauge steam locomotives for the mining industry. In 1899 he met an American businessman, Frank L. Gardner, whose fortune stemmed from gold mining in Australia, and who was a financial backer of Walter Arnold's Anglo-French Motor Carriage Company, who held British manufacturing rights to Benz engines and automobiles. Gardner agreed to finance Serpollet in a new venture, operating in a factory in rue Stendhal, Paris, where Serpollet produced 200 steam cars of his latest design within a two-year period. He made a new flat-four expander, mounted underfloor in an inboard/rear location, with the boiler and condenser in the rear overhang. Gardner-Serpollet cars were made in three sizes, and in 1902 he built a streamliner which set a new speed record of 120 km/h during the Nice Speed Week. When Gardner withdrew his support in 1905, Serpollet obtained backing from Alexandre Darracq, who ordered car production phased out in order to build steam-powered buses. The inventor died in Paris on February 11, 1907.

Serre, Charles Edmond (1882–1959)

Technical director of Renault Frères from 1903 to 1908; technical director of Automobiles Renault from 1908 to 1922; technical director and administrator of Société Anonyme des Usines Renault from 1922 to 1944; technical director of the Régie Nationale des Usines Renault 1945–46.

He was born in Paris in 1882 and attended a local grade school. He studied engineering at Ecole Colbert but dropped out because he had to work and earn his keep. He found employment as a draftsman with Maison Durand, gear makers in rue Oberkampf in Paris. Louis Renault hired him on March 1, 1899, to run the drawing office. He designed the 1900

Type C and the 1901 Types D and E, all with front-mounted water-cooled single-cylinder engines (Aster or de Dion–Bouton), three-speed gearboxes, and shaft drive to the rear axle.

In 1902 he created six new cars from 8 to 24 CV, as Renault began making its own parallel-twin and four-cylinder engines designed by M. Viet who was a brother-in-law of Georges Bouton. For the four-cylinder models H and K, Serre put radiators on both sides. The first Renault with the radiator behind the engine was the 1904 Type OB. Serre expanded the engineering staff, hiring Georges de Ram in 1903 and engaging Alexander Rothmüller as chief engineer in 1904. There was a proliferation of new models, as they went through the alphabet from M to S in 1903 and reached X in 1905.

The first six-cylinder model appeared in 1908, the 54 CV 9½-liter Type AR, and at the same time Renault began series production of the 8 CV 1.2-liter vertical-twin Type AX which was to enjoy a long and profitable career. After World War I he created Type IG, a 2120-cc four-cylinder touring car which evolved into the KZ, and Type JP with a six-cylinder 9123-cc engine. From this time to 1939, all Renault passenger-car chassis had transverse leaf springs on the rear axle. The 1922 model lineup included 12 types. He had overdone the proliferation, and cut back to six types for 1923. Production of the 6 CV Type NN began in 1924, and the KZ, updated at intervals, was built up to 1933. Radiators were moved to the front, beginning with the 1928 Type RM, an eight-cylinder 7125-cc luxury car. In April 1928, Serre and Emile Tordet accompanied Louis Renault on a factory tour of U.S. industry.

Model names were adopted as the 1927 RY became Monasix. The 1931 Primaquatre was an offshoot of the KZ, as was the 1932 Vivaquatre. The 1932 lineup comprised three four-cylinder models, three sixes, and two straight-eights. Chassis design, steering and brakes evolved at a slower pace during the thirties, and Renault's engines were not trendsetters. From 1939 on, Serre was nominally in charge of the 4 CV project, but did not get deeply involved. In 1946 he was sidelined, serving until 1949 as head of Renault's farm-tractor drawing office. He lived in retirement until his death in 1959.

Seze, Guy De (b. ca. 1880)

Consulting engineer to Peugeot from 1934 to 1937; research engineer of Peugeot from 1937 to 1944; director of research for Peugeot from 1944 to 1950.

He was born about 1880 and graduated from the Ecole Centrale des Arts et Manufactures in 1913. He was on military duty during World War I, and afterwards began to study chassis and suspension engineering on his own. He wrote a thesis on directional stability and steering phenomena, which he submitted to Professor Beghin of the Ecole Centrale. Prof. Beghin encouraged him to continue his work, and he conducted a thorough analysis of the various types of independent suspension then on the market in France and Germany. He opened a consulting engineer's office in 1931, and from 1934 onwards worked almost exclusively for Peugeot. In 1937 he made an exhaustive study of front-wheel drive, with particular emphasis on the work of H.G. Röhr, and became Louis Dufresne's assistant. He argued in favor of maximum-rigidity frames and long-travel suspension, which led him to look deeper into body engineering and structural rigidity. He felt that frames no longer had a place in the modern car, and prevailed on Louis Dufresne and Lucien Godard to adopt unit-construction bodies. Guy de Seze made the fundamental calculations and did much of the preliminary design work on the body shell for the Peugeot 203, which went into production at Sochaux in the fall of 1948.

He retired in 1950.

Seznec, Hubert (1930–1988)

Chief engineer in charge of Renault's passenger cars from 1977 to 1987; director of Renault's research division in 1987–88.

He was born in 1930 in Paris, attended the Lycée Carnot and graduated from the Ecole Polytechnique in 1952. He worked briefly for Renault and then joined Citroën's engineering staff in 1953. One of his first tasks was to develop the semi-automatic transmission for the DS-19. In 1958 he was named assistant to the aging Georges Sallot so as to prepare for taking over the position as head of the drawing office. His career turned instead to an appointment as head of new concepts and development. Most of the projects he worked on were unrelated to any future Citroën production model. He left Citroën in 1967 to join Renault as a test and development engineer. He was part of a team that continually upgraded the R-16 and helped develop the R-5 and the R-30. He directed the engineering of the R-9 and R-11 economy cars, the R-25 prestige car, the Super-Cinq and the Espace chassis, the R-19 and the R-21. He started the Clio program when he was sidelined into the research division. He died on May 28, 1988.

Shorter, Leo Joseph (1885–1965)

Chief designer of Humber, Ltd., from 1920 to 1924; chief engineer of Calcott Brothers, Ltd., from 1924 to 1927; chief engineer of Singer Motors, Ltd., from 1937 to 1944; and its technical director from 1944 to 1956.

He began his career with Rudge-Whitworth and later worked with the British importers of Duryea cars. From 1907 to 1913 he was a draftsman on Humber car projects, and he spent the next five years as a draftsman on Sunbeam aircraft engines. After a brief period was an aircraft-engine designer with Arrol-Johnston, he returned to Humber in 1920 and created the 8/18, 11.4 hp, and 9/20 models with F-head four-cylinder engines. His designs for Calcott, the 10/15, 12/24 and 16/50, were entirely conventional. He opened a consulting-engineer's practice in Coventry in 1928. One of his clients was Singer, who asked him to update the Junior engine. He did, and other tasks followed, leading to an invitation to join Singer's engineering staff, which he accepted in 1932. He created the Bantam for 1936, the Super Nine for 1937, Super 10 and Popular 10 for 1938. He discontinued the 14/6, modernized the Sixteen, and designed a new Twelve for 1937.

After World War II, he directed the engineering of the SM 1500, introduced in 1948, which evolved into the Hunter in 1954. He resigned when Singer Motors, Ltd., was taken over by the Rootes Group in 1956.

He died on November 26, 1965.

Siddeley, John Davenport (Lord Kenilworth) (1866–1953)

Head of the Siddeley Autocar Company from 1902 to 1905; general manager of the Wolseley Tool & Motor Car Company, Ltd., from 1905 to 1909; managing director of the Deasy Motor Car Manufacturing Company, Ltd., from 1909 to 1912; managing director of the Siddeley-Deasy Motor Car Company, Ltd., from 1912 to 1919; managing director of the Armstrong-Whitworth Development Company, Ltd., from 1919 to 1927; managing director of Armstrong-Siddeley Motors, Ltd., from 1919 to 1936; chairman of the Armstrong-Siddeley Development Company, Ltd., from 1927 to 1936.

He was born on August 5, 1866, at Cheadle Hume, Manchester, as the son of a hosier, and had no more than a high-school education. He was a keen bicyclist, raced for the Humber team, and went to work for the Humber Cycle Company in 1892. Harvey Ducros hired him in 1893 to run the Dunlop tire factory in Belfast, and in 1894 he set up the Dunlop Cycle Company there. He returned to Coventry in 1895 as a Dunlop representative, founded the Clipper Tyre Company in 1899 and led it profitably until 1901, when he withdrew to take on the Peugeot franchise for Great Britain. The first Siddeley cars of 1902 were pure Peugeot copies, built in Coventry. He prepared the specifications for a car of his own conception, and with the backing of Lionel Rothschild, approached Vickers, Sons and Maxim about taking on its production, which began at a Vickers factory in Crayford, Kent, in 1903.

In 1905, Siddeley and Rothschild sold the Siddeley Autocar Company to Albert Vickers and his brother Douglas, who were majority owners of the Wolseley Tool & Motor Car Company, Ltd. They were dissatisfied with the cars Wolseley was building, and wanted to make new models based on the Siddeley. Siddeley made up the specifications, but the actual designs were made by A.A. Remington and A.J. Rowledge. The products were good, but the company lost money every year under Siddeley's management. His relationship with Albert Vickers turned sour, and in 1909 Siddeley left to join a much smaller company headed by Captain H.H.P. Deasy. The Deasy car soon became the Siddeley-Deasy. In 1912 he privately founded the Stoneleigh Motors, Ltd., to make light vans and cars and bought the Burlington Carriage Company, coachbuilders in London, which he moved to Coventry. During World War I, Stoneleigh built military trucks, and Siddeley-Deasy took up aircraft production, also preparing to manufacture aircraft engines. In 1919 he merged the company with the aircraft division of Armstrong-Whitworth, and their car divisions were combined to form Armstrong-Siddeley Motors, Ltd. In the following years, Stoneleigh and Burlington ware absorbed by Armstrong-Siddeley. In 1927 he won independence from Armstrong-Whitworth for the Development Company, while the former partner continued its decline and ended up in a takeover by Vickers.

He did not meddle in car design beyond laying down the fundamental policy which (after the Stoneleigh was discontinued in 1924) kept the make firmly anchored in the high-class market. In 1935 he sold his interests to T.O.M. Sopwith and retired from active participation in company affairs a year later. He was elevated to the peerage as Baron Kenilworth in 1937 and after the end of World War II he retired to Jersey, where he died on November 3, 1953.

Sidgreaves, Arthur F. (1882–1948)

Managing director of Rolls-Royce, Ltd., from 1929 to 1946.

He was born in 1882 and began his career in the motor trade at the age of 20 by associating himself with S.F. Edge. That led to a position in the sales department of D. Napier & Son, Ltd., where he was active from 1904 to 1914. He joined the Royal Naval Voluntary Reserve when World War I broke out, but a year later was drafted by the Royal Navy Air Service. In 1917 he was given an appointment in the Air Ministry as a member of the aircraft-engine procurement staff. This brought him in direct contact with some of the suppliers, including Rolls-Royce, Ltd., who produced 4000 Eagle V-12 engines, notably for powering flying boats and Vickers Vimy bomber planes. At the end of the war, he went back to Napier, but left in 1920 to join Rolls-Royce as export sales manager. He was sent on a tour of U.S. industry and participated in the organization of an American Rolls-Royce factory at Springfield, Massachusetts, in 1921. He became general sales manager in 1926 and managing director in 1929. He negotiated Rolls-Royce's purchase of the assets of Bentley Motors from the receivers for £125,000 on November 13, 1931, and moved Bentley car production to Derby.

He authorized production of the Phantom III and headed Rolls-Royce's effort in getting aircraft engine production started in two government-financed Shadow factories, one at Crewe, Cheshire, the other at Glasgow.

He led the company through World War II and resigned in 1946. He died in June 1948.

Sidney, Edgar Harold (b. 1897)

Director and works manager of AC Cars, Ltd., from 1935 to 1940; engineering director of AC Cars, Ltd., from 1945 to 1962.

He was born on August 7, 1897, in Coventry and educated at Bablake School, Coventry. He worked as a draftsman with Beardmore of Glasgow from 1919 to 1923, joined Riley in 1924 as a draftsman and was soon appointed a design engineer. Leaving Riley in 1928, he spent the next seven years as works manager of the fuel-injection department of Bryce of Hackbridge. In 1935 he took charge of product development and production at the small AC factory at Thames Ditton. AC built no more than 75 cars a year, yet produced their own engines (an overhead-camshaft 1991-cc six dating back to 1919). He softened the leaf springs a little, fitted smaller wheels, and adopted

the Moss four-speed gearbox for 1936. He supercharged the engine for the 16/90 roadster of 1939 and prepared a new model based on the Flying Standard six-cylinder chassis. Towards the end of the war he designed a proposed postwar car, with a 2½-liter short-stroke six and independent front suspension, but the Hurlock brothers nixed it because production costs would be too high. The actual postwar model had the old engine, rigid axles, and a modern-looking body. He was enthusiastic about the Bristol-powered Tojeiro roadster after test-driving it, and smoothed its way into production as the AC Ace in 1953. A coupé version, the Aceca, followed in 1955. The four-door sedan was phased out in 1958, and tests with Ford V-8 engines in the Ace began in 1961.

Sidney suffered a cerebral hemorrhage in 1962 and went into retirement before the AC Cobra became a reality.

Siebler, Oskar (b. 1895)

Chief engineer of Horch from 1933 to 1938; director of Auto Union's technical center from 1938 to 1945; head of product development for DKW from 1949 to 1956; deputy director of Auto Union GmbH from 1956 to 1964.

He was born on May 10, 1895, and educated as an engineer. He began his career as a draftsman with Daimler Motoren Gesellschaft in 1924, then about to merge with Benz & cie. He left Daimler-Benz in 1928, and a year later moved to Saxony as chassis engineer of Horch-Werke. His first contribution was to put Type 400 on a lower frame for 1930. He adopted independent front suspension for the 830 BL in 1935 along with rear swing-axles and then put a de Dion–type rear end on the 853 in 1938.

The technical departments of Horch, Audi, DKW and Wandered were combined into one central Auto Union engineering office under Siebler's direction in 1938. He interrupted Audi's use of front-wheel drive and made the 1939 Audi a miniature Horch, assembled in the Horch plant in Zwickau. He directed the design team for Type 40, a Horch military reconnaissance and command car with V-8 power and four-wheel drive, which was produced in a variety of versions up to 1943.

He escaped to the West in 1945 and in 1949 helped in the establishment of Auto Union GmbH at Ingolstadt. He modernized the pre-war DKW into an F 89 P with an all-steel body, and developed the F 91 Sonderklasse with its three-cylinder two-stroke engine for 1953.

He created the DKW F 11 Junior, with a 660-cc two-cylinder two-stroke engine, later enlarged to 750 cc, which went into production in 1959, and directed the engineering of the DKW F 102, the first DKW with unit body construction, introduced in 1963. He retired on January 1, 1964.

Silva, Walter Maria De' *see* De' Silva, Walter Maria

Simms, Frederick Richard (1863–1944)

Director of the Daimler Motoren Gesellschaft from 1890 to 1892; founder of the Daimler Motor Syndicate, Ltd., in 1893; member of the supervisory board of Daimler Motoren Gesellschaft from 1895 to 1902; technical consultant to the Daimler Motor Company, Ltd., from 1896 to 1901.

The roots of the Simms family lay in Warwickshire, but he was born in Hamburg on August 12, 1863, as his grandfather ran a fishing-fleet supply company in the main port on the river Elbe. He grew up bilingual, attending schools now in Britain, now in Germany. He studied electrical engineering and at the age of 27 invented an aerial cableway for crossing rivers and valleys. In 1888 he saw a Daimler engine and arranged to meet Gottlieb Daimler, with whom he formed a fast friendship. He secured exclusive rights to Daimler's patents for the United Kingdom and the British Empire (except Canada) and in 1893 he set up the Daimler Motor Syndicate, Ltd., in London to promote and exploit the patents as well as to import and market Daimler engines.

He sold the syndicate in 1895 for almost twice the original investment to H.J. Lawson,

and a year later was named technical consultant to Lawson's Daimler Motor Company, Ltd. Privately he operated as Simms & Co., Engineers, to exploit his own patents and finance his inventions, and later began trading as the Motor Carriage Supply Company, Ltd., which imported two Daimler trucks from Cannstatt in 1899.

Some of his inventions were military. In 1899 he exhibited a Daimler car with an armored steel shell-like turtle with an open top, and a motor quadricycle carrying a Maxim repeater gun. He made the connection between the Daimler Motor Co., Ltd., and the electric streetcar manufacturers G.F. Milnes of Wellington in Shropshire, resulting in the production by Milnes from 1902 to 1914 of Daimler motor trucks designed at the Berlin-Marienfelde plant.

He had invented a low-tension magneto ignition system whose production was handled by Robert Bosch in Stuttgart. He left Daimler because the Coventry works would never standardize the Simms-Bosch ignition. The Simms-Bosch collaboration came to an end in 1907, and he formed Simms Magnetos, Ltd., to continue production in London. He invented the Simms Motor Wheel, easily attachable to powerless vehicles, and a pneumatic bumper which he mounted on Simms-Welbeck cars in 1905. He invented a new type of vacuum cleaner and a power mower with gasoline engine that was put in production by Ransome, Simms & Jeffries. In 1902–03 he designed and built an aircraft engine for the Spencer airship. In 1913 Simms Magnetos, Ltd., grew into Simms Motor Units, Ltd., and began to produce fuel gauges, lighting sets, and horns. He led this company until his retirement in 1935.

He never stopped inventing things. During World War II he began studying metallurgy, seeking knowledge he felt he needed for some new idea. He died at his home in Stoke Park, Coventry, on April 22, 1944.

Simpson, George (1942–)

Managing director of the Rover Group from 1989 to 1991; chief executive of the Rover Group from 1991 to 1993.

He was born on July 2, 1942, at Dundee, Scotland, and trained to be an accountant. He began his career as an accountant with the Scottish Gas Board in 1965, and in 1969 joined British Leyland in the central finance office. He served as finance director, Leyland Vehicles, Heavy Duty, from 1978 to 1980 and managing director of Coventry Climax from 1980 to 1984. After two years as managing director of Freight-Rover, he became managing director of Leyland Trucks in 1986, and held the office of chief executive, Leyland–DAF from 1987 to 1989. The British government sold the Rover Group to British Aerospace in March 1988, and he was picked to run the car company in 1989. In 1992 he also became deputy chief executive of British Aerospace and was knighted as Lord Simpson of Dunkeld. He left Rover in April 1993 and was named chairman of Lucas Industries, Ltd., in May 1993. British Aerospace sold its construction business, got rid of its executive jets, cut aircraft production capacity and sold Rover to BMW on January 31, 1994. He left Lucas at mid-year 1996 to take the chairmanship of the General Electric Company, Ltd. He sold off a lot of assets, renamed the company Marconi, and turned it from a wide-ranging electrical conglomerate into a telecommunications supplier, just in time for the telecom market to collapse. The value of Marconi went from £8 billion to "junk" status and he was ousted in September 2001. In October 1999 he had also invested several hundred thousand pounds of his personal fortune in a planned rebirth of the Jensen car, with a purpose-built Merseyside side factory to build the Jensen SV8 with a retail price of £38,000. The scheme collapsed.

Singer, George (1847–1909)

Managing director of the Singer Cycle Company, Ltd., from 1896 to 1899; chairman of Singer & Company, Ltd., and its subsidiary, Singer Motor Company, Ltd., from 1903 to 1909.

He was born on January 26, 1847, at Stanford in Dorset as the son of a farmer. He had some technical schooling, went to London, and was hired as a mechanic by John Penn & Sons, marine engineers at Lewisham. In 1859, he joined Newton, Wilson & Co. of Holborn, London, as a sewing-machine mechanic, and in 1861 moved to Coventry to go to work for James Starley, who has been a maintenance engineer with John Penn & Sons. Starley was just setting up the European Sewing Machine Company (Coventry Sewing Machine Company since 1857) at Cheylesmore. Singer became foreman in 1868, but left in 1870 to become manager of the Paragon bicycle department of Skidmore's Art Manufacturers and Constructive Iron Company in Alma Street, Coventry. He made his own prototype bicycle in 1875 and resigned from Skidmore's. In 1876 he set up his own agency, selling National, Xtra Ordinary, and Challenge bicycles. With the financial backing of his brother-in-law, he took over the Alma Street works and began production of a "safety" bicycle (equal-size wheels, chain drive to rear wheel) based on H.J. Lawson's Crocodile. Singer bicycles became popular and in 1891 he put up a shop on Canterbury Street, Coventry. He sold his company to Ernest T. Hooley in June 1896, staying on as managing director.

He began to take an interest in motor vehicles, and wanted to be free of day-to-day responsibilities so he could find time to make new plans. He kept a seat on the Cycle Company's board of directors, and in 1900 secured a license for the Perks and Birch motor-wheel patent. The motor wheel was designed to fit in the front fork of bicycles and tricycles. Also in 1900 he organized the production of the Compact motorcycle with its engine carried on the rear wheel, and within three years, motorcycles had replaced bicycles as the company's principal product. He also began production of a motorized three-wheeler, based on the Singer Tandem Tricycle. The cycle company was wound up after a disastrous fire in the Canterbury Street works, leading to the establishment of Singer & Company, Ltd. He introduced a new three-wheeled car, with a single driving wheel at the rear. He made arrangements with Richard Lea for manufacturing rights to the Lea-Francis automobile, which Singer started producing in 1905, in

8-hp and 12-hp models, with under floor two- and three-cylinder engines. New models with vertical front-mounted four-cylinder engines appeared in 1907; and the 1909 range included four-cylinder cars of 12/14, 16 and 20/25 hp rating.

He died on January 4, 1909.

Slaby, Rudolf (1880–1963)

Director of the S.B. Automobil-GmhH from 1920 to 1924; technical director of the motor car department of Zschopauer Motorenwerke J.S. Rasmussen AG from 1925 to 1928; technical director of the Deutsche Automobil- and Getriebe-Fabrik from 1928 to 1933.

He was born in 1880 in Berlin and educated as an engineer. He made postgraduate studies at the University of Berlin and lectured with the title of professor. During World War I he held a position on the engineering staff of Daimler Motoren Gesellschaft in Berlin-Marienfelde. In 1919 he formed a partnership with Dr. Hermann Beringer, a colleague at the University of Berlin, to explore new construction materials and new concepts in the making of car bodies. He invented a way to make a plywood structure strong enough to obviate the need for a metal frame and built a prototype with battery-electric drive in 1920. The S.B. minicar was put into production in a small factory at Berlin-Markgrafendamm in 1922. Two years later their company was bought by J.S. Rasmussen, who was eager to replace its battery-electric power train with his two stroke DKW engine. A small number of DKW-powered S.B. minicars were built in Berlin, while Slaby undertook the design of a practical-size small car with frameless construction, plywood body structure, and patent-leather "skin." He was able to find production facilities in the idled plant of Deutsche Industrie-Werke at Berlin-Spandau, and directed the production of the first DKW car, introduced in May 1928. The two-stroke parallel-twin engine and three-speed gearbox, designed by Hermann Weber, were manufactured at Zschopau. In 1929–30, the Spandau plant also built a DKW coach with a two-stroke V-4 engine. He remained a consultant to J.S. Rasmussen and designed the 1931 Audi P-Type, powered by the four-stroke 1122-cc engine from the Peugeot 201, assembled at Zwickau, Rasmussen reverted to steel frames for the front-wheel-drive DKW, and Slaby stayed at Spandau, adding an adjacent plant in 1930. In 1933 he returned to the University and continued to lecture on structural engineering and pursuing his research effort into construction principles and materials.

Slevogt, Carl (1876–1951)

Design engineer with Cudell Motor Company from 1903 to 1905; chief engineer of Laurin & Klement AG from 1905 to 1907; technical director of Johann Puch AG from 1907 to 1910; chief engineer of Apollo-Werke AG from 1910 to 1914 and 1919 to 1924; chief designer of Selve Automobilwerk from 1924 to 1927.

He was born in 1876 in Bohemia and educated as an engineer. Max Cudell of Aachen engaged him in 1903 to design new two- and four-cylinder engines. He also designed Cudell's Phönix model with a 45-hp 6.1-liter four-cylinder engine. Early in 1905 he joined Laurin & Klement of Jungbunzlau in Bohemia and created their Voiturette A with its 1005-cc V-twin engine, separately mounted three-speed gearbox and shaft drive. He also designed Type B with a 1349-cc V-twin and Type C with a 2042-cc V-twin, plus the 1906 Type E with a 4½-liter four-cylinder engine. His 1909 Puch "Prince Henry" had an overhead-camshaft four-cylinder 70-hp 4-liter engine, and his Puch 14/40 Alpine with its 3560-cc side-valve four-cylinder engine became a by-word for robustness. He designed the 1911 Apollo Type B with a 960-cc overhead-valve four-cylinder engine, a car that he drove himself in many rallies. He also designed the 2-liter Apollo Type R, 2.6-liter Apollo Type K and 3½-liter Apollo Type L, built from 1912 to 1914. He was engaged on military projects in World War I and returned to Apolda in 1919. He had a prototype 12/50 car with an overhead-valve V-8 ready in 1921, but the managing director, Artur Ruppe, would not agree to letting it go into production. Ruppe wanted low- and medium-priced cars, so

Slevogt designed a hemi-head 960-cc four-cylinder 4/20 and a side-valve 2.6-liter 10/40. He left Apolda in 1924 for Hamelin to design new models for Walther von Selve and hurriedly made up a new chassis for the Basse & Selve 2090-cc F-head 8/40 engine, already in production at Altena. In 1925 he drew up a new chassis for the six-cylinder 2850-cc Basse & Selve 11/45 engine and completed the designs for the Selve Selecta luxury car, powered by a 3.1-liter Basse & Selve six-cylinder engine.

In 1931 he created an experimental three-wheel economy car with a 200-cc motorcycle engine which he drove for eight laps of the AVUS[1] speed track between Potsdam and Berlin at an average speed of nearly 60 km/h with a fuel consumption of 4.25 liters per 100 km (55.3 miles per gallon).

1. Automobil Versuchs- und Uebungs-Strasse = automobile test and practice road.

Smith, A. Rowland (1888–1988)

General manager of the Ford Motor Company, Ltd., from 1929 to 1941; managing director from 1941 to 1950; chairman from 1950 to 1956.

He was born in 1888 in Gillingham, Kent, and educated at the Mathematical School, Rochester. He was apprenticed to Humber, Ltd., in 1904, and became a jig-and-tool designer. He began his career with a trading company that sent him to India in 1912, assigned to a local subsidiary, Russa Engineering, which handled Ford business in large parts of Asia from its Calcutta headquarters. He became manager of Russa Engineering during World War I, took charge of Ford distribution in Bengal, and when returning to Britain in 1924, joined Ford at Trafford Park, Manchester. He felt he had been passed over for promotion to assistant general manager in 1927 and promptly left to join the Standard Motor Company in Coventry as works manager. P. L. D. Perry brought him back into the Ford fold in 1929, just before construction work began at Dagenham. Smith supervised the building and equipment of the Dagenham plant, which came on stream in 1931. For ten

years, he was Lord Perry's closest assistant, and in World War II held chief-executive responsibilities for all Ford activities in Britain. He managed the return to civilian production and was named chairman in 1950.

He retired in 1956 and lived on until 1988.

Smith, F. Llewellyn (1909–)

Chairman of the Rolls-Royce Motor Car Division from 1968 to 1972.

He was born in 1909 at Wood Green, London, as the son of an insurance engineer. He was still a boy when the family moved to Rochdale in Lancashire, where his father's duties vis-à-vis the textile industry required his constant presence. He attended Rochdale High School and enrolled at Manchester University, obtaining a master's degree in engineering. He went next to Balliol College, Oxford, where he won praise for postgraduate studies on motor fuels and combustion, discussions of the effects of lead and incomplete combustion, and devising methods to make gas analyses at various stages of the combustion process.

He began his career as an aircraft-engine test engineer with Rolls-Royce at Derby in 1933, but was transferred to the motor car department a year later. He became a member of the team in charge of developing the 3½-Litre Bentley. They tried many approaches to improving its performance, including the installation of a centrifugal supercharger (which was discarded due to poor low-end torque). Instead, they enlarged the Bentley engine to 4¼ litres. He conducted an analytical study of a proposed straight-eight engine for Bentley and examined the possibilities of modernizing the sleeve-valve engine (recently abandoned by Daimler).

In 1936 he became an assistant to E.W. Hives, made an analysis of the Ford V-8 engine, and a wrote a study of possible layouts for a 24-cylinder aircraft engine. He also led an investigation into the service problems of the V-12 engine in the Phantom III, and clutch problems encountered on both Rolls-Royce and Bentley cars. In 1939, he was dispatched to the Shadow Factory at Crewe to find ways of increasing the production rate of the Merlin V-12 aircraft engine, before being named deputy works manager of a Rolls-Royce plant at Hillingdon near Glasgow. After the war, he returned to the motor car engineering staff in an administrative capacity. In 1962 he became chairman of Rolls-Royce's diesel-engine division, which he led until 1968. The Crewe plant was then building the Silver Shadow and Bentley T-Type in annual numbers exceeding 3000 units, and throughout his tenure, the car division was profitable. When Rolls-Royce was declared bankrupt in 1971, it was blamed on the aircraft-engine division, which had lost £65 million on failed tests of the RB 211 turbojet engine.

He resigned in 1972.

Smith, Frank (ca. 1880–ca. 1968)

Managing director of Clyno Engineering Company Ltd. from 1922 to 1929, works manager of Star Engineering Company Ltd. from 1929 to 1932.

He was born about 1880 at Thrapston in Northamptonshire, where he grew up and opened a mechanical workshop in partnership with his cousin Alwyn Smith. In 1896 they invented a variable-ratio pulley drive for motorcycles and began production. They assembled their first complete motorcycle in 1900, with a single-cylinder Stevens (later AJS) engine. In 1901 they moved into a factory on Pelham Street in Wolverhampton and built motorcycles with a choice of two-stroke 2.75-hp and four-stroke 5-hp Stevens engines.

In 1909 he adopted the Clyno name, founded the Clyno Engineering Company Ltd., and began producing his own V-twin engines. In 1913 he added a second plant in Brick-kiln Street, Wolverhampton, won financial backing from Thomas de la Rue & Co. for a car project, and hired S.C. Poole to design a light car with a four-cylinder engine. A prototype was running in 1914, but all plans were shelved when Great Britain entered the war. The factory made motorcycle and sidecar combinations equipped with a machine gun, produced Vickers machine guns at a rate of 20 to 25 a week, and won a contract for the manufacture of the ABC Dragonfly nine-cylinder air-cooled radial aircraft engine. Thomas de la Rue & Co. withdrew their support for the car project in 1919 and it took over a year to organize new sources of capital.

He had a new car designed by A.G. Booth, hired Henry Meadows as works manager, and began Clyno car production late in 1922 using Coventry Climax four-cylinder engines. When Henry Meadows left to start his own company, Leslie Munn became works manager. Assembly operations were quite efficient and nearly 5,000 Clyno cars were built in 1925. The following year over 13,000 were turned out and construction of a more modern factory began at Bushbury, coming on stream in 1928. But the market for Clyno cars had been eroding since 1927 and the company was put in receivership early in 1929. Smith appealed to Rootes Ltd., sole exporters of Clyno, to take over, but they refused. AJS took over the production of the Clyno 9 hp car, and the plant was sold to Alfred Herbert Ltd., machine-tool manufacturers and importers. Smith joined Star Engineering Company Ltd., which declared bankruptcy in 1932. Smith moved to London in 1933 and opened the Dome Garage on the Great West Road, Ealing, which he ran for years. During World War II he retired to Devon, where he died about 1968.

Smith, Frederick Robertson (1882–1930)

Chief engineer of Siddeley-Deasy from 1912 to 1918; chief engineer of Armstrong-Siddeley from 1919 to 1929.

He was born in June 1882 at King's Lynn, Norfolk, attended the local technical school and later studied engineering at the University School in London. After spending five years and six months as a trainee with Savage Brothers, builders of steam-powered trucks at King's Lynn, he joined Thornycroft as a draftsman in the Chiswick works, London, where the marine and motor-vehicle engines were produced.

In 1909 Montague Stanley Napier offered him an engineering position and he worked on motor car and aircraft engines at the Acton plant for two years. He became chief designer and general manager of Aberdonia Cars, Ltd., in 1912 but resigned to join J.D. Siddeley at Siddeley-Deasy in Coventry. He redesigned the six-cylinder BHP[1] aircraft engine, which was then renamed Puma. He also designed the Tiger aircraft engine. In 1918 J.D. Siddeley had bought a Marmon which he offered to Smith as a "guide" for the type of postwar car he wanted. What became the Armstrong-Siddeley 30 was not free of Marmon influence, yet was quite different. In 1922, he designed an 18-hp six-cylinder companion model, adding a 14-hp four-cylinder car in 1924. Towards the end of 1929, J.D. Siddeley wanted his son Ernest to take over design responsibility for the cars, and transferred Smith to another office with the new title of technical manager for aircraft engines and special projects. He died on June 15, 1930.

1. Bentley-Halford-Pullinger.

Smith-Clarke, George Thomas (1884–1960)

Chief engineer of Alvis from 1922 to 1950.

He was born on December 23, 1884, and educated at Bewdley National and secondary schools. He studied engineering at the Regent Street Polytechnic and joined the Great Western Railway as a trainee, filling a succession of assignments before ending up as chief draftsman of the GWR's motor car department. In 1914 he joined the Royal Flying Corps as a captain in charge of engine procurement in the Coventry district. After the war he joined the Daimler Company as assistant works manager, but left in 1922 to accept a position as chief engineer and works manager of Alvis, who had begun car production in 1920. He developed the existing 12/40 into a 12/50 and by modernizing it every year, ensured its production up to the end of 1932. He directed the design of the front-wheel-drive four-cylinder sports models of 1928–29 and the fabulous front-wheel-drive Grand Prix car with its dual-overhead-camshaft supercharged 1500-cc straight-eight engine. He switched the production models to six-cylinder engines (with rear-axle drive), beginning with the 14/75 of 1927 and continuing with the Silver Eagle, Crested Eagle, Silver Crest, Speed Twenty, Speed Twenty-Five, and the 4.3-Litre. During World War II he supervised aircraft engine production in the Alvis factories. He prepared the postwar 14-hp range and the six-cylinder 3-Litre before retiring in 1950. He died on February 28, 1960.

Snowdon, John Mark (1944–)

Product planning director of Austin-Morris from 1977 to 1979; director of product development for Austin-Morris from 1979 to 1981; managing director of Austin-Rover new product development from 1985 to 1987.

He was born on February 5, 1944, and educated at Consett Grammar School and University College, London. He joined Ford at Dagenham in 1966 to work in product planning and marketing, but left in 1971 to take classes in economics at London Business School. He joined Leyland Cars, Ltd., in 1973 and was instrumental in pushing the Austin Metro into production. When it became clear that Austin-Morris needed an outside partner, Snowdon led the study of competing makes that would determine Leyland's choice. Snowdon recommended Chrysler–UK as a partner, but was surprised when Peugeot took it over in 1979. His second choice was Honda, which had fine products and was looking for a production base in Great Britain. It led to a new generation of Morris, Triumph and Rover cars having direct counterparts in the Honda range, and an interesting setup of joint manufacturing. He served as commercial director of Leyland Cars, Ltd., from 1981 to 1983, and managing director of the Austin-Rover sales department from 1983 to 1985. He resigned from Austin-Rover at mid-year 1987 to become a vice president of Booz Allen & Hamilton, a top firm of "head-hunters."

Sporkhorst, Johann Friedrich August (1870–1940)

Technical director of Hansa Automobil-GmbH from 1905 to 1914; director of Hansa-Lloyd from 1914 to 1921; managing director of

Hansa Auto and Vehicle Works[1] from 1921 to 1929.

He was born on June 17, 1870, in Dortmund, studied engineering in Darmstadt and Berlin, and graduated with a degree from the Brunswick[2] Technical University in 1896. He began his career as technical director of Taemeling & Stoeve (weaveries) at Varel in Oldenburg. His private residence was practically next door to that of Dr. Robert Allmers, who had dreams of building cars, and was looking for an engineer. They founded Hansa Automobil-GmbH in 1905. His first Hansa car was a close copy of the Alcyon, powered by a 720-cc single-cylinder de Dion–Bouton engine. He laid out and equipped an engine plant at Varel, and hired Otto Garbe away from the Société d'Electricité at Charleroi in Belgium to design a four-cylinder Hansa power unit. The 12 hp Hansa went into production as a 1907 model. It was followed by a 2.7-liter touring car, Type A 20. Sporkhorst drove one himself in the 1908 Prince Henry Trials and finished without loss of points. He introduced the 6/16 in 1909, 7/20 in 1910, and 8/20 in 1911, committed to medium-size, medium-priced touring cars.

After the merger of Hansa with the makers of the Lloyd[3] in 1914, he became a member of the Hansa-Lloyd management board, and when they split in 1921, he took over as managing director of the Varel factory. Beginning in 1921 he also sat on the supervisory board of the newly established Rembrandt Karosseriewerke AG in Bremen.

In 1922 Hansa introduced the 8/36 car, adopting a one-model policy which lasted until 1927. In 1924, when Hansa-Lloyd came under new ownership, he was invited to join the management board of the Bremen-based enterprise. He accepted, and held office in both the Hansa and Hansa-Lloyd companies until they were taken over by Borgward in 1929.

1. Hansa Automobil- & Fahrzeugfabrik AG.
2. Braunschweig.
3. Norddeutsche Automobil- und Motorenfabrik AG of Bremen (NAMAG).

Spurrier, Henry (1898–1965)

General manager of Leyland Motors, Ltd., from 1942 to 1949; managing director of Leyland Motors, Ltd., from 1949 to 1957; chairman of Leyland Motors, Ltd., from 1957 to 1963.

He was born on June 16, 1898, as the son of Henry Spurrier, founder of the Lancashire Steam Motor Company, Ltd., and designer of the 1896 Leyland truck. He attended Repton School and received his technical training in the Leyland machine shops and foundries. In 1916 he volunteered for service in the Royal Flying Corps and became a pilot, serving in India and Mesopotamia. After World War I he returned to Leyland as an engineer, holding a succession of appointments on the production side, and reaching executive level in 1936. When he became chairman, the company was strictly a maker of heavy transport vehicles with no ties to the passenger-car business. He changed that by taking control of Standard-Triumph International in 1961. He provided fresh funds for new-model development and new plant construction, beginning with a stamping and body plant on Speke Road outside Liverpool. He retired in December 1963 due to ill health and died from a brain tumor on July 17, 1965.

Squire, Adrian Morgan (1910–1940)

Managing director of Squire Motors, Ltd., from 1931 to 1937; managing director of the Square Car Manufacturing Company, Ltd., from 1934 to 1936.

He was born in 1910 and educated at a Roman Catholic school at Downside. He was only 16 years old when he designed his first complete car. It was never built, however. He studied electrical engineering at Faraday House in London, but left before graduating in order to start an apprenticeship with Bentley Motors, Ltd. In 1929 he joined the MG Car Company, Ltd., as a draftsman, and proved to be a clean, precise, and very thorough designer. He left MG in 1931 to start his own business, selling, tuning and repairing sports cars in a garage at Remenham Hill, Henleyon-Thames. He designed the Squire roadster, an elegant, low-slung sports car with obvious racing potential, powered by a supercharged 1½-liter British Anzani. It was a top-quality machine with a list price of £995 in 1935, compared with £340 for an SS I and £275 for a Triumph Southern Cross. The Squire was also offered as a long-wheelbase four-passenger tourer at £1,195. The company built five roadsters and two tourers before being liquidated in July 1936. A year later he sold his interest in the garage business and joined W.O. Bentley at Lagonda Motors, Ltd. In 1939 he joined Roy Fedden's engineering staff at Bristol Aeroplane Company, Ltd., at Filton, Bristol. He was killed in a Luftwaffe raid on the Bristol area in September 1940.

Stanbury, John Vivian (1914–)

Chief designer, Rolls-Royce Motor Car Division, from 1957 to 1965.

He was born in 1914, held an engineering degree, and became a project engineer with the Hawker Engineering Company in 1931. He worked in the technical office at Kingston-on-Thames when the Hawker Fury and Hart aircraft were developed, and in later years pushed the Hurricane, Typhon and Tempest through to the production stage. He was named chief project engineer of the Hawker Aircraft Company in 1944 and led the studies for the Sea Hawk and Hunter airplanes. He joined Rolls-Royce in 1946 as a body engineer and structural analyst. He directed the design of the Silver Cloud body, which was mounted on a rugged frame. For the Silver Shadow, the company wanted a unit-construction body, for which he had to draw deeply on his aircraft fuselage experience. He left the company when the car went into production in 1965.

Stanzani, Paolo (1936–)

Technical director of Automobili Lamborghini from 1968 to 1971; managing director of Automobili Lamborghini from 1971 to 1976; executive engineer of Bugatti Automobili S.p.A. from 1987 to 1990.

He was born on July 20, 1936, in Bologna and graduated with a degree in mechanical engineering from the University of Bologna in 1961. He presented a postgraduate thesis on hydraulic servo mechanisms that was rated at 92 points out of a possible 100. He worked as an assistant professor at the University of Bologna until he was engaged by Ferruccio Lamborghini in 1963. He worked as Dallara's assistant on the chassis for the 350 GTV and the Miura, led the design team for the Islero, Espada, and the 500 LP Countach, and laid out the 250 LP Urraco. He resigned in 1976 to set up his own consulting business in Modena, which he named Pro. He built up Pro into an important development firm for the earth-moving vehicle industry. He was the originator of the Bugatti 110 project in 1987, bankrolled by Romano Artioli and scheduled for production in a brand new factory at Campogalliano near Modena. He was ousted during a shareholders fight in July 1990.

Starkloph, Franz (1863–1926)

Chief designer of Adler cars from 1898 to 1903; head of the automobile department of Polyphon Musikwerke from 1904 to 1908; director of the Siemens-Schuckert automobile department from 1908 to 1915.

He was born on May 13, 1863, at Gotha and educated as an engineer. He began his career as a tool designer in the Karlsruhe Maschinenfabrik, but in 1898 he met Heinrich Kleyer and was signed as works manager and chief designer of Adler Fahrradwerke in Frankfurt am Main. From the very beginning, he put the engine in the front of the frame, with shaft drive to the rear axle. The first Adler cars were powered by single-cylinder de Dion–Bouton engines, but Starkloph designed a 12-hp vertical-twin in 1902. Leaving Adler in 1903, he joined Polyphon MusikWerke in Leipzig-Wahren, makers of musical instruments who had acquired a license to produce the "curved-dash" Oldsmobile. It was sold as the Polymobil, but in reaction to dwindling demand, the management decided in 1907 to replace it with more modern designs. In 1908 Starkloph joined Siemens-Schuckert at Berlin-Nonnendamm to design battery-electric vehicles. At the same time, Siemens took over the Protos Motorenbau GmbH in Berlin-Reinickendorf and transferred Protos car production to Nonnendamm, under Starkloph's authority. He designed some four-cylinder models himself, and engaged Ernst Valentin to run the design office. The Nonnendamm works switched to war materiel production in 1915, and when Protos car production was revived in 1920, Starkloph had left the company.

Stieler von Heydekampf, Gerd (1905–)

Member of the NSU-Werke AG management board from 1953 to 1959; chairman of the management board of NSU Motorenwerke AG from 1959 to 1971.

He was born on January 5, 1905, in Berlin, and studied mechanical engineering at the Technical Institute of Brunswick, graduating in 1929. He worked for a year as assistant to the professor in charge of material strength and vibration studies at the Institute, and then went to America where he held engineering positions with Babcock & Wilcox (steam turbines) and the Baldwin Locomotive Works. On his return to Germany in 1933, he was engaged by Adam Opel AG as an assistant purchasing agent. Within two years he was Opel's director of purchasing, and became a deputy board member in 1936. He became a full member of the management board of Adam Opel AG in November 1939. He left Opel in 1942 to take over as managing director of Henschel & Sohn, truck manufacturers in Kassel, but in 1943 Hitler called on him to replace Dr. Ferdinand Porsche as head of the Panzer Commission, a military engineering office with responsibility for combat vehicle design. He held that position until the collapse of the Third Reich. He was interned, first by the British, then by the American occupation forces, and underwent a "denazification" process which lasted until 1948. Walter E. Niegtsch, chairman of NSU-Werke AG in Neckarsulm, engaged him in an executive capacity. He built NSU up to one of the world's leading motorcycle manufacturers, with a successful racing team which went on to win the 125-cc and 250-cc world championship in 1953 and 1954.

He anticipated the end of the motorcycle boom, sold the NSU motorcycle branch to Marshal Tito's government in Yugoslavia, and approved plans for going into the automobile business. The company had adequate resources to finance the development of the Prinz car and the building and equipment of a car-assembly hall, in parallel with the ongoing Wankel-engine program. NSU made money on the cars as long as they all belonged in the same basic category: rear-mounted transverse air-cooled engines across the board. With the addition of the front-wheel-drive Ro-80 in 1967 and K-70 in 1969, one with a Wankel engine and the other with a water-cooled front-mounted four-cylinder engine, the disparity caused problems. Each model demanded a number of specific parts, and different assembly methods and sequences. Rational production became impossible. NSU Motorenwerke AG was heading for ruin when Kurt Lotz of Volkswagenwerk AG stepped in and bought a 60 percent stake in NSU early in 1969. Stieler stepped down in 1971 but was given a seat on the Audi-NSU Auto Union GmbH supervisory board.

Stockmar, Jurgen (1941–)

Engineering director of Steyr-Daimler-Puch AG from 1981 to 1988; director of Audi car development from 1988 to 1990; technical director of Adam Opel AG from 1994 to 1997; director of worldwide research and development, Magna International, Inc., beginning in 1998.

He was born on December 29, 1941, at Wittichenau and graduated from Hanover Technical University in 1968 with a master's degree in mechanical engineering. He began his career with Deutsche Vergaser Gesellschaft (later Pierburg), where he spent four years on fuel-mixture systems. From 1972 to 1976 he was editor of a monthly car magazine, *Auto Zeitung*, and then joined Audi as head of suspension systems engineering. He led the chassis design for the 1977 Audi 100 and the 1978 Audi 80, but left Ingolstadt in 1981. He spent the next four years designing four-wheel-drive off-road vehicles for Steyr, and then led all of Steyr's research and development work for three years. He returned to Audi in 1988 as management board member for technical affairs, but resigned in September 1990 after an argument with Ferdinand Piëch over the replacement model for the Audi 100. He joined Opel at Rüsselsheim, where he succeeded Peter Hanenberger as technical director in 1994. In 1998 he joined Magna International, which was to take over Steyr Automobiltechnik, produce all-wheel drive systems, and assemble some models for Daimler Chrysler AG.

Stoewer, Bernhard Alfred Rudolf Karl (1875–1937)

Technical director of Stoewer Brothers[1] from 1898 to 1914; technical director of Stoewer Werke AG from 1926 to 1934.

He was born on October 18, 1875, in Stettin, as the second son of Bernhard Stoewer (1834–1908), a local industrialist. He attended the Marienstift College prior to serving an apprenticeship in the shops of the family enterprise, and later studied mechanical engineering at the Hanover Technical University and the Berlin Technical University at Charlottenburg. He learned business administration at the Berlin Commercial College and then made a journeyman's tour, holding temporary jobs in factories in several European countries. He designed and built a motor tricycle in 1897 and road tested his first four-wheeled car in 1899, a two-passenger voiturette with a rear-mounted single-cylinder engine. The first Stoewer production model was a high-chassis four-passenger vehicle with a horizontal two-cylinder engine mounted in the rear. He moved the engine to the front in 1901. He designed the 1902 20/40 PS model with a 7.4-liter four-cylinder engine and added an 8-liter 30/45 touring car in 1904. The last of the two-cylinder models, a 9/18 2.3-liter, was produced through 1907 and the first six came on the market in 1906, a luxury model with a 35/60 hp 8.8-liter engine. Stoewer's best-seller was Type G4, a 12-liter 6/12, with sales of 1270 cars in 1908–09.

In August 1914, he volunteered for service with the 13th Hussar Regiment, which was sent to the Western Front. He was badly wounded in action, and taken to a hospital in Germany. He returned to the factory in 1918 but due to health problems stemming from his war wounds, he retired to his country estate in Pomerania to breed horses. He returned to the factory in 1926, determined to create a new range of high-grade automobiles. With Fritz Fiedler as his assistant, he designed the Superior, Gigant, Marschall and Repräsentant, with straight-eight engines from 2.2 to 4.9 liters displacement. They hit the market just as the demand for luxury cars dried up, and caused heavy losses. He resigned in 1934, and a year later signed up with General Motors as a consulting engineer, based at the Opel headquarters in Rüsselsheim am Main. He went to Berlin for the 1937 Auto Show, but died there suddenly on February 27, 1937.

1. Gebrüder Stoewer, Fabrik für Motorfahrzeuge and Fahrradbestandteile, Stettin.

Stoewer, Emil Ludwig Karl (1873–1942)

Managing director of Stoewer Brothers[1] from 1898 to 1916; managing director of Stoewer-Werke AG from 1916 to 1932.

He was born on January 18, 1873, as the elder son of a local industrialist, Bernhard Stoewer of Stettin. He studied law and business administration, and served as sales manager as well as managing director when Stoewer cars were first put on the market. The car factory in Falkenwalder Allée, Stettin-Neutorney, was formerly known as the Stettiner Iron Works, established by his father in 1896, and presented as a gift to his sons. Emil was a founder-member of the Pomeranian Athletic Club and a speed-loving motorist who drove Stoewer cars in the Herkomer trials, the Prince Henry trials, and minor events. He was completely unprepared for the changes occurring in the car market in the wake of the Wall Street crash in 1929, and Stoewer-Werke AG was declared bankrupt in 1931. He secured refinancing of the company from the municipality of Stettin and local banks in 1932, but the Stoewer family had lost control, and he resigned. He lived in retirement in Pomerania until his death in 1942.

1. Gebrüder Stoewer, Fabrik für Motorfahrzeuge and Fahrradbestandteile, Stettin.

Stokes, Donald (b. 1914)

Managing director and vice chairman of Leyland Motors, Ltd., from 1963 to 1969; chairman of Standard-Triumph International, Ltd., from 1963 to 1969; chairman and managing director of British Leyland Motor Corporation from 1969 to 1975.

He was born on March 9, 1914, at Bexley, Kent, and educated at Blundell's School. He joined Leyland Motors, Ltd., as an engineering apprentice in 1930, and held a variety of appointments within Leyland Motors, Ltd., from 1934 to 1939. He was drafted into the Royal Army, served first with a transport group, and later as technical service engineer for Leyland-built tanks. He held the rank of lieutenant-colonel when he was discharged in 1946. Returning to Leyland, he was named export development director, and later promoted to export director. He served as sales- and service-director for Scammell Lorries, Ltd., from 1949 to 1953, and was elected to the Leyland Motors, Ltd., board of directors in 1953. It was strictly a truck-and-bus company until taking control of Standard-Triumph International, Ltd., in 1961. He built up the Triumph brand, adding the 1300 and 2000 sedans in 1963, and discontinued the Standard name plate. In 1966 Anthony Wedgwood Benn, industry minister in Harold Wilson's Labour government, leaned on him to buy shares of Rootes Motors, Ltd., in an effort to keep Chrysler Corporation from taking control. But Stokes kept Leyland out of it. He was more receptive when Benn talked to him about the losses of British Motor Holdings, Ltd., and its lack of prospects. Benn brought him together with George Harriman, which led to their merger on January 17, 1969. British Leyland Motor Corporation was formally established on May 14, 1969. That gave him full executive powers over Jaguar, which had been brought into British Motor Holdings, Ltd., in 1966, as well as Rover, which was merged into Leyland Motors, Ltd., in 1967. He bravely announced a five-year model-renewal and standardization plan, and ordered heavy investment in the plants. He abandoned the Riley badge in 1969 and eliminated the Wolseley badge in 1975. In return, he brought out several unprofitable Austin, Morris, and Triumph cars. Jaguar became a financial headache, and though successful, Rover's earnings were too small to offset all the losses. In December 1974, he applied to the government to rescue the corporation, and while offering loan guarantees up to $121 million, Anthony Wedgwood Benn refused direct aid without collateral. Benn's mind was set on nationalizing British Leyland Motor Corporation. Harold Wilson, prime minister and Denis Healey, chancellor of the Exchequer, decided to get rid of Stokes. British Leyland Motor Corporation was dissolved on June 27, 1975, and all its assets were assigned to a new state-owned entity named British Leyland, Ltd. Stokes resigned, and became chairman of Jack Barclay, London distributors for Rolls-Royce.

Storero, Luigi (b. 1869)

Technical director of Storero Fabbrica Automobili from 1912 to 1919.

He was born in 1869 in Turin as the son of the owner of a coachbuilding enterprise originally established in 1850. He served an apprenticeship with Officine Martina in Turin and joined the parental firm in 1884. He began to transform the business from carriage making to building bicycles and horse-drawn streetcars. In 1896 he secured manufacturing rights for the Phénix engine. In 1899 he built his first motor quadricycles, some with Phénix and others with de Dion–Bouton power. By 1901 he was no longer producing Phénix engines. He became a Fiat dealer and opened a big garage on Corso Massimo d'Azeglio in Turin. It was so successful, he wanted a garage in every town. And in 1905 he founded Società Garages Riuniti Fiat-Alberti-Storero and began knitting a nationwide network of Fiat sales and service points. It grew so big that Fiat decided to take it over in 1909. A year later he joined Cesare Scacchi, a former works manager of Fiat, in building the Scacchi car at Vercelli. They were powerful cars with four-cylinder engines of 4.4 and 5.3-liter size. In 1912 he returned to Turin and leased factory space at Chivasso to make Storero cars. They were a lot like Scacchi cars, but his 1912 lineup included a smaller 3.3-liter model. Car production was suspended in 1916, and for the next two years, the Storero plant build trucks for the military. Then the factory was closed and the company liquidated.

Straker, Sidney (ca. 1870–1929)

Consulting engineer with the Daimler Motor Company, Ltd., from 1899 to 1901; technical director of Sidney Straker & Squire, Ltd., from 1905 to 1913; managing director of Straker-Squire, Ltd., from 1913 to 1929.

He was born about 1870 in London and educated at King's College as a pupil of Professor Shelley. He made a journeyman's tour of industry on the Continent and returned to London in 1895. He built an automobile in 1895–96 and then began to design a steam-powered truck. He received a patent for a compound steam engine, and in 1899 formed a partnership with Edward Bayley to build Bayley-Straker steam vehicles at Stoke Newington, London. At the same time he was designing passenger cars for Daimler in Coventry, whose management was keen to evolve from the primitive chassis of the Cannstatt Daimlers. In 1901 he was a partner with L.R.L. Squire in founding the Straker Steam Vehicle Company with premises in Bush Lane, London, where some 200 trucks and buses were built over a five-year span. In 1905 they began importing Büssing gasoline-engine truck chassis and Cornilleau passenger cars, marketed with the Straker-Squire name. They won the financial backing of John P. Brazil, an Irish businessman who sat on the board of directors of several industrial firms, and offered factory space in Lodge Causeway, Fishponds, Bristol. Here they began to produce a new car designed by Roy Fedden, initially called Shamrock, but in 1908 just Straker-Squire 14/16. In 1913 the company title was simplified and the following year they bought the Twickenham, Middlesex, plant where Burford & Van Toll had formerly produced the New Orleans car. Straker-Squire truck (no longer pure Büssing) production was transferred to Twickenham. They made about 550 cars a year at Fishponds until January 1915 when the Admiralty requisitioned the plant and retooled it to make Rolls-Royce parts and complete aircraft engines. At the end of the war, John P. Brazil withdrew and the Fishponds works were taken over by the American-owned Cosmos group. Straker-Squire Ltd. bought an idled National Projectile plant at Edmonton, North London, where they began to produce a new six-cylinder 24/90 priced at £1350 for the bare chassis. He saw very quickly that they could not survive in that market without a supporting line of lower-priced cars and brought out the four-cylinder 15/20 W.2 in 1922, listing the complete car at £850. A 10/12 priced at £450 was added in 1923. More and more of the factory space was reserved for trucks and buses, including the Straker-Cloud trolleybus, and the last cars were made in 1926. He was killed in a hunting accident in 1929.

Strobel, Werner K. (1904–)

Assistant chief engineer and head of product development with BMW from 1937 to 1941; advanced projects engineer of Ford-Werke AG from 1948 to 1954; assistant (and deputy) chief engineer of Adam Opel AG from 1962. to 1969.

He was born on October 6, 1904, and graduated with a degree in mechanical engineering in 1928. Almost immediately he joined Horch-Werke AG and in 1929 was placed in charge of engine and driveline testing. Two years later he became head of the drawing office for power units, gearboxes, and rear axles. He designed the 3½-liter side-valve V-8, introduced overdrive transmissions, and designed the final drive units for the swing-axles when Horch abandoned the live rear axle. At Fritz Fiedler's invitation, he left Horch in 1937, to take charge of production-model development with BMW. He was associated with the creation of the 335, but in 1941 he was delegated to military projects.

His postwar career began in 1948 when he joined Ford at Cologne, and for the next five years, his time was divided between Dearborn and Cologne. He was mainly responsible for the 1952 Taunus 12M, following the trends set by the 1950 Ford Consul.

Arriving in Rüsselsheim in 1954, he ably seconded Karl Stief in developing an entirely new range of cars, starting with the Rekord and Kapitan introduced in 1958.

He designed the 993-cc engine for the 1962 Kadett and was named head of the engine design office, where he created new cylinder heads carrying camshafts while opening the valves by short pushrods and rocker arms. After his retirement in 1969, he became a lecturer on engine design at Darmstadt Technical University.

Strobl, Gottlieb M. (1916–)

President of Audi NSU Auto Union AG from 1975 to 1978.

He was born on October 14, 1916, in Munich, attended local schools and graduated from high school in 1938. That summer he was accepted as an engineering trainee at the Chemnitz headquarters of Auto Union AG. He escaped to the West in 1945, and in 1950 joined the newly founded Auto Union GmbH at the Düsseldorf plant, where the F 89 P was assembled. Starting as a purchasing agent, he climbed through the ranks, and in 1971 was promoted to director of purchasing and procurement of Audi NSU Auto Union AG. In 1973 he was given a seat on the management board of the parent organization, Volkswagenwerk AG. His first actions as president of Audi NSU Auto Union AG were to close the body shop at Heilbronn and the gearbox factory at Oehringen, transferring their activities to Neckarsulm. He moved the Audi range upmarket, giving up the Audi 50 (which went into production as the VW Polo), and introducing a bigger Audi 80 in 1978. He arranged assembly of the Porsche 924 at the Neckarsulm plant, powered by a 125-hp Audi engine, beginning in 1976.

He retired in 1978.

Taylor, George (1920–)

Joint managing director of Automobiles Peugeot from 1966 to 1973; managing director of Automobiles Peugeot from 1973 to 1977; president of Automobiles Citroën from 1974 to 1979.

He was born on December 22, 1920, in Romania to a Scots father and French mother, who brought him to Paris on his father's death. He studied languages, mathematics, and philosophy at Sainte-Croix de Neuilly, graduated from the HEC[1] at the age of 20, acquired a degree in Law, and received his CPA diploma before he was 30. He joined Peugeot at Sochaux in September 1941, then building trucks for the Wehrmacht. He was sent to the Bordeaux plant, making aircraft parts, as administrative secretary, and then transferred to the engineering office at La Garenne. After 1944 his responsibilities grew, and in 1949 he was selected to attend business school at Dartmouth College, New Hampshire, where he spent the year 1950–51. He was promoted to secretary-general of Peugeot in 1952, and was named managing director of the Indenor diesel-engine factory in Lille in 1956. He was plant director at Sochaux in the 1963–66 period, when the front-wheel-drive 204 was put in production, returning to the Paris headquarters, in higher office. When Peugeot bought Citroën, François Gautier put him in charge of the new subsidiary, where he introduced himself by telling the personnel "I don't want to be Peugeot's man at Citroën. I will be Citroën's man at Peugeot."

The CX had been launched, he phased out the D-series, stopped the Wankel-engine development, and abandoned Maserati, The Citroën LN, introduced in 1976, was an ill-disguised Peugeot 104. For Citroën, profits looked as distant as they did before. Taylor retired in the beginning of 1979.

1. Ecole de Hautes Etudes Commerciales.

Thomas, John Godfrey Parry *see* Parry Thomas, John Godfrey

Thomas, Miles Webster (1898–1980)

Director and general manager of Wolseley

Motors, Ltd., from 1938 to 1940; managing director of Morris Motors, Ltd., from 1940 to 1947 and vice-chairman of The Nuffield Organization from 1942 to 1947.

He was born in 1898 in London and had a general and commercial education. He was editor of a weekly magazine, *The Light Car*, from 1920 to 1924, when W.R. Morris hired him to run Morris Press, Ltd., which published his house organs, sales literature, service manuals and parts catalogs. He was promoted to sales manager and director of Morris Motors, Ltd., in 1927. His success was assured by the arrival of the new Minor with its 847-cc side-valve engine, and a price tag of only £119 for a two-door sedan. He was sent to rescue the van and truck operations, which were making heavy losses, and served as director and managing director of Morris Commercial Cars, Ltd., in 1934–36. He was then placed in charge of Wolseley, where he cut costs by sharing power trains and many other components with Morris models. He led Morris Motors, Ltd., and later the entire Nuffield Organization through World War II and prepared the wholesale renewal of its product lines. He coped well with diversionary proposals. When he was informed by the Ministry of Supply that the Volkswagen factory could be purchased under the War Reparations Scheme, he discussed it with the Nuffield board which quickly determined that there was "no virtue" in such a project. When he was approached by Harry Ferguson about taking on production of his light farm-tractor, Lord Nuffield gave him a flat "No." The aging chairman became opposed to most new projects, including the "Mosquito" which evolved into the 1948 Morris Minor, which Thomas had pushed from the beginning. Their quarrels rose in frequency and intensity, and Thomas walked out in December 1947. He had been knighted in 1943, becoming Lord Thomas of Remenham, and sat on the board of the Colonial Development Corporation. In 1948 he was appointed deputy chairman of British Overseas Airways Corporation, and its chairman on July 1, 1949. He left BOAC in 1955 to become chairman of Monsanto Chemicals. When he

retired in 1971, he was named a life peer and was active in the House of Lords. He also held a seat on the board of the Thomson Group (electrical equipment) and became chairman of Britannia Airlines.

He died in February 1980.

Thornley, John William Yates (1909–1994)

Assistant general manager of MG Car Company, Ltd., from 1948 to 1952; general manager of MG Car Company, Ltd., and a member of its board of directors from 1956 to 1969.

He was born on June 11, 1909, in London and educated at Ardingly College. He graduated from the London School of Economics and began his career at Peat, Marwick & Mitchell, a firm of accountants and auditors in London. In 1930 he bought an MG M-Type and became the secretary of the newly founded MG Car Club. He joined MG Car Co., Ltd., at Abingdon-on-Thames in 1931 and was named service manager in 1933. He became involved with the MG racing team, which was dissolved in 1935, but he privately took on the management of the Three Musketeers and the Cream Cracker trials teams. He was called up for military service in 1939, became an officer in the Royal Army Ordnance Corps, and was discharged in 1945 with the rank of lieutenant-colonel. He returned to his office as service manager of MG and in 1948 was promoted to join the management. He approved the TD for production and helped prepare the TF. The formation of British Motor Corporation in 1952 restricted MG to use Austin engines exclusively, and he was encouraged to make use of BMC components wherever possible. He had a supervisory role in the creation of the MGA and subsequently the MGB. He retired due to ill health in 1969 and died in July 1994.

Tiberghien, Pierre (1928–)

Director of product development for the Régie Nationale des Usines Renault, automobile division, from 1968 to 1975; director of Renault's automobile research and development from 1975 to 1981; joint managing director of Renault's au-

tomobile division from 1981 to 1982; managing director of the car division from December 1982 to October 1984.

He was born on December 9, 1928, at Tourcoing and educated in local schools. He held an engineering degree from the Ecole Centrale des Arts et Manufactures and joined Renault in 1951. From 1954 to 1962 he was engaged in several new-car projects in the areas of suspension, road behavior and acoustics. He was head of car testing from 1962 to 1968, when the R 16 was developed. He was involved with the R 12, R 15 and R 17, as well as the R 5 and R 30. His star was still rising until 1982. After his first year as head of the car division, the board of directors judged him as having been promoted beyond his level of competence, and he was moved back to the technical ranks, accepting early retirement in October 1987.

Tomaso, Alejandro De *see* De Tomaso, Alejandro

Toutee, Henri (1884–1943)

Chief engineer of Chenard-Walcker from 1922 to 1927; chief engineer of Ariès from 1927 to 1936.

He was born at Bléneau (Yonne) on August 18, 1884, as the son of an army general. After receiving his diploma as a mining engineer, he joined Chenard-Walcker in 1909 as a draftsman at the invitation of his former schoolmate, Henri Walcker. He urged the adoption of aluminum pistons in 1911 and participated in the design of the 1913 15 CV car. He was the principal designer of the new 1919 models, Type TT 10 CV and Type U 15 CV. He designed a 2-liter overhead-camshaft engine for the 1923 T3 sports model and incorporated Ricardo's turbulent combustion chamber for the 11 CV T4 which replaced the TT in 1923. He adopted the Hallot servo brake system for some 1923 models, and together with André Lagache, developed a freewheel device. He was responsible for the tanklike sports/racing cars of 1923–29 and designed the 1926 3-litre model. Leaving Chenard-Walcker in 1927, he was engaged by Baron Charles Petiet to renew the Ariès model range, with the emphasis on commercial vehicles. He designed the 10/12 CV Type CB, which was the only Ariès car made in 1930–32, adding the CB 4B in 1933. But the last Ariès cars were made in 1934. He designed a front-wheel-drive van, but it never went into production, as it demanded an industrial investment beyond the company's means. In 1936 he went into semi-retirement, concentrating his work on suspension-system innovations such as torsion-bar springing and variable-rate springs. He died on November 13, 1943.

Towers, John (1948–)

Managing director of Land Rover from 1991 to 1994; chief executive of the Rover Group from 1994 to 1996; president of MG–Rover since 2001.

He was born on March 30, 1948, and graduated from. Bradford University with a degree in mechanical engineering. He joined Perkins Engines Ltd. of Peterborough as a student apprentice in 1966, and climbed through the ranks to become president of Perkins. In 1987 he was named managing director of Massey-Ferguson, which owned Perkins. When Massey-Ferguson was absorbed by Varity Corporation, he was named Varity's vice president of international services. He left Varity in 1988 to take over as Land Rover's director of manufacturing, serving subsequently as head of Land Rover product development, purchasing director, and managing director. British Aerospace PLC sold the Rover Group to BMW in February 1994, and despite the title, Towers never held full executive powers. W. Reitzle, chairman of Rover since 1995, felt that Towers represented "all that was wrong with British industry" and forced his resignation, effective June 1, 1996. He became chief executive officer of Concentric, an engineering company and automotive supplier in the Midlands. He led the Phoenix consortium to win control of Rover back from BMW, with the backing of Abbey National, First National Finance, Mayflower, Lola Engineering, and an association of Rover dealers. In May 2000,

Phoenix bought Rover for £10. BMW kept the rights to the new Mini and the Cowley factory, extending a £500 million loan to Phoenix. Towers became president of the newly formed MG–Rover, having moved Rover 75 assembly from Cowley to Longbridge, Birmingham. He resumed production of the MG-F and in June 2001, MG Rover purchased the Qvale Mangusta operations in Modena for £815 million.

Towns, William (1937–1993)

Chief designer of the Aston Martin Lagonda styling studio from 1966 to 1969; head of an independent styling studio from 1970 to 1993.

He was born in 1936 near Guildford, Surrey, and began his career as a designer with the Rootes Group in 1955. He complained about interference from the Rootes executives and joined Rover in 1963. He designed the bodies for the BRM-chassis gas turbine car of 1965. He left Rover and designed the body for the Chrysler V-8 powered GKN FF (Ferguson) car. He designed the Aston Martin DB-S body and proposed a convertible which, after some delay, was built as the Volante. He was offered the position of chief stylist for Triumph, but decided to remain independent. He served as a consultant to Triumph for about five years, though few of his ideas were adopted. He designed the body for the Jensen-Healey roadster and in 1973 displayed the Minissima prototype on Austin Mini chassis. He designed an angular sedan body of exaggerated length for Aston Martin in 1978 and two years later created the Bulldog coupé for Aston Martin. From 1980 onwards he was closely linked with Derry Mallalieu Ltd., and in 1987 created the Tracer TCX prototype. He restyled the Reliant Scimitar in 1990 and tried to revive the Railton in 1991–92, using the Jaguar XJ-S chassis as the basis for a convertible.

He lost his fight against cancer and died on June 6, 1993.

Tranié, André P. (1911–2001)

Member of the Simca board of directors from 1948 to 1949; technical director of Panhard & Levassor from 1949 to 1965.

He was born on September 27, 1911, at Clermont-Ferrand. After graduation from the Ecole Polytechnique, he went in for a military career and served as an artillery officer in the French Army from 1933 to 1947. He fought against the German invasion in 1940 and in the liberation of Italy and France in 1943–45.

He entered the automobile industry when H.T. Pigozzi invited him to the board of Simca in 1948. Paul Panhard engaged him in 1949 as technical director, finance director, and secretary-general. He had his finger on the pulse of every aspect of the company's operations, and was directly responsible for the production of passenger cars, trucks, and military vehicles. The factory was then producing the EBR[1] armored car, an eight-wheeler with drive on the front and rear axles, and a set of slave wheels in the center. The engine was a flat-16 made up of eight Dyna-Panhard air-cooled flat-twins. It was in production until 1960.

Tranié organized the production of the Dyna 54 sedan, switching and supervising the body-supply source from Facel to Chausson. He supervised the development of the PL-17 sedan which replaced the Dyna 54 in 1961, and the AML 4 x 4 armored car.[2] He also championed the CD[3] sports/racing car and the production-model 24 CT which was built from 1963 to 1967.

He resigned from Panhard when the company came under Citroën's control. From 1965 to 1969 he served as managing director (and later, honorary chairman) of a mining-exploration group[4] and financial advisor to the management of the Lorraine coal mines.[5] In 1969 he became involved with the Jarret brothers and their electronic vehicle, sort of a high-technology golf cart, whose development was sponsored by Schneider SA, leaving in 1972. He was chairman and chief executive of TEJ[6] from 1976 to 1979, when he was named head of the supervisory board of Moteurs Leroy-Somer (electric motors), an office he held till 1986. He died on October 16, 2001.

1. Engin Blindé de Reconnaissance (armored scout machine).

2. Auto-Mitrailleuse Blindée (armored self-propelled machine gun).
3. Charles Deutsch.
4. Compagnie Française de Prospection Sismique.
5. Houillères de Lorraine.
6. Techniques Electriques Jarret.

Trefz, Karl (1890–1947)

President of Stoewer-Werke AG from 1936 to 1940.

He was born on March 27, 1890, in Ludwigsburg and educated in local schools. He began his career as an office clerk with Daimler Motoren Gesellschaft in 1912, but was drafted into military service in 1914. In 1922 he joined the sales department of Stoewer-Werke AG and did a great job of bringing cash into the company coffers for six years. He left Stettin in 1928 to accept the position of sales director of C.D. Magirus, fire-engine builders in Ulm on the Danube. Magirus held key patents on extension ladders and turn-tables, with lucrative license agreements. The company also made garbage-disposal vehicles and medium-duty trucks, mainly with Deutz diesel engines.

He resigned when Magirus became part of Klöckner-Humboldt-Deutz AG in 1935, and the following year made his return to Stettin. Stoewer had been under caretaker management by bankers and city officials for four years and its main product, the Greif Junior family car, was a Tatra design acquired from the defunct Röhr Automobil-AG. He eliminated the loss-making front-wheel-drive models and the Greif V-8, and put the small engineering staff to work on two conventional car projects, the four-cylinder Sedina and the six-cylinder Arkona. He survived another World War but died in 1947.

Trier, Vernon (18??–19??)

Technical director of Light Cars, Ltd., from 1910 to 1915; technical director of Waverley Cars, Ltd., from 1915 to 1931.

He was the inventor of the Trier & Martin carburetor and worked as a sales engineer for T.B. André & Co., Ltd., importers of Ballot engines, Malicet & Blin frames and axles, and Hartford friction-dampers. In 1909 Trier was a member of the team that organized assembly of the Marlborough car, designed by Malicet & Blin who also supplied the major components, in London. With backing from T.B. André & Co., Ltd., he set up Light Cars, Ltd., in Trenmar Gardens, Willesden, to produce a cyclecar of his design. It had a JAP[1] V-twin engine, two-speed gearbox, and no differential. Not many were built, for T.B. André had little faith in the cyclecar market. Trier then designed the Waverley, a conventional 10-hp light car, with a four-cylinder Chapuis-Dornier engine and major components from Malicet & Blin in Paris. Waverley production from 1910 to 1914 totaled 1200 cars. New models appeared in 1919, with a wider choice of power units: Coventry Simplex, Tylor, Aster and Chapuis-Dornier. In 1923–24 a few Waverley cars were also made with Burt-McCollum single-sleeve-valve engines. The first six-cylinder Waverley went into production in 1925 with a 1991cc Coventry Climax overhead-valve 16/50 engine. Trier also had new ideas for economy-car design, and drew up a small four-passenger vehicle with a rear-mounted V-twin and friction drive, but only three test cars were built. When Waverley car production came to an end in 1931, Trier was named managing director of T. B. André & Co., Ltd., and later became managing director of Silentbloc, Ltd.

1. John A. Prestwich, Tottenham, London.

Trippel, Hanns (1908–)

Founder of Trippelwagen special developments[1] at Tuttlingen in 1938; head of Amphicar Corporation from 1959 to 1968.

He was born on July 8, 1908, and his main goal in life was to make cars that could run on land and water with equal ease. In 1932 he converted a DKW chassis to carry the hull of a boat, creating an amphibious vehicle. New Trippel test vehicles appeared in 1935 and 1936, which he built at Homburg in the Saar, powered by Adler and Opel engines. In 1938 he moved into a factory in Tuttlingen, Württemberg, where he developed a military "swimming car" which he proposed to the Wehrmacht. After the fall of France in 1940, the Wehrmacht offered him space in the Bugatti factory at Molsheim and gave him a contract for delivery of amphibious vehicles powered by the six-cylinder 2½-liter Opel engine. Approximately 1000 units were delivered up to 1945. Taken prisoner by the French authorities, he served 36 months of a five-year sentence for "war crimes," obtaining his release in 1948. He promptly set up a new institute, Protek, in Stuttgart and in 1950 displayed the SK 10 streamliner, powered by a rear-mounted 600-cc Horex motorcycle engine. It was a road car with no amphibious capability. He sold about 20 of them.

Next he developed the Amphicar, for which he saw two different markets: one recreational, the other for rescue operations. The Amphicar was powered by a Triumph Herald engine giving it a road speed of 100 km/h and a cruising speed on water from 10 to 12 km/h. Production began in 1961 at the Lübeck and Berlin-Borsigwalde factories belonging to the Quandt group's IWKA[2] machinery affiliate, staffed by ex–Borgward and ex–Mercedes-Benz mechanics. About 3000 Amphicars were produced up to 1969. In 1968 he joined Motoporter AG in Zurich to develop a small utility vehicle powered by an air-cooled flat-twin motorcycle engine, but it never came on the market. In 1980 he retired to Erbach with his drawing board, and was still trying to promote new Amphicar projects as late as 1994.

1. Trippelwagen Sonderentwicklung.
2. Industrie-Werke Karlsruhe-Augsburg.

Turcat, Léon (1874–1953)

Technical director of Société des Ateliers de Construction d'Automobiles Turcat, Méry & Cie from 1899 to 1902 and from 1918 to 1921. Chief engineer of Lorraine-Diétrich from 1902 to 1905, technical director of Lorraine-Diétrich from 1905 to 1914.

He was born on December 7, 1874, the son of a well-to-do merchant in Marseilles and educated as an engineer. He visited Peugeot in 1895 to see how automobiles were built and ordered a Peugeot car. He also bought a Panhard-Levassor, to have a second source of inspiration. He began design and experimental

work on a car in 1896 and three years later formed a partnership with his brother-in-law, Simon Méry, to go into production.

The first Turcat-Méry car bore obvious Peugeot influence, but later models were designed with greater originality. Their widely acclaimed 1900 model had a four-cylinder vertical engine with chain drive to the rear wheels. The gearbox had two reverse and five forward speeds. In 1902 Adrien de Turckheim invited them to come to Lunéville and take charge of the car engineering department. They accepted and the management of the Turcat-Méry factory in Marseilles was delegated to Louis Méry, Simon's brother, while Paul Engelhardt took charge of the design office. For several years the Lorraine-Diétrich cars were near-identical with concurrent Turcat-Méry models. Turcat masterminded the transfer of car production from Lunéville to Argenteuil near Paris in 1907 and laid out the new plant.

By 1909 the plant was turning out eight different models, from a little two-cylinder 10 hp car to a six-cylinder 70 hp tourer. The company prospered, but in 1914 Turcat and Méry resigned, returned to Marseilles and tooled up their factory for war production. In 1918, Turcat designed a moderately priced 2.4-liter four-cylinder car which he planned to produce at a rate of 1000 units a year. That required heavy industrial investment, which depleted the company's cash reserves. It was declared bankrupt in 1921 and ownership passed into the hands of Arthur Provenzale, Berliet dealer in Nice, Louis Mouren, a former member of the office staff, and Andre Bridonneau. Turcat resettled in Toulon and opened an office, as an insurance broker. He died in Toulon in 1953.

Turckheim, Adrien G. de (b. 1866)

Managing director of Société De Dietrich & Cie from 1897 to 1906; managing director of Société Lorraine des Anciens Etablissements De Dietrich & Cie beginning in 1906.

He was born on August 19, 1866, in Alsace, son of Baron Edouard de Turckheim and nephew of Eugène De Dietrich. His ancestors had been associated with the De Dietrich metalworking enterprises since 1806. In 1880 he was a co-founder with Eugène De Dietrich and his brother Eugène de Turckheim in setting up a branch factory at Lunéville in Lorraine for making railroad rolling stock, and in 1890 he was named general manager of the Lunéville plant. It was his idea that the company should have a sideline in automobiles, and in 1896 he went to Le Mans for talks with Amédée Bollée, Jr., about manufacturing rights to his car design. He demanded a demonstration run to the Mediterranean shore and back, and in February the two pioneers set off together, stopping in Lyon and Monaco before heading for Lunéville. The license agreement was signed on March 6, 1897, and production got under way, not at Lunéville but at Reichshoffen in Alsace in July 1897 in a plant that had opened in 1848 to make railroad wagons. Twenty cars, with engines shipped complete from Le Mans, were completed by the end of the year. In the summer of 1898, the Lunéville plant began production of Amédée Bollée, Jr., cars with rear-mounted horizontal parallel-twin engines of 6, 12, and 18 hp. During 1899, some 40 cars a month were built at Lunéville.

At Eugène De Dietrich's initiative, the Reichshoffen plant began making 9 CV single-cylinder car under Vivinus license in 1899, marketed with a DAK name plate, in addition to the Amédée Bollée, Jr., models. DAK production was given up after a year, and in 1900 the Reichshoffen plant was offering a range of 6 CV, 9 CV, 11 CV and 18 CV cars, still with engines coming from Le Mans. Up to the end of 1901, De Dietrich had made over 500 cars in Alsace. In 1901 a representative of De Dietrich attended an industrial fair in Milan, where a four-cylinder Bugatti prototype was displayed. He sent a favorable report to the head office, and Eugène De Dietrich asked Ettore Bugatti to come and see him for talks about a license agreement. The designer, still a minor, arrived with his father. They signed a contract with per-car royalties and a salary for taking charge of the drawing office at Reichshoffen. The Bugatti-designed car went into production in 1902.

Adrien de Turckheim went to Marseille in February 1902, to see the Turcat-Méry partners and signed a license agreement for their up-market touring cars. Léon Turcat and Simon Méry moved to Lunéville to take charge of engineering and production. The Amédée Bollée, Jr., models, were phased out and in 1903 the Lunéville plant built 311 Turcat-Méry type cars. In 1904 Eugène De Dietrich dismissed his Italian designer and discontinued car production in Alsace.

In 1905 Adrien de Turckheim and his uncle had an argument which provoked the end of their business relationship. He negotiated the purchase of the Lunéville plant and organized a new company, financed by Henri Estier, chairman of Turcat-Méry, André Lebon, chairman of Crédit Foncier d'Algérie, Léopold Renouard of the Banque de Paris et des Pays-Bas, and Hubert de Portales, who became directors. Payment for the plant was made in bonds issued by the new company. The products were no longer De Dietrich, but Lorraine-Dietrich vehicles. About 650 cars were produced at Lunéville in 1906. The 1908 model range included four four-cylinder cars and a six: Type FS 18/24 hp, Type FM 28/35 hp, Type FO 40/50 hp, Type FER 60/70 hp and Type FO/6 70/80 hp. Construction of a new motor-vehicle plant in the Paris suburb of Argenteuil began in 1907 and car production at Lunéville was phased out in 1910–11. In 1908 Renouard, Lebon and Estier resigned from the board of directors, and Adrien de Turckheim reorganized the company in 1910, with no change in its title. Some 800 cars were made at Argenteuil in 1913, and in 1915 the plant was retooled to build aircraft engines. During World War I, the Lunéville plant produced trucks, machine-gun carriers, and railroad wagons.

Turnbull, George Henry (1926–1992)

General manager of Standard Triumph International from 1963 to 1968; managing director of Austin-Morris and deputy chairman of Standard-Triumph from 1968 to 1973; managing director of British Leyland Motor Corporation from May to September 1973.

He was born on October 17, 1926, and educated at King Henry VIII School, Coventry. He won a scholarship to Birmingham University where he obtained an honors degree (B.Sc.) in engineering. He joined Standard Motor Company, Ltd., as a trainee in 1950 and held a series of appointments in engineering and production. He was a daring policy maker and the company was profitable under his management. Once he came to Austin-Morris, his hands were no longer quite free. The Austin Allegro and Morris Marina were not successful in the market. Still, he was promoted, and widely regarded as the future BL chairman. But he fell out with senior management and resigned in 1973.

In 1974 he moved to Korea as vice president of Hyundai Motor Company, and at the end of his three-year contract, signed up with the Iran National Motor Company in Teheran as deputy managing director, running the plant that assembled the Paykan (Hillman Hunter). Returning to Britain in 1979, he was named chairman of Chrysler UK, then in the process of being sold to Peugeot SA. He stayed on as chairman and managing director when the company was renamed Talbot UK, phased out the Chrysler products, closed the Linwood plant in Scotland, and started assembly of Peugeot cars.

In 1984 he joined Inchcape, worldwide motor distributors, and was named chairman of Inchcape in 1986. But he was forced into retirement by ill health in 1991 and died in December 1992.

Turner, Edward (1901–?)

Managing director of the Automotive Division of BSA,[1] technical director of the Daimler Company from 1956 to 1960.

He was born in London on January 24, 1901, and educated as a mechanical engineer. He joined Ariel Motors Ltd. in 1927 and was co-designer with C.Y. Sangster of the Ariel square-four engine which went into production in 1934. He also designed the Ariel Red Hunter. In 1936 he became chief engineer of the Triumph Engineering Company Ltd. and created the Triumph Tiger and the 500 cc Triumph Speed Twin.

His association with BSA began in 1951 when BSA bought the Triumph Engineering Company Ltd. for £2.5 million. He was appointed a director of BSA Cycles while also serving as a director of Triumph. The creation of the BSA Automotive Division in 1956 put the Daimler Company under his authority.

He designed a modern 2½-liter V-8 engine with hemispherical combustion chambers that went into production in 1958 for installation in the Daimler Dart SP 250 sports car. Shortly after the sale of the Daimler Company to Jaguar Cars Ltd. in June 1960, it became apparent that the SP 250 was produced at no lower cost than the Jaguar E-Type, and had the drawback of a plastic body. It was taken out of production in 1964. Its V-8 engine continued and was installed in the small Jaguar sedan, rebadged as a Daimler 250 and built from 1962 to 1968. He resigned in 1966 when Jaguar Cars Ltd. was merged into the British Motor Corporation.

1. Birmingham Small Arms Company, Ltd., Small Heath, Birmingham.

Turrell, Charles McRobie (ca. 1872–?)

Secretary of the British Motor Syndicate, Ltd. in 1895–96, general manager of the Coventry Motor Company, Ltd. in 1898–99, chief engineer of Accles-Turrell Autocars Ltd. in 1900–01, technical consultant to the Pollock Engineering Company in 1901–02.

He was born about 1872 in Coventry and began an apprenticeship with J.K. Starley & Company at the age of 14. After a number of years with the Coventry Gas & Electrical Engineering Company, he met H.J. Lawson and joined the British Motor Syndicate Ltd. in November 1895. Lawson set up a subsidiary in Hertford Street, Coventry, to make the Coventry Motette in partnership with J. Russell Sharp. It began as a direct copy of the Léon Bollée "motor tandem," but Turrell made many modifications. Independently of Lawson, he had designed a single-cylinder 3½ hp light car and a prototype was built by the Pollock Engineering Company in 1896. He formed a partnership with Frederick Henry de Veulle of Coventry and they filed a number of joint patent applications, including a speed-change system for a constant-mesh gearbox, a motorcycle gearbox, and an air-cooled single-cylinder engine. He designed the engine for the 1898 Humber Motor Sociable and supervised its production by the Coventry Motor Company Ltd.

He went into partnership with J.G. Accles in the formation of Accles-Turrell Autocars Ltd., in January 1900 at Holford Works, Birmingham. Thomas Pollock and J.G. Bedson were also directors of the company. He designed the Accles-Turrell cars and obtained a patent for a hydraulically operated clutch mechanism. In November 1901, he joined Thomas Pollock at Ashton-under-Lyme in Lancashire and designed a 10/15 hp car with an inboard/rear-mounted engine, which was produced up to 1910.

Ueber, Max (1904–?)

Chairman and managing director of Ford-Werke AG and a director of Ford of Europe, Inc., from 1967 to 1971, member of the Ford-Werke AG supervisory board from 1971 to 1977.

He was born on January 24, 1904, at Freiburg in Breisgau and joined Ford Motor Co. AG in Berlin in May 1929 as a clerk. In 1936 he was transferred to the headquarters in Cologne in a minor executive position. He served as personnel director of Ford-Werke AG from 1947 to 1953 and director of domestic sales from 1953 to 1956. In 1956 he was named general sales director and in the following nine years Ford's share of the German market climbed from 7.8 percent to 18.5 percent. He was promoted to managing director in 1967.

Much of Ford's success was due to the popularity of the original Taunus 17M, built from 1960 to 1964. Its heavier 17M and 20M successors found a less eager market and the smaller front-wheel-drive Taunus 12M and 15M met a lot of sales resistance. The sales curve shot up again with the launching of the Escort in 1968 and the Capri in 1969. He was succeeded by John Banning in 1971 and held a seat on the supervisory board until 1977.

Uhlenhaut, Rudolf (1906–1989)

Technical manager of the Mercedes-Benz racing department from 1937 to 1939; manager of the passenger-car testing department from 1949 to 1959; chief engineer for passenger-car development from 1959 to 1972.

He was born on July 15, 1906, in London, where his father was an official of the local branch of Deutsche Bank. His mother was English. He began his schooling in Hampstead, but the family had to leave British soil at the outbreak of World War I. His schooling continued in Bremen where he later spent six months as a trainee with Hansa-Lloyd. He graduated from high school and got his driver's license in 1926, and then enrolled at the Technical University of Munich. He graduated in 1931 and was engaged as a test engineer with Daimler-Benz AG. He revealed himself as a superb driver with an uncanny ability to translate road-behavior quirks in a car into precise engineering modifications. In September 1936 he was placed in charge of racing car construction and testing. In 1939 he was assigned to the development of military vehicles for arctic climates, and in 1943 became manager of a group of small aircraft-engine plants dispersed around the countryside. From 1945 to 1948 he helped run a transport and forwarding business, returning to Daimler-Benz AG on February 1, 1948, as head of an experimental department.

While the factory was being rebuilt, he found and salvaged a supply of small-diameter steel tubes. He found use for them in the space-frame of a sports car which was to appear in 1952 as the 300 SL. He took on additional duties with the racing team in 1954–55 and then concentrated on developing the W-111 which appeared in 1959 as the 220-series. He made valuable contributions to the 300 SE, the 230 SL, the 600 and the 450 SEL 6.3, and forced the technical director to adopt semi-trailing arm suspension in place of the swing-axles in 1967.

He was due to retire in July 1971 but was given a one-year extension. He died on May 8, 1989.

Urach, Prince Wilhelm von (1897–1957)

Staff engineer with Daimler-Benz AG from 1927 to 1945; head of Daimler-Benz AG's technical research department from 1945 to 1957.

He was born on September 27, 1897, as the son of Duke Wilhelm von Urach, into a branch of the royal Hohenzollern dynasty. He might have been expected to choose a military career, but his interested lay in automobiles. He studied engineering at the Technical University of Stuttgart, graduating in 1922. He joined Walter Steiger in Burgrieden as a design engineer and later held an engineering position with Fritz Gockerell in Munich. In 1925–26 he was a member of the Bugatti drawing office in Molsheim. In 1927 Friedrich Nallinger invited him to join the engineering department of Daimler-Benz AG, and he accepted. In the summer of 1940 he was one of three Daimler-Benz officials placed in charge of military work in the Renault factories at Billancourt. They allowed no car production, but wanted specific types of trucks, and converted big halls to repair shops for armored vehicles and tanks. He took a lenient attitude to the French engineers' experiments with a small economy car which eventually saw production as the Renault 4 CV. Whenever a test car was observed, he would call in the Frenchmen for a reprimand, ending with "Keep it out of my sight!"

After 1945 he led the technical research efforts of Daimler-Benz with great competence and served as Germany's representative on the Technical Committee of the FIA[1] which set the rules and regulations for international motor racing and speed records.

He never retired and died, still at work, in 1957.

1. Fédération Internationale de l'Automobile.

Uren, Wilhelm Gustav Heinrich (1873–1937)

Technical director of Motorfahrzeug-Fabrik Köln, Uren, Kotthaus & Co. from 1903 to 1908, chief engineer of Aachener Stahlwarenfabrik from 1908 to 1919, chief engineer of Fafnir Werke AG from 1919 to 1926.

He was born on June 4, 1873, in Hanover and educated as an engineer. He joined Kölner Motorwagenfabrik GmbH vorm. Heinrich Brunthaler in 1897 and designed a small car with a 6 hp single-cylinder engine, spring-mounted in the chassis, with belt drive to the rear wheels. It was named Priamus and went into production in 1900.

He bought a stake in the company in 1903, changed its title, and moved into a bigger factory in Köln-Sulz. He designed a 1½-liter 8/10 hp two-cylinder model and in 1905 created the first four-cylinder Priamus, with a 3-liter engine, three-speed gearbox and shaft drive. It was enlarged to 3.8-liter size in 1907 and given a smaller 9/16 hp four-cylinder companion model. But the company was in financial trouble in 1908 and M. Molineus sold it to Dr. Emil Rehe and Bernhard Bölefahr. Wilhelm Uren left and joined Carl Schwanemeyer's Aachener Stahlwarenfabrik in Bachstrasse, Aachen (Aix-la-Chapelle), makers of Fafnir industrial, vehicular and marine engines.

He designed the first Fafnir cars in 1909, using current four-cylinder F-head engines. He designed the Type 384 with an L-head 2½-liter engine for 1910. The Type 472 8/22 hp 2-liter model of 1912 and the Type 466 6/16 1½-liter car of 1913 also had four-cylinder L-head engines. Car production was suspended during World War I and on March 8, 1919, the company was reorganized as Fafnir-Werke AG. The prewar models were put back into production, to be replaced by the Type 476 9/30 hp 2.2-liter touring car in 1924. He also created a super-sports model with Zoller-supercharged engine, the Type 471K, in 1925. But the company was making heavy losses and by the end of 1925 had piled up debts of RM 1.8 million. It was placed in receivership, and car production was halted in 1926. The factory survived for a short time on auto repair work, but in 1930 it was sold to Englebert Fils et Cie and converted to tire production.

Valentin, Ernst (1874–1950)

Head of Gobron-Brillié vehicle engineering from 1897 to 1900; chief engineer of the Berliner Motorwagenfabrik from 1900 to 1905; technical

director of Nagant Frères from 1905 to 1908;
chief engineer of the Protos branch of Siemens-
Schuckert from 1908 to 1911; chief engineer of
Rex-Simplex from 1911 to 1920.

He was born on September 18, 1874, in
Berlin. As soon as he had obtained his engineering degree, he sailed to America, where
he worked for several years in the machine-
tool industry. Returning to Europe in 1897,
his ship landed at Le Havre. He came to Paris
and took charge of car design for Gobron-
Brillié. In 1900 he returned to Berlin and
signed up as chief engineer of the Berliner
Motorwagenfabrik at Tempelhof, which
wanted to build trucks. He designed all their
trucks until he left in 1905, moving to Liège,
where he designed a new generation of Nagant
cars. He was contacted by representative of
Russo-Baltic, who were making cars under
Fondu license in Riga, Lithuania, and designed a new model range for Russo-Baltic,
going into production in 1909. At the invitation of Siemens-Schuckert, who had just
taken over the Protos Automobil-Werke, he
returned to Berlin in 1908 to design new Protos cars. His first creation was a six-cylinder
rally car for the Kaiser Preis, which became
the basis for a much-respected touring car, the
18/45 Type F, made from 1910 to 1912.

After three years with Protos, he joined
Richard Hering in Ronneburg, Saxony, as
chief engineer for their Rex-Simplex cars and
trucks. He also arranged for Russo-Baltic to
build copies of some Rex-Simplex models. He
spent World War I with Richard & Hering,
but withdrew from the automobile industry in
1920 when the company sold its motor vehicle division to Elite Motoren Werke AG of
Brand-Erbisdorf.

He lived on until 1950.

Valletta, Vittorio (1883–1967)

Managing director of Fiat from 1928 to 1943;
managing director of Fiat S.p.A. from 1943 to
1946; chairman of Fiat S.p.A. from 1946 to
1966.

He was born on July 28, 1883, at Sampier-
darena near Genoa as the son of a Sicilian railway executive and a lady of noble birth from

the Valtellina. He graduated from the Som-
meiller Technical Institute in Turin in 1900
with top marks and a gold medal from the
Chamber of Commerce. At the tender age of
17 he became the manager of a paper mill in
the Lanzo Valley, simultaneously taking evening classes in mathematics at the Turin University and business administration at a commercial college. After graduation, he took up
teaching, first in evening classes for factory
workers, later in professional training, and
eventually at Turin University, where he lectured on management, industrial law, banking
and economics.

At the beginning of World War I he was
factory manager for Antonio Chiribiri's aircraft and motor-vehicle enterprise in Turin,
and from 1915 to 1918 he served in the Military Aviation Corps of Engineers. In 1921 Giovanni Agnelli hired him as a part-time management consultant, giving him time to
continue his academic work. But his true calling was Fiat, where his first title was "central
manager." He was named general manager in
1925 and managing director in 1928. Since the
death of Edoardo Agnelli, son of the founder,
in 1935, Valletta was Fiat's de facto chief executive.

He realized before anyone that the Lingotto
plant would soon become inadequate, secured
land in a west-side suburb for plant construction, where the vast Mirafiori industrial complex was opened in 1939. After World War II,
he got the Italian plants restarted with a $10-
million loan from the Bank of America, and
revived car production at Heilbronn (Deutsche
Fiat). He authorized an across-the-board
model renewal, beginning with the 1400 in
1950 and the 1100/103 in 1953, following up
with the 600 in 1955 and the Nuova 500 in
1957. He put Fiat into a joint venture with
Spanish banks and the Franco government to
produce Seat cars in Barcelona, beginning in
1950. In 1955 he made Fiat a partner with
Giuseppe Bianchi and Pirelli (tires and cables)
in founding Autobianchi to make Fiat-based
cars at Desio. He later arranged for the production of Fiat cars by Crvena Zastava in Za-
greb, El Nasr in Cairo, Egypt, FSO in War-

saw and FSM in Bielsko-Biala. His plan to help the Soviet Union set up a plant near Kuibyshev on the lower Volga to make 600,000 Lada-version Fiat 124 cars a year had to be submitted to the U.S. Congress to make sure that no trade sanction would be applied against Italy. Approval followed in 1966. That was the crowning point of his career. He retired shortly afterwards to Le Focette di Pietrasanta near Lucca, where he died on August 9, 1967.

van Basshuysen, Richard *see* Basshuysen, Richard van

Van Doorne, Hubertus Josephus "Hub" (1900–1979)

Technical director of Van Doorne's Aanhang-wagen-Fabriek from 1932 to 1950; technical director of Van Doorne's Automobielfabriek NV from 1950 to 1965; technical director of Van Doorne's Transmissie BV from 1972 to 1979.

He was born on January 1, 1900 in a small Limburg village as the son of a blacksmith. He left school at 12 and worked for years as a bicycle repairman, motorcycle mechanic, and chauffeur. He did such a good job of rebuilding the engine of a local brewer's Stearns-Knight that the brewer financed him in setting up his own workshop in 1928. He began by making metal door frames, stepladders, and furniture, in partnership with his brother Willem "Wim" (1906–1978). He designed his first trailer in 1930 and two years later the brothers set up a factory at Eindhoven to build trailers and semi-trailers on an industrial scale. A tireless inventor, "Hub" designed and patented trailers with welded frames, automatic couplings, and swing-axle suspension. He also patented the "Trado" system of converting a two-axle truck to 6 × 4 and 6 × 6 configuration. In 1939 he was also granted a patent for an all-wheel-drive military six-wheel truck. During World War II he began drawing up a stepless transmission, using a cogged belt and variable-diameter pulleys. He filed patent applications for most of the details, and by 1945 he had tested a model that worked. He called it Variomatic and contin-

ued its development. The brothers also set up a sideline in making "new" trucks from war surplus Ford and GM components. In 1950 they started building trucks of their own design, from five to eight tons payload, powered by Leyland diesel engines. In 1956 the Variomatic was road tested in a Lloyd car, with results that convinced them to start car production. The first DAF built for sale came off the line on March 23, 1959, at Eindhoven, in the retooled trailer plant, for trailer production had been moved to Geldrop. The DAF car had a front-mounted air-cooled 600-cc flat-twin and rear wheel drive. It was aimed at the low end of the market, but sales were slow. DAF made its 10,000th truck in May 1955, and began building Leyland diesel engines under license in 1961. DAF-designed diesel engines went into production in 1972. "Hub" was given an honorary Doctor of Engineering degree in Technical Science by the Delft University of Technology in 1953. He retired in 1965 after having designed the DAF Pony, a military air-droppable platform with pedestrian controls and later set up a company for the development of the Transmatic (using a multi-link metal belt) and licensing. He died on May 23, 1979, having held 166 patents in his lifetime.

Van Eugen, Charles Marie (1890–1980)

General manager of the motor car division of Lea-Francis from 1922 to 1934, chief engineer of Autovia from 1934 to 1937, chief engineer of Wolseley from 1937 to 1943.

He was born in 1890 in the Netherlands, attended local schools and was apprenticed to Simplex in Amsterdam. He held a number of positions with Dutch companies until he went to Coventry in 1913, worked briefly for the Standard Motor Company, and then joined the engineering staff of the Daimler Company in July 1913. He became involved with Daimler's war-production program, including Daimler-Foster artillery tractors, complete aircraft designed by the Royal Aircraft Factory and 105 hp tank engines. When the war ended, he was head of the development team for the Dragonfly aircraft engine.

He left Daimler in 1919 to become works manager of Briton Cars Ltd. of Wolverhampton. In 1920–21 he held engineering positions with the Royal Ruby Cycle Company and Clyno Engineering Company, before joining Swift of Coventry in 1922. The Swift management released him later that year, after he had been asked to come to the rescue of Lea-Francis. He quickly introduced the 9 hp C-Type then undergoing testing and designed the 10 hp D and E-Types. He chose a 12 hp Meadows engine for the H and I-Types and began to experiment with supercharging.

In 1927 he designed a new, lower chassis which was used for the 12/50 Type 0, the S-Type Hyper Leaf with its twin-cam 1496 cc supercharged Meadows engine, and the 14/40 T-Type Light Six. In 1930 he designed the overhead-camshaft "Ace of Spades" 16 hp six-cylinder engine, which went into production in the Lea-Francis factory on Lower Ford Street, Coventry. The "Ace of Spades" was the mainstay of Lea-Francis car production up to 1935. At Victor Riley's invitation, he left Lea-Francis in 1934 to design a new car named Autovia, powered by a 2.9-liter V-8 engine conceived as two four-cylinder Riley engines placed at 90°, sharing a crankshaft. It was built in small numbers from 1936 to 1938, its existence ending when Riley fell under control of the Nuffield Organisation. Miles Thomas persuaded Van Eugen to take over the engineering responsibilities for Wolseley. He withdrew from the Nuffield Organisation in 1943 and lived on until 1980.

Varlet, Amédée (1863–1938)

Technical director of Delahaye from 1900 to 1938.

He was born on July 23, 1863, in Paris, as the son of a mechanic with Périn, Panhard & Cie. He was educated as an electrical engineer and joined the French Edison company in 1881. He worked on the installation of the power network in Paris and drew the wiring diagram for the central station in Avenue Trudaine. He also directed the installation of elec-

tric lighting in the Opera, Odéon, Gaité and other theaters. In 1892 he joined Etablissements Ollier and installed an underground generator station for the Olympia theater. In 1896 he was invited by a former Edison director to join him at the Etablissements de Quillac at Anzin and organize the production of automobiles. Varlet designed the Raouval car, an advanced design with a pressed-steel frame, dished steering wheel on a raked column, and variable-ratio steering car. The engine was an 8-hp two-cylinder Pygmée. The Raouval was produced from 1897 to 1900 and was awarded a gold medal at the World's Fair[1] in Paris in 1900.

Delahaye had just moved from Tours to Paris. Emile Delahaye was retiring, and Varlet was engaged as technical director. He designed new cars and won some minor races. Delahaye-powered speed boats were highly successful. His technical prowess was incredible, for in 1904 he designed a 300-hp four-cylinder marine engine with dual overhead camshafts, two spark plugs and six valves per cylinder.

He was also a glutton for work and introduced Delahaye trucks, taxicabs, fire engines and military vehicles. The touring-car range for 1909 included ten different models. He created the world's first V-six engine, fitted in the 1911–14 Delahaye 18/24 CV Type 44. His first inline six was a 4426-cc 18/22 CV L-head design, mounted in the 1919 Type 82, which was produced until the end of 1925. About 1927 he began to concentrate his thinking on fire trucks and military vehicles, leaving passenger-car design to his assistant (Jean François). In 1936 he designed an articulated twin-engine vehicle with a gun-turret on the bridge between the front and rear units, but no production was undertaken. He died unexpectedly in his home at Pomponne on the river Marne during the night of July 10, 1938.

1. Exposition Universelle.

Ventre, Philippe (1934–)

Renault engineer since 1957; safety project manager from 1976 to 1982; vice president of quality control for American Motors in 1982; director of the AMC engineering office in

Veraldi 328

1983–84; director of Renault vehicle engineering from 1985 to 1987; joint chief engineer from
1987 to 1989; chief engineer of product engineering from 1989 to 1995; senior vice president
of Renault engineering from 1995 to 1998.

He was born in 1934 at Annecy in the Savoy
Alps and attended local schools up to the age
of 18. His father wanted him to study literature, but he had his mind set on engineering
and graduated in 1955 from the Ecole Supérieure de Technique Aéronautique et Construction Automobile in Paris. After a year
with Simca, he joined Renault on the first day
of 1957 as a research and development engineer. He became involved with crash-testing
in 1959 and developed measuring equipment
and methods for accident analysis. In 1972–75
he created the Basic Research Vehicle as a parallel program with the R 30 production
model.

He led the design and development of the
Renault 21 and Renault 19, and later held
overall responsibility for the Clio, Safrane, Laguna and Mégane.

He retired in 1998.

Veraldi, Lewis C. "Lou" (1930–1990)

*Vice president of product development, Ford
of Europe, from 1973 to 1976, mastermind of
the Ford Fiesta program.*

He was born in Detroit on July 16, 1930, as
the son of southern Italian parents. He held a
degree in mechanical engineering from Lawrence Institute of Technology, studying while
on leave from Ford. He had originally joined
Ford at the age of 19 as a filing clerk in the engineering office, and showed such aptitude
that the company sent him to college. From
1955 to 1970 he advanced slowly through a
variety of assignments. He directed the engineering of Mustang II and was named chief
assembly engineer for Ford's automotive assembly division in 1972. A year later he was
transferred to the Technical Center at Brentwood, Essex, to take over the Bobcat project
from Eric Reickert. The Bobcat evolved into
the Fiesta, and production began at Almusafes
in Spain on October 18, 1976. It secured Ford's

position in a lower market segment and was
produced without major change until 1984.

Veraldi returned to Dearborn in 1976 as
vice president of advanced vehicles development. In 1988 he was named head of the product engineering staff. He retired due to ill
health on November 1, 1989, and died on October 13, 1990, from a heart condition aggra-
vated by diabetes.

Vermorel, Victor (1848–1927)

*Chairman of Etablissments V. Vermorel from
1890 to 1927.*

He was born in 1848 at Beauregard (Ain) as
the son of a watchmaker who moved to Villefranche-sur-Saône in 1853 and opened a general engineering shop. He went to work in the
paternal shop and designed equipment for
vineyards. He began making small industrial
engines in 1889 and started drawing a car. A
prototype with a single-cylinder rear-mounted
engine and tiller steering was built in 1895. In
1898 he built a more modern car designed by
François Pilain, with a front-mounted two-
cylinder engine, chain drive and a steering
wheel. Series production of cars began in 1908
in rue François Giraud at Villefranche, with a
four-cylinder T-head 2.2-liter shaft-driven
model designed by his son Edouard (1873–
1957). After an abortive 18 CV with a 3.3-liter
engine in 1910, they decided on a small range
of moderately priced family cars with four-
cylinder T-head engines from 1½ to 2 liter
size. Vermorel also made 200 trucks for the
military in 1913–16. Car production was revived in 1919 with the prewar 15 CV Type LO,
which was replaced in 1921 by the Type S with
a 2.2-liter L-head engine and a Vermorel-
patented cantilever rear-axle suspension. Type
AA introduced in 1923 had an overhead-valve
version of the same engine. The most popular models were the 1922–27 Types X and Z,
the former with an L-head 1.7-liter engine,
and the latter with overhead valves on the
same block. Victor and Edouard then decided
to test the economy-car market and brought
out the 6 CV 1132-cc L-head Type AG in
1926, but it was no great success, and it was

given a bigger (1327cc) engine. Edouard took over the reins when his father died and split off the car company as SA des Autobiles Vermorel. But the last Vermorel cars were made in 1930. The total output in 22 years of car-making has been estimated at 7,800 units.

Violet, Achille Narcel (ca. 1888–ca. 1972)

Tireless advocate of two-stroke engines, designer of light cars including the Violet-Bojey and Sima-Violet, designer of the 1930 Mathis Type PY.

He was born about 1888 at Armentières in northern France and graduated from the Ecole des Arts et Métiers in Lille with a degree in mechanial engineering in 1907. His first car, built in 1908, was a tandem-seat four-wheeler with the passenger's seat in front, above the front axle, and a 500 cc Quentin engine mounted under the driver's seat.

Financed by Count de Chevigné, he founded Société La Violette in 1909 and produced Violette cars for four years. The Violet-Bojey appeared in 1912, built in Paris with other business partners, and had a ten-year production life. In 1922 he founded SIMA[1] with backing from R. Legras and a factory at Courbevoie, to produce the Sima-Violet car with a 500 cc air-cooled two-stroke flat-twin engine and two-speed gearbox. Over 5,000 Sima Violet cars were produced up to 1928, when he left the company to join Deguingand as an engine designer.

In the following years, Violet held a position on the engineering staff of Mathis in Strasbourg, did some consulting work for Peugeot, Bucciali and Somua, and designed the AVA two-stroke flat-four aircraft engine. In 1938 he submitted to the Ministry of Defense plans for a military motorcycle and a small forward-control military vehicle with a rear-mounted two-cylinder engine — but they were ignored. After World War II he went to work for Bernardet and designed engines for motorcycles and scooters. He died about 1972.

1. Société Industrielle de Materiél Automobile.

Virgilio, Francesco Di *see* Di Virgilio, Francesco

Vitger, Erhard (1898–1991)

Managing director of Ford-Werke AG from 1945 to 1958, chairman of Ford-Werke AG from 1958 to 1969.

He was born on October 11, 1898, at Kong in Denmark and educated in accounting and business administration. He joined Ford's Danish branch office in Copenhagen in 1920 and was promoted to the Berlin office in 1925. Ford was then assembling the Model T in Berlin. A manufacturing plant at Cologne was opened in 1931 and he went there in an executive position. He was elected to the management board of Ford-Werke AG in 1935 and set up Credit AG to finance installment-plan sales. He spent World War II in Cologne and in 1945 was named managing director. He led the plant reconstruction and planned future expansion.

The US parent company agreed to finance the purchase of all available land adjacent to the existing plant, north and south on the left bank of the Rhine, but due to Dearborn's policy Ford-Werke AG had to finance all new plant construction out of its own earnings. That delayed work until 1950, when a new body plant and assembly hall were erected. At the end of the war, the tooling was not badly damaged and truck production resumed in 1945. But it was only in 1948 that the prewar Taunus car was put back in production.

As for product policy, the heads in Dearborn instructed Vitger not to compete head-on with Volkswagen. For the rest, they encouraged the Ford companies in Britain, Germany, and France to produce unrelated models, with the aim of gaining market share everywhere. That was reversed in the mid–1960s, when the Escort and Capri were jointly planned by Ford of Britain and Ford-Werke AG. As for plant financing, Dearborn relented in 1958, allocating funds for Ford-Werke AG to build a new power station, an engine plant and forges, a new stamping plant and a new final assembly line. In 1962 Ford-Werke AG bought land at Genk on the Albert Canal in Belgium for a major final-assembly plant that would add 50 percent to its production capacity. From 80,000 cars in 1956, output soared to 300,000 in 1963, and in 1965 Ford-Werke AG was turning out 2200 cars a day. In 1969 Vitger retired and settled in Lugano, where he died on October 29, 1991.

Vivinus, Alexis (1860–1929)

Director of the Société des Atéliers Vivinus from 1899 to 1911.

He was born on July 5, 1860, at Stenay (Meuse) in the French Ardennes, studied civil engineering and began a career in the French government's Department of Bridges and Highways. After seven years in public service, he joined a company that made boats and naval equipment. Moving to Belgium in 1886, he worked as a designer of railway equipment for Usines Rollin at Braine-le-Comte, and two years later became a tool designer for Atéliers Bouton in Brussels. He left Bouton in 1893 and started a factory to make bicycles. As a sideline, he opened a dealership for Benz engines and cars in 1895. He designed and built his first motorcycle in 1896, and two years later designed a four-wheeled automobile. He showed his drawings to Count Jacques de Liedekerke, pioneer motorist, who agreed to finance the production of his car. They founded Société des Atéliers Vivinus in February 1899, with a factory in rue du Progrès at Schaerbeek, Brussels. The first Vivinus car was a voiturette with a 785-cc single-cylinder engine and belt drive. Vivinus completed 152 cars in 1900–01. His design was also built under license by Georges Richard in Paris as the Poney, by De Diétrich of Niederbronn as the DAK and by Burford & Van Toll in London as the New Orleans. This business proved so lucrative that Vivinus was able to move into bigger premises in rue Destouvelles, Brussels, in 1904. The first four-cylinder Vivinus was a 15/18 CV model introduced in 1902, and he built a few six-cylinder cars from 1906 to 1909. The Vivinus range for 1911 included three models with four-cylinder engines: 10/12 CV, 16/20 CV and 24/30 CV. But the company was losing money, and declared bankruptcy in 1911. He began to import the Clément-Bayard for the Belgian market, but joined the engineering staff of Minerva in 1914. After 1920,

he helped design many Minerva cars, trucks and buses. His crowning effort was the 1929-model AP, with a 4-liter straight-eight sleeve-valve engine.

He died in Antwerp in 1929.

Voisin, Gabriel (1880–1973)

Director of Société des Aéroplanes Voisin from 1911 to 1945.

He was born on February 5, 1880, as the son of a foundry engineer at Belleville-sur-Saône and educated as an architect at the Ecole des Beaux-Arts in Lyon. He went to Paris in 1899 was engaged by Godefroy & Freynet, architects. But his passion was not houses, it was flying. He had been experimenting with model gliders ever since boyhood, and in 1902 he began to build an aeroplane. He left the architects to join Ernest Archdeacon, head of a group known as the Aviation Syndicate. In 1904 he was joined by his brother Charles Voisin (1882–1912) and Louis Blériot in taking over the Surcouf Aviation factory at Billancourt, and by 1907 they were flying planes capable of controlled flight for any distance (as long as the fuel load permitted). Blériot left them in 1908, and they formed Voisin Frères to continue the business. They moved to a bigger factory at Issy-les-Moulineaux in 1911, forming a new company. In 1914 the Minister of War, Alexandre Millerand, selected the Voisin biplane as the standard design for the French air force. The Voisin factory was too small to fill the orders, and planes designed by Voisin were also built by Breguet, Nieuport, Esnault-Pelterie and others. Voisin also built big bomber planes with Panhard-Levassor engines.

Though he had owned Rochet-Schneider, Lorraine-Diétrich, Mercedes, Métallurgique, and Panhard cars, Voisin had no natural interest in building cars. But when his factory was left without aircraft orders in 1919, he opted to convert the plant into making automobiles. He did not design engines nor chassis, but had a lot of original ideas of using aircraft techniques in body design. The first Voisin car was ready-made chassis, designed for André Citroën by Louis Dufresne and Ernest Artaut, which became available when Citroën did not want it. In 1923 the engineering functions were split between André Lefebvre for the up-market models and Marius Bernard for the lower end. Voisin's body designs, for many years, made no attempt at streamlining. His aeroplanes, in fact, had not been streamlined, and Voisin had no precise idea of aerodynamic drag forces. His main contribution to body design was lightness. The 1934 Voisin Aéro-dyne, was not his design, but André Noel's. He was no businessman and lost control of the company to Belgian investors in 1935. His mind was occupied by innovations in car de-sign. He created a prototype with two six-cylinder engines in line, making a straight-12. He also designed a six-passenger sedan with wheels in lozenge formation (two driving wheels, right and left, aft-of-center, with steering on both the front and rear (single) wheels. The driver sat in front, with a pas-senger on each side, slightly behind, and a three-abreast rear seat ahead of the engine. The company was taken over by Gnome & Rhône in 1938, but he was kept on as a direc-tor until 1945 when Gnome & Rhône was in-corporated into SNECMA.[1]

He designed a minicar he called the Bi-scooter. When SNECMA rejected it, he of-fered it to Autonacional SA of Bercelona, who produced over 5000 of them from 1949 to 1958. He retired to Tournus in 1960, but kept his mind busy with ideas for innovations in aircraft and automobiles. He died on Decem-ber 25, 1973.

1. Société Nationale des Etudes et de Construc-tions de Moteurs d'Aviation.

Volanis, Antoine (1948–)

Head of the Matra automobile styling studio from 1971 to 1980.

He was born on August 6, 1948, at Saloniki in Greece and came to France with his parents in 1953. He studied engineering and industrial design. He began his career as a designer with Renault but left after one year to join Peugeot as a body engineer in 1968. At the time, Simca controlled Matra Sports, and Matra was work-ing on sports car projects for Simca. He ap-plied direct to Matra and was engaged as chief stylist. He designed the Bagheera, the Mu-réna, and the Rancho. He also made the orig-inal styling proposal for a multi-purpose van, intended for Simca, but which later evolved into the Renault Espace, produced by Matra. He resigned from Matra and set up as an in-dependent styling consultant in November 1980. Over the years, he worked under con-tracts from Peugeot, Volkswagen, and Seat, plus non-automotive clients.

Vollmer, Joseph (1871–1951)

Automotive consulting engineer; head of the Deutsche Automobil-Konstruktions-AG in Ber-lin from 1907 to 1932.

He was born on February 13, 1871, in Baden-Baden and was educated as an engi-neer. He began his career by designing a car, the Orient Express, for Bergmann's Industrie-Werken in Gaggenau in 1894. He stayed on at Gaggenau, supervising car production and improving the product, until 1897, when he returned to Berlin. He designed a motorized fore-carriage for Kühlstein Wagenbau in Berlin-Charlottenburgs which was produced in a variety of models, with battery-electric drive or gasoline engines, in important num-bers up to 1902. Vollmer then designed a 12-hp car for NAG[1] which led to a five-year con-tract with NAG. He also designed a truck for Paul Heinrich Podeus of Wismar in 1902. In 1907 he incorporated his office in Berlin and hired an engineering staff, and began adver-tising his ability to supply complete designs for cars and trucks, industrial engines, trac-tors, and armored vehicles. He attracted new orders from Podeus and Lloyd in 1908, and in 1910 designed three tractors for Hanomag (one wheeled on-road type, one wheeled off-road type, and a crawler) with a line of diesel and spark-ignition engines. In 1911 he began a long relationship with Hille of Dresden, During World War I he served in the vehicle engi-neering section of the German War Office and accomplished much for the standardization of motorized fighting equipment. Throughout the war he also acted as a technical consultant to Ludwig Loeb, and sold military-equipment designs to the Berlin-Erfurter Maschinenfabrik. In 1918 Ludwig Loeb merged his Berlin works with Carl Rüttger's who had built army trucks at Hohenschönhausen to form the Dinos Automobilwerke. Initially Dinos was to build a Vollmer-designed basic-transport vehicle, but in 1920 Vollmer brought out a popular 8/35 PS four-cylinder model and a powerful 16/72 PS tourer. But in 1922 Ludwig Loeb sold Dinos to Hugo Stinnes, and Vollmer was on his own. He designed new tractors for Hanomag in 1923–24. He also went to Elbing and rearranged Franz Komnick's factory as well as selling him new car designs. From 1925 to 1932 he designed trucks for Wumag of Görlitz. He then liquidated his company and disbanded the staff, but continued consulting work up to 1941. He died in 1951.

1. Neue Automobil-Gesellschaft, a subsidiary of AEG (Allgemeine Elektrizitäts-Gesellschaft = General Electricity Company).

von Eberhorst, Robert Emmerich Manfred Eberan *see* Eberan von Eberhorst, Robert Emmerich Manfred

von Falkenhausen, Alex *see* Falkenhausen, Alex von

Von Falkenhayn, Fritz *see* Falkenhayn, Fritz Von

von Heydekampf, Gerd Stieler *see* Stieler von Heydekampf, Gerd

von Kuenheim, Eberhard *see* Kuenheim, Eberhard von

von Selve, Walther von *see* Selve, Walther von

von Urach, Prince Wilhelm *see* Urach, Prince Wilhelm von

Walcker, Henri Etienne (1873–1912)

Co-founder of Chenard, Walcker et Cie; technical director of Chenard-Walcker from 1899 to 1912.

He was born on August 7, 1873, at the Château d'Argenteuil as the son of a retail shopkeeper, Adolphe Guillaume Walcker and his wife, Maria Ludowica née Lückfiël. He attended the Lycée Condorcet in Paris and the Lycée de Versailles before enrolling at the Ecole des Mines in 1895, graduating in 1898 with a degree in mining engineering. He was a keen cyclist during his student days and became familiar with the two and three-wheelers made by Ernest Chenard. They went into partnership in January 1899.

He was the chief designer of the first Chenard-Walcker automobile, completed in September 1900, and later designed their first engines and gearboxes. His first four-cylinder engine went into production in the spring of 1903, and a year later he invented a unique rear suspension with open drive shafts, the rear wheel hubs being connected by a beam axle, which was used up to 1914, though the first models with live rear axles came on the market in 1911. In 1906 he designed a taxicab which was produced in significant numbers, and updated in 1908 and again in 1911. He never designed a six-cylinder engine, but his biggest four-cylinder unit was the 30/40-hp Type N of 5880 cc, appearing in 1907.

He had an appendectomy in mid–June 1912 and died in the clinic on June 20, 1912.

Walkinshaw, Thomas Dobbie Thomson "Tom" (1947–)

General manager of Tom Walkinshaw Racing from 1976 to 1996; chairman of the TWR Group since 1996.

He was born in 1947 into a Scots family of gentleman farmers, played rugby in his youth, and became a horse breeder. In 1974 he opened a garage, acquired a clientele fore race- and rally-preparation, participated successfully in rallies himself, and secured several sales-and-service franchises. He won a contract with Jaguar to build an endurance-racing car with a 700-hp 7-liter V-12 engine, and the XJR won at Le Mans in 1988 and 1990. The engine came from Jaguar and he engaged Tony Southgate to design the chassis. Its direct derivative, the XK-220, became a limited production model. In 1991 he allied himself with Flavio Briatore in taking over Ted Toleman's racing team and turning it into Benetton Formula One. He became engineering director of the Benetton racing team and built it up to world championship status. He held a 35 percent stake in Benetton Formula One and an equal share in the Ligier Grand Prix stable, while the Benetton family holding "21 Investimenti" took a 50 percent interest in Tom Walkinshaw Racing. In January 1995 he became a 51 percent partner with AB Volvo in founding Auto-Nova with a factory at Uddevalla in Sweden, to produce the C-70 Coupe and Cabriolet.

During 1996 he severed his connection-with Benetton, repurchasing their shares in Tom Walkinshaw Racing, which now became the TWR Group. His shares in Ligier were sold to Alain Prost. He moved his headquarters from Milton Keynes to Leafield some 20 miles west of Oxford, and put his chain of retail dealers into a new corporate unit called Ixion Motor Group. He bought the Arrows racing team from Jackie Oliver and hired Frank Dernie to prepare a new Arrows car with Yamaha power. He also engaged Geoff Goddard, a racing engine designer from Cosworth. The TWR group created the chassis for the Aston Martin DB7, and also the Nissan 8390 (by Tony Southgate) as well as managing the Nissan team during 1997. In August 1999 he sold the Ixion Motor Group to Frank Sytner, formerly head of BMW's touring-car race team.

Walzer, Peter (1937–)

Technical director of Seat[1] from 1990 to 1993; vice president and deputy managing director of Seat 1993–94.

He was born on May 4, 1937, in Stuttgart. After studying aeronautical engineering at the Stuttgart Technical University, he took an economics course at the University of Aachen. He began his career with Dornier (aircraft) and spent five years in the drawing office for wings and fuselage. Next he took a teaching appointment and became assistant professor of turbomachinery at the Aachen Institute of Technology. He joined Volkswagenwerk in 1971, working in the research division at Wolfsburg. Among other projects, he worked on gas turbine development from 1971 to 1976. He was head of a power-train research group from 1979 to 1983 and then took charge of the gas-flow research branch, which included internal and external air flow, and aerodynamics. VW built its own wind tunnel for these studies. He served as director of VW's research department from 1987 to 1989, when he was delegated to Seat to oversee their new-model planning and development. He restructured Seat's industrial plant and was responsible for laying out the Martorell factory. The Seat Toledo was launched in 1991 and a new Ibiza in 1993, followed by the Cordoba in 1994. After a few months as interim president of Seat in 1994, he returned to Wolfsburg but resigned before the end of the year to become general manager of FEV Motorentechnik in Aachen, an engine-consulting business with capability from updating an old unit to producing an all-new engine from scratch.

1. Sociedad Española de Automoviles de Turismo.

Wankel, Felix (1902–1988)

Inventor of the rotary engine.

He was born in Lahr, Bavaria, son of a ranger in the Black Forest. Wankel showed an early interest in mathematics and engineering. He received his first patent for a rotary piston engine in 1929, five years after he first hatched the idea. He could find no funding until 1936, when Hermann Goering invited him to set up the Wankel Test Institute. The Institute worked on a number of aircraft projects, but not a rotary engine. In 1951 Wankel was contacted by the German company NSU, which was having problems with the rotary valves on its racing motorcycles. He developed very efficient superchargers for use in record-breaking motorcycles, and in 1957 the first rotary engine ran at the NSU works. The single rotary-powered Sport Spyder went into production in 1964, and was followed in 1967 by the twin-rotor Ro80 saloon, which was made up to 1977. Meanwhile Mazda had bought rotary patents from NSU, and began production of rotary-engined cars in 1967. Ultimately many more rotary-engined Mazdas were made than NSUs — more than a million Mazdas (one model is still in production today) to just under 40,000 NSUs. Although a prominent name in automobile engineering, it is believed that Wankel never drove a car.

Ware, Peter (1918–2000)

Director of engineering, Rootes (Motors) Ltd., passenger-car division, from 1958 to 1966.

He was born on May 14, 1918, as the son of Sydney Ware, factory manager of Straker-Squire and the inventor of a carburetor. He joined the Royal Navy as a cadet and went on to the Navy's engineering college. Discharged from the Royal Navy for medical reasons in 1939, he went to work as an engine development engineer for the Bristol Aeroplane Company, Ltd., under the orders of Roy Fedden. He accompanied Fedden to Whitehall and was appointed technical secretary to the committee on motor boats in the Ministry of Supply. In 1945 he was a member of the Bristol engineering team that studied the KdF-Wagen as a candidate for production in England, but left Bristol that year to join Roy Fedden in the pursuit of his own car projects. When they failed, he went into farming in Somerset. He returned to the motor industry in 1949, working on fuel-injection systems for CAV[1] and Leyland Motors, Ltd. He joined Humber, Ltd., to take over the top technical office and directed the Hillman Hunter program. He also held top responsibility for the Hillman Imp. He resigned in 1966 and became managing director of the Dunlop Wheel Company. He retired in 1983 and died at the end of February 2000.

1. Motor accessories makers and traders, founded by Charles Anthony Vandervell.

Waseige, Charles (1884–1943)

Chief engineer of Société Fernand Charron from 1911 to 1914; chief engineer of Société H. & M. Farman from 1919 to 1935.

He was born on July 24, 1884, into a middle-class family in Paris. Proving himself a bright and conscientious pupil, he was encouraged by his teacher to seek a career in arts and handicrafts. He applied to Ecole Lavoisier and was accepted, but was expelled after participating in a student protest march and went to work in a printing shop. This involved a lot of blueprint copying, which brought him in contact with the world of engineering. He quickly learned to read blueprints and landed a position as a tracer in Clément-Bayard's drawing office. He worked his way up to draftsman. In 1902–03 he accompanied Marius Barbarou to Mannheim, helping to design new Benz cars. Returning to Paris, Barbarou and his team of young engineers were engaged as a group by Delaunay-Belleville, where Waseige learned to work independently on components and complete chassis. He met Fernand Charron and designed four- and six-cylinder Alda cars for him. The Alda factory in Courbevoie was requisitioned for war work in 1914. He stayed on and organized the production of artillery shells. In 1919 Henry and Maurice Farman engaged him to design a luxury car. He created a high-class chassis with a 6.6-liter overhead-camshaft six, four-speed gearbox, torque-tube drive and cantilever rear springs. He also designed V-12 and W-18 aircraft engines for Farman, and a remarkable straight-eight with three-speed supercharger drive.

He left Farman in 1935 and took over the Société Viet at Bois-Colombes, which he reorganized as Société Air Equipment to manufacture landing gear, auxiliary power groups and compressed-air installations for the aircraft industry. He retired to Blois in 1937 and spent his last years toying with two-seater minicar projects. He died on April 27, 1943.

Watson, William George (1889–1971)

Chief engineer of Invicta Cars, Ltd., from 1924 to 1934; chief engineer of Invicta Car Development Company (London) Ltd. from 1945 to 1950.

He was born in 1889 in London and was apprenticed to the Thames Iron Works of Millwall and Greenwich, which built Thames cars from 1906 to 1911. Privately he built his own motorcycle, with a Fafnir engine and

Chater-Lea components. He was commissioned into the Royal Naval Auxiliary Reserve in World War I. In 1919 he met Noel Macklin and Hugh Orr-Ewing, and became involved in the design of the Eric-Campbell, Silver Hawk, and Invicta cars. He created a succession of Invicta sports cars with 2½-liter, 3-liter and 4½-liter Meadows engines, and in 1931 a lower-priced Invicta with a 1½-liter Blackburne six. Leaving Invicta in 1934, he joined the engineering staff of Wolseley Motors, Ltd. He later did some design work for Ford at Dagenham before joining W.O. Bentley at Lagonda Motors, Ltd. He spent the war years with Lagonda but left in 1945 and became one of the instigators of the revived Invicta. He designed the Invicta Black Prince, with a 3-liter twin-cam six, Brockhouse Turbo-Transmitter automatic transmission, and all-independent torsion-bar suspension. When the Invicta venture failed, he joined Aston Martin and was responsible for the Lagonda V-12 engine project, which was abandoned. He went to Jaguar in 1958 and worked on gearbox design until he retired in 1967.

He died on November 3, 1971.

Weaver, William Arthur (1888–1970)

Managing director of Coventry Victor Motor Company, Ltd. from 1920 to 1926 and its chairman from 1926 to 1970.

He was born on October 8, 1888, in Peterborough and educated at private schools and Salford Technical College, Manchester. In 1904 he went into partnership with T.E. Morton to take over the textile machinery business of William Gardner in Cox Street, Coventry, which they reorganized as Morton & Weaver Ltd., machine-tool manufacturers. He designed and built his first flat-twin and flat-four engines in 1905, and in 1906 created the Weaver Ornithoplane — not an ornithopter but a well-conceived monoplane which could have become a trendsetter but did not. The engines became the company's main product line and in 1911 the corporate title was changed to Coventry Victor Motor Company Ltd. He was a prolific engine designer with original

ideas for keeping costs down by means of modular design. He added single-cylinder and flat-six units. The engine line-up included air and water-cooled versions of the same basic designs, and cylinder heads with overhead valves as well as L-head types. He designed the first Coventry Victor motorcycle in 1919, powered by an air-cooled 688 cc side-valve flat-twin. The water-cooled version of this engine was installed in the 1926 three-wheel runabout, with chain drive to the single rear wheel. The engine was enlarged to 749 cc in 1928 and became available with overhead valves in 1929. He discontinued motorcycle production in 1931. The three-wheeler was produced up to 1938, when he decided to concentrate on offering a range of small industrial engines. After World War II, when the market for marginal transport flourished, he took steps to re-enter the car business and displayed his Venus prototype with a 474 cc flat-four engine in 1949. At the same time he began production of the first Coventry Victor diesel engine, an air-cooled vertical single-cylinder named Zephyr, which came into great demand and led to a family of water-cooled diesels as well. The Venus was quietly shelved.

Webster, Henry George "Harry" (1917–?)

Director of engineering, Standard Motor Company Ltd. from 1957 to 1959, director of engineering for Standard-Triumph International Ltd. from 1959 to 1968, director and chief engineer for Volume Car and Light Commercial Vehicle Division, British Leyland Motor Corporation from 1968 to 1974.

He was born on May 27, 1917, in Coventry and educated at Welshpool County School and Coventry Technical College. He joined the Standard Motor Company, Ltd. at Canley, Coventry, as an engineering apprentice in 1932 and became an assistant technical engineer in 1938. In 1940 he was appointed Deputy Chief Inspector and from 1946 to 1948 he worked as assistant technical engineer on the Vanguard program. He held the title of chief chassis engineer from 1948 to 1955 and chief engineer from 1955 to 1957.

Triumph had lost its supplier of unit-construction bodies when the British Motor Corporation took control of Fisher & Ludlow in 1953. This prompted Webster to a new kind of structure, with a "back-bone" frame and bolt-on body panels, first applied to the 1959-model Triumph Herald and subsequently to the Spitfire, GT-G and Vitesse. He led the design team for the 1300 sedan, a front-wheel-drive car intended as a Herald replacement and for the Triumph 2000 sedan, both introduced in September 1963. He was transferred to Longbridge, Birmingham, in 1969 and took over the ADO-67 project, which went into production in 1973 as the Austin Allegro. In 1971 he began design work on the ADO-71, intended to replace the Austin 1800/2200 models, with transversely mounted four and six cylinder engines and all-independent Hydragas suspension, it was marketed as the Princess, beginning in 1975. He resigned from British Leyland in June 1974 to become technical director of Automotive Products, where his first major development was a narrow-profile automatic transmission for front-wheel-drive cars with transversely mounted engines.

Weigel, Daniel M. (18??–19??)

Managing director of Clement Talbot Ltd. from 1903 to 1905; managing director of Weigel Motors (1907) Ltd. from 1907 to 1909.

He was managing director of the British Automobile Syndicate, importers of Clément-Bayard cars to Britain, and was chosen to run the Clement Talbot operations in 1903. The Talbot factory in Barlby Road, Ladbroke Grove, London, was placed in the hands of C.R. Garrard who had formerly worked for Adolphe Clement in Paris. Weigel resigned in 1905, and within two years had organized a company to make Weigel cars. The Weigel cars were assembled from a mix of British, Belgian and Italian components, beginning in a garage in Goswell Road, London, and later moved to Olaf Street, Latimer Road, Shepherds Bush, West London. The chassis frames were finished by the Wilkinson Sword Company and the bodies were built by the English

branch of J. Rothschild et Fils. Engines came from Pipe, Itala, and the local industry. At the end of 1909, his company was taken over by Crowdy, Ltd., who continued to produce cars in the Latimer Road factory.

Wenderoth, Hans Georg (1925–)

Director of testing and deputy research director of NSU from 1961 to 1966; deputy technical director of NSU from 1966 to 1971; chief development engineer of Volkswagenwerk AG from 1971 to 1973.

He was born on June 23, 1925, in Berlin and served in the German Navy during World War II, going on many submarine raids. He studied at the Berlin Technical University from 1946 to 1951 and earned an engineering degree. From 1952 to 1956 he worked for Shell Oil Co. in Hamburg, and then returned to Berlin as an instructor at the Technical University. He pursued his own studies at the same time, and was given a doctor's degree in engineering in July 1960. He began his career as technical director of H. Trüller AG of Celle, near Hanover, but eagerly joined NSU Motorenwerke AG in Neckarsulm in July 1961. He worked on the development of the Prinz 4 and its derivatives, and became project manager for the Ro-80. He designed the K-70 which was later given a VW badge, and was brought to Wolfsburg towards the end of 1971. He led the design and development of the Scirocco and Golf, which went into production in 1974. He had resigned from VW, however, in March 1973 and joined Continental Gummiwerke AG in Hanover. By 1976 he was production manager of Continental, and spent the rest of his career with the tire-and-rubber products company.

Wennlo, Sten (1925–)

General manager of the Saab car division from 1976 to 1984; executive vice president of Saab-Scania AB from 1984 to 1987; senior vice president of Saab-Scania AB from 1987 to 1990.

He was born on May 25, 1925, and began his career as a journalist. He joined Scania in 1957 as a public-relations officer for ANA[1] and was promoted to sales director of Saab–ANA in 1965. He realized that the small capacity of the Trollhättan plant prevented the Saab car from being cost-competitive in the small-car market to which the company had been committed. He argued for moving the Saab product into higher price classes, with wider per-car profit margins, where limited production volume was no handicap. His recommendations led to the planning of the Saab 99 which went into production in November 1967. It did not immediately replace the smaller Saab 96 sedan and 95 station wagon, which were transferred out of Trollhättan for assembly by a joint venture with Oy Valmet at Nystad, Finland. He was strongly opposed to a merger with AB Volvo, as proposed by Peter Wallenberg, who held majority control of Saab. Wennlo rallied a number of executives to his side, and also won the support of the labor unions. Wallenberg changed his plans and merged Saab with truck builder ScaniaVabis. Wennlo initiated plans for cooperation with Lancia, and Saab dealers in the Nordic countries began selling the Lancia Y-10 subcompact car. The Saab 9000 project began as a Swedish version of the Lancia Thema, and a further step up-market. The production-model 9000 ended up sharing nothing more than a few sheet-metal parts in the under structure, and never met its cost targets. It was also 150 kg heavier than the Thema. He retired from full-time service with Saab-Scania in the beginning of 1987, but sat on committees and continued to wield some influence in the marketing and product-planning areas.

1. Aktiebolaget Nyköpings Automobilfabrik, Nyköping.

Werbin, Dan (1944–)

Director of Volvo Car Corporation product planning from 1978 to 1983, executive vice-president of Volvo Car Corporation with responsibility for long-range product planning from 1984 to 1987.

He was born on October 1, 1944, at Ljungby and went to school in Linköping, where from 1960 to 1964 he built 23 "Reva" sports cars, all with different specifications. He graduated from Chalmers University of Technology in Gothenburg in 1969 with a master's degree in science and joined Volvo in 1970 as a product specialist for the VESC (Volvo Experimental Safety Car). He was named manager of Volvo's car product program in 1972 and served as a director of Volvo North America from 1973 to 1978. Werbin returned to Gothenburg and took charge of preparing new models, beginning with the 760 and 740. He planned the 960 and 940, which were launched in 1989–90 and took on new assignments in marketing and sales organization. He became project manager for the Volvo S.4 (a sister model to the Mitsubishi Carisma) in 1994, the two being built on the same line at the Born plant in the Netherlands. He went to Tokyo in 1996 as president of Volvo Cars of Japan and in 2001 was promoted to president and chief executive officer of Volvo North American Sales.

Werlin, Jakob (1886–1965)

Member of the Daimler-Benz AG management board from 1933 to 1942, director of Gezuvor[1] from 1937 to 1939.

He was born in 1886 in Austria and served an apprenticeship with Puch-Werke AG in Graz. He joined the Benz sales organization and in 1919 became general manager of the Benz branch in Munich, handling not only distribution throughout Bavaria but also exports to the Balkan countries. In 1921 he sold Adolf Hitler a second-hand Mercedes limousine and became one of Hitler's political backers. After Benz' merger with Daimler Motoren Gesellschaft in 1926, he was general manager of the Daimler-Benz AG branch in Munich.

When Hitler came to power in 1933, Wilhelm Kissel made Werlin a director of the parent company, correctly calculating that Werlin could bring government business and perhaps provide protection from the SS, Gestapo, etc. That proved only too true, for twice Heinrich Himmler tried to get Wilhelm Haspel dismissed, as Mrs. Haspel was half–Jewish, but Werlin intervened with Hitler, and Haspel stayed on the job.

As early as 1932 Hitler talked to Werlin about his ideas for a low-cost "people's car."

An artist/painter by profession, Hitler had even made some drawings of what such a car might look like. Hitler asked Werlin if he could get Daimler-Benz to produce it. He put it before the board, who turned it down. Then Hitler mentioned Porsche. They had met in May 1933, at the instigation of Baron Klaus von Oertzen of Auto Union AG, who sought state subsidies for the cost of building and racing the P-Wagen. Werlin organized another meeting between Porsche and Hitler in May 1934, to discuss the "people's car." It led directly to a contract. And in 1936 Daimler-Benz AG agreed to build a series of 30 test cars. Gezuvor was established in May 1937, the directors being Bodo Lafferentz, an aide to Robert Ley of the Deutsche Arbeitsfront,[2] Ferdinand Porsche, and Jakob Werlin. Werlin enlisted 200 test drivers from the SS ranks. In the summer of 1937 he was a member of a delegation visiting Detroit on Gezuvor business. And throughout this period he lobbied to secure orders for military vehicles, aircraft engines etc. for Daimler-Benz AG. World War II soon gave him other duties. In 1942 Hitler named him SS Standartenführer and sent him to Minsk as superintendent of a huge repair works for tanks, halftracks and trucks.

1. Gesellschaft zur Vorbereitung des deutschen Volkswagens. (Company to prepare the German People's Car).

2. Deutsche Arbeitsfront was to finance the People's Car venture through its leisure time organization, KdF (Kraft durch Freude = Strength through Joy).

Werner, Helmut (1931–)

Chairman of the Mercedes-Benz AG management board from 1993 to 1997.

He was born on September 2, 1931, in Cologne as the son of a bank manager. He attended local schools and graduated from college with a degree in business economics. He began his career with Englebert & Co. GmbH of Aachen in 1961, working in marketing and finance. He was transferred to the Englebert corporate headquarters at Liège in 1970 as general product manager, and was named head of all product groups of Uniroyal Europe in 1977. A year later he became managing director of Uniroyal Europe. He became a board member of Continental Gummiwerke AG in 1979 when Uniroyal sold its European tire-making activities to Continental, and served as chairman of Continental Gummi from 1982 to 1987. He joined Daimler-Benz AG as director of the truck division in 1988, and from 1992 he was in charge of the corporation's planning, finance and purchasing. He became chairman of the Mercedes-Benz AG management board on January 1, 1993, succeeding Werner Niefer. He said Mercedes-Benz cars were over-engineered and proceeded to take some of the cost out of them. Some of the quality went out too, and the 1995 E-Class cars became known for engine problems, braking problems, and rust problems. He also argued that Mercedes-Benz could not grow by restricting itself to the three sizes of luxury cars that accounted for the bulk of its sales. He started new ventures in small and niche market products such as the A-Class, the M-Class, the MCC Smart, and the SLK sports models.

He resigned in January 1997 when Schrempp merged Mercedes-Benz AG into the Daimler-Benz AG parent organization.

Werner, William (1893–1975)

Technical director of Auto Union AG from 1935 to 1945; director of production, Auto Union GmbH from 1949 to 1962.

He was born on November 7, 1893, in New York City, to immigrant German parents from the Freiberg area of Saxony. He was barely 14 when his parents chose to return to the Vaterland. After high school, he joined Multigraph GmbH in Berlin as a mechanic while taking evening classes in mechanical engineering. After obtaining his degree, he held technical positions in machine-tool design and production methods with Bergmann-Borsig, Schuchart & Schütte, and Ludwig Loewe. He became chief engineer of Schiess AG, machine builders in Düsseldorf. In 1925 he joined Horch Werke AG at Zwickau as plant director, and was credited with raising the quality of Horch cars to a higher level. He was named production manager of Auto Union AG in 1932, with responsibility for the car factories building Audi, DKW, Wanderer and Horch cars. In 1935 he was promoted to technical director and took over J.S. Rasmussen's seat on the management board of Auto Union AG. That gave him full authority over product engineering as well as production, plus responsibility for the Auto Union Grand Prix racing cars. He remained at Zwickau throughout World War II and escaped to the West in 1945. He set up an independent consulting office in partnership with Gerhard Müller, and in 1949 joined the newly formed Auto Union GmbH in Ingolstadt, to plan and supervise the production of DKW motorcycles, cars, and commercial vehicles.

The Ingolstadt plant was bursting at the seams, with motorcycles, the Vespa scooter (under license from Piaggio SA) and delivery vans. No adjacent land was available for expansion. Werner leased a Rheinmetall-Borsig factory in Düsseldorf in 1950 and got DKW car production started. The Düsseldorf plant became Auto Union GmbH property in 1954. Production of the DKW Munga 4x4 military vehicle began at Düsseldorf in 1953, and in 1958 Auto Union GmbH began construction of a greenfield plant outside Ingolstadt, after selling the motorcycle branch to Nürnberger Zweirad Union. Werner laid out the new plant, organized the tooling and methods, and assembled the first DKW Junior there in August 1959. He retired in 1962 and died in 1975.

White, David McCall *see* McCall White, David

Whittaker, Derek (1930–?)

Managing director of British Leyland Motor Corporation's Body and Assembly Division from 1973 to 1975, managing director of Leyland Cars Division from 1975 to 1978.

He was born in 1930, the son of a Ford Motor Company Ltd. production manager and joined Briggs Motor Bodies as a trainee in 1946. When Ford took over the Briggs factory in 1954, he became controller of Ford's transmission and chassis division. Leaving Ford in 1967, he joined the General Electric Company and in 1969 he was named managing director of the London Electric Wire Group.

John Barber recruited him to British Leyland, along with several other ex–Ford men, and he became manager of the Oxford assembly and body works at Cowley, recently retooled to produce the Morris Marina. When he took command of the BL car division in 1975, Austin had been building the Allegro at Longbridge for nearly two years and was getting ready to produce the Princess. Rover's SD-1 (3500) was in the final stages of preparation. He invented the Jaguar Operating Committee to restore morale at Browns Lane and the Radford engine plant in the wake of Geoffrey Robinson's departure. He resigned in anger shortly after Michael Edwardes' taking office as chairman of British Leyland, leaving in January 1978.

Wiedeking, Wendelin (1952–)

Chairman of Dr. Ing. h.c. F. Porsche K-G beginning in September 1992.

He was born in 1952 and studied at universities in Rhineland-Westphalia. He first joined Porsche in 1983 as a production engineer, but left five years later to become chairman of Glyco Metallwerke. He then returned to Porsche in 1991 as board member with responsibility for production and materials. The company had made 55,000 cars in 1985 but produced only 14,000 in 1992 and was close to bankruptcy. In September 1992 he was named chairman of the management board, and within a month he had eliminated 850 production jobs. He hired Japanese consultants on production systems, Shen-Gijutsu led by ex-Toyota man Yoshiki Iwata to reorganize the factory layout, manufacturing methods and the assembly setup. Within two years he slashed production costs by 30 percent and lowered the breakeven point from 29,000 cars to 21,000. The company returned to profits at the end of 1995. In 1995–96 he authorized spending $835 million on design, development, and industrial investment for the Boxster and 996 (revised 911) and established Porsche Engineering Services GmbH at Bietigheim-Bissingen to handle contracts with outside clients. When the Zuffenhausen plant could not keep up with the order flow for the Boxster, he rented assembly capacity for the Boxster at a Valmet plant in Finland.

Wilding, James Armstrong (1870–?)

Technical director of Clement Talbot Ltd. from 1913 to 1919, technical director of Swift of Coventry Ltd. from 1919 to 1931.

He was born about 1870 at Warrington, Cheshire, left school at 16 and served an apprenticeship with William Muir & Co., a Manchester engineering firm. He graduated from Manchester Technical College. One of his first assignments in industry was to head the jig and tool department for Hiram S. Maxim, inventor of the machine gun mass-produced by Vickers. He also spent some time with a typewriter company. In 1898 he was responsible for the design of a battery-electric vehicle for A. Hall and was awarded a patent, as co-inventor with General Kelly-Kelly, of a motorized ambulance. Wilding was named technical director of Clement Talbot, Ltd. in 1913, when the company was producing a 15 hp and a 25 hp four-cylinder touring car. In 1914 he became responsible for converting the Ladbroke Grove, London, factory to war production. He resigned in 1919 when it became clear that the Earl of Shrewsbury and Talbot, the majority shareholder, had no intention of resuming car production.

He signed up with Swift of Coventry Ltd., then building cars at the Quinton Road Works and making bicycles at Cheylesmore. It became his task to transfer car production to the Cheylesmore Works. In 1920 Swift was offering three four-cylinder models of 10, 12 and 15 hp rating. The 15 was taken out of production in 1921, while the 12 was replaced by a new 12/25 for 1926 and the 10 was built through 1931. He designed a new 14/40 2-liter four-cylinder model for 1927. But production costs were too high for an output of less than 1000 cars a year, and the company was losing money. An 8 hp Cadet with an 847 cc Coventry Climax engine was introduced in September 1930, but had no impact in the market. The factory was closed in April 1931.

Wilfert, Karl (1907–1976)

Head of structural research, Daimler-Benz AG, body engineering section, from 1949 to 1954; director of body styling and research from 1954 to 1976.

He was born on July 1, 1907, in Vienna and attended schools in his home town. He graduated from the Technical Vocational Federal School of Vienna in 1926 and was promptly engaged by Steyr-Werke AG as a draftsman in the body engineering department. In 1929 he joined Daimler-Benz AG in its Vienna office and was transferred to Stuttgart in 1931. In 1934 he became an assistant to the chief body designer (Ahrens) and in 1936 was named head of testing for the body engineering section at Sindelfingen. He became a designer in 1940, but spent the next four years on aircraft-engine projects. From 1945 to 1949 he made a living in the Stuttgart area, doing temporary jobs, privately and for small business enterprises. He returned to Sindelfingen in 1949 and was immediately entrusted with greater responsibilities. He worked very closely with the inventor Bela Barenyi in the advanced-engineering group, and the two of them introduced countless innovations in the safety and occupant-protection area. They made Mercedes-Benz the world's leader in automotive safety. Wilfert did most of the body engineering for the 1960 220-series and its derivatives, the 300 SE and 230 SL. He was given a seat on the management board in 1956 and spent the rest of his career as the company's top authority on body structure development and the evolution of body shapes. In 1968 he designed the body for an experimental streamliner that became known as the C-111. In 1973 he was awarded an honorary doctorate in engineering by the Vienna Technical University.

He died from sudden heart failure on February 26, 1976.

Wilkes, Peter (1921–1972)

Technical director of Rover from 1964 to 1971.

He was born in 1921 in Birmingham and served his apprenticeship in the machine-tool industry from 1937 to 1940. He spent five years on active service with the Royal Air Force and joined the Rover engineering staff in 1946.

Privately, he created the Marauder sports car in 1950, in partnership with Spencer King and George Mackie. The Marauder made use of Rover powertrain and chassis components, which made it heavy and uncompetitive as a sports car. Only about 15 Marauders were completed.

In 1952 he joined J.W. Gethin, Ltd., in Birmingham as service manager, returning to Rover in 1954 as production manager for gas turbines. In 1956 he was named assistant to the chief engineer for passenger cars, working on the P5 (3-Litre) and P6 (2000). In 1967–70 he laid down the basic design for the proposed P10 (relabeled RT-1 in 1971) which was promptly shelved in favor of Spencer King's SD-1 design.

He retired at the end of July 1971 on his doctor's advice and died from heart failure in 1972.

Wilks, Maurice Fernand Cary (1904–1964)

Chief engineer of Rover from 1930 to 1956; technical director of Rover from 1956 to 1964.

He was born on August 19, 1904, on Hayling Island as a younger brother of Spencer B. Wilks. He attended Malvern College, graduated with a degree in mechanical engineering and went to Detroit to get inside experience of the American motor industry. He held engineering positions with General Motors from 1926 to 1928. On his return to Britain, he joined Hillman in Coventry as a planning engineer and was a member of the group that laid down the basic specifications for the 1931 Wizard and the first Minx.

One of his first projects at Rover was the little Scarab with its rear-mounted engine. It was not put in production. He created a new six-cylinder Pilot for 1932, prepared a much-improved Ten for 1934, and brought out a new Sixteen for 1937. His engineering was conservative, as all prewar models had rigid axles front and rear, mechanical brakes, and engines

set well back in the chassis. But he was fond of little luxuries, such as rheostat-controlled instrument lighting, free-wheeling, vacuum-assisted clutch pedal, and Bijur automatic chassis lubrication.

In 1943 he thought that Rover would need a small car for the postwar market, and prepared the M-Model two-passenger coupe. But the cars that went into production in 1945 were prewar models. Independent front suspension and brand-new F-head engines were adopted in 1948. The P3 of 1949 introduced a modern silhouette. He took personal charge of the Land Rover and supported the experimental gas turbine car projects. In 1952 Rover began building an optional diesel engine for the Land Rover. He adopted unit construction (with a front sub-frame) for his last design, the P5 (3-litre) introduced late in 1958, but his involvement in the P6 was purely supervisory. He died unexpectedly while vacationing on the isle of Anglesey on September 8, 1964.

Wilks, Spencer Bernau (1891–1971)

Managing director of the Hillman Motor Car Company, Ltd., from 1919 to 1928; managing director of the Rover Company from 1933 to 1962; chairman of the Rover Company from 1957 to 1962.

He was born at Rickmansworth in Hertfordshire on May 26, 1891, as the son of a tanner. After graduating from Charterhouse School, he studied law from 1909 to 1914, and served as a captain in the Royal Army from 1914 to 1918. He joined Hillman as joint managing director in 1919, his brother-in-law John Black holding the same title. Under their management, Hillman was strictly a one-model manufacturer, with an 11-hp car giving way to a 14-hp car in 1926. When Hillman came under the control of Rootes, Ltd., in 1928, he resigned, and joined Rover as works manager in 1929. Rover ran up considerable losses in 1930–31, and the management blamed Wilks for excessive production costs. With a chartered accountant at his side, he was able to demonstrate that the losses did not stem from the factory but from poor financial control by management. The Rover board appointed Wilks as managing director, and his accountant, H.E. Graham, was named finance director.

Wilks put a stop to Rover's economy-car experiments, restricted the model range to four- and six-cylinder family cars higher-priced than their direct rivals, cultivated a quality image for Rover, and kept the name out of racing.

He authorized wise spending on plant and tooling in 1935, and by the end of 1937, Rover reported record profits. During World War II Rover operated government owned Shadow factories at Solihull and Acocks Green, making aircraft engines and parts, while Rover's works at Tyseley and Coventry were converted to build airframes, wing structures, and military vehicle bodies. After the war, Wilks had the idea of a four-wheel-drive go-anywhere utility vehicle which went into production as the Land-Rover in 1948. He was involved in merger talks with Standard-Triumph, first in 1955 and again in 1959, but broke them off because he never found that terms were acceptable to Rover. He retired in 1962 to the whisky-producing Isle of Islay where he died in 1971.

Williams, Thomas Lawrence (1890–1964)

Chairman and managing director of Reliant Engineering Company (Tamworth) Ltd. from 1934 to 1963; chairman and managing director of Reliant Motor Company, Ltd., from 1963 to 1964.

He was born on March 15, 1890, at Tamworth in Staffordshire and educated at the Birmingham Technical Day School. He began his career as a trainee in the heavy steam-vehicle business, but joined the Triumph Cycle Company in Coventry in 1916 as a motorcycle designer. In 1924 he moved to Sheffield, Yorkshire, to join Dunford & Elliott, where he created a new range of Dunelt motorcycles. Six years later he went to Raleigh Cycle Company, Ltd., of Nottingham, as head of the motor department. He designed the Raleigh Safety Seven three-wheeled van, which had a steering wheel inside the cab but motorcycle-type fork suspension for the single front wheel. It was powered by a 600-cc air-cooled J.A.P.[1] V-twin driving the rear wheels.

When Raleigh decided to close the motor department in 1933, Williams resigned, purchasing the rights to the Safety Seven, returned to Tamworth, and started to build the vans in his back garden. In 1934 he took over the Twin Gate factory on Watling Street, Tamworth, and in a three-year period made some 3000 three-wheelers with the Reliant name plate.

A passenger-car version appeared in 1935, and in 1937 he adopted the 747-cc four-cylinder water-cooled Austin Seven engine. A year later, Reliant began producing its own version of the Austin engine. Though periodically updated, the same basic vehicle was produced up to 1950. Williams introduced the Regal three-wheeler in 1951, with an Austin A 30 engine and aluminum-panel body over an ash frame. Two years later, the Regal was redesigned with a plastic body.

The first four-wheeled Reliant was a 1958 station wagon named Sussita, developed for Autocars Limited of Haifa, Israel. It led to production of the Sabre sports coupé and convertible at Tamworth in 1960, both powered by a Ford Consul engine. The Sabre Six, with a Ford Zodiac engine, followed in 1961. The Robin three-wheeler replaced the Regal in 1962, with a new aluminum-block Reliant engine produced in a new plant at Shenstone. By 1964, Reliant was turning out 15,000 cars a year.

Wilsgaard, Jan Einar (1930–)

Chief designer of Volvo car bodies from 1966 to 1988.

He was born on January 23, 1930, in Brooklyn, New York, to Norwegian parents who later returned to their homeland. He started out as an interior architect, graduating from the Art Industry School in Gothenburg in 1950. Later that year, he joined Volvo in the body design department. He was one of many who worked on projects to modernize the PV 444, and he was the principal designer of the 1956 Amazon (later renamed 121-122S). He was given leave of absence to study industrial design and spent the years 1964–66 at the Art Center college of Design in Los Angeles. On his return to Sweden, he was named chief designer and styled the Volvo 244. The bigger the Volvo cars grew, the more intent the management became on a boxy look, and Wilsgaard found his only styling freedom in the details. He was responsible for the 164 in 1968, the 240 and 260 for 1975, and 1977 262 C which was produced by Bertone. He styled the 760 for 1982 and the 740 for 1984. He began the studies for the 940/960 series, but retired in 1988.

Winter, Bernard B. (1894–?)

Technical director of the Rootes Group from 1938 to 1959.

He was born in 1894 and had not completed his engineering studies when he was called up for military service in 1914. After his discharge in 1919 he was employed in the auto repair business until he joined Rootes Ltd. as chief service executive in 1923. In 1929 he was transferred to Coventry as service engineer with Humber Ltd., which was then building about 4,000 cars a year. In 1934 he was promoted to executive manager of the Humber-Hillman engineering department.

His responsibilities were administrative rather than creative, though he found opportunities for creativity when Sir William Rootes told him to coordinate the Humber, Hillman, Talbot and Sunbeam model ranges into a rational grouping. In this task he was seconded by Arthur G. Booth, formerly of Clyno, A.J.S. and Singer. Car production was discontinued in the Sunbeam works at Wolverhampton, which were converted to trolleybus production. A new Hillman Minx was introduced in October 1935, and its chassis was shared with the new Talbot Ten. The new Hillman 16 and Hawk were the same car with different engines, both side-valve sixes, the 2576 cc unit of the Hillman 16 being a smaller-bore version of the 2731 cc Humber 18 power unit. The old Humber 12 was phased out in 1937 and the Hillman 16 was rebadged Humber 16 for 1938. The 1938 Talbot 3-liter shared the Humber

Snipe chassis and body shell, but had a 3181 cc version of the Humber 18 engine, while the Snipe had been given a 4086 cc version for 1936.

Throughout World War II Winter was responsible for the Group's military vehicles, with an engineering staff that included A.C. Sampietro, A.C. Miller, W.T. Oliver, Donald M. Healey, Leslie Dyer, Benjamin G. Bowden and A.M. Kamper, all highly talented men. After the war, all Rootes Group passenger car production was combined in the former Shadow Factory at Ryton-on-Dunsmore, outside Coventry, and the Talbot factory in London became a service depot. The Group's number of body shells was whittled down to four, and the engine program streamlined to just two sizes of four-cylinder units and two sizes of sixes. When the Rootes Group acquired Singer Motors Ltd. in 1956, he competently accommodated new Singer models into the line-up and eliminated Singer's engines. He retired in 1959.

Wishart, Thomas D.

Design engineer of Crossley Motors, Ltd., from 1919 to 1939.

He joined Crossley Motors, Ltd., at the Napier Street, Gorton, Manchester, works in 1919, assisting A.W. Reeves and G. Hubert Woods in directing an engineering staff. He designed the four-cylinder T-head 3.7-liter engine for the 1920 19.6-hp model and its four-speed gearbox. The 19.6 and its 20/70 Sports model were produced until mid-year 1926. In 1921 he obtained a patent for combining inlet and exhaust manifolds into one single casting. He designed the first six-cylinder engine, the 2.7-liter 18/50 with overhead valves, in 1925, and engines for two companion models, the 1927 3.2-liter 10.9-hp and the 1928 15.7-hp 1991-cc "Shelsley." Crossley and Willys-Overland engineers started a joint project, the X-Car, in 1925, but the Americans backed out of it in 1929. Wishart took it over and turned it into the Crossley Ten, powered by an 1122-cc Coventry Climax engine, with Moss gearbox and an ENV rear axle. The company sold only just over 1000 Tens in its three-year production span. He introduced a 1½-liter engine for the Regis Sports in 1935. A.W. Reeves and G. Hubert Woods left the company when the production of Crossley cars was discontinued in 1937, but he stayed on until 1939. His last project was a 26/90 L-head six based on the 3.3-liter Studebaker Dictator.

Wiss, George (1868–1928)

Managing director of Bergmann's Industriewerke GmbH from 1893 to 1905, chairman of the Süddeutsche Automobil-Fabrik GmbH from 1905 to 1909.

He was born on August 6, 1868, at Kleischmalkalden in Thuringia and graduated from the Commercial College of Antwerp. He traveled widely, to England and the U.S.A. to gain business experience, and upon his return to Germany became a co-founder of Waggonfabrik Fuchs in Heidelberg. He joined Theodor Bergmann and played a part in his makeover of the Eisenwerke Gaggenau AG[1] in 1893 and the formation of its subsidiary, Bergmann's Industriewerke GmbH, to manufacture small steam engines, retail automats, and compressed-air guns and pistols. A year later he started production of the Orient Express car, designed by Joseph Vollmer. By 1902 the Orient Express was obsolete and the sales curve dropped. He introduced new cars with the Bergmann nameplate, front-mounted engines and chain drive, and business revived.

In 1905 he bought the automobile department from Theodor Bergmann and reorganized it as the Süddeutsche Automobil-Fabrik GmbH. The single-cylinder 600 cc Liliput, designed by Willi Seck, was built from 1904 to 1907. Bigger models (12, 18, 22, 24, and 40 hp) were sold with an SAF badge. A 60 hp 9-liter four-cylinder overhead-camshaft model was added in 1909. The Gaggenau plant had also built trucks under contract from Benz & Cie since 1907 and in 1909 Benz & Cie bought the company from Wiss. It was reorganized as Benz-Werke Gaggenau GmbH and Wiss was elected to the Benz & Cie board of directors. He remained active in the company's affairs until the merger with Daimler Motoren Gesellschaft in 1926, and died in 1928.

1. Originally established in 1680 by the Markgraf of Baden with a drop-forge in Gaggenau and a nail-smithy at Mittelbach, later reorganized as the Murgtaler Eisenwerke and in 1888 as Eisenwerke Gaggenau AG which came under Bergmann's control in 1889.

Wolff, Ernst (1868–?)

Technical director of NAG[1] from 1912 to 1926, chairman of Pluto Automobil-Fabrik in 1927–28.

He was born on August 30, 1868, in Berlin and educated as an engineer. He began his career with Krupp in Essen and later joined AEG[2] in Berlin. In 1912 he was transferred to an AEG subsidiary which had been building NAG automobiles since 1900, NAG trucks since 1904, and aircraft engines (under Wright license) since 1907. In 1913 the NAG factory at Berlin-Oberschöneweide was offering a range of five models from 6/18 to 33/75 PS, plus electric vehicles and taxicabs. Annual production was about 2000 cars.

During World War I the factory built NAG trucks and Benz aircraft engines, and new buildings were erected in 1917. He directed the reconversion of the plants to production for civilian markets in 1919, adopting a one-model program for the car side with the 2-liter four-cylinder C4, but broadening NAG's range of commercial vehicles, to the extent of adding a three-axle truck in 1924, and taking up the production of semi-trailers. In 1926 NAG bought Protos Automobil-Werke from Siemens-Schuckert and a year later took over Presto-Werke AG in Chemnitz and Dux Automobilwerke AG in Leipzig-Wahren. He resigned in 1926 and purchased a factory at Zella-Mehlis from Ehrhardt-Werke, which he retooled for series-production of the Pluto 4/20 and 5/30 light cars, both under license from Amilcar. But the business failed within a year.

1. Neue Automobil-Gesellschaft mbH evolved into NAG Aktiengesellschaft in 1912 and was renamed Nationale Automobil-Gesellschaft AG in 1915.

2. Allgemeine Elektrizitäts-Gesellschaft = General Electric Company.

Woodwark, Christopher John Stuart "Chris" (1946–)

Managing director of Cosworth Engineering from 1993 to 1994; chairman of Rolls-Royce Motor Cars from 1995 to 1998.

He was born on October 9, 1946, and educated at Felsted School. He studied economics and was given a diploma in marketing. He continued at business school until he joined the Rootes Group as a management trainee in 1964. He worked in sales and marketing for Chrysler–UK from 1967 to 1971, and then joined British Leyland in international marketing. He worked in France and the Benelux countries in 1974–5 and then in Africa, from Kenya to Zimbabwe, from 1975 to 1982. On his return to Britain, he was assigned to Leyland Trucks. He worked on Land Rover exports in 1985–86 and served as sales director of Austin Rover from 1986 to 1993. He was recruited by Vickers to run Cosworth Engineering in 1993 and in January 1995 was chosen to succeed Peter Ward as chairman of Rolls-Royce Motor Cars, as Ward left in protest over the agreement to install BMW engines in Rolls-Royce and Bentley cars. Woodwark took the manufacturing contract for Rolls-Royce body shells away from Rover (also under BMW ownership) and gave it to the Mayflower Group. The Mayflower Group tried to buy control of Vickers in November 1997, a time when both Volkswagen and BMW were bidding for Rolls-Royce Motor Cars. The Mayflower bid failed, and Woodwark left Vickers in January 1998. In May 1998 he became chief executive officer of Stavely Industries (machine tools) but his strategy led to the dismemberment of the group. In 1999, he linked up with Jon Moulton of Alchemy Partners, who was trying to buy the MG name when BMW put Rover up for sale. He was named chief executive of MG Cars in March 2000, an appointment that was cancelled when BMW sold MG–Rover to the Phoenix Consortium in May 2000.

Woollard, Frank Griffiths (1881–1956)

Works manager of Morris Motors (Engines Branch) Ltd. from 1924 to 1926, general production manager and a member of the board of Morris Motors Ltd. from 1926 to 1931.

He was born in 1881 in London into a family from Badmondisfield Hall, near Newmarket. He was educated at Goldsmith's and Birkbeck College, served an apprenticeship with the London & South Western Railway and graduated from the City & Guilds engineering college in South Kensington, London.

In 1910 he joined E.G. Wrigley & Company Ltd. gearbox and axle manufacturers of Soho, Birmingham, where he rose to the position of chief draftsman. In 1912 Wrigley got the first of many orders from WRM Motors Ltd. Woollard became chief engineer of the company and later, assistant managing director. In 1923, E.G. Wrigley & Company Ltd. was taken over by Morris Motors Ltd. and reorganized as Morris Commercial Cars Ltd., producing vans and light-duty trucks. Woollard was transferred to the ex–Hotchkiss engine plant in Coventry, which had been taken over by Morris Motors Ltd. in 1923 and incorporated as Morris Motors (Engines Branch) Ltd. He introduced new methods of flow-production, raised capacity and reduced unit cost. In 1926 W.R. Morris put him in charge of all production plants. He raised efficiency and boosted output, with significant reductions in unit cost. But in 1931 he quarreled with W.R. Morris and left in protest. He became managing director of Steele Griffiths & Company, Austin distributors in London, and later joined Rudge-Whitworth as managing director.

After World War II he became a director of Birmingham Aluminum Castings (1903) Ltd. which evolved into the Birmid Industries Group. He also served as chairman of the executive committee of the Council of Zinc Alloy Die Casters Association and a private consultant on automation and materials flow.

Wright, Geoffrey P. (1936–)

Head of product development Volvo Car BV in 1980–81, executive vice-president Volvo Car BV from 1981 to 1992.

He was born on March 21, 1936, in the Midlands. He studied at Coventry Technical College and the Lanchester College of Engineering, graduating in 1963 with a degree in mechanical engineering. He began his career with the Rootes Group in 1957 and designed gearboxes and chassis components from 1961 to 1966. He was named project manager for Sunbeam Tiger and Alpine V production in 1966, and served as manager of safety and reliability for Chrysler UK from 1968 to 1970. A year later he was promoted to head of Chrysler UK engineering operations and technical director, development and production engineering, cars and trucks. When Peugeot SA took over Chrysler UK, he became director for light and medium trucks, PSA Europe.

He left Peugeot SA in 1980 to join Volvo at the Born plant in the Netherlands. He designed the Volvo 480 Aerodeck, pushed the 480 Turbo and the 330 into production, and supervised the 460 program. He resigned after Mitsubishi Motors Corporation took a one-third stake in the company.

Wyer, John Leonard (1909–1989)

Technical director of Aston Martin from 1955 to 1956; general manager of Aston Martin from 1957 to 1963.

He was born on December 11, 1909, at Kidderminster. In 1927 he joined Sunbeam as an apprentice engineer and at the end of a six-year term at Wolverhampton, he went to Solex carburetors in 1933 as a technician. During World War II he was engaged in production-engineering for Solex. In 1947 he became a director of Monaco Engineering, dealers in high-performance cars with a racing department. He left when Monaco was taken over by Spurling in 1950 and accepted a six-month contract with David Brown as racing manager of Aston Martin. David Brown put him on the permanent staff in August 1950, and from 1951 to 1955 he served as a development engineer as well as being head of the racing team.

Seeing the limitations of the DB 2 series, he helped formulate the specifications for the DB 4 and directed its testing and development. He planned and supervised the development of its derivatives, DB 5, DB 6 and DBS. Leaving Aston Martin, he became associated with Ford's GT-40 racing program in 1963 and led it until the end of 1966.

He formed a partnership with John Willment to take over the racing department of Ford Advanced Vehicles and merge it with John Willment Racing, Ltd., forming J.W. Automotive Engineering Ltd. on January 1, 1967. Wyer served as its general manager and racing manager, campaigning for the "left-over" Ford GT 40s and improving them. By April 1967 they were no longer Ford but Mirage cars. Sponsored by Gulf Corporation of Pittsburgh, Pennsylvania, the JW Mirage team won at Le Mans in 1968 and 1969 with supposedly outdated cars. The JW team then acquired Porsche 917 cars but never achieved the same success. John Wyer retired at the end of the 1975 season and retired to Menton on the French Riviera. He subsequently moved to Scottsdale, Arizona, where he died on April 8, 1989.

Zahn, Joachim (1914–)

Finance director of Daimler-Benz AG from 1958 to 1965; spokesman for the Daimler-Benz management board from 1965 to 1971; chairman of the management board from 1971 until the end of 1979.

He was born on January 24, 1914, in Wuppertal as the son of a lawyer, and studied law at the universities of Tübingen, Cologne and Königsberg. After graduation, he went into government service and by 1945 held the title of Councillor.[1] From 1947 to 1957 he was a director of a government-licensed organization of trustees for industrial property.[2] He ran a cellulose plant for a private company[3] in Hessen from 1957 to 1958 when he moved to Stuttgart at the insistence of Hermann Abs, chairman of the Deutsche Bank, then a big shareholder in Daimler-Benz AG. By the end of 1958, Daimler-Benz chairman Fritz Kö-necke had stopped worrying about finances. As a result of his accounting techniques, the bottom line had a way of satisfying all interests. Under his rule, Daimler-Benz AG enjoyed a stable growth rate of at least 10 percent a year. The company became Germany's third-biggest industrial enterprise. He concentrated the resources on passenger-car production, which increased by 126 percent from 1965 to 1978. He retired at the end of 1979.

1. Regierungsrat.
2. Deutsche Treuhandgesellschaft.
3. Aschaffenburger Zellstoffwerken.

Zerbi, Tranquillo (1891–1939)

Technical director of Fiat from 1928 to 1939.

He was born in 1891 in Saronno, just north of Milan, and educated in Swiss and German schools. In 1912 he graduated from the Mannheim Technical University and proceeded to gain experience in engine design and development in the works of Sulzer Frères in Winterthur and Franco Tosi in Legnano. He joined Fiat as an engine designer in 1919. His first complete engine design for Fiat was the Type 404, a 112-hp 1991-cc twin-cam in-line six for the 804 racing car of 1922. He spent the next several years in the aircraft engine design office, and in 1928 succeeded Carlo Cavalli in the highest technical office in the Fiat organization.

The first models created under his direction, the 1932 Belilla and Ardita, were of conventional engineering, but he saw the need for progress and encouraged advanced ideas. He embraced the styling innovations of Mario Revelli de Beaumont and the frame-construction and suspension patents of André Dubonnet. He involved the best brains in the aircraft engine department (Fessia and Giacosa) in the planning of new cars. At his orders, Fiat also began testing the semi-automatic Cotal transmission in 1936 and automatic Hayes transmission in 1937.

He personally led the design of the Fiat AS 2 and AS 3 aircraft engines which powered the planes that won the Schneider Cup in 1926

and 1927, and later the A 22, A 22 T, A 22 R, Type 50 and the remarkable AS 6 which combined two V-12 engines in such a way as to turn two coaxial propellers in opposite directions.

He died unexpectedly in 1939.

Zerbst, Fritz (1891–1958)

Chief engineer of DKW from 1934 to 1943; technical director of Auto Union GmbH in Ingolstadt from 1949 to 1956.

He was born on August 18, 1891, at Guesen in West Prussia, as the son of a landowner. Having little inclination to bookish learning, he trained as a mechanic's apprentice in Breslau and followed up by volunteering for unpaid practice in the shops of the Adler, Opel, Mercedes and NAG factories. He eventually earned his engineering degree by taking evening classes at the technikum in Berlin, and began his paid career by joining NAG as a test engineer in 1913.

He also drove NAG cars in rallies and races, and in 1919 went to the United States for a long study tour, with stints of working in the Packard, Ford and Lincoln factories. Returning to Germany in 1924, he joined Horch at Zwickau, assigned to the drawing office and working under Paul Daimler, and later Fritz Fiedler. He was transferred to DKW in 1934 and directed the development of both the front-wheel-drive and the rear-wheel-drive models up to 1943. In 1945 he was placed in charge of rebuilding the Horch plant at Zwickau, but escaped to the West. He prepared drawings for a light utility vehicle with a two-stroke engine and front-wheel drive, which became the DKW F 89 L Schnellaster, going into production at Ingolstadt in August 1949.

He developed the 1950 F-89 Meisterklasse sedan, powered by a 700-cc parallel-twin two-stroke engine, and the 1953 F-91 Sonderklasse sedan, with a new 750-cc three-cylinder two-stroke power unit. He followed up with the F-93, SP-1000 sports coupé and prepared the F-94 before his retirement in 1956. He died on January 7, 1958.

Zetsche, Dieter (1953–)

Chief engineer, vehicle engineering and development for Mercedes-Benz passenger cars from 1992 to 1995.

He was born on May 5, 1953, in Istanbul, Turkey, to German parents. He attended high school in Frankfurt and obtained a master's degree in electrical engineering from the University of Karlsruhe in 1976. He began his career with Daimler-Benz AG in the research department in 1976 and in 1981 became a development assistant in the commercial-vehicle branch. In 1982 he presented a thesis which won him a doctor's degree in mechanical engineering from the Paderborn Technical University and two years later he was placed in charge of truck development for the overseas branch factories. In 1986 he became senior staff manager for off-road vehicle design and development and in 1987 went to South America as chief development engineer for trucks and buses with Mercedes-Benz do Brasil. In 1988 he was transferred from the technical ranks to the management side and in 1989 became chief executive officer of Mercedes-Benz Argentina. He was president of Freightliner Corporation in Portland, Oregon, in 1991–92.

On his return to Stuttgart, he was assigned to passenger cars. His contribution was not so much in design concepts or engineering innovation as in streamlining the process of putting a new model on the market. The mid-range W-124 launched in 1984 had taken 54 months from the first stroke on the drawing board to the first production model off the line. He speeded up the development of the W-210 (E-Class) cutting its time-to-market to 38 months. It went into production at Sindelfingen and Rastatt in 1996.

Zetsche returned to truck development engineering in 1995 and then worked as a truck sales executive until he was named president of Chrysler in Detroit in November 2000.

Index

A. Darracq & Cie *see* Darracq
A. Darracq & Co. [1905] Ltd.
  see Darracq
A. Harper Sons & Bean Ltd.
  (ABC Motors [1920] Ltd.) 47
A. Horch & Co. *see* Horch
AB Volvo *see* Volvo
Abadal 141
Abarth, Carlo 5
Abarth (Abarth & Co) 5, 36, 69
Abarth & Co. *see* Abarth
ABC (Aeroplane Engine Com-
  pany, All-British Engine Com-
  pany, ABC Road Motors Ltd.)
  47
ABC Motors [1920] Ltd. *see* A.
  Harper Sons & Bean Ltd.
ABC Road Motors Ltd. *see*
  ABC
AC Cars Ltd. 73
Accles, James George 5
Accles & Pollock Ltd. *see*
  Accles-Turrell
Accles & Shelvoke *see* Accles-
  Turrell
Accles Limited *see* Accles-
  Turrell
Accles-Turrell (Accles Limited,
  Accles & Pollock Ltd., Accles
  & Shelvoke Accles-Turrell Au-
  tocars Ltd., Grenfell & Accles)
  5, 6
Accles-Turrell Autocars Ltd. *see*
  Accles-Turrell
AC-Delco General Motors 102
Ace (Baguley Cars Ltd., Salmon
  Motor Co.) 16
ACMA *see* Renault
Adam Opel AG 210
Adam Opel AG *see* Opel
Adams, W.L. 47
Ader (Société Ader) 6
Ader, Clément 6
Adler-Werke AG 18, 84, 121, 160
ADO-88 101

AEG (Allgemeine Elektrizitäts-
  Gesellschaft, General Electrical
  Company) 49, 52, 119
Aeroplane Engine Company *see*
  ABC
AF-Grégoire 242
AFM 109
AFN Ltd. (GN) *see* Frazer Nash
AGA (Aktien-Gesellschaft für
  Automobilbau, Berlin-
  Lichtenberg) 114
Agnelli, Gianni 7
Agnelli, Giovanni 6, 12, 27, 28,
  219
Agnelli, Umberto 7, 8
Ahrens, Hermann 8
Ainsworth, Henry Mann 9, 29
Akar, Emile 215
Aktiengesellschaft für Motor-und
  Motor-fahrzenghau vorm
  Cudell & Co. *see* Cudell &
  Co.
Albany Carriage Company (Park
  Ward, H.J. Mulliner) 46
Albion Motor Car Company
  Ltd. 38, 200
Alda (Automobiles Alda) 63, 64
Alden, John 9
Aldington, Harold J. 10
Aldington, William H. 10
Alessio, Antonio 176, 253
A.L.F.A. (Anonima Lombarda
  Fabbrica Automobili, Alfa
  Romeo) 210
Alfa Corse *see* Alfa Romeo
Alfa Lancia Industriale SpA *see*
  Alfa Romeo
Alfa Romeo (Alfa Romeo SpA,
  Alfa Lancia Industriale SpA,
  Alfa Corse, ARNA) 7, 12, 23,
  31, 36, 56, 58, 62, 69, 70, 78,
  87, 89, 90, 94, 202, 209, 240,
  253
Alfa Romeo SpA *see* Alfa
  Romeo

Alfieri, Giulio 10, 78
Alford & Alder 89
All-British Car Co. 208
All-British Engine Company *see*
  ABC
Allard (Allard Motor Co. Ltd.)
  11
Allard, Denis 11
Allard, Sydney Herbert 11
Alldays & Onions *see* Enfield-
  Allday
Allemano 5
Allen, Michael 62
Allen, Nigel 62
Allen-Bowden Ltd. 46
Allgemeine Elektrizitäts-
  Gesellschaft *see* AEG
Allmers, Robert Anton 11, 12
Alpine (Burkhard Bovensiepen
  K-G) 64, 155
Aluminium Company of Amer-
  ica (formerly Alcoa) *see* Alu-
  minium Manufacturers Ltd.
Aluminium Français 65
Aluminium Manufacturers Ltd.
  (Aluminium Company of
  America/Alcoa) 245
Alvis (Alvis Cars, Alvis Car &
  Engineering Co. Ltd., Alvis
  Ltd.) 39, 96, 157, 161, 216,
  231
Alvis Car & Engineering Co.
  Ltd. *see* Alvis
Alvis Cars *see* Alvis
Alvis Ltd. *see* Alvis
Alzati, Eugenio 12
American Automotive Company
  see Berliet
American Motor Car Company
  (Central Motor Car Company,
  Perry, Thornton & Schreiber)
  136, 234
American Nash 110
Amigo 71
Amilcar 103, 129, 215

André Citroën Ingénieur-
Constructeur *see* Citroën
Andreau, Jean 12, 204
Andren, Bertil T. 13
ANF 110
Ansaldi, Cavaliere Michel 61
Ansaldo 61
Antem 42, 204
Anthony Crook Motors Ltd. 73
Anzani Engine Company Ltd.
17
Appleby, William Victor 13
Aquila Italiana 58
Ardie-Ganz 121
Ardie Motorenwerk AG 120
Argus Motorenwerke, Berlin
223
Argyll Motors Ltd. 57, 234
Ariès 211
Armstrong-Siddeley (Armstrong-
Whitworth) 25, 47, 76, 232
Armstrong-Whitworth *see* Arm-
strong-Siddeley
ARNA *see* Alfa Romeo
Arnott & Harrison Ltd. 59
Arrol, Sir William 38
Arrol-Johnston (Arrol-Johnston
Ltd., New Arrol-Johnston Co.
Ltd.) 33, 51, 208, 231, 231
Arrol-Johnston Ltd. *see* Arrol-
Johnston
Artaut, Ernest 37, 96
ASA (Autocostruzioni S.p.A)
37
Ashcroft, Timothy 127
Associated Equipment Co. Ltd.
see Daimler
Aster 61, 64
Aston Martin (Aston Martin
Ltd., Aston Martin Lagonda
Ltd., Company Developments
Ltd., David Brown & Sons
[Huddersfield] Ltd., Lagonda
Motors Ltd.) 14, 15, 20, 21,
30, 32, 33, 41, 50, 51, 79, 98,
99, 199, 201, 224
Aston Martin Ltd. *see* Aston
Martin and David Brown &
Sons [Huddersfield] Ltd.
Aston-Martin Lagonda Ltd. *see*
Aston Martin, David Brown
& Sons [Huddersfield] Ltd.,
Lagonda Motors Ltd., and
Company Developments Ltd.
Atcherley 50
Ateliers de La Fournaise 33
Ateliers M.A. Julien 164
Ateliers Schneider 50
ATS (Automobili di Turismo a
Sport S.p.A) 37
Audi (Audi AG., Auto Union

GmbH, Audi NSU Auto
Union AG 19, 22, 23, 52, 86,
95, 240, 250
Audi AG *see* Audi and Auto
Union
Audi NSU Auto Union AG *see*
Audi
Audibert & Lavirotte 87
Audibert, Lavirotte & Cie *see*
Berliet
August Horch Automobilwerke
see Horch
Austin (Austin Motor Company
Ltd., Austin-Rover, Rover-
Italia, Austin-Morris) 14, 15,
16, 18, 19, 31, 51, 63, 78, 79,
100, 199, 217, 223, 235, 247
Austin, Sir Herbert 14, 15, 41,
51, 100, 247
Austin-Healey 103
Austin-Morris *see* Austin
Austin Motor Company Ltd. *see*
Austin
Austin-Rover Group *see* Austin
Austro-Daimler 160, 165, 169
Austro-Fiat *see* Fiat
Autem 42
Auto Avia 202
Auto Project 71
Auto Union (Auto Union AG,
Audi, Comdoit, Comotor,
DKW, Horch, NSU, Wan-
derer) 8, 27, 36, 41, 52, 77,
78, 98, 137, 211, 212, 249
Auto Union AG *see* Auto Union
Auto Union GmbH *see* Audi
Autoar SA Industrial, Commer-
cial, Financiera y Mobiliara
98
Auto-Avio Construzioni Ferrari
SpA *see* Ferrari
Autobloc *see* Claveau
Autocostruzioni S.p.A. *see* ASA
Autocrat 9
Automobil fabrik Komnick *see*
Komnick
Automobile Club of Great
Britain & Ireland (later Royal
Automobile Club) 162
Automobile Lamborghini *see*
Volkswagen
Automobile Palace Ltd. 118
Automobiles Alda *see* Alda
Automobiles Bignan *see* Bignan
& Picker [Engines]
Automobiles Bugatti 224, 241
Automobiles Causan 59, 60
Automobiles Cottin & Des-
gouttes (Société d'Automobiles
Pierre Desgouttes & Cie) 87
Automobiles Delage *see* Delage

Automobiles E. Bugatti *see*
Bugatti
Automobiles et Cycles Peugeot
see Peugeot
Automobiles Georges Irat SA *see*
Georges Irat
Automobiles Hotchkiss *see*
Hotchkiss
Automobiles Impéria 211
Automobiles La Licorne *see*
Bugatti
Automobiles Michel Irat SA
155
Automobiles Mors *see* Mors
Automobiles Peugeot *see* Peu-
geot
Automobiles Talbot *see* Sun-
beam
Automobiles Zédel of Pontarlier
see Donnet Zédel
Automobil-fabrik Komnick AG
see Komnick
Automobilfabrik Zella Mehlis
GmbH 102
Automobili di Turismo a Sport
S.p.A *see* ATS
Automobili Lamborghini *see*
Lamborghini
Automobil-Technischer Kalender
56
Automobil-Technisches Hand-
buch 55
Automotive Products 63, 23
Autosports Products *see* De
Tomaso
Autostar, Livorno (Bizzarrini
SpA., Prototipi Bizzarrini SpA)
36
Autovaz Lada 212
Aveling-Barford 101
Axe, Royden 5

Babs (speed record car) 232
Bache, David 15
Baguley, Ernest Edward 16, 91
Baguley Cars Ltd. 16
Bahnsen Uwe 16, 174
Bailey, Claude Walter Lionel 17
Bailey, Ernest 154
Ballot, Ernest 237
Bamford, J.C. 71
Bamford, Robert 201
Bamford & Martin Ltd. 201
Band, C.J. 118
Banque, Charles Victor 64
Banque Lazard 66
Barbarou, Marius J.B. 17, 26,
106
Barber, John Norman Romney
18, 231
Barenyi, Bela 18

Barker & Co. (coachbuilders) 20, 107
Barry, Sir John Wolfe 205
Bartlett, Charles John 19
Bashford, Gordon 19
Basshuyson, Richard 19
Bastow, Donald 20
Baur, Karl 36
Bazzi, Liugi 69
Beach, Harold 20
Bean Industries of Tipton, Birmingham 13, 67, 89
Beardmore, Sir William 51
Beardmore Aero Engines 47
Becchia, Walter 21
Beckett, Terence "Terry" Norman 21
Bedford 19
Bell, F.R. 168
Bellanger, Robert 22, 23
Bellentani, Vittorio 23, 69
Belsize-Bradshaw see Belsize Motors Ltd.
Belsize Motors Ltd. (Belsize-Bradshaw) 7, 47
Bendix, Vincent 234
Bendix Brake Corporation 234
Benelli 88
Bensinger, Jörg 23
Bensinger, W.D. 23, 24
Benteler, Helmut 146
Bentley (Bentley Motors Ltd.) 15, 24, 25, 39, 79, 241
Bentley, Horace 24
Bentley, W.O. 20, 24, 25
Bentley Motors Ltd. see Bentley
Benz (Benz & Cie) (Benz & Cie, Rheinische Gas Motorenfabrik, Carl-Benz Söhne) 6, 17, 25, 26
Benz, Carl Friedrich Michael 25
Benz, Eugen see Carl-Benz Söhne
Benz, Richard see Carl-Benz Söhne
Benz & Cie see Benz
Benz & Cie, Rheinische Gas Motorenfabrik see Benz
Bercot, Pierre 26, 27, 220, 227, 251
Beretta (Fabbrica d'Armi Pietro Beretta, Gardons Val Trompia) 37
Berge, Ernst 217
Berger, Dr. Arthur 221
Bergmann & Co. see Bergmann
Bergmann, Sigmund 27
Bergmann Elektrizitäts-Werke AG (Bergmann & Co., Bergmann Elektromotor und Dynamo Werke AG,

Bergmann Métallurgique, Deutsche Métallurgique Gesellschaft) 27
Bergmann Elektromotor und Dynamo Werke AG see Bergmann
Bergmann Métallurgique see Bergmann
Berliet (American Automotive Company, Audibert, Lavirotte & Cie, SA des Automobiles Marius Berliet 27, 28, 94, 252
Berliet, Marius Maximin François Joseph 27, 75, 87
Bernard, Marius 96
Bernardi, Enrico Zeno 28, 59
Berriman, Algernon Edward 29
Bertarione Vincent 21, 29, 67
Bertelli, Augustus Cesare 30, 143
Berthetto, Adrien 30
Berthon, Peter 73
Bertodo, Roland 31
Bertone, Giuseppe "Nuccio" 31
Besse, Georges 31
Bettman, Siegfried 32, 118, 231
Bez, Ulrich 32
Biada, Elizalde y Cia see Elizalde
Bianchi, Edouardo 210, 209
Bielefelder Maschinenfabrik AG vorm. Dürkopp & Co. see Dürkopp
Biggs, Theodore James 33, 51, 252
Bignan, Jacques 33, 79
Bignan & Picker [engines]
Bilbie, Barrie 140
Binks, Charles 34
Binks Carburettors see Charles Binks Ltd.
Bionier, Louis 34
Bira, Prince 105
Birfield Engineering 20
Birkigt, Marc 35
Birkin, Sir Henry 105
Birmingham Small Arms Company Ltd. see BSA
Biscasetti di Ruffia, Roberto 6
Bitter, Erich 35
Bitter & Co. Automobile GmbH see Bitter Automobile Company
Bitter & Co. Automobile GmbH of Schelm see Bitter Automobile Company
Bitter Automobile Company (Bitter & Co. Automobile GmbH, Bitter & Co. Automobile GmbH of Schelm, Bitter Automobile GmbH & Co.

K-G., Bitter Automobiles of America, Rallye Bitter) 35, 36
Bitter Automobile GmbH & Co. K-G see Bitter Automobile Company
Bitter Automobiles of America see Bitter Automobile Company
Bizzarrini, Giotto 36
Bizzarrini SpA see Autostar, Livorno
BL Light Machine Division see BMC
Black, John (Captain) 144
Black, Sir John 37, 89
Blackburne 117
Blackwood Murray, Thomas 38
Blatchley, John 38, 107
Blériot 110
Blitzkarren 43
Blum, Charles 180
BMC (British Leyland 214, 228, 230, 235, 231, 235
BMC (British Motor Corporation, British Leyland, BL Light Machine Division, Leyland Motors, Leyland-Innocenti SpA., Rover Group) 14, 15, 16, 18, 71, 81, 88, 89, 96, 100, 197, 200, 232, 235
BMS (Brescia Motor Sport) 78
BMW (Baverian Motor Works [Bayerische Motorenen Werke GmbH], Brandenburg Motor Works, Carl F.W. Borgward GmbH, Goliath Werke Borgward & Co. GmbH, Hansa Lloyd & Goliath Werke Borgward & Tecklenborg GmbH, Hansa-Lloyd-Goliath Werke Carl F.W. Borgward, Goliath Werke GmbH, Lloyd Machinenfabrik GmbH [Lloyd Motoren Werke GmbH], Borgward-Hansa) 10, 12, 22, 41, 42, 43, 44, 47, 48, 222, 241, 243, 246, 247, 252
BMW (Baverian Motor Works) see Borgward
Boano, Giampolo 87
Böhler Fidelis 245
Boillot, Georges 239
Boillot, Jean 39
Bollée (Amedee Bollée of Le Mans, Usines Bollée, Usines Léon Bollée, Etablissements Léon Bollée) 39, 40, 41, 57, 82
Bollée, Amédée, Jr. 39, 82, 237
Bollée, Amédée, Sr. 40
Bollée, Léon 40, 41, 57, 80

Böning, Alfred 41
Bonisson 110
Bonnet, René 27, 41, 42, 89, 131, 174
Bono, Gandenzio 42, 152, 213
Bönsch, Helmut Werner 134
Booth, A.G. 55
Bordino, Petro 219
Borgward, C.F.W. 12, 212
Borgward-Hansa see BMW
Borisch, Helmut Warner 43
Bosch 44, 248
Bott Helmut 23, 33, 44, 48
Boulanger, Pierre Jules 26, 44, 251
Bouton, Georges-Thadée 45, 83
Bowden, Benjamin G. 46, 139
Boyle, Robert W. 19, 46
Bradshaw, Granville Eastwood 47
Braess, Hans Hermann 47, 48
Brandenburg Motor Works see BMW
Branitzky, Heinz 48
Braq, Paul 46
Brasier, Charles Henry 48
Brasier, Henri 215
Brasier, Richard 49, 94
Breda, Ernesto 58
Breitschwerdt, Werner 49, 150, 222
Bremer Kühlefabrik Borgward & Co 43
Brennabor 8
Brillié, Eugène 126
Briscoe 22
Brissonneaux & Lotz 47
Bristol (Bristol Aeroplane Co., Bristol Cars Ltd., Bristol-Siddeley Engines) 10, 20, 31, 73, 76, 233
Bristol Aeroplane Co. see Bristol
Bristol Cars Ltd. see Bristol
Bristol-Siddeley Engines see Bristol
British Airship R-100 55
British Electric Traction Company 73
British Leyland see BMC
British Motor Corporation see BMC
British Motor Holdings 136
British Motor Syndicate 64
Brixia-Züst 120
BRM (Rubery Owen) 216
Broadley, Eric 216
Bronotore 69
Brough, George 50
Brough Superior Cars 50
Brown, David 25, 50

Brown, George W.A. 51, 130, 252
Brown, R.J. "Tom" 51, 76
Brown, Bouverie & Cie 250
Brueder, Antoine 26, 251
Brun, Richard 52
Brune, L. Prideaux 30
Brush Engineering Company 73
BSA (Birmingham Small Arms Co., Ltd.) see Daimler
BSA (Birmingham Small Arms Company Ltd., BSA Cycles Ltd., BSA Group Research, Daimler Motor Co., Coventry, Lanchester Motor Company Ltd.) 9, 16, 20, 29, 47, 51, 52, 55, 90, 91
BSA Cycles Ltd. see BSA
BSÁ Group Research see BSA
Bucciali 60
Buckland Body Works Ltd. 153
Budge, John 118
Bugatti (Automobiles E. Bugatti, Automobiles La Licorne, Bugatti-Gulinelli Motor Co.) 52, 53, 54, 58, 60, 69, 241
Bugatti, Ettore Arco Isidore 53, 82, 203, 239
Bugatti, Gianoberto "Jean" Carlo Rembrandt Ettore 52
Bugatti Automobiles SAS see Volkswagen
Bugatti-Gulinelli Motor Co. see Bugatti
Buick (General Motors) 19
Bükken, Curt 118
Bullock, William Edward 54
Bureau d'Etudes et de Recherches Exloratoires (Berex) see Renault
Bureau d'Etudes et Recherches Techniques 66
Burney, Sir Charles Dennistoun 55
Burt McCollum Company 30
Burzi, Dick 15
Bussien, Richard 55
Büssing 141, 172
Busso, Giuseppe 56, 121
BW 228

Caccamo, Paolo 31
Cadillac 208
Caesar, Richard D. 157
Cahen, Emile 215
Calcott Bros., Ltd. see Sparkbrook Manufacturing Co. Ltd.
Calthorpe (Calthorpe Motors Company Ltd., G.W. Hands Motor Company Ltd) 135

Calthorpe Motors Company Ltd. see Calthorpe
Calvet, Jacques Yves Jean 39, 56, 115, 230
Cam Gears Ltd. 63
Campbell, Sir Malcolm 80
Camuffo, Sergio 90
Camusat, Maurice 57
Canaday, Ward M. 9
Canavese, Giovanni 233
Canham, Reginald 11
Cantavella, Paolo 57
Cappa, Guilio Cesare 21, 58
Caprioni, Gianni 113
Carden, John Valentine 58
Carden Engineering Company (Carden-Lloyd) 58, 59
Carden-Lloyd see Carden Engineering Company
Cardoux, Christian 233
Carl-Benz Söhne see Eugen and Richard Benz
Carrozzeria Bertone 31, 61, 62
Carrozzeria Ghia SpA see Ford
Carrozzeria Ghia 155
Carrozzeria Lotti 219
Carrozzeria Ocra 36
Carrozzeria Touring 21
Caspers, Albert 59
Castro, J. 35
Cattaneo, Giustino 59
Causan, Némorin 33, 59
Cav. Achicle Magliano 219
Cavelli, Carlo 60, 158, 202
CD-GRAC 89
CD-Panhard 88
CD-Peugeot 88, 89
CeComp of Turin 36
Ceirano (Fabbrica Junior Torinese di Automobili, Fabbrica Ligure di Automobili, Genova, Fratelli Cierano, Giovanni Ceriano Junior & Co, Matteo Ceriano e C, SCAT [Società Ceriano di Automobili Torino], SA Giovanni Ceirano Fabbrica di Automobili, Società Ligure Piemontese Automobili SA, Società Piemontese Automobili Ansaldi-Ceirano) 6, 7, 60, 61, 221
Ceirano, Giovanni 19, 60
Ceirano, Giovanni Battista 61
Ceirano, Matteo 61
Cena, Bruno 61, 233
CGV (Charron, Girardot et Voigt) 63
Chaffiotte, Pierre 62
Chaigneau, Jean 49
Chalmers Motor Company, Detroit 60, 198

Champion Automobilwerk 147
Chantiers de l'Horme et de La Buire 299
Chapman, Anthony "Colin" Bruce 62, 71, 95, 216
Chapron, Henri 129
Charbonneaux, Philippe 46
Charles, Hubert Noel 63, 78
Charles Binks Ltd. (Binks Carburettors [Amal]) 34
Charnwood, Lady 201
Charnwood, Lord 30
Charron, Ferdinand "Fernand" 63, 23, 110
Charron, Girardot et Voigt see CGV
Chassangay, Marcel 174
Chaussan 34, 42, 204, 220, 242
Chedru 94
Cheinisse, Jacques 64
Chelsea Motor Carriage Builders Company 46
Chénard, Ernest Charles Marie 64
Chénard, Lucien 65
Chénard-Walcker (Société Parisienne de Carrosserie Automobile) 65, 84
Chevrolet 198
Chiti, Carlo 37, 78
Chrysler (Chrysler Cars, General Motors) 13, 15, 18, 26, 88, 230, 233, 242
Chrysler Cars see Chrysler/General Motors
Cisitalia (Cisitalia Automobili SpA, Cisitalia Argentine SA, CIS-Italia [Consorzio Industriale Sportive Italia]) 5, 97, 98
CIS-Italia [Consorzio Industriale Sportive Italia] see Cisitalia
Cisitalia Argentine SA see Cisitalia
Cisitalia Automobili SpA see Cisitalia
Citroën (André Citroën Ingénieur-Constructeur, SA André Citroën, Société d'Engrenages Citroën) 10, 12, 21, 26, 27, 31, 34, 39, 42, 44, 45, 57, 65, 79, 95, 96, 98, 204, 211, 215, 219, 226, 229, 230, 250, 251
Citroën, André 12, 65, 79, 95, 96, 215
Citroën, David 84
City & Surburban Electric Vehicle Co. 162
Clarke, John 15
Claveau (Autobloc Claveau) 67

Claveau, Emile 66
Clayton-Dewandre 100
Clegg, Owen 67
Clément, Gustave Adolphe 64
Clément & Cie see Clément-Bayard
Clément-Bayard (Clément & Cie, Clément Gladiator & Humber [France] Ltd., Clément-Panhard, Clément Talbot Ltd., Société Annonyme des Anciens Etablissements Clément-Bayard), (SA des Vélocipedes Clément) 17, 46, 51, 63, 64, 67, 68, 92, 211, 231, 233, 237
Clément Gladiator & Humber [France]Ltd. see Clément-Bayard
Clément-Panhard see Clément-Bayard
Clément-Talbot see Sunbeam/Clément-Bayard
Clément Talbot Ltd. see Clément-Bayard
Clerget, Pierre 92, 242
Clyno Engineering Company Ltd. 246, 307
Coatalen, Louis 21, 67, 68, 75, 234
Cockerill, John 86
Colombo, Gioacchino 23, 69
Coman see Fiat
Commer Cars Ltd. 37
Comotor SA (Citroën/NSU Motorenwerke AG) 140
Compagnie Générale d'Electricite 129
Compagnie Lilloise des Moteurs see Peugeot
Company Developments Ltd. see Aston Martin
Constantinesco, George 69
Cooper, John 216
Cooper, Joseph 214
Cordiano, Ettore 70, 213
Cornacchia, Felice 70
Costin, Francis "Frank" Albert 70
Costin, Michael "Mike" Charles 71, 95
Costin Drake Technology see Costin Engineering Ltd.
Costin Engineering Ltd. (Costin Drake Technology, Costin Research & Development, Sanderson & Costin of Newbury, Timothy Research Ltd.) 70, 71
Costin Research & Development see Costin Engineering Ltd.

Cosworth Castings see Cosworth Engineering Ltd.
Cosworth Engineering Ltd. (Cosworth Technology, Cosworth Castings, Cosworth Racing) 71, 95
Cosworth Racing see Cosworth Engineering Ltd.
Cosworth Technology see Cosworth Engineering Ltd.
Cotal, Jean 71
Cottereau 84
Cottin, Cyrille 87
Coulet, Robert 89
Cousins, Cecil 71, 103
Coventry-Climax see Coventry Motor Company
Coventry Cycle Co. see Coventry Motor Company
Coventry Motor Company (Coventry Cycle Co., Coventry-Climax, Coventry Premier Ltd.) 20, 32, 51, 214, 216
Coventry Premier Ltd. see Coventry Motor Company/Singer
Coventry Repetition Company see Singer
Craig, Alexander 72
Craig, Joe 63
Crawford, Charles S. 226
Crespelle 33
Critchley, James Sidney 72, 74
Critchley-Norris 73
Crook, Thomas Anthony "Tony" Donald 73, 145
Crosio, Ambrogio 108
Crosio, Angelo 108
Crossley (Crossley Motors Ltd.) 47, 54, 55, 72, 73, 74, 208
Crossley, William John 73, 74
Crossley Motors Ltd. see Crossley
Crouch, John William "Billy" Fisher 74
Crouch Cars Ltd. see Crouch Motors Ltd.
Crouch Cavettes see John William Fisher Crouch
Crouch Motors Ltd. (Crouch Cars Ltd.) 74
Crowden, Charles 68
Crystal Palace Motor Exhibition 14
Cudell, Max 74, 75
Cudell & Co. (Aktiengesellschaft für Motor-und Motorfahrzenhau vorm Cudell & Co., Cudell & Co. K-G, Cudell Motoren GmbH) 74, 75

Cudell & Co. K-G *see* Cudell & Co.

Cudell Motoren GmbH *see* Cudell & Co.

Cunningham, Alexander A. 75

Cureton, Thomas 75

Curtice, Harlow 245

Cutler, Gerard Mervyn 76

Cutting, Edward 199

Cyklon Maschmenfabrik GmbH *see* Cyklonette

Cyklon Werke, Berlin *see* Cyklonette

Cyklonette (Cyklon Maschmenfabrik GmbH, Cyklon Werke, Berlin) 143, 154

D.Ing.h.c.F. Porsche K-G *see* Porsche

D. Napier & Son (D. Napier & Son Ltd) 218

Daewoo (Daewoo Motor Company) 31, 32, 33

DAF 81, 101

Dahlberg, Wesley P. 16

Daimler (Associated Equipment Co. Ltd., Birmingham Small Arms Co., Ltd., Daimler Motor Company Ltd., Coventry, Daimler Motor Syndicate, Lanchester Motor Co. Ltd., Motor Mills) 11, 16, 57, 72, 73, 77, 80, 90, 91, 201, 208, 245

Daimler (Benz & Cie, Daimler-Benz) 10

Daimler (Daimler-Benz AG, Daimler Motoren Gesellscaft, Austro-Daimler. Auto Union-/Daimler-Benz) 8, 18, 20, 22, 24, 29, 44, 45, 49, 52, 72, 73, 76, 77, 205, 207, 216, 217, 218, 221, 224, 246, 247, 248, 250, 252

Daimler, Gottlieb William 150, 160, 238

Daimler, Paul Friedrich 77, 206

Daimler-Knight 124

Daimler Motor Company Ltd., Coventry *see* Daimler/BSA

Daimler Motor Syndicate *see* Daimler

Daimler Motoren Gesellschaft *see* Daimler

Damiam, Don Matéu 35

Dangauthier, Marcel 78

Daniels, Jack 78, 157

Daninos, Jean Clément 79

Darl'Mat, Emile 79

Darracq (A. Darracq & Cie, A. Darracq & Co. [1905] Ltd.,

SA Darracq) 6, 67, 68, 69, 80, 225

Darracq Alexandre 41, 80, 251

David Brown & Sons [Huddersfield] Ltd., David Brown Holdings, David Brown Gear Industries, David Brown Tractors Ltd., David Brown Tractions Ltd. (Vosper-Thornycroft) 50

David Brown Gear Industries *see* David Brown & Sons [Huddersfield] Ltd.

David Brown Holdings *see* David Brown & Sons [Huddersfield] Ltd.

David Brown Tractions Ltd. *see* David Brown & Sons [Huddersfield] Ltd.

David Brown Tractors Ltd. *see* David Brown & Sons [Huddersfield] Ltd.

David Ogle Ltd. (Ogle Design Ltd.) 166

Davidse, A.V. 14

Dawfrey, Lewis Henry 80

Dawson, Alfred J. 80, 144

Dawson Car Company Ltd. 32, 80, 147

Day, Bernard 107

Day, Graham 81, 217

DB 88

de Arriortua, José Ignacio Lopez 241

Deasy (H.H.P. Deasy & Co., Deasy Motorcar Manufacturing Company Ltd) 81, 91

Deasy, Henry Hugh Peter 74, 81

Deasy Motorcar Manufacturing Company Ltd. *see* Deasy

de Castelet, Gaeton de Coye 30

Decauville 102

Deconinck, Rémi 81

De Cosmo (De Cosmo & Cie) 82

De Cosmo, Joseph 8

De Cosmo & Cie *see* De Cosmo

De Diétrich (Société De Diétrich & Cie) 40, 53, 82, 203

De Diétrich, Eugène 82

De Dion (de Dion–Bouton & Cie, De Dion–Bouton British & Colonial Syndicate, Société de Dion–Bouton Trepardeux) 23, 45, 61, 64, 68, 74, 75, 83, 84, 99, 170, 191

De Dion, Wandonne de Malfiance, Count Jules-Félix Phillipe Albert 45, 83

De Dion–Bouton & Cie *see* De Dion

De Dion–Bouton British & Colonial Syndicate *see* De Dion

De Duiche, Duc (Duke) 54

Degen 5

de Jong, Jacques 84

de Jong, Sylvain 83

Delagarde, Louis 84

Delage (Automobiles Delage) 13, 57, 59, 71, 85, 211, 233, 240

Delage, Louis 59, 85, 211

Delahaye (Emile Delahaye Ingénieur-Constructeur), (Emile Delahaye & Cie), 65, 71, 79, 82, 85, 211, 240

Delahaye, Emile 82, 85

Delamare-Deboutteville, Edouard 86

Delangere & Clayette 57

Delaunay-Belleville 17

Dellara, Giampaolo 77

De Lorean 63

De Luca Daimler 208

Demel, Herbert 86

De Montais, Roger 48

De Nora, Niccolo 37

Deroy (Deroy Sports Car Company) 228

Desgouttes, Pierre 87

De'Silva, Walter Maria 87

Desmarais, Léon 86

De Tomaso (Autosports Products, De Tomaso of America Inc., De Tomaso Industries Inc., De Tomaso (Modena S.p.A) Automobili, Ghia SpA., Vignale SpA) 202

De Tomaso, Alexandro 10, 78, 87, 231

De Tomaso Industries Inc.

De Tomaso (Modena S.p.A) Automobili *see* De Tomaso

De Tomaso of America Inc.

De Turkheim, Adrien 40

Deutsch, Charles 42, 88

Deutsche Automobilebank AG 75

Deutsche-Fiat *see* Fiat

Deutsche Métallurgique *see* Bergmann

Deutsche-Werke 8

Deutz (Gasmotorenfabrik Deutz of Cologne) 53, 76, 246

De Zuylen de Nyevelt de Haar, Baron 83

DF Automobiles *see* Doriot, Flandrin & Cie

DFP Automobiles *see* Doriot, Flandrin & Cie

Dialto 54, 68

Di Carpena, Count Belli 61

Dick, Alick Sydney 89

Diehl, Georg 26, 221
Di Giusto 89
Diligeon & Cie of Albert
(Somme) 40, 41
Di Virgilio, Francesco 90
Dixi of Eisenach 100, 102, 247
DKW see Auto Union
DKW (Zschopauer Motoren-
werke AG) 52
DKW-Auto Union 133, 149
Docker, Sir Bernard 90
Docker, Frank "Dudley" 16, 90
Dodson 50
Dombret, Emile 91
Dompert, Karl 125
Donald Healey Motor Company
Ltd. 139, 140
Donnet, Joseph Albert Jerome
12, 13, 92
Donnet-Zédel (SA des Automo-
biles Donnet-Zédel, SA des
Automobiles Donnet, Auto-
mobiles Zédel of Pontarlier)
92, 241
Doriot, Auguste 92
Doriot, Flandrin & Cie (Doriot,
Flandrin et Parant), (DF Auto-
mobiles), (DFP Automobiles)
92, 93
Doriot, Flandrin et Parant see
Doriot, Flandrin & Cie
Dorman Engine Company 47
Dornier 49
Douin, Georges 93
Douvrin La Française de Méca-
nique) 94
Dowson J.M.P. "George" 157
Dozekal, Felix 125
Drew, Harold 93
Dreyfus, Pierre 94, 130
Dubonnet, André 94
Duckworth, David Keith 71,
95
Du Cros, Harvey 14
Duesenberg 54, 94
Dufresne, Henri 95
Dufresne, Louis 95
Dumaine, Emile 92
Dunn, Michael Donald David
96
Dunn, William 96
Dupin, Jean 129
Durin, Michael 96
Dürkopp (Dürkopp-Werke
GmbH), (Dürkopp-Werke
AG), (Bielefelder Maschinen-
fabrik AG vorm. Dürkopp &
Co.), (Karosseriefabrik Wie-
mann & Co.) 26, 97
Dürkopp, Ferdinand Robert
Nikolaus 97, 250

Dürkopp-Werke AG see
Dürkopp
Dürkopp-Werke GmbH see
Dürkopp
Dusio, Piero 5, 97, 248
DVL (Deutsche Versuchsanstalt
für Luftfahrt [German Test
Establishment for Aviation])
24
D-Wagen see Deutsche-Werke
Dyckhoff, Otto Eduard Maria
98
Dyfresbem Gebru 78
Dyna-Panhard 79, 85, 88

E.E.C. Mathis see Mathis
E.G. Wrigley & Company Ltd.
167
Earl, Harley 163
Eberan (Professor) 20
Eberan, von Eberhorst, Robert
119
Eberhard, von Kuenheim 134
Edge, Stanley Howard 14, 100,
218
Edison, Thomas Alva 40
Edmond Gentil (Société Gentil
& Cie, Société des Automo-
biles Alcyon) 121
Edwardes, Michael 100
Edwards, "Joe" 101
Egan, John Leopold 102
Ehrhardt, Heinrich 102
Ehrhardt-Szawe 102
Elizalde (Biada, Elizalde y Cia,
Fabrica Española de Auto-
moviles Elizalde) 103
Elizade y Rouvier, Arturo 103
Elliott, A.G. 107
Elliott & Sons of Reading 46
Elswick Works 103
Elva 138
Emile Delahaye & Cie see Dela-
haye
Emile Delahaye Ingénieur-
Constructeur see Delahaye
ENA (Ecóle Nationale d'Admin-
istration) 56
Enever, Sidney "Sid" 103, 139
Enfield-Allday (Alldays &
Onions) 30
Engelbach, Carl R.F. 101, 103
Engelhardt, Paul 104
Engellau, Gunnar 104, 133
England, Frank "Lofty" Ray-
mond Wilton 105
ENI (Entre Nationale Idocar-
buri) see GEPI
Ennos, Clive 105
Enrico, Giovanni 106, 212
Enzo Ferrari 202

Epron, Luc 106
ERA (English Racing Automo-
biles Ltd) 138, 145, 216
Erfiag 121
Eric, Campbell & Co., Ltd. (In-
victa Cars, Railton Cars, Silver
Hawk Motors Ltd.) 198
Eric-Campbell see Eric, Camp-
bell & Co., Ltd.
Erle, Fritz 17, 26, 221
Esser, Heinrich 203
Esslinger, Friedrich Wilhelm 25,
26
Etablissement H. Précloux 34
Etablissements André Dubonnet
94
Etablissements de Construction
d'Automoteurs Système E.
Mathieu see Mathieu
Etablissements Léon Bollée see
Bollée
Etablissements Rolland et Pilain
see Rolland-Pilain
Everden, Harold "Ivan" Freder-
ick 39, 107
EX 135 (speed record car) 103
Eyston G.E.T. l3

F.W. Berwick & Co., Ltd. 47
Fabbrica Automobile Nazzaro
120
Fabbrica Automobili Züst 120
Fabbrica d'Armi Pietro Beretta,
Gardons Val Trompia see
Beretta
Fabbrica Italiana Automobili
Torino see Fiat
Fabbrica Junior Torinese di Au-
tomobili see Ceirano
Fabbrica Ligure di Automobili,
Genova see Ceirano
Fabrica Española de Automoviles
Elizalde see Elizalde
Fabrique Nationale d'Armes de
Guerre see FN Herstallez-
Liège
Fabry, Maurizio 219
Faccioli, Avistide 61
Facel (Forges et Atéliers de Con-
struction d'Euret-Loire Facel),
(Facel Metallon) 79
Facel Metallon see Facel
Fahrzeugfabrik Eisenach AG
102
Falchetto, Battista Giuseppe 108
Falkenhayn, Fritz von 109
Farkas, Armand 167
Farman, Dick 110, 188
Farman, Henry 64
Farman, Maurice 110
Farmer, George 89

Farmer, Lovedin "George" Thomas 110
Faulkenhausen, Alex von 108
Feeley, Frank 127
Felz, Jean-Martin 115
Fenaille Pierre 129
Ferrari (Ferrari Automobili, Auto-Avio Construzioni Ferrari SpA, Scuderia Ferrari) 7, 12, 23, 27, 31, 69, 77, 213, 282
Ferrari, Alfred "Dino" 159
Ferrari, Alfredo 3
Ferrari, Enzo 23, 36, 69
Fessia, Antonio 90, 108, 113
Fiala, Ernst 114
Fiat (Deutsche Fiat) 212, 213
Fiat (F.I.A.T.—Fabbrica Italiana Automobili Torino, Austro-Fiat, Coman, Deutsche-Fiat, Fiat SpA, Fiat Auto SpA, Fiat Automoveis, Fiat Argentine, Iveco N.V, Magirus Deutz, Polski—Fiat) 5, 6, 7, 8, 12, 18, 21, 29, 31, 42, 56, 57, 58, 60, 61, 62, 70, 73, 87, 88, 89, 94, 97, 98, 198, 199, 209, 213, 219, 233, 240, 241, 251
Fiat (Polski-Fiat) see Fiat
Fiat Argentine see Fiat
Fiat Auto SpA see Fiat
Fiat Automoveis see Fiat
Fiat SpA see Fiat
Fidelio 102
Fides-Brasier 106, 120
Fiedler, Fritz 10, 41, 233
Fikentscher, Fritz 149
Filipinetti, Georges 231
Fils des Peugeot Frères see Peugeot
Les Fils des Peugeot Frères 210
FIRE 1000 (Fully Integrated Robotized Engine) 158
Fisher, A.G.A. 228
Fisher & Ludlow 38
FLAG (Fabbrica Junior Torinese di Automobili) see Ceirano
Flandrin, Ludovic 92
Flegl, Helmut 48
Flick, Fritz 52
Florentia 59
FN Herstallez-Liège (Fabrique Nationale d'Armes de Guerre (FN) 82, 84
Fogolin, Claude 178
Follett, Charles 161
Ford (Carrozzeria Ghia SpA, Ford of Britain, Ford of Europe Inc., Ford Motor Co. Ltd., Ford-Werke AG) 7, 13, 16, 18, 19, 21, 22, 37, 46, 47, 55, 59, 79, 87, 88, 95, 96,

166, 198, 204, 212, 224, 228, 232, 234
Ford, Henry 14
Ford Motor Co. Ltd. see Ford
Ford of Britain see Ford
Ford of Europe Inc. see Ford
Ford-Werke AG see Ford
Forges et Atéliers de Construction d'Euret-Loire Facel see Facel
Fornaca, Guido 115, 126
Fornicon, Michel 116
Franchini, Enzo 70
Francis, A 51
Françon 94
Frankenborger, Victor 116
Fraschetti Andrea 23
Fraschini, Oreste 59
Fratelli Ceirano see Ceirano
Frazer Nash (AFN Ltd.) 10, 233
Frazer Nash, A. 10
Fremminville, Charles de 117
Freville, G.P.H. de 161
Frisinger, Haakan 117
Friswell, Charles 118, 205
Froede, Walter 19
Fry, David 157
Fry, Joseph G. 157
Fry, Tim 76
FSA Peugeot-Citroën see Peugeot
Fuhrmann, Ernst 119
Fulgura see Bergmann
Fulton, Norman Osborne 38, 162
Fuscaldo, Ottario 120, 219

G. Carde Fils & Cie see SA des Automobiles Autobloc
G.W. Hands Motor Company Ltd. see Calthorpe
Galileo Ferraris National Electrotechnical Institute 42
Gallion, George 36
Gandini, Marcello 31
Gangloff 53
Ganss, Julius 17, 26
Ganz, Joseph 120
GAP — Générale Automobile Rostique see Société Rostique des Vosges
Garcea, Giampaolo 121
Gasmotorenfabrik Deutz of Cologne see Deutz
Gautier, François 230
Gautier-Wehrlé 82, 180
GDA (Germainschaft Deutscher Automobilabriken) 11, 12
GEM (Société Générale d'Automobiles Electro-Mécaniques) 124

General Electrical Company 49, 52
General Electrical Company see AEG
General Motors (General Motors European Operations, GM Overseas Group) 198, 226, 241
General Motors European Operations see General Motors
Gentil, Edmond 121
Georges, Yves 121
Georges Irat (Automobiles Georges Irat SA, Irat & Cie, Irat Diesel, Irat Moteurs, Société Cherifienne d'Etudes des Automobiles Georges Irat, SA d'Exploitation des Automobiles Georges Irat 155, 156
GEPI (Gestrone a Participazioni Industriali), (Ente Nationale Idrocarburi) 88
Germainschaft Deutscher Automobilabriken see GDA
Gesellschaft zur Vorbereitung des deutschen Volkswagons (Gezvor) see Porsche
Gestrone a Participazioni Industriali see GEPI
Gevin 180
Ghia SpA see De Tomaso
Ghidella 8
Giacosa, Dante 97, 122, 212, 251
Gilchrist 9
Giovanni Ceriano Junior & Co. see Ceirano
Girardot, Léonce 64
Giugiaro, Giorgetto 233
Giugiaro, Giorgio 31
GK (Groupo Industriale Ghidella) 122
GKN (Guest, Keen, Nettlefolds, Uni-Cardan AG) 24
GLAS (Hans Glas GmbH, Isaria Maschinenfabrik Hans-Glas K-G) 124, 125, 171
Glas, Andreas "Anderl" 125
Glas, Hans 124
GM Overseas Group see General Motors
GN Ltd. see Godfrey & Nash
Gobbato, Pier Ugo 125
Gobbato, Ugo 125, 231
Gobran & Brillié 49, 50, 84
Gobron, Gustave 50
Godfrey, Henry Ronald 117, 127
Godfrey & Nash (GN Ltd., GN Motors Ltd.) 117, 1270
Goggomobil 125
Golden Arrow (speed record car) 156

Goliath-Werke GmbH 44, 211
Good, A.P. 25
Gordon 37
Gordon Bennett 106
Gorrini, Osvaldo 227
Goux, Jules 239
Grade Automobilwerke und Aviatik *see* Grade Motorenwerke GmbH
Grade Motorenwerke GmbH (Hans Grade-Werk GmbH, Grade Automobilwerke und Aviatik) 128
Grade, Hans 128
Gräf & Stift *see* Spitz
Graham-White Aviation 30
Granjean, A. 33
Great Horseless Carriage Co. *see* Horseless Carriage Company
Grégoire, Jean-Albert 128, 129, 164
Grenfell & Accles *see* Accles-Turrell
Greschen, Wilhelm H. 123
Griffin, Charles 129
Grillot, Pierre Marcel "Alphonse" 130
Grinham, Edward G. 130
Grosseau, Albert 130
Group Lotus Companies Plc *see* Lotus
Grylls, Shadwell Harry 131
Guédon, Phillipe 131
Guerzoni, Vittorio 23
Gulinelli, Count 53
Gutbrod, Wilhelm 121
Guy, Sidney Slater 32
Guy Motors Ltd. 132
GWK (Grice, Wood & Keiller) 200
Gyllenhammer, Pehr Gustave 118, 132

H.H.P. Deasy & Co. *see* Deasy
H.J. Mulliner 46
H.M. Hudson (Aircraft & Motor) Components Ltd. 245
Habbel, Wolfgang R 133
Hagemeier, Ernst 160, 170
Hahn, Carl 52
Hahnemann, Paul Gustav 134
Haja, Henry 36
Halford, Edward A. 127
Hall Engineering (Holdings) Ltd. 89
Hammesfahr, Ftiz 107
Hanau 226
Hancker, Oscar H. 160
Hancock, A.J.W. "Joey" 14, 15
Handley Page Ltd. 198
Hands, George William 135

Hanks, Reginald F. 135
Hanomag 98
Hanomag-Henschel 237, 245
Hanon, Bernard 136
Hans Glas GmbH *see* GLAS
Hans Grade-Werk GmbH *see* Grade Motorenwerke GmbH
Hansa (Hansa Automobil GmbH, Hansa Automobil-Werke AG, Hansa Automobil- *see* Hansa/Borgward
Hansa Automobil GmbH *see* Hansa/Borgward
Hansa Automobil und Fahrzeug-Werke *see* Hansa
Hansa Automobil-Werke AG *see* Hansa
Hansa-Lloyd Werke *see* Hansa/Borgward
Harper, R.O. 221
Harriman, George 102
Harry Ferguson Research Ltd. 143
Hartmann, George 97
Hartwich, Günther 137
Harvey-Bailey, R.W. 137
Haspel, Wilhelm 137, 172, 193
Hassan, Walter "Wally" 20, 137, 170, 216
Hauk, Franz 138, 155
Hault, Franz 19
Hawker, Harry 47
Hayter, Don 139
Healey, Donald 46
Healey, Geoffrey Carroll 140
Heess, Albert 221
Heinrich, Jean 140
Heinrich Ehrhardt Automobil-werke AG 102
Helbe 85
Heldé (Leveque & Dobenrieder) 250
Hennessy, Sir Patrick 21
Henning, R. 109
Henry, Ernest 239
Henze, Paul 74, 140, 211
Henzel, Nikolaus 97
Herbert, Alfred 19, 144
Herbert, William 144
Hercules 62
Hereil, George 141
Hermès 53
Heron, Samuel D. 24
Heynes, Walter M. 17
Heynes, William Munger 142, 170
Hezemans, Toine 36
Hieronymus, Otto 142
Hill, Claude 143
Hillier, Karl "Gustave" 143
Hillman, William 80
Hillman-Coatalon 68

Hillman, Herbert & Cooper 32
Hillman Motor Company Ltd. 37, 46, 68, 76, 80
Hills-Martini Ltd. 81
Hirst, Ivan 144
Hispano-Suiza (La Hispano-Suiza, Fabrica de Automoviles, Barcelona) 18, 49, 61, 62, 64, 94, 234
Hitzinger, Walker 144
Hives, E.W. 55
Hobbs, Dudley Erwin 145
Hodkin, David 146
Hoeppner, Ernst 119
Hofbauer, Peter 146
Holbein, Hermann 146
Holbrook, Claude Vivian 147
Holden *see* General Motors
Holls, David R. 36
Holste, Werner 147
Honda (Honda Italia di Arresa) 11, 101
Honda Italia di Arresa *see* Honda
Honsig, Anton 160
Hooven, John L. 148
Horch *see* Auto Union
Horch (A. Horch & Co., August Horch Automobilwerke, Horche-Werke AG) 8, 77, 98, 149
Horch, August 148
Horche-Werke AG *see* Auto Union
Hörnig, Rudolf 150
Horrocks, Raymond 150
Horseless Carriage Company (Great Horseless Carriage Co., Motor Manufacturing Company Ltd.) 137, 154, 181, 182
Hotchkiss (Automobiles Hotchkiss) 8, 9, 29, 79, 80, 91, 243
Houdaille, Maurice 203
Hounsfield, Leslie Hayward 150
Hounslow, Alex 103
Hoyal Bond Building Corporation 46
HRG Engineering Company Ltd. 127
Hruska, Rudolph 151
Hübbert, Jürgen 152
Hudson 50
Hull, Graham 15
Hulse, F.W. 52
Humber (Humber Ltd.) 33, 37, 46, 51, 84, 251, 252
Humber, Guntram 152
Humber Ltd. *see* Rootes Group
Hunt, Gilbert A. 152
Hupmobile (Hupp Motor Car Company, Detroit) 198

Hurlock, Charles Fleetwood 153
Hurlock, William Albert Edward 153
Hurtu (Hurtu, Hautin et Diligeon) 203
Hüttel, Franz Louis 154

Iacocca, Lee A. 13, 87
Iden, G.W. 74
Iden Motor Company 81
IFI (Instituto Finanzierio Industriale) 7
Ignez-Vox (Automotive gas-generator works) 5 Innocenti (Leyland Innocenti SpA, SpA Nuova Innocenti) 10, 87, 88
IMI (Instituto Mobiliari Italiano) see GEPI
INKA 212
Innocenti Ferdinando 155
Instituto per la Riconstruzione Industriale (Instituto Mobiliari Italiano) 88
Invicta see Macklin
Iota 178
Irat, Georges 155, 237
Irat, Michel 155
Irat & Cie see Georges Irat
Irat Diesel see Georges Irat
Irat Moteurs see Georges Irat
IRI (Instituto per la Ricostruzione Industriale) see GEPI
Irving, J.S. 13
Isaria Maschinenfabrik Hans Glas K-G. see GLAS
Iso Rivolta 31, 37, 78
Isotta, Cesare 59
Isotta-Fraschini 7
Issigonis, Sir Alec 52
Ital Design (Giugiaro) 199
Itala (Itala Fabbrica Automobili) 16, 58, 61
Itala Fabbrica Automobili see Itala/Ceirano
IVECO 101
Iveco N.V. see Fiat

J. & A. Niclausse 237
J.A. Grégoire 9, 22, 34, 59, 60, 65
J.R. Engineering 20
Jackson, Reg 103
Jackson, Robin 157
Jacobus 207
Jacoponi, Stephano 157
Jaguar (Jaguar Cars Ltd., SS Cars Ltd.) 16, 17, 36, 138, 142, 170, 196, 197, 216
Jaguar Cars Ltd. see Jaguar
Jahre, Karl 217

James Young 46
Jamieson, T. Murray 14
Janecek, Frantisek 158
Jano, Vittorio 69
Jaray, Paul 159
Jarrott, Charles 73, 74
Jarrott & Letts 74
JAWA (Zbrojovka Ing. F. Janecek AS) 158
Jeep 9
Jellinck, Count Emile 77, 159, 160, 206
Jem, Marsh 71
Jenbacher Werke 229
Jenscheke, Karl F. 160
Jensen (Jensen Motors Ltd) 62, 140
Jensen, Frank "Alan" 160
Jensen, Richard Arthur 161
JET 1 (experimental gas-turbine car) 168
John, Thomas George 127, 161
John Newton Fabbrica Automobil see NB
Johnson, Claude Goodman 162
Johnston, George 162, 208
Jones, David Brynmor 163
Jordan, M. Charles 36
José Maria Vallet y Cia 103
Joseph, Paul 164
Jowett (Jowett Cars Ltd.) 20, 228, 246
Jowett Cars Ltd. see Jowett
Julien, Maurice François Alexandre 164

Kales, Josef 164
Kalkert, Werner 165
Kamm, Professor W.A. 228, 245
Karcher, Xavier 165
Karen, Tom 166
Karl Baur Karosseriefabrik 36
Karmann, Wilhelm 166
Karmann, Wilhelm, Jr. 166
Karosseriefabrik Wiemann & Co. see Dürkopp
Kayser, Frank 14
Keighley Gear Company 50
Kennington, Eric 163
Kia Motors of Korea 168
Kieffer, Joseph-Nicolas 166
Kimber, Cecil 63, 72, 78
Kimberley, Michael "Mike" John 167
King, Charles "Spencer" 168
Kingham, Chris 157
Kissel, Wilhelm 137, 168, 217, 247, 249
Klaue, Hermann 223
Klement, Vaclav 169, 181
Kleyer, Erwin 170

Kleyer, Heinrich 169
Knight, Robert J. 170
Knight & Kilbourne 201, 229
Knipperdolling see Dürkopp
Knyff, René de 173, 229
Koch, Hans C. 171
Kolbing, Leo 74
Komenda, Erwin 171
Komnick, Bruno 171
Komnick, Karl "Franz" 171
Komnick, Otto 171
Komnick (Automobilfabrik Komnick, Automobil-fabrik Komnick AG, Maschinenfabrik Komnick) 171, 172
Korn, Alois 180
Korn & Latil (Latil et Cie, Société Industrielle des Automobiles Latil) 180
Kornecke, Friz 172
Kraus, Ludwig 22, 117, 139
Krebs, A.C. 60
Kreiger, Louis 173
Kronpriz AG 43
Küchen, Richard 109
Kuenheim, Berhard von 173
Kynast, Fritz 43

L.G. Motors (Staines) 127
Labourdette (Coachbuilders) 13
La Buire 243
Lacoba (Lacoste & Battmann) 57
La Cuadra, Emilio 35
Lafferentz, Bodo 98
Lago, A.F. "Tony" 21, 67
Lagonda (Lagonda Motors Ltd.) 20, 24, 25, 39, 50, 51, 99
Lagonda Motors Ltd. see Aston-Martin
La Hispano-Suiza, Fabrica de Automoviles, Barcelona see Hispano-Suiza
La Licorne 60
Lambert, Percy 51
Lamborghini (Automobili Lamborghini) 10, 12, 31, 37, 77, 78, 241
Lamborghini, Ferrucio 175
La Mouche (Teste, Moret & Cie) 251
Lampredi, Aurelio 176, 251
Lamy, Joseph 215
Lanchester (Lanchester Motor Company Ltd., Daimler Ltd., Coventry) 91
Lanchester, Frank 176
Lanchester, Frederick William 176, 201
Lanchester, George Herbert 176, 177

Lanchester Motor Company Ltd. *see* BSA/Daimler Ltd., Coventry
Lancia (Lancia e C) 7, 12, 31, 62, 78, 89, 90, 198, 231
Lancia, Vincenzo 108, 178, 179, 219
Land Rover 101
Lange, Karlheinz 179
Langen, Arnold 53
Langen, Eugen 76
Lanstad, Hans 179
LAP Engineering 175
La Perle 60
Lapine, Tony 174
Latil, Auguste "Georges" 180
Latil, Lazare 180
Latil et Cie *see* Korn & Latil
Laurin, Vaclav 180
Laurin & Klement AG Automobil-Fabrik 142, 169, 180
Lavalette, Count Henri de 94, 207
Lawson, Geoffrey 181
Lawson, Harry John 34, 41, 181
Lea-Francis 72
Learoyd, A.E. 34
Lefaucheux, Pierre 94, 130
Lefebvre, André 96, 220
Lefèvre, Louis 60
Legros, Austin 85
Leiding, Rudolf 248, 250
Leigh, Lord 14
Leimer, René 176
Lemaire, Pierre 164
Léon Bollée 203
Leonhardt, Carl 149
Lestienne 60
Letts, William 73, 74
Levassor, Emile 173, 188, 206, 207, 228, 230
Lewis, E.W. 81
Lewis, Michael 168
Leyland Cars *see* BMC
Leyland-Innocenti *see* De Tomaso/BMC
Leyland Innocenti SpA *see* BMC
Leyland Motors *see* BMC
Leyland Motors Ltd. *see* BMC
Le Zèbre (Borie & Cie) 215
Liberty Motor Car Company, Detroit) 198
Liebieg, Baron von Theodor 141
Ligier 89
Lister, Brian 71
Lohr, Fritz 223
Lola Engineering 89
London General Omnibus Company 73

L'Orange, Prosper 217
Lord, Leonard Patrick (Lord Lambury). 102, 103, 136, 149, 157, 193, 215, 223, 228
Lorenz, Wilhelm 206
Lorraine-Diétrich 17, 58, 234
Loste, Ernest 241
Lotus (Lotus Engineering Company, Lotus Cars Ltd., Group Lotus Companies Plc, Lotus Engineering) 36, 62, 71, 95
Lotus Cars Ltd. *see* Lotus
Lotus Engineering *see* Lotus
Lotus Engineering Company *see* Lotus
Lotz, Kurt 148, 194, 250, 253
Lucas hybrid (electric experimental car body) 166
Lucchini, Giuseppe 78
Luftfahrzeutg Motorenbau GmbH 206
Luraghi, Giuseppe 151, 152, 195
Lutz, Bob 36
Lutzmann 225
Lyons, Sir William 105, 138, 168, 170, 196, 197

Macclesfield, Lord George Loveden William Henry Parker (Earl of Macclesfield) 214
Mackerle, Julius 197
Macklin, Noel Campbell 198
MacPherson, Earle Steele 198
Maggiora 36
Magirus Deutz *see* Fiat
Maier, Dr. Albert 146
Maikäfer 121
Maioli, Mario 89, 199, 233
Maladin, Léon 86
Malleret, Guy 227
Malthis, E.E.C. 53
MAN (MTU), (Motoren and Turbinen Union) 49
Manessius of Levallois 241
Mann & Overton 9
Mannesmann AG 236
Manufacture d'Armes et de Cycles 92
Manville, Sir Edward 201
March, Earl of 153
March Special Body Designing Consultancy 153
Marcos 71
Marek, Tadek 199
Marendaz, Donald M.K. 199
Markland, Stanley 200
Marsaylia Bank 58
Marseel (Marseal Engineering Company Ltd., Marseal Motors Ltd., D.M.K. Marendaz

Ltd., Marendaz Special Cars Ltd.) 199, 200
Marston, John 75, 200
Marta (Magyar Automobil Reszveny Tarsasag Arad, Benz & Cie) 217
Martin, Lionel Walter Birch 30, 201
Martin, Percy 201
Martineau, Francis Leigh 137
Martini 81, 95
Maschinenfabrik Augsburg-Nürnberg *see* MTU/MAN
Maschinenfabrik Komnick *see* Komnick
Maschinenfabrik Walter Steiger & Cie *see* Steiger
Maserati (Officine Alfieri Maserati) 10, 12, 16, 23, 27, 69, 71, 78, 87, 88, 220, 227
Masoero, Pilado 219
Mason, George W. 140
Massacesi, Ettore 202
Massey-Harris 89
Massimino Alberto 23, 202, 227
Mathieu, Eugène 202
Mathieu or U.S. (Etablissements de Construction d'Automoteurs Système E. Mathieu, Usines de Saventhem) 202, 203
Mathis (E.E.C. Mathis, Mathis & Cie, Mathis AG, Mathis SA) 84, 203 204, 210
Mathis, Emile Ernest Charles 203
Mathis AG *see* Mathis
Mathis & Cie *see* Mathis
Mathis SA *see* Mathis
Matra (GAP — Générale Automobile Plastique) 42, 56, 89
Matteo Ceirano e C *see* Ceriano
Mattern, Ernest 204, 205
Maudslay, Reginald Walter 38, 197, 205
Maudslay Motor Company 72
Maybach (Maybach Motorenbau GmbH) 22, 115, 206, 207
Maybach, August "Wilhelm" 205
Maybach, Karl Wilhelm 206, 208
Maybach, Wilhelm 76, 77, 160
Maybach Motorenbau GmbH *see* Daimler-Benz
Mayola 155
Mays, Raymond 73
McCall White, David 208
McCormack, Arthur J. 208
McEvoy, Michael (Colonel) 144
McNeil, A.F. 38

Meadows Henry 17, 117, 198
Mediobanca 88
Mellde, Rolf Wilhelm 209
Mercédès (Société Française d'Automobiles) 159
Mercédès-Benz 8, 18, 216, 221, 224
Merosi, Giuseppe 209
Mersheimer, Hans 210
Méry Simon 104
Messerschmidt, Willy 108, 111, 112
Metcalfe, C.H.N. 127
MG Car Company Ltd. 63, 71, 78, 167, 233
Michaux, Gratien 210
Michelat, Arthur "Leon" 57, 211
Michelin 66, 79, 164
Michelin, André 251
Michelin, Edouard 45
Michelin, François 27
Michelin, Marcel 44
Michelin, Pierre 45
Michelotti 36, 42
Mickl, Josef 171
Mickwausch, Günter 211
Micot 110
Miglan, Lee 36
Milde-Krieger (Milde) 173
Mills, G.P. 51
Mimran, Patrick 11
Minerva (Minerva Motors SA, S. de Jong & Cie) 32, 83
Minerva Motors SA see S. de Jong & Cie
Misurata, Count Giovanni Volpi di 202
Mitsubishi Motors 133
Mo-Car Syndicate 38
Moglia, Edmond 67
Momberger, August 211
Momo, Cesare 212
Montabone, Oscar 70, 212, 213
Montanari, Vittorio 131, 213
Mooney, James D. 226
Morane, Georges 86
Morel, André 215
Moret, Jules 251
Morgan, Henry Frederick Stanley 213
Morgan Motor Company 11, 213
Morris (Morris Motor Engines Ltd., Morris Garages [MG], Morris Motors Ltd., Oxford Automobile & Cycle Agency, Hotchkiss, E.G. Wrigle, Osberton, Holilck & Pratt, Pressed Steel Company, Léon Bollée, Le Mans, MG Car Company, Wolseley Motors Ltd., Riley Motors Ltd.) 9, 15,

18, 19, 27, 46, 51, 63, 72, 76, 78, 214, 216, 223, 228
Morris, William Richard (Baron Nuffield/Viscount Nuffield) 9, 57, 91, 205, 209, 214
Morris Garages [MG] see Morris
Morris Motor Engines Ltd. see Morris
Morris Motors Ltd. see Morris
Mors (Automobiles Mors, Société d'Electricite et d'Automobiles Mors, Société de Construction d'Automobiles Trèfle à Quatre Feuilles, Société des Automobiles Brasier, Société Nouvelles des Automobiles Mors, Société de Trèfle à Quatre Feuilles 48, 65, 84, 95, 160, 215
Mors, Emile 48, 215
Mors, Louis, Jr. 215
Moteurs Salmson 237
Motor Guzzi 88
Motor Manufacturing Company Ltd. see Horseless Carriage Company
Motor Mills see Daimler
Motoradwerk Max Thun 5
Motoren and Turbinen Union see MAN
Motoren und Turbinen Union see MTU
Moyet, Edmond 215
MTU see MAN
MTU (Motoren und Turbinen Union, MAN [Maschinenfabrik Augsburg-Nürnberg]) 222
Müller, Josef 216
Mulliner Ltd. (Mulliner Park Ward) 46, 89, 235
Mulliner Park Ward see Mulliner Ltd.
Mundy, Harry 138, 216
Munn, Leslie 307
Musgrove, Harold 15, 217
Mussolini 7, 59, 241
Mutel 73

NAG-Voran 56
Nagant, Maurice 126
Nallinger, Friedrich "Fritz" 217, 218, 249
Napier (D. Napier & Sons) 25, 145, 182
Napier, Montague Stanley 218
Narval 94
Nash Healey see Nash-Kelvinator Corporation
Nash-Kelvinator Corporation 140

Nash Metropolitan see Nash-Kelvinator Corporation
Nathan, Roger 71
National Enterprise Board 101
Nationale Automobil-Gesellschaft AG see Hansa
Nazzaro, Felice 120, 219
Nazzaro e C. Fabbrica di Automobili see Tipo
NB — Newton-Bennett (John Newton Fabbrica Automobili, Newton & Bennett Ltd.) 220
Ne, Jacques 219
Nebbi, Bartolomeo 60
Nedcar BV see Volvo
Neue Kraftfahrzeug GmbH 154
Neumeyer, Hans Friedrich "Fritz" 220, 249
New Arrol-Johnston Co. Ltd. see Arrol-Johnston
New Castro 35
New Leader Cars 34
New Triumph Cycle Company see Triumph Limited
Newton & Bennett Ltd. see NB — Newton-Bennett
Newton, Bennett & Carlisle Ltd. 221
Newton, Noel Banner 220
Nibel, Hans 107, 217, 221
Nichols, Frank 216
Niefer, Werner 221
Nissan Motor Company 202
Norcros Ltd. 161
Nord Deutsche Lloyd see Hansa
Norddeutsche Automobil & Motoren AG, Bremen see Hansa
Nordoff, Heinz 222, 248
North American Automotive Operations see General Motors
North American Automotive Operations [Chevrolet, Chrysler, Holden, Opel, Pontiac, Vauxhall] 14, 15, 19, 70, 75, 93, 94
NSU see Auto Union
NSU-Wankel KKM 119
Nuffield, Lord 52, 63
Nuffield Organization 19, 63, 71, 78
Nutting, J. Gurney 38

Oak, Albert Victor "Vic" 51, 223
Oberhaus, Herbert 223
Oberle, Georges 223
Obländer, Kurt 224
Officine Alfieri Maserati 202
Officine Alfieri Maserati see Maserati

Ogier, John 166
Ogle, David 166
Ogle Design Ltd. *see* David Ogle Ltd.
Oldfield, John 224
OM of Brescia (Officine Meccaniche, originally Brixia-Züst) 7
Omer, Orsi 227
Opel, Carl von 224
Opel, Friedrich "Fritz" 225, 226
Opel, Heinrich 225
Opel, Wilhelm von 225, 226
Opel Adam Opel AG (Adam Opel KG, Opel Brothers GmbH) 9, 19, 84, 222, 223, 224, 225, 226
Opel Brothers GmbH *see* Opel
Orio, Stefano 209
Orio & Marchand 209
Orr-Ewing, Hugh Eric 198
OSCA (Officine Specializzate Constuzioni Automobili) 69, 88
Osmond, Jacques 49
Osswald, Bernhard 228
Otto, Gustav 246
Otto & Langen 73

Pallavicino di Priola, Marquis 58
Palmer, Gerald Marley 228
Panhard, Adrien "Hippollyte" François 228
Panhard, Jean Joseph Léon 229
Panhard, Joseph "Paul" René 42, 229
Panhard, Léon 229
Panhard, Louis François René 230
Panhard, René 189, 229, 238
Panhard & Levassor (Société de Constructions Militaires Panhard-Levassor) 27, 34, 42, 54, 59, 60, 64, 68, 73, 77, 84, 85, 96, 97, 210, 220, 236, 237
Parant, Alexandre 93
Parant, Jules-René 93
Parayre, Jean 174
Parayre, Jean-Paul 39, 230, 233
Park, Alex 230
Park Ward *see* Mulliner Ltd.
Parker, George 15
Parkes, J.J. 231
Parkes, Michael "Mike" John 76, 231
Parnell, Frank Gordon 100, 231
Parry-Jones, Richard 232
Parry Thomas, John Godfrey 232
Patchett, George W. 158

Paulet, Léon 211
Paulin, Georges 79
Payton, Ernest 104
Pennington, Edward Joel 137, 182
Penny, Noel 168
Perkins Engines Ltd. 146
Perlo, Giuseppe 232
Peronnin, Jean 233
Perrett, John Bertram 233
Perrot, Henri 48, 234
Perrot Freins 234
Perry, Percival Lea Dewhurst (Lord Perry of Stock Harvard) 234
Perry, Richard "Dick" William 235
Persson, Jan Christer Bernhard 235
Peter, Wolfgang 235
Petiet, Baron Charles 236
Petit, Emile 156, 237
Petri, Helmut 237
Peugeot (Automobiles Peugeot, FSA Peugeot-Citroën, Fils des Peugeot Frères, Peugeot Canada, SA des Automobiles Peugeot, SA des Société Peugeot Frères, Peugeot & Cie, Cycles Peugeot, Automobiles et Cycles Peugeot, Compagnie Lilloise des Moteurs) 13, 15, 23, 26, 39, 46, 47, 53, 56, 57, 64, 78, 79, 80, 85, 92, 94, 95, 96, 97, 204, 210, 230, 238, 244
Peugeot, Armand Godefroy Pierre 40, 206, 210, 237
Peugeot, Emile 237
Peugeot, Jean-Pierre 238
Peugeot, Robert 204, 210, 239
Peugeot & Cie *see* Peugeot
Peugeot Canada *see* Peugeot
Pezon, André 242
Pfaender, Otto 206
Phänomen (Phänomen Fahrradwerke Gustav Hillier, Phänomen Werke Gustave Hillier AG) 143, 154
Phänomen Fahrradwerke Gustave Hillier *see* Phänomen
Phänomen Werke Gustave Hillir AG *see* Phänomen
Phillips, Jack 131
Piaggi, Enrico 58
Picard, Fernand 122, 239
Piccone, Alessandro 240
Piëch, Anton 98
Piëch, Ferdinand 19, 22, 24, 240
Piedboeuf, Adrien G. 141, 211
Pierce-Arrow 245

Piganeau, Gerard 234
Pigozzi, Henri Théodore 7, 26, 92, 341
Pilain, Emile 242
Pilain, François 242
Pilain/SAP/SLIM (Société des Voitures Automobiles F. Pilain & Cie, Société des
Pinin Farina (Stabilimenti Farina) 15, 37, 42
Pininfarina, Serio 110
Pischotsrieder, Bernd 243
Piziali, Alfred P. 243
Place, René 244
Plastow, David Arnold Stuart 244
Platt, Maurice 244
Pollack, Thomas 6
Pollich, Karl 245
Pollock, Fergus 181
Polski-Fiat *see* Fiat
Polygon Engineering Works *see* Trojan
Pomeroy, Laurence Henry 29, 245
Poole, Stephen Carey 246
Popp, Franz Joseph 245
Poppe, Peter August 179, 247
Porsche (Dr. Ing. F. Porsche Konstruktions GmbH, Dr. Ing.h.c.F. Porsche K-G, Gezvo — Gesellschaft zur Vorbereitung des deutschen Volkswagons) 5, 23, 240, 247
Porsche, Ferdinand "Ferry" Anton Ernst 18, 77, 98, 160, 165, 192, 217, 240, 247, 248
Poulsen, Jean-Felix 164
Pourtout, Marcel 79
Powel & Matter 86
Praxl, Edwald 249
Pressed Steel Fisher Division *see* Pressed Steel Ltd.
Pressed Steel Ltd. (Pressed Steel Fisher Division) 39, 235
Prestcold 101
Pretôt 180
Prinetti & Stucchi 53
Prinz, Gerhard 250
Proctor, Stuart 127
Prodi, Romano 88
Prototipi Bizzarrini *see* Autostar, Livorno
Puch, Johann 250
Puch-Werke, Graz (Puchwerken AG of Graz) 160, 169, 250
Puchwerken AG of Graz *see* Puch-Werke
Puiseux, Robert 250
Puleo, Giuseppe 251

Pullcar (Pullcar Company, Preston) 137
Pullinger, Thomas Charles Willis 33, 75, 251

Quadrant 84
Quandt, Harold 252
Quandt, Herbert 252
Quaroni, Francesco "Franco" 253

R. Incerti e Cie 7
Rabag 54
Rabe, Karl 151, 165
Radermache, Karlheiz 254
Radford, Harold 16
RAF (Reichenberger Automobil-Fabrik) 140, 141, 169
Railton see Macklin
Railton, Reid Anthony 232, 254
Raleigh Cycle Co. Ltd. 32, 33
Rallye Bitter see Bitter Automobile Company
Randle, James "Jim" Neville 255
Rapid (Società Torinese Automobili Rapid) 121, 158
Rapier (Rapier Cars Ltd.) 127
Rasmussen, Jörgen Skafte 52, 149, 255
Ravelli de Beaumont, Mario 21
Ravenel, Raymond 256
Ravigneaux, Pol 256
Raworth 24
Rayner, Herbert 257
RDA (Reichsverband Deutscher Automobilindustrie) 12
Rebling, Arthur 257
Rédélé, Jean 64, 257
Reeves, A.W. 74
Régie Nationale des Usines Renault see Renault
Reichstein, Carl 258
Reitzle, Wolfgang 243
Reitzle, Wolfgang 243, 259
Reliant (Reliant Engineering Company) 166
Remington, Alfred Arthur 259
Remington, Phil 154
Renault (ACMA, Berex — Bureau d'Etudes et de Recherches Exploratoires, Regié Nationale des Usines Renault, RET, RIET — Renault Industrie Equipement et Techniques, RMO — Renault Machines-Outils) 15, 26, 30, 31, 47, 57, 62, 64, 230, 233, 239
Renault, Louis 236, 240, 259, 260, 261
Renault, Marcel 110
Renault Industrie Equipement et

Techniques (RIET) see Renault
Renault Machines-Outils (RMO) see Renault
Reno, Emile 33
Renwick, William Somerville 78, 143
Renwick & Bertelli 143
RET see Renault
Reuter, Edzard 222, 261
Revelli de Beaumont, Mario 261
Riancey 180
Ribeyrolles, Paul 262
Ricardo & Co., Ltd. 231
Ricart, Wilfredo Pelayo y Medina 56, 263
Richard, Carl Hermann Ludwig 263
Richard, Georges 49, 264
Richard-Brasier 34, 106
Richardson, Percy 265
Ridlington, James 20
Riecken, Christian 265
Rigolly, Louis-Emile 50, 126
Rigoulot, Louis 266
Riley (Riley Motors Ltd.) 72, 79
Riley, Percy 266
Riley, Stanley 267
Riley, William "Victor" 267, 268
Riolfo, Auguste 249, 268
Rivolta, Renzo 37, 269
Robert Bosch GmbH 86
Robins, Guy H. 127
Robinson, Geoffrey 105, 269
Robot Gears Limited 233
Robotham, William Arthur "Roy" 269
Rochet, Edouard 270
Rochet-Scheider 104
Roesch, Georges Henri 46, 67, 271
Roger, Emile 26, 272
Röhr, Hans Gustav 121, 272
Rolland, François 242
Rolland, Lucien 242
Rolland-Pilain (Etablissements Rolland et Pilain, SA des Etablissements Rolland & Emile Pilain, SA des Etablissements Emile Pilain) 242
Rolls, Charles Stewart 162, 273
Rolls-Royce Motor Cars Ltd. 20, 25, 38, 39, 55, 96, 162, 235, 241, 244
Rolls-Royce Motors Ltd. see Rolls-Royce Motor Cars Ltd.
Rolt, APR 143
Rombo, Società Anonima Brevetti Fuscaldo 120
Romeo, Nicola 158, 273

Romiti, Cesare 122, 274
Ronayne, Michael 275
Rootes, Geoffrey 153
Rootes, William Edward (Baron Rootes of Ramsbury) 156, 275
Rootes Group 15, 19, 46, 76, 231
Roots & Venables 103
Rose, Max Caspar 25, 26
Rose, Raymond "Hugh" 276
Rosengart, Lucien 23, 239, 276
Rosetti, Henri-Georges 176
Rothschild (Coachbuilders) 7
Rouge, Francis 277
Rover (Rover Company Ltd., Rover Group) 15, 16, 19, 28, 46, 55, 67, 72, 79, 81, 101, 243, 247
Rover Company Ltd. see Rover
Rover Group see British Leyland/ BMC
Rover-Italia see Austin
Rowan Controllers Inc. see De Tomaso
Rowledge, Arthur J. 278
Royce, Frederick "Henry" 107, 278
Roydale Engineering Company Ltd. 34
Rubery Owen see BRM
Rucker, M.D. 251
Rudd, Anthony "Tony" Cyril 279
Rudd, Ken 154
Rumpler, Edmund 56, 221, 279
Ruppe, Oskar "Berthold" 280
Ruppe, Paul "Hugo" 281
Rybeck, Martin 235
Rykneid Motor Co., Ltd. 16

S. de Jong & Cie see Minerva Motors SA
SA André Citroën see Citroën
SA Ansaldo Autobili 212
SA Ariès see Société La Cigogne
SA Automobili Nazzaro see Tipo
SA Darracq see Darracq
SA des Automobiles Autobloc (G. Carde Fils & Cie) 92
SA des Automobiles Belanger Frères 22
SA des Automobiles Donnet see Donnet Zédel
SA des Automobiles Donnet-Zédel see Donnet Zédel
SA des Automobiles Hispano Suiza, Bois Colombes see Hispano-Suiza
SA des Automobiles Marius Berliet see Berliet
SA des Automobiles Peugeot 237

SA des Automobiles Peugeot *see* Peugeot
SA des Automobiles Talbot 175
SA des Etablissements Emile Pilain *see* Pilain
SA des Etablissements Rolland & Emile Pilain *see* Rolland-Pilain
SA des Société Peugeot Frères *see* Peugeot
SA des Vélocipedes Clément *see* Clément-Bayard
SA d'Exploitation des Automobiles Georges Irat *see* Georges Irat
SA Giovanni Ceirano Fabbrica di Automobili *see* Ceirano
SA Piaggio 176
SA Reggiane 176
SAAB Aktiebolaget 36, 70, 209
Sacco, Bruno 281
SAFRAR (Sociedad Anonima Franco-Argentina de Automotores G y F) 165
SAI (Società Assicurazione Industriale) 8
Sailer, Max 281
Saintigny, Henri 282
Sainturat, Maurice 282
Salamano, Carlo 213, 219
Salmon Motor Co. 16
Salmson, Emile 64, 79, 92, 283
Salomon, Jules 96, 215, 283
Sampietro, "Sam" Achille C. 139, 284
Sanderson & Costin of Newbury *see* Costin Engineering Ltd.
Sangster, Charles 284
Sangster, John "Jack" Young 32, 90, 285
Sankey 214
Saoutchik 94
Sarazin, Edouard Auguste 285
Sarre, Claude-Alain 286
Sartorelli, Sergio 87
Satta Puliga, Orazio 286
Satta, Grazio 56
Saunders, Ron 166
SAVA (Société Anversoise pour la Fabrication de Voitures Automobiles) 84
Savey, Dominique 286
Savonuzzi, Giovanni 97, 287
Savoye, Raymond 287
Scaglioni, Franco 31
Scammell Lorries Ltd. 228
Scarisbrick, Herbert 44
SCAT (Società Ceirano Automobili Torino) 221
SCAT (Società Ceirano di Automobili Torino) *see* Ceirano

Schaefer, Herbert 288
Schaeffer, Rodolfo 288
Schapiro, Jakob 11, 247, 288, 289
Schaudel, Charles 91
Scheele, Nicholas "Nick" V. 289
Scheibler, Friz 290
Scherenberg, Hans 290
Schleicher, Rudolf 108
Schlumpf Collection 229
Schmücker, Toni 290
Schneider, Théodore 49, 291
Schrempp, Jürgen 292
Schulte, Mauritz J. 32
Schulz, Otto 167
Schürmann, Gustav 292
Schutz, Peter 48, 293
Schwarz, Georg 293
Schweitzer, Louis 294
Schwenke, Robert 294
Scolari, Paolo 57, 70, 294
Scuderia Ferrari *see* Ferrari
Scuderia Filipinetti 231
Scuderia Italia 78
Seaman, Dick 105
SEAT *see* Volkswagen
Seck, Willy 295
See, Marcel 295
Segre, Luigi 296
Seidel-Naumann 84
Seiffert, Ulrich 297
Selve (Selve Automobilwerke AG) 140, 186
Sénéchal, Robert 297
Sensaud de Lavaud, Dimitri 298
SERA (Société d'Etudes et Realisations Automobiles) 89
Serpollet, Léon Emmanuel 298, 237, 243
Serre, Charles Edmond 299
Seze, Guy de 300
SF Edge Ltd. 219
Shaw, Angus 75
Shelby, Carroll 154
Shelvoke, G.E. 6
Shepherd, John 157
Shorter, Leo Joseph 55, 300
Siddeley, John Davenport (Lord Kenilworth) 14, 301
Siddeley-Deasy 91, 199, 200
Sidgreaves, Sir Arthur E. 127, 301
Sidney, Edgar Harold 302
Siebler, Oskar 302
Siegfried Bettman 147
Sigliarini, Armande 5
Sima-Standard 92
SIMA-Violet 121
Simca (Société Industrielle de Mécanique et de Carosserie

Automobile) 5, 7, 26, 31, 65, 79, 212, 233, 241
Simms, Frederick Richard 77, 303
Simonsson, Svante 2009
Simpson, George 303
Singer (Singer & Co., Ltd., Coventry Premier Ltd., Coventry Repetition Company) 72, 246
Singer, George 82, 304
Singer, Paris E. 162
Singer & Co. Ltd. *see* Singer
Sir W.A. Bailey & Co. 34
Sir William Beardmore & Co. 51
Skoda (Skoda Werke) 134, 169, 181, 241
Slaby, Rudolf 305
Slevogt, Carl 74, 250, 305
SMAE (Société de Mécanique Automobile de l'Est) 140
Smith, A. Rowland 306
Smith, F. Llewellyn 306, 307
Smith, Frederick Robertson 307
Smith, Mabberley 75
Smith-Clarke, George Thomas 308
SNCASE (Société de Constructions Aeronautiques du Sud-Est [later Sud-Aviation]) 141
Snowdon, John "Mark" 308
Società Faccioli Ferro Rampone 108
Società Italiana Bernardi 28
Società Ligure Piemontese Automobili SA *see* Ceirano
Société Ader *see* Ader
Société Alsacienne de Constructions Mécaniques 53
Société Anonyme des Anciens Etablissements Clément-Bayard *see* Clément-Bayard
Société Anonyme des Anciens Establissement Panhard & Levassor (Société des Constructions Mécaniques Panhard & Levassor, Société Panhard & Levassor, SA Panhard & Levassor) 228, 229, 230
Société Anonyme Française de l'Automobile 128
Société Anversoise pour la Fabrication de Voitures Automobiles *see* SAVA
Société Cherifienne d'Etudes des Automobiles Georges Irat *see* Georges Irat
Société d'Automobiles Nanceiennes 126
Société d'Automobiles Pierre

Desgouttes & Cie *see* Automobiles Cottin & Desgouttes
Société de Construction d'Automobiles Trèfle à Quatre Feuilles *see* Mors
Société de Constructions Militaires Panhard-Levassor *see* Panhard & Levassor
Société De Diétrich & Cie *see* De Diétrich
Société de Dion–Bouton Trepardeux *see* De Dion
Société d'Electricite et d'Automobiles Mors *see* Mors
Société d'Engrenages Citroën *see* Citroën
Société de Trèfle à Quatre Feuilles *see* Mors
Société des Ateliers de Construction de Lavalette 207
Société des Automobiles Alcyon *see* Edmond Gentil
Société des Automobiles Alpine Renault 64
Société des Automobiles Ariès 236, 237
Société des Automobiles Brasier *see* Mors
Société des Automobiles Brasier 234
Société des Automobiles François Pilain *see* Pilain
Société des Automobiles René Bonnet 41
Société des Moteurs Zürcher 121
Société des Voitures Automobiles et Moteurs [F. Pilain & Cie] *see* Pilain
Société des Voitures Automobiles F. Pilain & Cie *see* Pilain
Société Générale des Automobiles Porthos 166, 167
Société Gentil & Cie *see* Edmond Gentil
Société H. & M. Farman 110
Société Industrielle des Automobiles Latil *see* Korn & Latil
Société Janvier 33
Société La Cigogne (SA Ariès) 34
Société Nouvelle des Automobiles Mors *see* Mors
Société Parisienne de Carrosserie Automobile *see* Chénard-Walcker
Société Peugeot Frères 237
Société Rostique des Vosges (GAP — Générale Automobile Rostique) 42
Society for Technical Progress 18
Sopwith, T.O.M. 47

SP — Società Piemontese Automobili Ansaldi-Ceirano *see* Ceirano
SPA (Società Ligure Piemontese di Automobili) 212
SpA Nuova Innocenti *see* Innocenti
Sparkbrook Manufacturing Co., Ltd. (Calcott Bros. Ltd.) 54, 55
Speedex Castings & Accessories Ltd. 71
Speer, Albert 12
Spitz (Gräf & Stift) 142
Spitz, Arnold 142
Sporkhorst, Johann Friedrich "August" 11, 308
Spurrier, Henry 309
Spyker, Hendrik 207
Squire, Adrian Morgan 78, 309
SS 1 (Swallow Sidecar Company, Swallow Coachbuilding Company, SS Cars Ltd.) 197
SS Cars Ltd. *see* Jaguar
Stabilimenti Farina *see* Pinin Farina
Stanbury, John Vivian 310
Standard (Standard Motor Car Company Ltd., Standard Motor Co., Ltd., Standard Triumph International Ltd.) 7, 13, 14, 19, 37, 38, 72, 80, 81, 89, 101, 161, 200, 205, 246
Standard Motor Car Company Ltd. *see* Standard
Standard Motor Co. Ltd. *see* Standard
Standard Superior (Deutschen Volkswagen) 121
Standard Triumph International Ltd. *see* Standard
Stanguelini 202
Stanzani, Paolo 310
STAR (Société Torinese Automobile Rapid) 61
Star Engineering Company 307
Starkloph, Franz 310
Steiger (Maschinenfabrik Walter Steiger & Cie, Walter Steiger AG Automobilefabrik) 140
Steiger, Walter 141
Steinway, William 76
Stevens, H.C.M. 55
Stewart & Ardern 214
Steyr (Steyr, Daimler-Puch AG, Steyr-Daimler-Puch AG, Steyr-Werke AG) 18, 36, 70, 160, 165, 169, 171, 182, 183, 211, 249
Steyr, Daimler-Puch AG *see* Steyr

Steyr-Daimler-Puch AG *see* Steyr
Steyr-Werke AG *see* Steyr
Stieler von Heydekampf, Gerd 311
Stockmar, Jürgen 311
Stoewer, Bernhard Alfred Rudolf Karl 114, 312
Stoewer, Emil Ludwig Karl 312
Stokes, Sir Donald 18, 182, 313
Stoneleigh 91
Storero, Luigi 313
Straight, Whitney 105
Straker, Sidney 314
Straker-Squire 231
Stratton, Desmond 154
Straussler, Nicholas 162
Streamline Cars Ltd. 55
Strobel, Werner 314
Strobl, Gottlieb M. 315
Sunbeam (Automobiles Talbot, Clément Talbot, Sunbeam Motor Car Co. Ltd., Sunbeam-Talbot-Darracq Motors, Ltd.) 13, 15, 17, 21, 25, 29, 34, 39, 63, 67, 68, 71, 114, 131, 233, 242
Sunbeam-Lotus 63
Sunbeam-Mabley (Sunbeamland Cycle Company, Sunbeam Motor Car Company Ltd.) 200, 201, 251
Sunbeam Motor Car Company *see* Sunbeam-Mabley/Marston
Sunbeam Motor Car Co. Ltd. *see* Sunbeam
Sunbeam-Talbot-Darracq Motors Ltd. *see* Sunbeam
Sunbeamland Cycle Company *see* Marston
Sunderland, Sir Arthur 30
Supra (Automobilbau Simson & Co.) 141
Swallow Coachbuilding Company *see* SS1
Swallow Sidecar Company *see* SS 1
Sykes, A.A. 232
Szabo & Wechselmann 102

T.B. Barker & Company 176
T.G. John Ltd. 161
Talbot (Talbot Darracq) *see* Sunbeam
Talbot Motors Co., Ltd. *see* Sunbeam
Tamplin, Edward A. 59
Tata Engineering and Locomotive Company 172
Tatra (Tatra-Werke) 197
Tavoni, Romolo 37

Taylor, George 315
Tecklenborg, Wilhelm 43
Teste Auguste 251
Thomas, Miles Webster 167, 315
Thomson & Taylor 138, 232, 233
Thornley, John William Yates 71, 316
Thornycroft 13
Thrupp & Maberley 107, 153
Thunderbolt (Land/Speed Record Car 1938) 13
Tiberghien, Pierre 316
Tickford Ltd. 50
Timothy Research Ltd. *see* Costin Engineering Ltd.
Tipo (Nazzaro e C. Fabbrica di Automobili, SA Automobili Nazzaro) 219
Tojeiro, John 154
Torre, Pierluigi 155
Touring 151
Toutee, Henri 317
Towers, John 317
Towns, William 318
Toyota Motor Company 63, 115
Tracta (Automobiles Tracta) 128, 129
Tranié, André P. 318
Trefz, Karl 319
Trepardeux, Charles Armand 45, 83
Treser, Walter 24
Trier, Vernon 319
Trippel, Hanns 319
Triumph *see* Standard
Triumph (Triumph Limited, New Triumph Cycle Company, Triumph Cycle Co. Ltd., Triumph Engineering Ltd, Triumph Motor Company Ltd.) 80, 81, 101, 118, 139, 147, 231
Triumph Cycle Co. Ltd. *see* Triumph
Triumph Engineering Ltd. *see* Triumph
Triumph Limited *see* Triumph
Triumph Motor Company Ltd. 147
Trojan (Polygon Engineering Works, Trojan Commercial Vehicles, Trojan Ltd.) 150, 151, 233, 234
Trojan Commercial Vehicles *see* Trojan
Trojan Ltd. *see* Trojan
Tucker, Samuel 36
Tufarelli, Nicola 122
Turcat, Léon 104, 310

Turcat-Méry 104
Turckheim, Adrien G. de 320, 321
Turnbull, George Henry 321
Turner, Alan 154
Turner, Edward 322
Turrell, Charles McRobie 5, 322

Ueber, Max 323
Uhlenhaut, Rudolf 323
Unic (Automobiles Unic) 7, 203
Urach, Prince Wilhelm von 324
Uren, Wilhelm Gustav Heinrich 324
U.S. (Usines de Saventhem) 203
Usines Bollée *see* Bollée

Valentin, Ernst 324
Valletta, Vittorio 126, 325
VALT (Vetture Automobili Leggere Torino) 221
Valveless 50
Van Doorne, Hubertus "Hub" Josephus 326
Van Eugen, Charles Marie 326
Vanwall 71
Varlet, Amédée 327
Vauxhall (Vauxhall Motors Ltd., General Motors) 9, 13, 18, 19, 75, 93, 147, 157, 228, 244, 245
Vauxhall Motors Ltd. *see* Vauxhall/General Motors
Ventre, Philippe 327
Veraldi, Lewis "Lou" C. 328
Verdet, Lucien 210
Vermorel, Victor 328
Vernandi 60
Vickers, Alfred 208
Vickers-Armstrong *see* Vickers Ltd.
Vickers Ltd. (Vickers Sons & Maxim, Vickers-Armstrong) 14, 95, 233, 244
Vickers Sons & Maxim *see* Vickers Ltd.
Vignale SpA *see* De Tomaso
Villiers 17
Violet, Achille 329
Vischer, Gustav 160
Vitger, Erhard 329
Vivinus, Alexis 82, 330
Voigt Emile 64, 124
Voisin Gabriel 120, 330
Voitures Automobiles et Moteurs [F. Pilain & Cie] (Société des Automobiles François Pilain) 242, 243

Volanis, Antoine 331
Volksauto (Zündapp Gesellschaft für den Bau von Spezialmaschinen GmbH, Zündapp-Werke GmbH) 220
Volkswagen (Automobile Lamborghini, Bugatti Automobiles SAS, SEAT, Volkswagen AG, VW do Brasil, VW Group Power Unit Development Center, Volkswagenwerk 12, 18, 24, 86, 87, 114, 117, 119, 134, 137, 138, 144, 146, 147, 164, 165, 166, 183, 186, 194, 222, 240, 243, 249, 250
Volkswagen AG *see* Volkswagen/ Porsche
Volkswagenwerk *see* Volkswagen
Vollmer, Joseph 75, 180, 331
Volpi, Count 37
Volvo (AB Volvo, Volvo Cars BV, Nedcar BV) 81, 104, 132, 133, 117, 209, 235
Volvo Cars BV *see* Volvo
von Falkenhausen, Alex 41
Von Fischer, Friedrich 26
von Selve, Walther 141, 186, 297
Voran Automobiles AG 55, 56
Vorster, Teddy 109
Vosper-Thornycroft *see* David Brown & Sons (Huddersfield) Ltd.
VW do Brasil *see* Volkswagen/ Porsche
VW Group Power Unit Development Center *see* Volkswagen
Vykoukal, Rudolf 158

W.J. Smith & Sons 161
Wagner, Max 217, 221
Wakefield, Charles Cheers 156
Walcker, Henri 64
Walkinshaw, Thomas "Tom" Dobbie Thomson 332
Wallenberg, Peter 133
Walmsley, William 197
Walter Steiger AG Automobilfabrik *see* Steiger
Walzer, Peter 333
Wanderer *see* Auto Union
Wankel 19, 24, 97, 116, 117, 140; *see also* NSU/Auto Union
Wankel, Felix 119, 333, 250
Ward & Avey Ltd. 58
Ware, Peter 334
Wartburg 102
Waseige, Charles 334
Watson, R. Gordon 127
Watson, William George 334

Weaver, William Arthur 335
Webster, Henry "Harry" George 335
Weigel, Daniel M. 336
Weissach 48
Welleyes 61, 107, 175
Wenderoth, Hans Georg 336
Wennlo, Sten 336
Werbin, Dan 337
Werlin, Jakob 337
Werner, Helmut 338
Werner, William 338
West, Ronald 131
West Bromwich Motor & Carriage Works 161
White, Sir George 10, 73
White & Poppe 179, 214, 247
Whittaker, Derek 150, 139
Wiedeking, Wendelin 48, 339
Wiegand, Heinrich 11
Wilde, Alfred Herbert 9, 205
Wilding, James Armstong 340
Wilfert, Karl 8, 46, 340
Wilhelm Karmann GmbH 166
Wilkes, Peter 340
Wilkinson Sword Company Ltd. 82

Wilks, Maurice Fernand Cary 15, 46, 341
Wilks, Spencer Bernau 144, 145, 168, 341
Williams, Thomas Lawrence 342
Willys Overland 9, 97
Wilsgaard, Jan Einar 342
Wilson-Pilcher 103, 104
Winter, Bernard B. 343
Wishart, Norman 157, 232
Wishart, Thomas D. 343
Wiss, George 344
Wolff, Ernst 344
Wolseley (Wolseley Sheep Shearing Machine Co., Wolseley Tool & Motor Car Company) 129, 135, 136, 208, 223, 246
Wolseley Sheep Shearing Machine Co. see Wolsely
Wolseley Tool & Motor Car Company see Wolsely
Woods, G. Hubert 74
Woodwark, Christopher "Chris" John Stuart 345
Woollard, Frank Griffiths 345

Wormwarld, Arthur 145
Worthington Automobile Company 41
Wright, Geoffrey P. 346
Wright, Orville 41
Wright, Wilbur 41
Wright Aeronautical Corp. 47
Wyer, John 21
Wyer, John Leonard 346

Youngren, Harold T. 198

Zahn, Joachim 250, 346
Zborowski, Count Louis Vorow 201, 232
Zedal 121
Zenith Carburetors 63
Zeppelin, Count 54, 206
Zerbi, Tranquillo 202, 347
Zerbst, Fritz 347
Zetsche, Dieter 248
ZF (Zahnradfabrik Friedrichshafen) 146
Zimmermann, Hans 149
Zoller, Arnold 120, 219
Zucarelli, Paul 239
Züst, Robert 59